The Geometry of Ecological Interactions: Simplifying Spatial Complexity

The concept of invasion fitness, defined as the initial per capita growth rate of a rare mutant in the environment set by the resident types, lies at the heart of adaptive dynamics theory. Current research seeks to provide techniques for determining measures of invasion fitness in different ecological settings. These measures are well established for populations without spatial structure. However, for spatially heterogeneous populations, the patterns that typically arise from short-range ecological interactions often decisively influence invasion fitness. This first volume of the Cambridge Studies in Adaptive Dynamics provides systematic introductions to the modern tools available for describing ecological and evolutionary change in spatially structured populations.

ULF DIECKMANN is Project Coordinator of the Adaptive Dynamics Network at the International Institute for Applied Systems Analysis (IIASA) in Laxenburg, Austria.

RICHARD LAW is Reader in Biology at the University of York.

JOHAN A.J. METZ is Professor of Mathematical Biology at the Institute of Evolutionary and Ecological Sciences at the University of Leiden, and Project Leader of the Adaptive Dynamics Network at IIASA.

Cambridge Studies in Adaptive Dynamics

Series Editors

ULF DIECKMANN
Adaptive Dynamics Network
International Institute for
Applied Systems Analysis
A-2361 Laxenburg
Austria

JOHAN A.J. METZ
Institute of Evolutionary
and Ecological Sciences
Leiden University
NL-2311 GP Leiden
The Netherlands

The modern synthesis of the first half of the twentieth century reconciled Darwinian selection with Mendelian genetics. However, it failed to incorporate ecology and hence did not develop into a predictive theory of long-term evolution. It was only in the 1970s that evolutionary game theory allowed the consequences of frequency-dependent ecological interactions to be analyzed. Adaptive Dynamics extends evolutionary game theory by describing the dynamics of adaptive trait substitutions and by analyzing the evolutionary implications of complex ecological settings.

The *Cambridge Studies in Adaptive Dynamics* highlight these novel concepts and techniques for ecological and evolutionary research. The series is designed to help graduate students and researchers to use the new methods for their own studies. Volumes in the series provide coverage of both empirical observations and theoretical insights, offering natural points of departure for various groups of readers. If you would like to contribute a book to the series, please contact Cambridge University Press or the series editors.

1. *The Geometry of Ecological Interactions: Simplifying Spatial Complexity*
   Edited by Ulf Dieckmann, Richard Law, and Johan A.J. Metz

   In preparation:

2. *The Adaptive Dynamics of Infectious Diseases: In Pursuit of Virulence Management*
   Edited by Ulf Dieckmann, Johan A.J. Metz, Maurice Sabelis, and Karl Sigmund

3. *Elements of Adaptive Dynamics*
   Edited by Ulf Dieckmann and Johan A.J. Metz

The Geometry of Ecological Interactions: Simplifying Spatial Complexity

Edited by

Ulf Dieckmann, Richard Law, and Johan A.J. Metz

CAMBRIDGE UNIVERSITY PRESS
Cambridge, New York, Melbourne, Madrid, Cape Town, Singapore, São Paulo

Cambridge University Press
The Edinburgh Building, Cambridge CB2 2RU, UK

Published in the United States of America by Cambridge University Press, New York

www.cambridge.org
Information on this title: www.cambridge.org/9780521642941

First published 2000
Reprinted 2001
This digitally printed first paperback version 2005

A catalogue record for this publication is available from the British Library

ISBN-13 978-0-521-64294-1 hardback
ISBN-10 0-521-64294-9 hardback

ISBN-13 978-0-521-02209-5 paperback
ISBN-10 0-521-02209-6 paperback

Contents

page

Contributing Authors   xiii

1   Introduction   1
    Richard Law, Ulf Dieckmann, and Johan A.J. Metz

A   Empirical and Statistical Background:
    A Plant Ecological Perspective   7

2   A Neighborhood View of Interactions among Individual Plants   11
    Peter Stoll and Jacob Weiner
    2.1   Introduction   11
    2.2   Competition Mechanisms   12
    2.3   Moving from the Population to the Individual Level 18
    2.4   What is a Plant's Neighborhood?   19
    2.5   Challenges for a Neighborhood Perspective of
          Plant Interactions   24
    2.6   Suggestions for Modelers   26

3   Spatial Interactions among Grassland Plant Populations   28
    Jonathan Silvertown and J. Bastow Wilson
    3.1   Introduction   28
    3.2   Methods for Measuring Competition in the Field   29
    3.3   Results of Field Experiments   32
    3.4   Competition Matrices   38
    3.5   Community Consequences of Spatial Interactions   42
    3.6   Concluding Comments   46

4   Spatio-temporal Patterns in Grassland Communities   48
    Tomáš Herben, Heinjo J. During, and Richard Law
    4.1   Introduction   48
    4.2   Spatio-temporal Patterns in Plant Communities 48
    4.3   Externally versus Internally Generated Spatial Patterns . . 52
    4.4   Concepts in Spatio-temporal Processes in
          Plant Communities   54
    4.5   Ergodic and Non-ergodic Communities   60
    4.6   Concluding Comments   64

5   **Statistical Modeling and Analysis of Spatial Patterns**   65
    David R. Cox, Valerie Isham, and Paul Northrop
    5.1   Introduction   65
    5.2   Descriptive Analysis   66
    5.3   Stochastic Models   70
    5.4   Model Fitting   80
    5.5   Concluding Comments   88

B   **When the Mean-field Approximation Breaks Down**   89

6   **Grid-based Models as Tools for Ecological Research**   94
    Christian Wissel
    6.1   Introduction   94
    6.2   Grid-based Simulation Models   95
    6.3   Spread and Control of Rabies   97
    6.4   Dynamics of a Dwarf Shrub Community   104
    6.5   A Generic Forest Fire Model   109
    6.6   Concluding Comments   114

7   **Coexistence of Replicators in Prebiotic Evolution**   116
    Tamás Czárán and Eörs Szathmáry
    7.1   Introduction   116
    7.2   Metabolic Replication: A Cellular Automaton Model   . .   119
    7.3   The Phenomenology of Coexistence   123
    7.4   Spatial Pattern and the "Advantage of the Rare" Effect   . .   127
    7.5   Resistance to Parasites and the Evolution of
          Community Size   129
    7.6   Toward a Dynamical Theory of Surface Metabolism   . . .   133

8   **Games on Grids**   135
    Martin A. Nowak and Karl Sigmund
    8.1   Introduction   135
    8.2   One-round Games   137
    8.3   Repeated Games   145
    8.4   Extensions and Related Work   149
    8.5   Concluding Comments   150

9 The Interplay between Reaction and Diffusion 151
Mikael B. Cronhjort
9.1 Introduction 151
9.2 The Models: Cellular Automata versus
Partial Differential Equations 153
9.3 Spiral and Scroll Ring Patterns 159
9.4 Cluster Dynamics 163
9.5 Concluding Comments 169

**10 Spirals and Spots: Novel Evolutionary
Phenomena through Spatial Self-structuring** 171
Maarten C. Boerlijst
10.1 Introduction 171
10.2 A Spatial Hypercycle Model 173
10.3 Spirals and Spots 174
10.4 Local versus Global Extinction 175
10.5 Resistance to Parasites 178
10.6 Concluding Comments 180

11 The Role of Space in Reducing Predator–Prey Cycles 183
Vincent A.A. Jansen and André M. de Roos
11.1 Introduction 183
11.2 Individual-based Predator–Prey Models 184
11.3 A Deterministic Model of Two Coupled Local Populations 187
11.4 Larger Spatial Domains 193
11.5 The Spatial Rosenzweig–MacArthur Model 196
11.6 Concluding Comments 199
11.A Stability Analysis of a Multi-patch System 200

C Simplifying Spatial Complexity: Examples 203

12 Spatial Scales and Low-dimensional Deterministic Dynamics 209
Howard B. Wilson and Matthew J. Keeling
12.1 Introduction 209
12.2 Two Models from Evolutionary Ecology 210
12.3 Identifying Spatial Scales 213
12.4 Dynamics, Determinism, and Dimensionality 219
12.5 Concluding Comments 225
12.A Singular Value Decomposition 225

13 Lattice Models and Pair Approximation in Ecology   227
Yoh Iwasa
13.1 Introduction 227
13.2 Plants Reproducing by Seed and Clonal Growth 228
13.3 Forest Gaps 236
13.4 Colicin-producing and Colicin-sensitive Bacteria 243
13.5 Limitations, Extensions, and Further Applications 247

14 Moment Approximations of Individual-based Models   252
Richard Law and Ulf Dieckmann
14.1 Introduction 252
14.2 Spatial Patterns and Spatial Moments 253
14.3 Extracting the Ecological Signal from
Stochastic Realizations 256
14.4 Qualitative Dependencies in a Spatial Logistic Equation . 261
14.5 Exploration of Parameter Space 267
14.6 Concluding Comments 269

15 Evolutionary Dynamics in Spatial Host–Parasite Systems   271
Matthew J. Keeling
15.1 Introduction 271
15.2 Dynamics of the Spatial Host–Parasite Model 272
15.3 A Difference Equation for the Dynamics of
Local Configurations 279
15.4 Evolution to Critical Transmissibility 282
15.5 Concluding Comments 288
15.A Mathematical Specification of the PATCH Model 289

16 Foci, Small and Large: A Specific Class of Biological Invasion   292
Jan-Carel Zadoks
16.1 Introduction 292
16.2 Epidemic Orders 293
16.3 A Theory of Foci 298
16.4 Generalizations 312
16.5 Concluding Comments 315
16.A Quantitative Applications of Models for Spatial
Population Expansion (by Johan A.J. Metz) 315

17 Wave Patterns in Spatial Games and the Evolution of Cooperation   318
Régis Ferrière and Richard E. Michod
  17.1 Introduction 318
  17.2 Invasion in Time- and Space-continuous Games 319
  17.3 Invasion of *Tit For Tat* in Games with
      Time-limited Memory 323
  17.4 Invasion of *Tit For Tat* in Games with
      Space-limited Memory 329
  17.5 Concluding Comments 332

D Simplifying Spatial Complexity: Techniques   337

18 Pair Approximations for Lattice-based Ecological Models   341
Kazunori Satō and Yoh Iwasa
  18.1 Introduction 341
  18.2 Pair Approximation 344
  18.3 Improved Pair Approximation 349
  18.4 Improved Pair Approximation with Variable Discounting . 355
  18.5 Concluding Comments 357

19 Pair Approximations for Different Spatial Geometries   359
Minus van Baalen
  19.1 Introduction 359
  19.2 The Dynamics of Pair Events 364
  19.3 Average Event Rates 368
  19.4 Pair Approximations for Special Geometries 372
  19.5 Pair Approximations versus Explicit Simulations 379
  19.6 Invasion Dynamics 382
  19.7 Concluding Comments 385

20 Moment Methods for Ecological Processes in Continuous Space 388
Benjamin M. Bolker, Stephen W. Pacala, and Simon A. Levin
  20.1 Introduction 388
  20.2 Moment Methods 389
  20.3 A Spatial Logistic Model 391
  20.4 A Spatial Competition Model 400
  20.5 Extensions and Related Work 403
  20.6 Concluding Comments 405

20.A Mean Equation 406
20.B Covariance Equation 408
20.C Analyzing the One-species System 409
20.D Analyzing the Two-species System 410

21 Relaxation Projections and the Method of Moments 412
Ulf Dieckmann and Richard Law
21.1 Introduction 412
21.2 Individual-based Dynamics in Continuous Space 418
21.3 Dynamics of Correlation Densities 425
21.4 Moment Closures and their Performance 438
21.5 Further Developments and Extensions 447
21.A Derivation of Pair Dynamics 452

22 Methods for Reaction–Diffusion Models 456
Vivian Hutson and Glenn T. Vickers
22.1 Introduction 456
22.2 Continuous Models 459
22.3 Linearized Stability and the Turing Bifurcation 466
22.4 Comparison Methods 471
22.5 Traveling Waves 475
22.6 The Evolution of Diffusion 479
22.7 Concluding Comments 481

23 The Dynamics of Invasion Waves 482
Johan A.J. Metz, Denis Mollison, and Frank van den Bosch
23.1 Introduction 482
23.2 Relative Scales of the Process Components 483
23.3 Independent Spread in Homogeneous Space:
       A Natural Gauging Point 485
23.4 Complications 497
23.5 The Link with Reaction–Diffusion Models 504
23.6 Dispersal on Different Scales 507
23.7 Concluding Comments 512

24 Epilogue 513
Johan A.J. Metz, Ulf Dieckmann, and Richard Law

References 517

Index 553

Contributing Authors

Maarten C. Boerlijst (boerlijst@bio.uva.nl) Population Biology Section, Institute for Biodiversity and Ecosystem Dynamics, University of Amsterdam, Kruislaan 320, NL-1098 SM Amsterdam, The Netherlands

Benjamin M. Bolker (ben@eno.princeton.edu) Department of Ecology and Evolutionary Biology, Princeton University, Princeton, NJ 08544-1003, USA

David R. Cox (david.cox@nuffield.oxford.ac.uk) Nuffield College, Oxford OX1 1NF, United Kingdom

Mikael B. Cronhjort (btc@egi.kth.se) Theoretical Biophysics, Department of Physics, Royal Institute of Technology, SE-100 44 Stockholm, Sweden

Tamás Czárán (czaran@ludens.elte.hu) Research Group in Theoretical Biology and Ecology, Department of Plant Taxonomy and Ecology, Eötvös University, Ludovika ter 2, H-1083 Budapest, Hungary

André M. de Roos (aroos@bio.uva.nl) Population Biology Section, Institute for Biodiversity and Ecosystem Dynamics, University of Amsterdam, Kruislaan 320, NL-1098 SM Amsterdam, The Netherlands

Ulf Dieckmann (dieckman@iiasa.ac.at) Adaptive Dynamics Network, International Institute for Applied Systems Analysis, A-2361 Laxenburg, Austria

Heinjo J. During (J.J.During@boev.biol.ruu.nl) Section of Vegetation Ecology, Department of Plant Ecology and Evolutionary Biology, University of Utrecht, P.O. Box 800.84, NL-3508 TB Utrecht, The Netherlands

Régis Ferrière (Regis.Ferriere@snv.jussieu.fr) Laboratoire d'Écologie, École Normale Supérieure, CNRS-UMR 7625, 46, rue d'Ulm, F-75230 Paris Cedex 05, France & Adaptive Dynamics Network, International Institute for Applied Systems Analysis, A-2361 Laxenburg, Austria

Tomáš Herben (herben@site.cas.cz) Institute of Botany, CZ-252 43 Pruhonice, Czech Republic

Vivian Hutson (V.Hutson@sheffield.ac.uk) Department of Applied Mathematics, University of Sheffield, The Hicks Building, Sheffield S3 7RH, United Kingdom

Valerie Isham (valerie@stats.ucl.ac.uk) Department of Statistical Science, University College London, Gower Street, London WC1E 6BT, United Kingdom

Yoh Iwasa (yiwasscb@mbox.nc.kyushu-u.ac.jp) Department of Biology, Faculty of Science, Kyushu University, Fukuoka 812-8581, Japan

Vincent A.A. Jansen (vincent@einstein.zoo.ox.ac.uk) NERC Centre for Population Biology, Imperial College at Silwood Park, Ascot, Berkshire SL5 7PY, United Kingdom. Present address: Department of Zoology, University of Oxford, South Parks Road, Oxford OX1 3PS, United Kingdom

Matthew J. Keeling (matt@zoo.cam.ac.uk) Zoology Department, University of Cambridge, Downing Street, Cambridge CB2 3EJ, United Kingdom

Richard Law (RL1@york.ac.uk) Department of Biology, University of York, York YO10 5YW, United Kingdom

Simon A. Levin (simon@eno.Princeton.edu) Department of Ecology and Evolutionary Biology, Princeton University, Princeton, NJ 08544-1003, USA

Johan A.J. Metz (metz@rulsfb.leidenuniv.nl) Section Theoretical Evolutionary Biology, Institute of Evolutionary and Ecological Sciences (EEW), Leiden University, Kaiserstraat 63, NL-2311 GP Leiden, The Netherlands & Adaptive Dynamics Network, International Institute for Applied Systems Analysis, A-2361 Laxenburg, Austria

Richard E. Michod (michod@u.arizona.edu) Department of Ecology and Evolutionary Biology, University of Arizona, Tucson, AZ 85721, USA

Denis Mollison (denis@ma.hw.ac.uk) Department of Actuarial Mathematics and Statistics, Heriot Watt University, Edinburgh EH14 4AS, United Kingdom

Paul Northrop (paul@stats.ucl.ac.uk) Department of Statistical Science, University College London, Gower Street, London WC1E 6BT, United Kingdom

Martin A. Nowak (nowak@ias.edu) Program in Theoretical Biology, Institute for Advanced Study, Olden Lane, Princeton, NJ 08540, USA

Stephen W. Pacala (steve@eno.princeton.edu) Department of Ecology and Evolutionary Biology, Princeton University, Princeton, NJ 08544-1003, USA

Kazunori Satō (sato@sys.eng.shizuoka.ac.jp) Department of Systems Engineering, Faculty of Engineering, Shizuoka University, Hamamatsu 432-8561, Japan

Karl Sigmund (ksigmund@esi.ac.at) Institute for Mathematics, University of Vienna, Strudlhofgasse 4, A-1090 Vienna, Austria & Adaptive Dynamics Network, International Institute for Applied Systems Analysis, A-2361 Laxenburg, Austria

Jonathan Silvertown (J.Silvertown@open.ac.uk) Ecology and Conservation Research Group, Department of Biology, The Open University, Milton Keynes MK7 6AA, United Kingdom

Peter Stoll (stoll@sgi.unibe.ch) Institute of Geobotany, University of Berne, Altenbergrain 21, CH-3013 Berne, Switzerland

Eörs Szathmáry (szathmary@zeus.colbud.hu) Department of Plant Taxonomy and Ecology, Eötvös University, Ludovika ter 2, H-1083 Budapest, Hungary & Collegium Budapest, Institute for Advanced Study, Szentharomsag utca 2, H-1014 Budapest, Hungary

Minus van Baalen (mvbaalen@snv.jussieu.fr) Institut d'Écologie, CNRS-FR3, Université Pierre et Marie Curie, Bâtiment A, 7me étage, 7, quai St.-Bernard, F-75252 Paris Cedex 05, France

Frank van den Bosch (frank.vandenbosch@ztw.wk.wau.nl) Department of Mathematics, Wageningen Agricultural University, Dreijenlaan 4, NL-6703 HA Wageningen, The Netherlands

Glenn T. Vickers (g.vickers@sheffield.ac.uk) Department of Applied Mathematics, University of Sheffield, The Hicks Building, Sheffield S3 7RH, United Kingdom

Jacob Weiner (jawej5@staff.kvl.dk) Botany Section, Department of Ecology, Royal Veterinary and Agricultural University, Rolighedsvej 21, DK-1958 Frederiksberg, Denmark

Howard B. Wilson (h.b.wilson@ic.ac.uk) Biology Department, Imperial College at Silwood Park, Ascot, Berkshire SL5 7PY, United Kingdom

J. Bastow Wilson (Bastow@Otago.ac.NZ) Botany Department, University of Otago, P.O. Box 56, Dunedin, New Zealand

Christian Wissel (wissel@pinus.oesa.ufz.de) Department of Ecological Modeling, Center for Environmental Research Leipzig-Halle, PF2, D-04301 Leipzig, Germany

Jan-Carel Zadoks (JCZadoks@User.DiVa.NL) Department of Phytopathology, Wageningen Agricultural University, P.O. Box 8025, NL-6700 EE Wageningen, The Netherlands

1

Introduction

Richard Law, Ulf Dieckmann, and Johan A.J. Metz

> Species form different kinds of patches; these patches form a mosaic and together constitute the community. Recognition of the patch is fundamental to an understanding of structure. Patches are dynamically related to each other. But there are also departures from this inherent tendency to orderliness. At any given time, therefore, structure is the resultant of causes which make for order, and those that tend to upset it. Both sets of causes must be appreciated.
>
> *Abbreviated from Watt (1947, p. 2)*
> *Pattern and process in the plant community*

A sea change has come over theoretical ecology in the past 10 years. The era of the simple general model that tries to capture the elusive essence of an ecological community is rapidly fading from sight. This is the age of the individual-based, spatially explicit, computer-based model (Huston *et al.* 1988; DeAngelis and Gross 1992; Judson 1994).

Why has this transformation taken place? First there is the simple matter of practicality: desktop computing power has reached a level at which it is quite feasible to simulate individuals as they move across a landscape, interact, reproduce, and die. Second is the issue of language: for many ecologists, rules encoded in computer algorithms are much more accessible than the formal mathematical language of dynamical systems. Third is the appreciation that important ecological intricacies, such as the mechanisms by which organisms interact in communities, often cannot be incorporated sufficiently faithfully into simple models. Fourth is an awareness that the simple models traditionally used in ecology have not always proved very successful in accounting for phenomena observed in natural systems.

Individual-based simulations are most realistic when they encompass the randomness of individuals in births, deaths, and movements (e.g., Pacala *et al.* 1996). Our computer screens then give realizations of complex spatio-temporal stochastic processes. The simulations have their own intrinsic interest; they can be a valuable aid in defining and characterizing the processes involved and can lead to the discovery of new and interesting phenomena. But we should not to infer too much from a few realizations of a process: it is not the location and behavior of each individual that matters, since the stochasticity will ensure that every realization is different at least in detail. It is the gross properties of the stochastic process that are likely to be of interest in the long run.

Helpful though simulations are, they can be no more than a step toward understanding properties of the stochastic process. When you next look at such a realization as it unfolds on the screen, ask yourself the following questions:

- Can you distinguish between the random variation intrinsic to any stochastic process and the ecological signal that characterizes the system's representative behavior?
- What spatial and temporal patterns come about in the long run? In other words, can you characterize the asymptotic states of the system?
- Can you identify different kinds of patterns that develop as the initial configuration of the community is changed? Are there alternative metastable states that depend on the starting conditions?
- Can you work out how many different kinds of patterns could develop from different starting conditions?
- Can you understand what happens when you change the environment in which the organisms live by altering the parameters of the process?
- How readily can you sample the parameter space and determine the effects of parameters on the qualitative and quantitative properties of your system?

These are important questions, but ones that are very difficult to answer from individual-based simulations. The heterogeneity of natural environments in time and space provides a strong imperative for such analyses, but dealing with such heterogeneity, always a major undertaking in ecology, is especially demanding in the context of a stochastic process.

As a result of developments in theoretical ecology over the past decade, enormously complex models have replaced simple ones. If you doubt this, imagine a community being modeled on a spatial lattice of, say, 100 cells. To keep matters simple, suppose population sizes in the cells are large enough for stochasticity to be ignored. If you have 10 species in the community, your dynamical system comprises 1000 equations. Can it really be that community dynamics need a state space of such staggering dimensionality? We believe that very often the dynamics can be adequately represented in a more parsimonious set of equations. It should be possible to project the dynamics into a low-dimensional space which carries the essential information. This is more than wishful thinking: Rand and

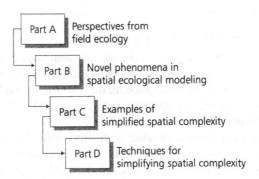

Figure 1.1 Structure of the book.

Wilson (1995) found a spatial resource–predator–prey model that has effective dynamics in a four-dimensional embedding space. The problem is, what simplified state space would be an appropriate target for projecting the dynamics? Can projections be found that properly hold in place the main effects of spatial structure?

Ecology needs new ideas and methods to deal with dynamics of processes in a spatial setting (Wiens *et al.* 1993; Hastings 1994; Levin *et al.* 1997; Tilman and Karieva 1997). This book points to and explains some possible ways forward (Figure 1.1). In the first place, there is obviously much to be learned from individual-based modeling of ecological communities. Such models can be motivated by direct observations of individuals in the field, and they force precise thinking about the processes involved. They help in developing intuition about how ecological systems behave. They show us repeatedly how new, unexpected phenomena emerge when spatial structure is introduced.

Spatially explicit, individual-based models contrast with models that lack spatial structure, widely used in theoretical ecology in the past. These earlier models make an assumption that the effects of neighbors are proportional to their density averaged across a large spatial domain (the so-called mean-field assumption, see Box 1.1). In communities where individuals interact with their neighbors, the presence of nonrandom spatial pattern, for which there is abundant evidence in nature, will most likely lead to major departures from the mean-field dynamics. The world is full of spatial structure, and this has fundamental consequences for many ecological processes. Individual-based models are an important step toward seeing what happens when the mean-field assumption is abandoned, and Part B gives some striking illustrations of the remarkable behavior that can then emerge.

Box 1.1 The mean-field assumption in ecology

At the heart of much ecological theory lies an assumption that individual organisms encounter one another in proportion to their average abundance across space. You find this assumption in, for instance, the Lotka–Volterra equations for two interacting species i and j, expressed as the product of their mean densities $N_i N_j$.

Before being applied to ecological problems, assumptions of this type were used in physics and chemistry (Weiss 1907). Examples are collisions between molecules in a well-mixed gas, the electrical field experienced by electrons within an atom, and the magnetic field around elementary magnets of a solid. In the last two cases, all electrons in an atom (all elementary magnets in a solid) are assumed to be locally surrounded by the same electric (magnetic) field, called the "mean field." This is why, even in ecology, the assumption is widely referred to as the mean-field assumption.

The mean-field assumption is most likely to hold as a good approximation when the physical environment of organisms is homogeneous and

- physical forces exist that cause strong mixing of organisms,
- organisms themselves are highly mobile, or
- organisms interact with others over long distances.

As conditions depart from those above, the mean-field assumption becomes less and less appropriate. A lack of mixing, whether due to the external environment or immobility of the organisms, generates neighborhoods around individuals that deviate from the spatial averages. Differences in local environmental conditions become especially important if organisms only interact over short distances (integrating over large neighborhoods can give spatial averages quite close to the mean field). The local environment organisms experience can then be quite different from the mean environment, averaged across the entire ecological habitat (see figure). Such departures from the mean field can feed through to the vital rates of individuals and can have fundamental effects on their dynamics.

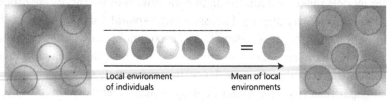

Local environment   Mean of local
of individuals   environments

Constructing the mean field. Spatial variation in environmental conditions (e.g., measuring a resource's availability) is shown by gray scales; individuals are depicted as points, and their local environments, as circles. Constructing the mean field by averaging over local environments shows why the mean-field assumption may break down: a population's response to a spatially heterogeneous environment is often different from its response to the mean field.

However, theoretical ecology needs to move on from simulations of individual-based processes to manageable approximations that give a better understanding of the generic properties of these processes. This is already being done with some success, as can be seen from the examples in Part C and the methodological chapters in Part D. When the patterns we are interested in have a large spatial extent, the methods include diffusion approximations using partial differential equations. These methods have been available for many years (Okubo 1980) but still have much to offer ecologists. In addition, novel methods, such as pair approximations and correlation dynamics, are being developed that concentrate on dynamics of small-scale spatial structure. Analysis of the resulting deterministic equations can deal with many of the issues left unresolved by stochastic simulations, including

- overall qualitative features buried in the processes;
- the attractors that are present and whether they correspond to spatially homogeneous systems or indicate the presence of spatial structures;
- the effect on eventual states of communities of changes in the environment and ecological interactions, using bifurcation analysis in moderately large parameter spaces;
- the fate of newly introduced mutants and immigrant species, whether they will invade or be driven to extinction by the resident system.

There is much for the theorist to do here and a great deal to challenge the ecologist. But a major factor hindering progress is the difficulty ecologists and theorists have had in developing an effective dialogue. We think it is essential to develop theory that is demonstrably relevant to real ecological systems and to show how it illuminates our understanding of ecology. We begin with several chapters in Part A that explain what ecologists have learned about spatio-temporal processes in ecological communities to provide some guidelines for developing theory.

The book covers a much wider span of knowledge from ecology to mathematics than is usual in a single textbook, and we recognize that you may not want to read it from cover to cover (although of course we hope you will!). But we hope that you will be encouraged to build bridges from the parts of the book that lie in your own area of expertise – whether ecology, computation, or mathematics – to other, less familiar parts and that the book will aid your understanding of these different areas.

In a sense, the path mapped out in this book – from field observations, to individual-based simulations, to deterministic approximations of stochastic processes, and back again – is how ecological theory might have developed in an ideal world. But progress in research has its own imperative, and scientists work on the problems that appear promising at the time. Although the simple models from an earlier age of theoretical ecology may now look somewhat *ad hoc*, their importance should not be underestimated: there certainly are circumstances where spatial structure is less important, and in these circumstances the earlier theoretical framework will prove helpful. Our focus on spatio-temporal processes tries to extend the formal framework of ecology, not to replace one paradigm with another. As theoretical ecology develops, the broader framework that emerges should place earlier theory in its proper context within the structure of our expanding understanding.

Acknowledgments This book was conceived while two of the editors, Ulf Dieckmann and Richard Law, were working with Tomáš Herben at the Institute for Advanced Study, Wissenschaftskolleg zu Berlin, in the academic year of 1995/1996. It became evident during the course of our discussions that there was a real possibility of moving on from individual-based models, currently at the center of much theoretical ecology, toward a more rigorous and elucidating treatment of spatial dynamics. The Institute for Advanced Study provided ideal conditions in which to develop our ideas, and we are very grateful to the staff for the welcome they gave us and for providing such a good working environment. Special thanks go to Wolf Lepenies, Joachim Nettelbeck, Hans Georg Lindenberg, and Andrea Friedrich.

Subsequent stages in development of the book took place at the International Institute of Applied Systems Analysis (IIASA), Laxenburg, Austria, where IIASA's current director Gordon J. MacDonald and former director Peter E. de Jánosi provided critical support. We organized two workshops in which authors were brought together to discuss their contributions to achieve as much continuity across subject areas as possible. The success of a book of this kind depends very much on cooperation of authors in dealing with the many points editors are bound to raise, and we thank our authors for their patience over the past two years. The book has benefited greatly from the support of the Publications Department at IIASA; we are especially grateful to Ellen Bergschneider, Anka James, Martina Jöstl, and Eryl Maedel for the work they have put into preparing the manuscript. Any mistakes that remain are our responsibility.

Part A

Empirical and Statistical Background: A Plant Ecological Perspective

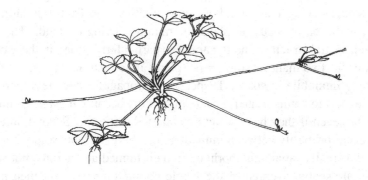

Introduction to Part A

What have ecologists learned about spatio-temporal processes in natural communities? The first part of the book gives some answers to this question and is intended to provide ecological background to which theorists can turn. We believe such a basis is necessary if we are to achieve a constructive and enlightening dialogue between ecologists and mathematicians. The information in Part A indicates how ecologists think about spatio-temporal processes and sets boundaries on the kinds of spatio-temporal models likely to be of lasting interest in ecology. This part of the book is a small step in an iterative process of mutual education of theorists and ecologists.

Ecologists study many kinds of communities, and from this large set we have chosen to focus on plant communities living on land. The link between spatial structure and dynamics is particularly strong in these communities for two main reasons. First, plants in terrestrial communities are relatively immobile in space. Dispersal of propagules does less than one might expect to compensate for such immobility because most seeds do not travel far, even if they have structures that aid dispersal. Second, interactions occur primarily between immediate neighbors. Plants respond to the state of a small spatial neighborhood in their immediate vicinity, not some large-scale spatial average of the whole community (the so-called mean field). If there are circumstances in natural communities under which local spatial pattern should be important, they are in these terrestrial plant communities. The variation in spatial pattern from one location to another is both sensed and partially generated by the plants; such plant communities are therefore an obvious place to start looking for dynamics in which space plays an important part. This is not to suggest that it is only in plant communities that spatial structure can play a major role – several chapters later in the book point to the importance of spatial structure in other ecological and evolutionary processes.

Part A concentrates on temperate grasslands. Here, nonrandom spatial pattern is particularly evident. One reason for this pattern is that many plant species in such communities reproduce by clonal growth, which greatly restricts dispersal of propagules and gives rise to clumps of conspecifics. What is less obvious, but equally important, is that these spatial structures are in a continual state of flux, and the time scale on which this flux occurs is short enough for changes in spatial pattern to be observed over a few years. Grassland communities are about as close as ecologists can get to systems in the field in which (1) spatial structure should play a major role and (2) turnover rates are great enough to make it feasible to study the dynamics.

The first three chapters follow a sequence from small to large spatial and temporal scales. The sequence starts in Chapter 2 at the microscopic scale, with the concept of a neighborhood around an individual plant within which local interactions take place, resulting in overlapping zones of influence. Stoll and Weiner consider the mechanisms of competition within this neighborhood through limited resources such as light and mineral nutrients. They discuss how ecologists have tried to characterize the neighborhood as an area around an individual plant and the assumptions implicit in this work. They also point out some of the main issues that remain open, such as how to deal with the modular structure that many plants have, and how to allow the neighborhood to expand as plants grow. Last but not least, they express a viewpoint about the role of theory in ecology, quite widely shared by ecologists.

Chapter 3 moves up a step in the scale of time to tackle the turnover of individuals as they interact in small neighborhoods. A detailed mechanistic view of neighborhoods of the kind described in Chapter 2 is difficult to distill from ecological data. Silvertown and Wilson take a more phenomenological approach, integrating over the known (and unknown) mechanisms of interaction by means of a single measure of neighborhood dependence, often given as a competition coefficient for a pair of species. They describe how such coefficients are estimated in the field and the information that can be gleaned from matrices of these coefficients for several coexisting species. Rather little has been done to develop these ideas explicitly in a spatial framework. The authors discuss the state of this art, together with the work that they and their colleagues have done in estimating parameters for spatial invasion of grasses in the field. Their work leads to a cellular automaton model of the spatial dynamics.

Chapter 4 moves up a step further in the scale of both time and space to the spatio-temporal patterns observed in grassland communities. In the long run, the success of spatio-temporal models of plant communities will be judged by their capacity to capture these macroscopic features from the underlying microscopic (neighborhood-dependent) processes of birth, death, and movement. Nonrandom spatial patterns are certainly typical of grassland communities. In Chapter 4, Herben, During, and Law discuss the extent to which these patterns can be said to be self-generating or imposed from outside by heterogeneities in the external environment. Several informal theories can be found in the ecological literature that foster an understanding of how spatial structure of plant communities develops through time. The models include cyclic sequences of community states (mosaic cycles), random sequences of states (carousel model), non-allowable states (guild proportionality), and community states that are "frozen" through time by space preemption.

The last chapter in Part A provides a link between what ecologists can do in the field and what theorists can do at their desks. It is essential to have in place a formal statistical framework for analysis of the spatio-temporal processes taking place in nature. Chapter 5, by Cox, Isham, and Northrop, gives an introduction to the armory of methods that statisticians have available for treating empirical data associated with spatio-temporal processes. First, the chapter covers descriptive methods for preliminary inspection of data collected over time, space, or both time and space together. Second, it deals with stochastic models for describing spatial data, in particular, Poisson-based models and Markov random fields. Third, it considers methods by which the parameters of such models can be estimated, notwithstanding the complex interdependencies that spatio-temporal data typically exhibit.

2

A Neighborhood View of
Interactions among Individual Plants

Peter Stoll and Jacob Weiner

2.1 Introduction

In no area of ecology is the role of space more fundamental than in the study of plant communities (Hutchings 1986; Crawley and May 1987). Individual plants are rooted in one place and their ability to move and occupy space is restricted to growth (Eriksson 1986). A plant cannot relocate from an unfavorable location to a more favorable one. Rather, it grows as well as possible where it finds itself or it dies. Basic plant biology suggests that plant–plant interactions are inherently local in nature. For example, individual plants do not experience global population density *per se*, but only interact with neighbors over restricted distances. The mobility of animals makes their spatial behavior potentially far more complex than that of plants, but, ironically, this ability to move can make the modeling of space for animal populations unnecessary in many cases. For example, because animals can "diffuse" in space from areas of higher density to areas of lower density, models based on mean spatial behavior or overall density may often be sufficient. Because a plant's ability to move is quite restricted (except during dispersal), local conditions are of much greater significance to plants than to animals. When feeding fish in a tank, it does not matter where on the water surface one places the food, because the fish will come to it. But when watering or fertilizing the garden one must make sure that the resource comes close to the plant – if one waters only half the garden the other half will not obtain sufficient water. Thus, while it is possible that the spatially averaged behavior of individuals may sometimes provide sufficient information for modeling some processes within populations and communities, this is much less likely to be the case for plants than for animals.

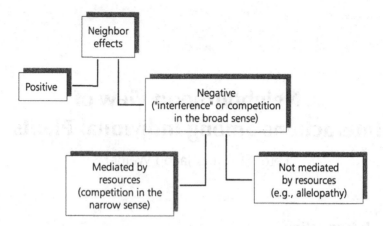

Figure 2.1 Classification of neighbor effects in plants.

In this chapter, we discuss the role of local spatial processes in plant communities, focusing on the concept of an individual's neighborhood. We emphasize competition among plants because it is thought to be one of the primary factors determining plant performance in the field and has therefore been the most studied ecological interaction in plant communities. However, the neighborhood approach taken here could also be applied to other ecological interactions, such as herbivory or pollination. Our goal in the context of the present volume is to foster much-needed communication between theoreticians and empiricists in ecology by providing modelers with an empirical perspective on local plant interactions, which we hope will be of use in building models and developing modeling techniques.

2.2 Competition Mechanisms

The study of interactions among plants in natural communities presents the ecological researcher with daunting complexity. The mechanisms by which plants interact are understood only at a general nonquantitative level (Bazzaz 1990), although we can be encouraged by a few robust patterns at the population level, such as the relationship between density and biomass (Silvertown and Lovett Doust 1993). In discussing interactions among plants, our conceptual framework follows that of Harper (1977). Plants are influenced by neighbors, and we call all such interactions "neighbor effects" (Figure 2.1). While many of these effects are negative ("interference" *sensu* Harper or competition in the broad sense), some are positive. Positive neighbor effects, such as protecting neighbors from excessive solar radiation and resultant water loss and providing mechanical support and

---

Box 2.1 Positive neighbor effects

Plants can have positive as well as negative effects on their neighbors. The classic example of positive neighbor effects is that of "nurse plants" in arid systems (e.g., Franco Pizana *et al.* 1996). Some desert plants can only establish themselves in close proximity to a larger plant, usually a shrub, because the shade of the larger plant provides protection from the intense solar radiation and resultant heat and transpiration that a seedling otherwise would experience. Plant establishment in deserts is largely determined by the negative effects of a superabundant plant resource – solar radiation – for which plants in other environments compete. There has been increased interest in positive plant–plant interactions. Below are two examples.

During drought periods, sugar maple (*Acer saccharum*) demonstrates "hydraulic lift," nocturnal uptake of water by roots from deep soil layers that is released from shallow roots into upper soil layers (Dawson 1993). Neighboring plants use from 3–60% of the hydraulically lifted water supplied by sugar maple trees. Hydraulic lift may not be limited to arid or semiarid environments where chronic water deficits prevail and might be important in relatively mesic environments when subjected to periodic soil water deficits.

Facilitation by neighbors may be quite common in wetlands. For example, emergent wetland plants often alleviate the effects of anaerobic soils on root respiration by transporting oxygen below-ground through continuous air spaces (aerenchyma) within the plant. Oxygen leaking from the roots into the rhizosphere may oxidize minerals in the soil or become available to other plants. Callaway and King (1996) investigated the ability of cattail (*Typha latifolia*), a widespread wetland plant with aerenchymous tissue, to aerate sediments and affect the growth of two neighbors, a willow (*Salix exigua*) and forget-me-not (*Myosotis laxa*). At lower temperatures, rooted willow cuttings survived only when planted with cattail and forget-me-not transplants grew significantly larger when planted with cattail. At higher soil temperatures, however, there was evidence of competition rather than facilitation.

---

protection from herbivores, may be more common than previously thought (Box 2.1; Aarssen and Epp 1990). In some ecosystems, particularly those of nutrient-poor or other extreme environments such as salt marshes (e.g., Bertness and Shumway 1993), positive effects may be as important as negative effects. It is important to remember that the net effect of one plant on another is the sum of positive and negative effects (Berkowitz *et al.* 1995). Because the relative importance, timing, and spatial structure of the numerous positive and negative mechanisms may vary, it is not easy to summarize the effects plants have on one another by using simple coefficients. Chapter 3 gives an overview of how plant ecologists have tried to obtain this information and the results accumulated so far.

Negative neighbor effects are usually more important than positive ones, because all plants require basically the same resources. If plants are growing in close proximity, it seems almost inevitable that they will eventually compete for some of these resources. Negative neighbor effects can be divided into those mediated by resources (competition in the narrow sense) and those mediated by other mechanisms or organisms (Figure 2.1). Indirect neighbor effects include changes in environmental conditions such as temperature, humidity, and wind velocity, and attraction or repelling of animals, which thereby affects predation, trampling, etc. (Harper 1977). Most plant ecologists consider competition for resources to be generally more important and more likely to be predictable than other neighbor effects, but there may be communities in which other mechanisms such as allelopathy play a major role (Rice 1984). Quantifying mechanisms such as allelopathy in a field situation is a distant goal, but ecologists have begun to study resource-mediated competition quantitatively (Tilman 1982, 1988; Fitter 1986; Keddy 1991).

It is important to distinguish between the effects an individual plant has on resources and how that plant responds to the preemption of resources by its neighbors (Goldberg 1990; Tremmel and Bazzaz 1993). Thus, the intensity of competition is determined by two processes: (1) the effects of neighbors on resource availability and (2) the ability of individuals to tolerate or compensate for these effects through plasticity and other "behavioral" responses (Box 2.2). Plasticity is the ability of a single genotype to develop into different phenotypes in different environments (Bradshaw 1965). For example, plants can change their growth form in response to neighbors – for instance, by putting more effort into height growth at the expense of lateral growth when they are shaded (Schmitt 1993; Schmitt and Wulff 1993) or can "expect" to be shaded in the near future (Ballaré *et al.* 1990). Resource acquisition depends on the placement of plant parts in relation to resources and the ability to take up these resources when they are encountered. The ways plants obtain resources, use them to obtain more resources, and consequently make these resources unavailable to neighboring plants, and the ways plants respond to reduced resource levels caused by neighbors can be considered the mechanisms of resource competition (Bazzaz 1990). The effects of plants on each other in the field are primarily the result of such mechanisms.

Box 2.2 Plant behavior: Clonal growth, foraging, and division of labor

Higher plants are composed of a repetitive branched system of units (e.g., ramets, modules), each consisting of a segment of stem, leaves, and axillary buds (meristems) with the potential to form a new unit. The whole shoot system is a population of such units that may be united in sharing a common root system. This unity is often lost during clonal growth – lateral spread by "vegetative reproduction" – of plants such as strawberries (Figure 2.2). When connecting horizontal parts of the shoot systems (stolons or rhizomes) die and rot away, rooted, physiologically independent offspring are left (Harper 1977). One consequence of the modular construction of plants is that leaves and root tips are located on branches that project them into habitat space (Bell 1984; Schmid 1990; Hutchings and de Kroon 1994). The architecture of resulting branching patterns can be described by variables such as spacer length, branching frequency, and branching angles. Morphological plasticity refers to changes in these architectural parameters in response to the plant's environment.

In an analogy to foraging in animals, Slade and Hutchings (1987a) defined "foraging" in plants as "the process whereby an organism searches or ramifies within its habitat in the activity of acquiring essential resources." According to this analogy, leaves and root tips (resource-acquiring structures) are "feeding sites," which are located at the ends of "spacers" (i.e., horizontal branches). Shortening spacers and increasing branching in response to high local abundance of resources in order to place more feeding sites into resource-rich microhabitats should increase resource acquisition from the habitat and enhance plant fitness (Hutchings 1988).

Spacers not only place feeding sites within the environment, they also perform bidirectional transport processes, thus providing intraclonal specialization and cooperation analogous to the economic principle of spatial division of labor between shoots in different patches within the environment (Stuefer et al. 1994; Stuefer 1995). Water provided through stolons from shoots growing in shaded microhabitats may be delivered to shoots growing in full sunlight, while shoots in full sunlight may provide assimilates to shoots growing in areas with lower light levels (Evans 1991, 1992).

Simply put, resource-mediated competition occurs when individual plants consume resources which are therefore not available to other individuals. If the lack of a resource limits the growth of an individual, then that individual has suffered from competition. One important difference between plants and animals is that animals of different species may have varying degrees of overlap in the resources they need. In some cases, two different animal species may use few, if any, of the same resources. Plants,

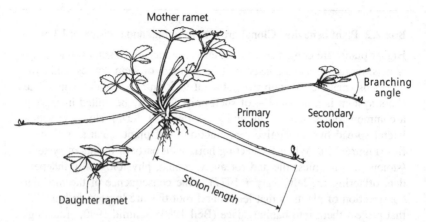

Figure 2.2 A strawberry (*Fragaria* sp.) as an example of clonal growth. The genetic individual ("genet") consists of numerous physiological/morphological individuals ("ramets") that may exchange resources if they remain connected or live independently if connections (stolons, i.e., horizontal shoots) are no longer functional. Clonal growth by plants can be seen as analogous to movement by animals.

on the other hand, all need the same basic resources: physical space, light, carbon dioxide, water, oxygen, and a suite of mineral nutrients. Thus, the ability of plants to avoid competition through niche differentiation is quite limited, although plants can use resources in different proportions (Tilman 1982), at different depths in the soil (Parrish and Bazzaz 1976, 1985), or at different times of the year (Eissenstat and Caldwell 1988).

Ecologists have begun to develop general models for local depletion and renewal of soil resources (Fitter and Stickland 1991; Fitter *et al.* 1991; Huston and DeAngelis 1994; Leadley *et al.* 1997), but it is not clear how far such generalizations can take us. Light, for example, is so fundamentally different from other resources that it is difficult to imagine how it can be treated similarly to soil resources. Light is unidirectional, cannot be stored (although the products of photosynthesis can), and does not diffuse from one point in space to another. In this sense, light as an energy source is inherently local: a plant cannot benefit from light that it does not intercept. Moreover, because plant movements are so limited, a plant's ability to move to areas of greater light availability is very restricted, although one can consider plants to be "foraging" through their growth patterns (Slade and Hutchings 1987b).

Although mineral nutrients are not unidirectional or renewable in the same sense as solar radiation, the diffusion of nutrients through the soil appears to be extremely limited. For example, plant roots can deplete local

---

Box 2.3 Competition at the microscale: Distribution and dynamics of neighboring plant roots

Although roots are difficult to observe, especially under field conditions, progress is being made. Caldwell and his coworkers (Caldwell *et al.* 1991, 1996) have used various techniques to investigate the deployment of roots in relation to neighboring plants and the availability of soil resources, which may vary both in space and time. For example, they found that root proliferation of the sagebrush (*Artemisia*) was considerably influenced by the presence of different grass species (*Agropyron* or *Pseudoregneria*). Root density of the shrub was generally two to three times higher with *Pseudoregneria* than with *Agropyron*, and there was a greater tendency for the roots of the shrub and *Pseudoregneria* to segregate (i.e., to avoid one another). Caldwell *et al.* (1991) interpreted these patterns as interference at the level of individual roots but only speculated about the possible mechanisms, such as resource pre-emption or allelopathy. In a subsequent experiment, Caldwell *et al.* (1996) found that shrub and grass roots tended to avoid each other at a scale of millimeters to centimeters, although there was no direct evidence of resource competition. While resource competition cannot be entirely dismissed, other mechanisms may have contributed to the species-specific relationships between shrub and grass roots. Growth inhibition of roots following contact with roots of other plants has been shown (Mahall and Callaway 1992; Krannitz and Caldwell 1995; Huber-Sannwald *et al.* 1996), and Caldwell *et al.* (1996) conclude that such species-specific and sometimes even genotype-specific responses strongly suggest a recognition mechanism.

---

nitrogen, which can diffuse from areas where it has not been depleted, but this occurs over very short distances and at the local level of neighboring fine roots (Box 2.3). Water is perhaps the most diffusible of plant resources, but the distances involved are still quite limited relative to the size of the plant. For example, in an experiment where water was supplied only to the outer root system (more than 10 cm from the center) of branch units of bunchgrass (*Bouteloua gracilis*) in containers, growth was significantly less than when water was applied only to the central root system (Hook and Lauenroth 1994).

Dispersal is one way that plants "move" extensively, and seeds can sometimes move great distances via wind and water, or with the help of animals. One function of dispersal may be to escape local competition. However, most studies show that, by far, most seeds end up very close to the mother plant. Seed dispersal can be modeled as a diffusion process, with fewer and fewer seeds at increasing distances from the plant (e.g., Pacala *et al.* 1996). Although long-distance dispersal is quite rare, it may be extremely important for plant community dynamics. Plant species diversity, for example, may often be limited by dispersal (Ricklefs and Schluter 1993; Tilman 1997). The few seeds that are dispersed far from their mother plant – that is, those that escape from the maternal plant's neighborhood – may be able to escape mortality due to seed predators (Janzen 1970) or diseases (Augspurger 1984) concentrated near the mother plant. We can think of plants interacting at a local scale as they survive and grow, and subsequently experiencing a more mobile phase during propagule dispersal.

2.3   Moving from the Population to the Individual Level

Because of the local nature of plant interactions, analysis and modeling of plant–plant interactions have moved from "mean-field approximations" toward explicit modeling of local interactions. Until the mid-1980s the study of density dependence focused on mean plant behavior (Bleasdale and Nelder 1960; Watkinson 1980; Vandermeer 1984). However, there has been an increase in the study of what one could call local density dependence, that is, the study of the performance of individual plants as a function of their local competitive conditions. As is often the case in ecology, models of local competition have been less successful in accounting for variation in the observed phenomena in the field than in stimulating new ideas and approaches to the study of plant–plant interactions, and more questions have been raised than have been answered. It has been demonstrated that

- plants do interact locally;
- local crowding reduces plant growth, reproductive output, and probability of survival;
- the effect of neighbors attenuates with distance (although the nature of this attenuation is not well understood);
- beyond a certain distance plants have no detectable effect on each other.

For example, Tyler and D'Antonio (1995) showed that, for seedlings of the shrub *Ceanothus impressus*, both survivorship and growth increased with increasing distance from near neighbors. Their study site was a burned area, and disturbances such as fire might preclude competition by releasing a flush of nutrients or by reducing biomass and thereby diminishing the consumption of resources. Thus, even after disturbance, when some resources are apparently abundant on a large scale, competition for resources may be important in determining small-scale patterns of seedling growth and survival. Survival of seedlings was reduced by the presence of neighbors up to a distance of 20 cm. More distant neighbors no longer had an effect on survival but still reduced growth. There are numerous studies that demonstrate negative neighborhood interactions (see Chapter 3). However, information on the relative importance of different mechanisms in different environments and evidence for the importance of the observed effects for population or community processes are rarely available.

2.4 What is a Plant's Neighborhood?

The study of local interactions begins with the question of what "local" means. We define a competitive neighborhood as an area within which a plant can be affected by local factors, such as the abundance of neighboring plants. There are two different approaches to the definition of a neighborhood. In one approach the neighborhood is defined as a patch of space within which plants interact. Interactions do not occur between patches, and all individuals within a patch can potentially interact. This framework comes from the study of environmental heterogeneity and has been further developed in the context of patch dynamics (Pickett and White 1985). Although such an approach has the virtue of simplicity, most plant ecologists find it insufficient to capture the spatial dynamics of plants. In contrast, a "plant-centered" view of neighborhoods does not aggregate all individuals within a patch of space but lets the individual define the neighborhood, usually thought of as a circular area around the individual. The study of local interactions among plants has moved in the direction of plant-centered neighborhoods.

In an ecological perspective that emphasizes plants' local neighborhoods, a major goal would be to describe the performance of a plant (its growth and reproductive output) as a function of the plant's genotype and local environment, broadly defined to include neighboring plants and other organisms. We are far from being able to describe such a function. In many studies, environmental heterogeneity, such as local variation in soil quality (Lechowicz and Bell 1991), seems to be more important than the local abundance of competing plants (Mitchell-Olds 1987). The question for modelers becomes, What would be an adequate description of a plant's neighborhood for a specific ecological purpose or research problem?

In the simplest plant-centered neighborhood approach, a neighborhood is defined by a radius from an individual plant's center, and the number of neighbors of different species within the area defined by the neighborhood radius is a measure of the local density. This approach has been extensively developed by Pacala, Silander, and their coworkers. Starting with populations of a single species of annual plant (Pacala and Silander 1985; Silander and Pacala 1985), they progressed to two-species models (Pacala 1986, 1987; Pacala and Silander 1987) that were fitted to field data (Pacala and Silander 1990). Predictors of individual plant performance, such as survival, growth, and reproduction, are functions of the number of neighbors in the neighborhood – a circular area around a subject plant circumscribing all other individuals that interact with the subject plant. In Pacala and Silander's models, the positions and sizes of neighbors within the neighborhood are not considered. They argue that when the "optimal" neighborhood size (i.e., the neighborhood radius of the circle that explains most of the variability in performance) was determined by statistical fitting, the number of neighbors within the neighborhood alone had almost equal predictive power as more detailed and complicated models. The assumptions of this approach are that the neighborhood can be considered internally homogeneous and that individuals of the same species can be considered equal, independent of their size. If these assumptions hold, then modeling and analyses of plant neighborhoods are quite tractable. In apparent contradiction to the emphasis on spatial dynamics in the recent literature (Durrett and Levin 1994b; Pacala and Deutschman 1995) and in this book, Pacala and Silander's model collapses into a mean-field model (Pacala and Silander 1990). This may be due to the fact that their local neighborhood model differs only in scale (local) and location (plant-centered) from mean-field models of density dependence (S. Thomas, personal communication). Pacala and Silander's local density model is just that: neighborhoods are defined around individuals and the local density (in the simplest sense, the number of individuals) is the independent variable.

Neighborhood competition studies have been criticized recently for several reasons. First, some researchers have questioned the implicit assumption that competition among nearby individuals is the primary determinant of observed dynamics. For example, Ellison et al. (1994) cite theoretical and experimental evidence showing that intrinsic variation in plant growth rate alone can give rise to hierarchical distributions of biomass or other metrics of plant size. Second, the statistical analysis of local competition is fraught with problems. For example, in most spatial models of plant population dynamics, plants are represented by mathematical points in space that occupy no area. For rosette-forming plants like Arabidopsis, individuals with a large diameter will necessarily have fewer neighbors than those

with a small diameter when plants are considered as points and neighborhoods are defined by a fixed radius. A plant with a tight rosette of leaves at
its base cannot have a neighbor above ground within the radius of the rosette
itself. Thus, it is then unclear whether a particular individual is small because it has many neighbors or has many neighbors because it is small. This
can be seen as another example of the general problem that it is usually inappropriate to look at plant performance (e.g., size) at time t as a function
of neighborhood conditions at time t because these two quantities are not
statistically or inferentially independent. Rather, we should take a dynamic
perspective and try to look at performance over the period t to $t + \Delta t$ as
a function of neighbor conditions at time t. By including appropriate controls and alternative hypotheses in replicating Silander and Pacala's (1985)
experiment with Arabidopsis, Ellison et al. (1994) concluded that neighborhood competition significantly affects population dynamics in plant monocultures because neighbors impair target plants' biomass, growth, and fecundity (relative to plants grown in the absence of competition). However,
in addition to the expected effects on fecundity mediated by biomass, there
were also neighbor effects on plant fecundity that were independent of the
effects on plant shape and biomass. This suggests that we cannot always
infer fecundity from plant size alone, which presents more obstacles to the
development of neighborhood models of plant performance.

Because of the indeterminate, modular nature of plant growth and the
resultant plasticity in plant size, which means that a neighbor may be a tiny
seedling or a huge adult, many researchers consider the number of neighbors alone an insufficient measure of local crowding. Therefore, several
researchers have attempted to describe the competitive neighborhoods of
plants more fully by looking at the distance and size as well as the number
and species of neighbors. Some researchers have adopted a more complete
physical Ansatz to describing a plant's neighborhood: the effect of a neighbor is proportional to its size and decreases with the square of its distance
from the subject plant (Weiner 1984). In such models different species can
still have different per-unit-biomass effects through the use of competition
coefficients. The amount of variation in individual plant performance that
such models can "explain" ranges from nearly 0% to almost 90%, depending on the model, the species, and the environment (Bonan 1993; Hara
and Wyszomirski 1994). The amount of variation accounted for is usually
quite low, possibly due to other factors such as genetic variation and microsite heterogeneity. For example, from 75 neighborhood analyses in 20
manuscripts Bonan (1993) derived an average of $42\% \pm 28\%$ (mean $\pm$ s.d.)

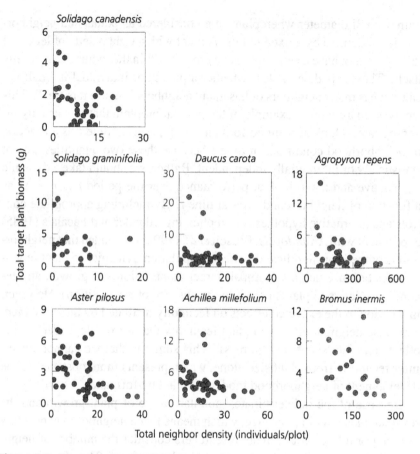

Figure 2.3 Biomass of target *Solidago canadensis* individuals versus local density for several species of neighbors. *Source*: Goldberg (1987).

of accounted-for variation. Sixty percent of these analyses accounted for less than half of the total variation in individual plant performance. These figures roughly correspond to an effect size of competition of 34% ± 5% (mean ± 95% confidence interval) given for producers in a meta-analysis of 73 studies (Gurevitch *et al.* 1992).

The relationship between plant performance (e.g., growth or size) and the abundance of neighbors is often triangular (see Figure 2.3), suggesting that local competition is more a limiting condition than a direct determinant of plant performance. When neighbors are very abundant, plants will be small; but when neighbors are scarce, plants can be either large or small. This type of relationship suggests that either our measures of local competition are inadequate or local competition is just one of many factors determining individual plant performance. When local crowding is severe, it limits plant growth; but when local crowding is not so severe, plant growth is often limited by other factors in the local environment, such as soil quality or water availability.

Because plant growth is indeterminate and modular, leading to huge plasticity in plant size, and because all plants have basically the same requirements, several researchers have proposed that the size or biomass of neighbors is perhaps more important than their number or species. Size is thought to be an important determinant of competitive ability (Keddy and Shipley 1989). If neighbor size is important, equal per-unit-biomass effects of neighbors is the appropriate null model for comparing the effects of different species of neighbors on a target species (Goldberg and Werner 1983). Of course, different species may have different per-unit-size effects, but the size of a neighbor may be the single most important factor determining its effect on another plant. Although some researchers think that an emphasis on the size of neighbors rather than their species may be a useful first approximation for describing and modeling neighborhoods (whereas other researchers think this would be "throwing out the baby with the bathwater"), even modeling size effects alone is not as straightforward as it might seem at first. The relative size of neighbors with respect to a target plant is often as important as their absolute size, because competition among plants is usually "size asymmetric," that is, larger plants have a disproportionate (for their relative size) effect on smaller plants. The inclusion of size asymmetry can result in a marked improvement in the performance of neighborhood models of competition (Thomas and Weiner 1989).

Recent work in forest ecosystems (Pacala *et al.* 1996) demonstrates how useful a neighborhood approach can be when combined with mechanistic models of resource competition, as long as a balance between the level of detail and generality can be found that meets the basic design criteria of simplicity, observability, and biological realism. However, this balance can only be found if field methods, statistical estimators, and model structure are designed simultaneously to ensure that parameters can be estimated from data collected in the field (Pacala *et al.* 1996). Many plant ecologists believe that such mechanistic neighborhood models are much more promising than earlier, purely phenomenological models.

2.5   Challenges for a Neighborhood Perspective of Plant Interactions

Another problem in developing a neighborhood-based view of plant–plant interactions is that the location of an individual, and often the individual it-self, usually is not easy to define. In certain cases where plants have an erect main stem, an individual's location can be adequately described as a point in two-dimensional space (Stoll *et al.* 1994). However, many plants do not have such a straightforward location or identity. Plants are often clonal (Figure 2.2; Box 2.2; see Chapters 3 and 4) and it is their open, "modular" architecture (Schmid 1990) that enables them to respond to their local en-vironment by "adjusting" birth, growth, and death of modules (Box 2.4). The modules (also called ramets if they are capable of living on their own) can remain connected through horizontal structures, and the extent and im-portance of transport processes and possible sharing of resources among modules are much debated in the literature. However, in contrast to mod-ules such as branches of trees, the ramets of clonal plants are also capable of living independently when the connections become severed; indeed, in many clonal plants the loss of connections after ramets have been placed and established seems to be the rule. Thus, clonal plants (and to some de-gree non-clonal plants) have a hierarchical structure in which the genetic individual (genet) is made up of smaller physiological units (ramets).

In the field it is usually quite difficult, if not impossible, to identify genets: what we see are ramets. In a ramet-based neighborhood view, many of an "individual's" neighbors may be (genetically) the individual it-self. Severe competition among genetically identical individuals decreases Darwinian fitness. From a sociobiological viewpoint, one can ask if in-dividuals are able to recognize "themselves," and there is evidence that plant roots react differently to contact with genetically identical roots than to roots of other genets (Huber-Sannwald *et al.* 1996). It could be that such recognition is quite limited, in which case we could build hierarchical neighborhood models with genets growing by iterating and placing ramets that maintain their connections for a given amount of time or under given conditions (Bell 1986). In fact, clonal growth can make the description of growth easier to the extent that we can describe size simply as the num-ber of ramets, either considering all ramets equal or having just a few size classes. The modeling of clonal growth can benefit from a neighborhood perspective (e.g., Cain *et al.* 1995). For example, most models of clonal spread are based on simple rules for internode length, branching frequency, and branching angle, and these parameters are assumed to apply to a whole

---

Box 2.4 Demography of plant parts and neighborhood interactions below the individual scale

It has been shown that modules of trees, like ramets of herbaceous plants, respond to their local environment (Jones 1985; Franco 1986; Jones and Harper 1987a, 1987b; Franco and Harper 1988). These studies used the demography of modules (see also Maillette 1982a, 1982b) to describe tree growth. The most comprehensive approach is to combine the study of allocation patterns with that of the demography of modules (Küppers 1994). It should be noted that even if modules within branches respond to their immediate local environment, branches within trees may still be integrated (Sprugel *et al.* 1991). For example, within shaded branches of trees that had some branches in full sunlight yearly growth increments were smaller than growth increments of branches of control trees growing completely in the shade (Stoll and Schmid 1998). This difference was interpreted as correlative inhibition, for example, resource allocation to branches in the sun that therefore inhibited the growth of shaded branches. Evidence for such an interpretation was demonstrated in pea plants (*Pisum sativum*) by Novoplansky *et al.* (1989). When they grew two connected shoots in different light conditions, the shaded shoot was inhibited and eventually even withered and died. It elongated and became etiolated only when the shoot in the stronger light was removed.

We know of only one model of plant competition that explicitly includes plasticity at a modular level (Sorrensen-Cothern *et al.* 1993). Including plasticity through modular foliage in their spatially explicit model of competition fundamentally changed the population structure. For example, if trees were equipped with plasticity through modular foliage, the whole stand had a greater leaf area index and individuals grew taller than without plasticity.

---

genet (Sutherland and Stillman 1988). A more recent view is that the general rules are not completely fixed for an individual, but vary locally in response to local conditions. A parallel can be drawn between the growth of individual plants responding locally to the environment and foraging animals (Hutchings and de Kroon 1994). The point here is simply that even if the individual plants we see above ground are genetically and sometimes physiologically parts of larger organisms, behavior can, to some degree, be explained by local neighborhood conditions. The definition of an individual and of its neighborhood is best determined in the context of the scientific question being asked.

The difficulties of spatially delimiting plant neighborhoods are made much worse because the size of a plant increases by many orders of magnitude during its growth from a seedling to a reproducing adult. Thus, we would expect the area in which a plant experiences and is experienced by neighbors to increase accordingly. Most neighborhood studies investigate plant performance over one time interval with one neighborhood definition. It is difficult for many field ecologists to imagine such an approach yielding sufficient information to enable us to predict the dynamics of the system. In many cases, it may be necessary to model plant growth over several intervals during which the neighborhood grows along with the plant. Therefore, we predict better performance of future models with dynamic neighborhoods as opposed to the *a priori* fixed neighborhoods used in cellular automata.

2.6   Suggestions for Modelers

We have taken an empirical approach in an attempt to communicate to theoreticians some of what we empiricists think are the essential aspects and central questions in the development of a neighborhood approach to interactions among plants. There has been an increasing emphasis on bridging the gap between modeling and empirical work in ecology, although the tendency for the two to go in totally different directions is still very strong (Weiner 1995). In the spirit of building this much-needed bridge, we make the following suggestions to our theory-oriented colleagues.

- Models are more likely to be useful in solving empirical problems if they are directed at observed patterns in nature rather than at very general, abstract questions (Grimm 1994; Weiner 1995).
- Each model should have a clear purpose. The model should not be the object of study, but merely a tool. If the "occupational hazard" of being a field ecologist is thinking that everything is important and therefore must be included, the occupational hazard of theoreticians is building general, abstract models without a clear goal other than exploring the dynamics of the model itself. The question of how to simplify the model (or, in other words, how to determine what can be ignored) is closely linked to the model's goal. If the purpose of a model is to predict community dynamics (i.e., the abundance of different species over time), then what is essential for the model may be very different from what is essential in a model concerning persistence (how many species can coexist, independent of their specific abundances) or the genetic diversity of some species.

Perhaps the most important potential role for modeling in ecology today could be to help direct empirical research. At this point in the development of ecology, models may be most useful not for what they can deliver in the near future in terms of prediction, but for elucidating the sorts of information that are most needed if we are to build models that can do what we want them to do. Models can help make empirical research more strategic and less haphazard. If a formal, systematic theoretical framework can serve as an alternative to trial and error in exploring empirical "parameter space," advances in mathematical theory will have contributed much to what future generations of scientists will (we hope) call the emergence of a mechanistic and predictive ecology.

Acknowledgments   We wish to thank E. Schreier for drawing Figure 2.2, and D.E. Goldberg and the Ecological Society of America for permission to use Figure 2.3. J.R. Porter and D. Newbery made valuable comments on an earlier version of the manuscript. This work was financially supported by grants from the Swiss National Science Foundation and the Roche Research Foundation.

3

Spatial Interactions among Grassland Plant Populations

Jonathan Silvertown and J. Bastow Wilson

3.1 Introduction

The neighborhood perspective on plant interactions discussed in Chapter 2 shows the richness of processes in plant neighborhoods and how difficult it can be to obtain a detailed mechanistic understanding of them. Faced with this dilemma, plant ecologists have sometimes adopted a more phenomenological approach, which integrates over the known (and unknown) mechanisms by which neighbors interact, using a few neighborhood-dependent measures of plant performance. These measures are interaction coefficients, usually thought of as relating to competition between pairs of species.

In multispecies communities, competition coefficients can summarize a lot of information on the effects species have on one another and help ecologists to understand the spatial structures that develop over the course of time. This chapter describes

- how plant ecologists have tried to investigate competition in the field,
- what has been learned about plant competition coefficients from this field work,
- the information available about the structure of matrices of these competition coefficients, and
- how sensitive the outcome of competition is to the initial spatial pattern.

We focus on grassland communities. These communities are almost ideal for studying the significance of spatial pattern for ecological processes because they are patchy and approximately two-dimensional. In fact, the appearance of patchiness is more complicated than it might seem at first. Borders between patches that look well-defined from a distance can turn out to be quite diffuse when viewed more closely. Patches of one species frequently overlap patches of another, and the canopies of different species can be intimately mixed (Watkins and Wilson 1992; Chapter 4), especially in species-rich communities. It is our impression that patches are more discrete and individuals overlap less in communities of modular organisms that lack roots, such as lichens and bryozoans. The lesson seems to be that if you want to be discrete, life must remain superficial!

One reason for the complexity of spatial structure in perennial grass-lands is that, strictly speaking, they are not two-dimensional. Function-ing connections in the soil can stretch between ramets (effectively, rooted branches) that appear unconnected above ground; structurally it is better to think of a perennial grassland as a forest buried up to its canopy in soil than as a two-dimensional surface. It is possible that annual communities (Wu and Levin 1994; Moloney and Levin 1996; Rees *et al.* 1996) fit a two-dimensional model better than perennial ones. However, in this chapter we only discuss the latter, and it should be kept in mind that an assumption of two-dimensionality can be no more than a rough approximation to such communities.

3.2   Methods for Measuring Competition in the Field

Definitions of competition tend to be contentious, but we shall use the op-erational definition that competition is an interaction between neighboring plants in which each suppresses the other's performance (i.e., how well it grows). It is helpful to distinguish between two components of a species' behavior when in competition with neighbors. Each species has an *effect* on its competitor and a *response* to competition from its neighbor (Miller and Werner 1987). The effect on its competitor is actually the same as the com-petitor's response, so it is enough to measure the responses of both species to determine their interaction. However, competitive effect and response of a single species when it interacts with a neighbor need not be correlated: an interaction may be asymmetric, such that one species exerts a much greater effect than its neighbor. Only when effect and response are both negative is the interaction a genuinely competitive one.

The experimental literature on interactions between herbs, many of which are grassland plants, is voluminous. However, the literature *relevant* to understanding the dynamics of spatial structure in plant communities is quite small for two reasons. First, most grasslands only remain grasslands if they are mown, grazed, or burned. Therefore the majority of the literature, which reports interactions between plant species in pots in glasshouses, ig-nores important aspects of reality in the field. Second, field experiments often have time scales so short that they deal only with the performance of individual plants, not with the effects of competition on population dynamics. At the very least, experiments need to be run long enough for some population turnover of plants in the experimental mixture to occur. Otherwise, one must make untested assumptions about how short-term effects of competition upon *performance* translate into longer-term effects upon population *dynamics* and spatial structure.

The standard method for measuring competition between grassland plants in the field is to remove one or more neighbors from around a "target" individual and to observe the target's response compared with that of a control individual whose neighbors are left in place. Alternatively, plants can be added to the neighborhood around a target; however, in practice this is much more difficult to achieve than neighbor removal, because planting is likely to damage the target's roots.

A target's response to the manipulation of its neighbors can be measured in a variety of ways. The most common method is to harvest target plants at the end of the experiment and compare their dry weight with that of control plants (see Box 3.1). This method does not explicitly measure how a target plant's occupancy of space changes in response to competition from neighbors, even though any increase in plant size due to neighbor removal is likely to mean the plant occupies more space.

If we are interested in how plants compete for space, then it makes sense to measure directly how occupancy of space is affected by competitors. However, a removal experiment is not necessarily the best method for determining this effect. The degree to which a target plant occupies the space left vacant by a neighbor that has been artificially removed is a measure of how well the target invades bare space, but invading bare space is not the same as capturing space from a neighbor. For example, white clover (*Trifolium repens*) is a creeping plant that is good at invading bare ground but poor at invading moderately tall grass (Thompson and Harper 1988). From the point of view of competition for space, what is really of interest is how well white clover can invade other species and how well it resists displacement by other invaders. These questions can be addressed by monitoring the spatial distributions of plants in a community at a fine scale over a long period and calculating species-by-species replacement rates (Law *et al.* 1997), or by experimentally creating interfaces between monospecific patches of different species and recording how plants move across the interface (Silvertown *et al.* 1994).

The advantage of measuring competition in terms of the space neighbors capture from each other is that the spatial consequences of the interaction

---

Box 3.1 Nonspatial measurement of competition

A variety of models have been used to estimate interspecific competition coefficients from plant performance in nonspatial experiments. If the total weight of plant material reaches an asymptote as density rises, then a simple relationship that often fits experimental data well is the reciprocal yield model (Wright 1981; Spitters 1983). For a mixture of two species i and j, the mean weights w_i, w_j of the plants under competition are

$$\frac{1}{w_i} = B_{i0} + B_{ii} N_i + B_{ij} N_j , \qquad (a)$$

$$\frac{1}{w_j} = B_{j0} + B_{jj} N_j + B_{ji} N_i , \qquad (b)$$

where B_{i0} and B_{j0} are the reciprocals of mean plant weight when competitors are absent (i.e., have been removed) for species i and j, respectively; B_{ii} and B_{jj} measure intraspecific competition; B_{ij} and B_{ji} measure interspecific competition; and N_i and N_j are the densities of species i and j, respectively. Competition coefficients are generally standardized so that the effect of species j on species i is expressed relative to the effect of species i on itself. Thus, competition coefficients (α, β) can be calculated from Equations (a) and (b):

$$\frac{B_{ij}}{B_{ii}} = \alpha \quad \text{and} \quad \frac{B_{ji}}{B_{jj}} = \beta , \qquad (c)$$

where α is the competition coefficient for species j's effect on species i and β is the competition coefficient for species i's effect on species j. The effect of one species is the response of the other.

In principle it is straightforward to extend this to a multispecies community, giving a matrix B of all the pairwise responses (and effects). Row i of this matrix describes the different responses of i to its neighbor species; column j gives the different effects of neighbor species j on each species.

---

are explicit. The main practical problem is that measuring invasion requires clearly demarcated boundaries between species, which can be difficult to engineer between mature plants in the field. There is also an interpretational problem because invading another species across a boundary is not necessarily the same as capturing space at its expense. Even when a species is at equilibrium density in monoculture, physical space exists between shoots that in theory can be penetrated by a smaller neighbor – in effect, this is invasion without displacement of the resident territory holder. If two species are able to interdigitate physically, then an invading species may occupy more space than the resident loses. Only when the competition coefficient (see Box 3.1) is equal to 1, and species are thus effectively identical to one another, do gains and losses of space balance. Ideally, then, one needs a measure of the space lost by the invaded species (which we term its *displacibility due to competition*) and a measure of that taken by the invader (the invader's *invasiveness*), because the two quantities may not be the same. (The loser's *invasibility*, however, is equivalent to the invader's invasiveness.) So few experiments on spatial competition have been conducted (see Section 3.3) that there is not yet a standard way to conduct or analyze this type of experiment; Box 3.2 makes some suggestions on how to proceed.

3.3   Results of Field Experiments

In this section, we look in turn at experiments where the effects of competition were measured in terms of performance and experiments that explicitly measured the effects of neighbors on each other's occupancy of space.

Experiments on effects of neighbors on performance

Goldberg and Barton (1992) reviewed field experiments on interspecific competition between plants. Among the 101 experiments they examined, by far the most common (76 cases) test of interaction between species involved removal of neighbors from around a target plant whose growth was then compared with that of control plants. In the other experimental designs, one or more species were planted in the field to manipulate relative or absolute densities of competitors. Some of the studies did not test responses for statistical significance, but 79% of those that did found neighbor presence or abundance to have a negative effect on the target, and 18% found some positive effects on the target.

The commonness of negative interactions is no surprise, as all plants require more or less the same resources and thus compete with one another when in physical proximity. A plant must acquire and hold onto space in order to gain access to resources such as light, water, and the mineral nutrients obtained from soil. Some partitioning of space into different niches might be possible – for example, if plants can divide space by placing their roots at different soil depths or if short species can tolerate conditions in the shade of taller ones – but there seem to be far fewer such niches than there are species in most plant communities.

Nonetheless, neighbors sometimes have positive effects on each other (Box 2.1), and the mechanisms that lie behind such interactions require further study (Callaway 1995). One such mechanism that is potentially

---

Box 3.2 Measurement of competition through occupation of space

There is not yet a consensus on how spatial competition experiments should be carried out. This box suggests a protocol based on monocultures of species i and j placed side by side so that each can invade the other. The basic arena comprises a strip of each species, and the region of interest is close to the boundary between them.

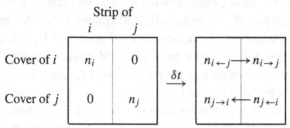

Here, n_i and n_j are the cover of species i and j, respectively, at the start; after a period δt has elapsed, $n_{i \to j}$ is the cover of species i in j's territory and $n_{i \leftarrow j}$ is the cover of i in its original territory (which changes as a result of invasion by j). Two simple measures of the interaction as experienced by species i are as follows:

$$\text{rate of invasion of } i \text{ into } j: \quad \frac{\delta n_{i \to j}}{\delta t} = \frac{n_{i \to j}}{\delta t},$$

$$\text{rate of displacement of } i \text{ due to invasion by } j: \quad \frac{\delta n_{i \leftarrow j}}{\delta t} = \frac{n_i - n_{i \leftarrow j}}{\delta t},$$

$$\text{with net rate of change of } i \text{ with } j: \quad \frac{\delta n_{i \to j} + \delta n_{i \leftarrow j}}{\delta t}.$$

Elaborations on this theme would entail controls for changes in cover that occur irrespective of the presence of the other species and measures expressed per unit cover. Because the spread into each other's territory takes place gradually, the boundary strip in which measurements are made should be made long and thin.

---

relevant in grasslands (because of the importance of herbivory) is that plant species that are well-defended against herbivores provide refuge for more palatable species growing in their immediate vicinity (Atsatt and O'Dowd 1976). There are some examples of such a mechanism operating with insect herbivores (Pfister and Hay 1988), but the mechanism runs counter to experience in grasslands grazed by vertebrates. Plants that are poisonous to vertebrates, such as ragwort *Senecio jacobaea*, thrive under the heaviest grazing pressures when all their competitors are removed: rabbits simply graze all other plants from around them. Sheep, too, can graze in a very fine-grained manner, but can the smell of a poisonous plant protect its neighbors from these grazers? An experiment by Launchbaugh and Provenza (1993) suggests that this is unlikely. They found that sheep would not persistently avoid food that smelled aversive if it did not also taste of a toxin. In other words, there is no refuge for a palatable species in the chemical neighborhood of an unpalatable one. Contrary results have been found with woody plants that have physical defenses against herbivory such as spines. Watt (1919) reported that spiny holly and brambles provided refuge from herbivory that facilitated the regeneration of oaks, and Hjalten *et al.* (1993) have described a case where hares attacked birch trees less often when the trees were mixed with less palatable woody species. It is probably too early to rule out the existence of associative defense from vertebrate herbivores among grassland plants because too few field experiments have investigated the combined effects of competition and herbivory in grasslands.

In view of the fact that most grasslands only continue to exist by virtue of mowing, grazing, or burning, it is important to design field studies of competition so that the effects of these processes can be determined. Yet, in their survey Goldberg and Barton (1992) found only eight studies where competition and herbivory were combined in a factorial design, none of which were performed in grasslands even though both these processes are important in this kind of vegetation. To this list we can add an experiment we carried out in a seminatural grassland at Little Wittenham, England (Bullock *et al.* 1994; Silvertown *et al.* 1994). We outline this experiment in some detail in the next subsection because the experiment deals explicitly with spatial invasion of one species by another and because the results are used in a simulation model described later in this chapter.

Experiments on space occupancy by neighbors

The seven studies we know of that have explicitly looked at mutual invasion between grassland species are listed in Table 3.1. All of them can be improved upon and all leave some important questions unanswered. For instance, a common design, used in more than half the studies, was a hexagonal plot sown as a monoculture and surrounded by a different species on each side. Hexagonal plots were usually closely packed, producing a tessellated arrangement. The problem with this design is that the plots are not statistically independent of each other, and two invaders of a third species may soon begin to interfere with one another. Cellular automaton models (Silvertown *et al.* 1992) suggest that such higher-order interactions among

Table 3.1 Summary of spatial competition experiments on grasses.

Study	No. of species (grasses/dicots)	Design	Invasiveness and invasibility correlated?	Competition transitive?	Ranks altered by treatment?
Thórhallsdóttir (1990b)	5/1	6 × 6 species in hexagonal plots	No	Yes	–
Turkington (1994)	6/0	6 × 6 species in hexagonal plots	Positively in 1 treatment	No	Yes
Wedin and Tilman (1993)	4/0	4/6 possible pairwise combinations	No	Probably	Yes
Marshall (1990)	7/0	1 × 6 species in hexagonal plots	Fourfold range in invasibility among 6 species	–	–
Van Andel and Nelissen (1981)	2/5	Hexagonal plots	No	Cannot tell	–
Stewart and Potvin (1996)	3/1	Mapped plants in mixture	Positively in some treatments	Yes	Yes
Silvertown et al. (1994)	4/0	All species pairwise	No	Yes	Yes

three or more species may be important, but in experiments like those listed in Table 3.1, where interactions are assumed to be pairwise, they could make interpretation difficult.

Our own experiment (Silvertown *et al.* 1994) was designed to avoid the problems of nonindependence and unplanned higher-order interactions and to determine the effect of grazing upon spatial competition. We performed the experiment in a seminatural grassland, examining the mutual invasion of all possible pairs of four perennial grass species. The competition experiment was nested within a replicated sheep grazing experiment that imposed two levels of grazing in each of three seasons of the year (winter, spring, summer), factorially combined in a randomized block design to give eight different field environments (grazing treatments) in each of two replicate blocks. Grazing levels in winter and spring were determined by the presence or absence of sheep. Summer grazing levels were determined by adjusting sheep stocking levels to achieve mean sward heights of 3 cm or 9 cm. The vegetation in the experimental area was dominated by the grasses *Lolium perenne* and *Agrostis stolonifera*, but also contained about a dozen other grasses and a very low percentage cover of some 40 dicotyledonous species. Our mutual-invasion experiment used the grass species *Festuca arundinacea* (A), *Festuca rubra* (R), *Lolium perenne* (L), and *Poa pratensis* (P), in all pairwise combinations, and involved transplanting pairs of small, monoculture turfs into the native grassland so that each pair shared a common boundary over which invasions could occur. Over a year later, we counted the number N_{ij} of tillers of each species i that had invaded the neighboring turf of species j. Net invasion rates were calculated for each permutation of species i, j as $N_{ij} - N_{ji}$. (In the absence of more detailed information, these net rates assume that i's gain in tillers in j's neighboring turf is equal to j's loss.) The rates are shown in Figure 3.1.

In so far as it is possible to generalize from this ragbag of experiments, three main results emerge. First, there is no clear tendency for the invasiveness and invasibility to be correlated (Table 3.1). The rare cases where there is a positive correlation suggest the existence of a gradient in mobility between species from the "stay-at-homes" (neither invasive nor invasible) to the "tramps" (invasive and invasible).

Second, competition is typically transitive, meaning that it is possible to rank the species by their net invasion rates in a strict pecking order. For example, in our experiment, the spring-grazed, 3-cm treatment shown in Figure 3.1 has the following pecking order of net invasion rates: L > R > A > P.

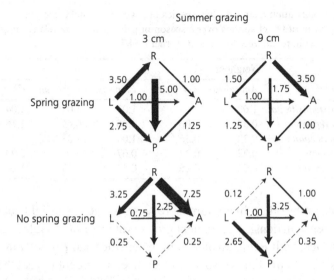

Figure 3.1 Net rates of tiller invasion between all pairwise combinations of four grass species under four different grazing regimes in a seminatural grassland. *Abbreviations*: A = *Festuca arundinacea*; R = *Festuca rubra*; L = *Lolium perenne*; P = *Poa pratensis*. *Source*: Silvertown *et al.* (1994).

Third, the rank a species has in a transitive competitive hierarchy depends on the grazing treatment. This is true, for instance, in our own experiment (Figure 3.1). Analysis of variance of square-root transformed tiller counts identified a significant interaction between spring and summer grazing treatments affecting invasion rates. For example, Figure 3.1 shows that net invasion rates between L and R were reversed by spring and summer grazing treatments.

Caveats

Clonal plants (see Box 2.2) vary greatly between and even within species in the degree to which they invade the space around them. Lovett Doust (1981) introduced the terms "guerilla" for plants that send out long stolons and "phalanx" for those that place new ramets near the parent. There is a continuum of variation between these two extreme types of clonal morphology.

It might seem inevitable that such differences in morphology will have major consequences on space capture in the field. But the following example of two herbs suggests that matters are not necessarily so simple. Schmid and Harper (1985) compared the responses of a phalanx species (*Bellis perennis*) and a guerilla species (*Prunella vulgaris*) to the presence

Table 3.2 Competition matrix for five species in a weedy old-field community. Numbers given are growth of target species over a season in pairwise mixture relative to growth of target in monoculture. *Source*: Miller and Werner (1987).

Target species	Neighbor species				
	Ambrosia	*Agropyron*	*Plantago*	*Trifolium*	*Chenopodium*
Ambrosia artemisiifolia	1.00	0.94	1.03	1.02	1.04
Agropyron repens	0.37	1.00	0.86	0.95	1.16
Plantago lanceolata	0.23	0.36	1.00	1.08	0.73
Trifolium repens	0.22	0.27	0.67	1.00	0.90
Chenopodium album	0.08	0.16	0.43	1.15	1.00

or absence of the other. In mixture there was evidence of a slight but significant difference in density-dependent growth (number of modules per plant) of the species, with *Bellis* favored at high densities and *Prunella* favored at low densities. In the light of such a difference, one might expect the planting arrangement of species to influence interspecific competition between them, but no such influence was found. This suggests that slight differences in how species capture space may have undetectable consequences in the field, where many other factors impinge upon interactions between species. The species differed in how they recolonized bare ground, with *Prunella* doing so almost exclusively by clonal growth and *Bellis* recolonizing twice as many gaps by establishment from seed as by clonal growth (Schmid 1985). The colonization of gaps is a significant spatial process in grasslands, but beyond the scope of this chapter.

3.4 Competition Matrices

Competition matrices provide a convenient summary of the information on pairwise interactions (including competition) obtained by field experiments (Box 3.1). An example of such a matrix is shown in Table 3.2; this matrix is based on five species at an early stage of succession in an old-field community (Miller and Werner 1987). The information in such matrices plays an important part in spatio-temporal modeling of communities, because it is these interactions that couple the dynamics of different species. Depending on the structure of the matrices, we may find spatial patterns developing that are homogeneous or heterogeneous in space, including exotic kinds of behavior such as the spirals described in Chapter 10. It is important for theoretical work to be properly informed about what is known (and not known) empirically about these matrices.

Competition matrices are few, and complete matrices for whole communities are unknown. However, the pairwise field competition experiments in the literature might be representative of the interactions one would find more widely within a single community. In any case, this is all the information there is to go on at present. Sparse though the data are, two features of competition matrices stand out: their transitivity (Shipley 1993) and the lack of difference between intra- and interspecific interaction strengths.

Transitivity

A competition matrix is transitive if each species can be unambiguously assigned a rank in a competitive hierarchy. Keddy and Shipley (1989) and Shipley (1993) surveyed studies where several species were combined in pairwise competition experiments and concluded that transitivity was the rule. However, Goldberg (1997) pointed out that this survey did not distinguish between hierarchies of effect of competition (ranking down the columns of the matrix) and hierarchies of response (ranking across the rows of the matrix). She surveyed field and non-field experiments, testing for concordance in the hierarchies of effects and responses. Under standardized environmental conditions, 80% of response hierarchies (12/15 cases) and 66% of effect hierarchies (14/21 cases) showed evidence of consistent transitivity; these findings are in approximate agreement with those of other authors. (The matrix in Table 3.2, for example, has consistent hierarchies, both for response and effect.) But hierarchies often changed from one environment to another: 41% of response hierarchies (9/22) and 50% of effect hierarchies (7/14 cases) were contingent upon environmental conditions.

One reason why transitive competition is important is that, in an unvarying environment, it should lead to the exclusion of all other species by the topmost species in rank. This is in contrast with intransitive networks of the form $A > B$, $B > C$, $C > A$, etc., in which coexistence of the species may be possible (Karlson and Jackson 1981). How plant species come to coexist is a long-standing issue in ecology (Silvertown and Law 1987), and it is evident that the transitivity of competition matrices does not help to resolve it. However, it may be premature to reach firm conclusions: the mosaic cycles described in Chapter 4, in which there is a cyclic sequence of states within patches of vegetation, although not quantified to the level of competition matrices, appear to be driven by intransitivities.

Intra- versus interspecific competition

The relative magnitude of interactions within and between species is also important in the context of species coexistence. In Lotka–Volterra competition models, the coexistence of competitors is expected if intraspecific competition is stronger than interspecific competition, particularly when species are aggregated. Yet, there is no clear indication of diagonal dominance in competition matrices. A lack of diagonal dominance is evident, for example, in Table 3.2: growth is in some cases much more limited by heterospecifics than conspecifics. This is in agreement with Goldberg and Barton's (1992) conclusion that "The very limited field evidence available for coexisting species thus suggests that conspecifics do not usually compete more strongly than heterospecifics."

Competition matrices determined from traits of species

Faced with the great difficulty of estimating competition matrices from manipulation experiments in the field, it would obviously be helpful to find other ways to proceed. Ultimately, one might hope to specify the structure of the competition matrix simply from knowledge of the relevant traits of the species in the community, but we are some way from being able to do so at the moment. Indeed, the first question to ask is whether it is reasonable to expect to be able to predict competitive outcome from species' traits (Goldberg 1997). Recent progress in the study of plant life histories suggests a qualified "yes" in answer to this question. Evidence is accumulating that life-history variation in plants has a single major axis with many traits correlated with lifespan (Silvertown et al. 1993; Condit et al. 1996; Franco and Silvertown 1996; Grime et al. 1997). If many important traits are strongly correlated with one another, the likelihood that they will include traits that influence competitive outcome is great, because competitive ability has important effects on fitness. If life-history traits including those influencing competitive ability have a simple correlation structure, then predicting competitive outcome should only require knowledge of a limited number of trade-offs.

Two obvious qualifications that could prevent the use of traits to predict competitive outcome are that (1) species are genetically variable for traits that affect competitive ability and (2) the fitness conferred by any trait is environment dependent. If trait correlations within species are the same as trait correlations between species, then the first problem can be circumvented simply by treating genotypes rather than species as the taxonomic unit. The second problem cannot be circumvented but can be accommodated by defining how the fitness of traits changes along environmental gradients.

A study by Sugiyama (in press) is interesting in this context. Sugiyama investigated how varietal differences within three species of grasses

(18 cultivars of *Dactylis glomerata*, 15 cultivars of *Lolium perenne*, 5 cultivars of *Festuca arundinacea*) influenced the outcome of competition against a standard cultivar of another species. When the cultivars were grown in monoculture, no differences among cultivars in yield per unit area were found. Although the cultivars differed in mean tiller densities, such differences were counterbalanced by tiller weights so that the yields were the same. However, when the cultivars were grown in interspecific mixture with a standard cultivar of another species, they varied hugely in their final yields. After two years in mixture with *Festuca*, the yield of *Dactylis* cultivars varied from 40–84% of the total yield of the mixture, and the yield of *Lolium* cultivars in mixture with *Dactylis* varied from 54–90%. *Festuca* also showed large differences in proportion of final yield among five cultivars (11–43%). In all three species there was a significant relationship between the mean tiller weight of cultivars and the relative contribution of that species to the yield of the mixture. However, the form of the relationship was different in each species: in *Lolium* it was linear and positive, in *Dactylis* it was parabolic with a maximum at 120 mg mean tiller weight, and in *Festuca* it was parabolic with a maximum at 50 mg tiller weight.

The differences Sugiyama observed between species could be explained in terms of two relationships involving tiller size: (1) regrowth after defoliation and (2) competitive ability. Defoliation took the form of periodic clipping of the plots. Regrowth after defoliation declines with tiller size, but competitive ability increases with tiller size. Under light clipping and high soil fertility, cultivars with large tillers were favored, but a complete competitive reversal (favoring small tillers) was achieved by Sugiyama and Nakashima (1995) through heavy clipping or low soil fertility. This work is the most convincing demonstration that we know of that competitive outcome might be predicted from the traits of the species concerned.

Aarssen (1983) suggested that genetic variation for competitive ability within species might lead to intransitive competitive relationships between species and thus provide a mechanism for coexistence. We know of no formal theoretical model of this mechanism, so it is not clear under what assumptions it would work. However, clearly one necessary condition is that competitive ability should vary greatly among genotypes, and Sugiyama's experiments as well as much earlier work (Charles 1964) suggest that it does. It is not so clear, however, whether there is sufficient genetic variation for competitive ability *within* neighborhoods (Taylor and Aarssen 1990), although if grassland plants are very mobile it is possible that the genotypic composition of neighborhoods changes with time.

Box 3.3 Estimation of transition matrices from an invasion experiment

Numbers of tillers N_{ij} of each species i invading each other species j were determined from a field experiment in a seminatural grassland described in Section 3.3. Tillers of the four species in the experiment were of different sizes and occupied different amounts of space, so values of N_{ij} were converted to the proportion of space occupied by species j taken by species i using the formula

$$a_{ij} = N_{ij} (Cn_i)^{-1} \ ,$$

where n_i is the density of tillers of species i in monoculture as determined at the start of the experiment and C is the area available for invasion, which we estimated to be a strip 10 mm wide and 80 mm long along the turf border. Note that this conversion from N_{ij} to a_{ij} assumes that the space occupied by species i must be equal to the space lost by species j, though this assumption is not necessarily correct (see Section 3.2). Values of a_{ij} for all i, j formed the elements of a transition matrix A of dimension (4,4) summarizing spatial interactions (replacements) between species sharing a common border. Elements (a_{ii}) in the leading diagonal of this matrix were set to 1.

The competitive relationships among the four species were transitive in all five transition matrices representing different patterns of grazing. That is, in each case the species could be unambiguously ranked in terms of the proportion of available area captured from each other in one iteration of the model with a random starting arrangement. These hierarchies are shown at the bottom of Table 3.3 and are different from those in Figure 3.1 [which are based on tiller numbers given by Silvertown *et al.* (1994)] because of the transformation of tiller numbers into proportion of area captured necessary for the model.

3.5 Community Consequences of Spatial Interactions

The transitivity of competition matrices that emerges from field experiments might be thought to imply that the highest-ranking species in a competitive hierarchy should always displace those of lower rank from the community. However, this ignores spatial structure: once the spatial component of community dynamics is taken into account, it is not so obvious that the highest-ranking species has an immediate advantage. The aggregation of species, and their pattern of juxtaposition, may radically alter the outcome of competition over the medium term (Silvertown *et al.* 1992).

Below, we show that the initial spatial configuration of a community matters greatly for the dynamics of communities with transitive competition matrices. We do this by means of a cellular automaton model, in which the transition probabilities of cell states are given by the results of our spatial invasion experiment, described in Section 3.3 and Figure 3.1.

A cellular automaton model

The model was based on a lattice of 40×40 square cells, with synchronous updating and a von Neumann (four-cell) neighborhood. Because it is more realistic to think of a finite spatial region for the patchy environments and communities found in nature, we used absorbing boundaries instead of the periodic boundaries often used in the theoretical literature.

At any time t a cell could be occupied by one of the four species in Figure 3.1. At time $t + 1$ the species in a cell had a chance of being randomly invaded and replaced by one of the neighbors present at time t in one of the four immediately adjacent cells. Replacements of one species by another occurred according to transition probabilities derived from invasion rates measured in the field experiment (see Section 3.3 and Box 3.3). Five transition matrices were used: an overall matrix based on mean invasion rates across the whole grazing experiment and transition matrices based on invasion rates in each of the four combinations of spring and summer grazing (Figure 3.1).

To provide some contrasting initial spatial patterns, we aggregated the species into monospecific bands and placed the bands in each of the 12 possible orders shown in Table 3.3; we also used a spatially random initial configuration. Each species started with the same total abundance of 400 cells regardless of the initial pattern. For each of these 13 patterns and 5 transition matrices, we ran 10 realizations of the cellular automaton model to predict the composition of the community after 500 iterations (notionally 500 years).

Effects of initial spatial configuration

For all five matrices, random initial arrangements of competitors led to the rapid extinction of all species except the first in rank (Table 3.3). In contrast, when species were initially aggregated in bands, the outcome depended strongly on how the bands were ordered. Replicate runs were reasonably consistent, thus by comparing how the survival of lower-ranked species varied with the initial arrangement of species, a set of parsimonious rules for survival in each grazing treatment can be deduced. Table 3.4 gives these rules in terms of the juxtaposition of species required for survival. The main features to note about the results are as follows.

Table 3.3 Results of a cellular automaton model on grasses. The results show initial banding arrangements of four species of grasses (R, L, A, P) and the species surviving after 500 iterations. See Figure 3.1 for species names. The model was based on an invasion experiment in a seminatural grassland described in Section 3.3. The most common result is given; the number of runs in which it occurred out of 10 is shown in parentheses. Surviving species are listed in the standard order RLAP, with each dash indicating an extinction. Grazing treatments are as follows: (1) summer grazing to mean sward height of 9 cm, no spring grazing; (2) summer grazing to mean sward height of 9 cm, spring grazing; (3) summer grazing to mean sward height of 3 cm, no spring grazing; (4) summer grazing to mean sward height of 3 cm, spring grazing.

	Species surviving in each grazing treatment				
Initial banding	Mean over whole experiment	(1)	(2)	(3)	(4)
RLAP	RLAP (10)	RLAP (10)	RLAP (9)	R - - P (10)	- LA - (7)
RLPA	RLA- (9)	RLA- (10)	RLA- (8)	R - AP (10)	- LA- (10)
RALP	RL-- (10)	RL-- (8)	RL-P (9)	R - - P (9)	RLA- (10)
RAPL	RL-- (10)	RLA- (5)	RL-- (8)	RL-P (10)	RLA- (10)
RPLA	RLA- (10)	RLA- (9)	RLAP (5)	RLAP (8)	- LA- (10)
RPAL	RL-- (10)	RL-- (6)	RL-- (7)	RL-P (9)	RLA- (10)
LRAP	RL-P (6)	RLAP (10)	RL-P (8)	R - - P (10)	- LA- (6)
LRPA	RL-- (7)	RLA- (8)	RLA- (6)	R - AP (8)	- LA- (8)
ARLP	RL-- (10)	RLA- (10)	RL-P (9)	R - - P (10)	- LA- (10)
ARPL	RL-- (10)	RLA- (10)	RL-- (6)	RL-P (10)	- LA- (6)
PRLA	RLA- (10)	RLA- (8)	RLAP (9)	R - - P (6)	- LA- (10)
PRAL	RL-- (10)	RL-- (9)	RL-P (10)	R - - P (6)	RLA- (10)
Random (unbanded)	- L - -	- L - -	R - - -	R - - -	- - A-
Competitive rank in transition matrix	L>R>A>P	L>R>A>P	R>L>A>P	R>P>L>A	A>L>R>P

Table 3.4 Results of a cellular automaton model on grasses. The results show initial banding arrangements needed for the survival of grass species after 500 iterations. See Figure 3.1 for species names. The term x denotes an allowable position for any species not identified by name. For example, xRxLxAx for target species A indicates that A will survive as long as the bands are ordered R, L, A, at the start; the fourth species P may be present anywhere (or absent altogether) without affecting the survival of A. In the column labeled "Ranks," species are replaced with their competitive ranks in the underlying transition matrix. See Table 3.3 for explanation of grazing levels.

Target species		Initial banding patterns by species	
Rank	Identity	Identities	Ranks
Overall matrix			
1	L	xxxx	xxxx
2	R	xxxx	xxxx
3	A	xRxLxAx	x2x1x3x
4	P	xxAP*	xx34
(1) Grazing: Summer to 9 cm, no spring grazing			
1	L	xxxx	xxxx
2	R	xxxx	xxxx
3	A	xxxA or xxAP	xxx3 or xx34
4	P	xxAP	xx34
(2) Grazing: Summer to 9 cm, spring grazing			
1	R	xxxx	xxxx
2	L	xxxx	xxxx
3	A	xRxLxAx	x1x2x3x
4	P	xxxP	xxx4
(3) Grazing: Summer to 3 cm, no spring grazing			
1	R	xxxx	xxxx
2	P	Not xxRP*	Not xx12
3	L	xRxPxLx	x1x2x3x
4	A	xRxPxA*	x1x2x4
(4) Grazing: Summer to 3 cm, spring grazing			
1	A	xxxx	xxxx
2	L	xxxx	xxxx
3	R	xLxAxRx*	x2x1x3x
4	P	xxAP*	xx14

*Survival does not occur in every run with these conditions.

First, the aggregation of conspecifics into bands clearly increases the length of time to extinction. Even though the competitive relationships were transitive, aggregation provided a spatial refuge for weaker competitors and permitted their survival in the medium term (Silvertown *et al.* 1992). Although ultimately weaker species were excluded, survival for an estimated 500 years suggests a delay that could be of considerable ecological significance. Few communities remain unperturbed by disturbance or environmental change for 500 years. In a community where species are finely mixed, and indeed overlapping (e.g., Roxburgh *et al.* 1993), indirect effects such as apparent mutualism may further delay the elimination of weak competitors (Lawlor 1979; Stone and Roberts 1991).

Second, because different grazing treatments alter the relative invasiveness and invasibility of species and hence their competitive ranks, they produced different outcomes from the same initial conditions. This is as one might expect from the differences in competitive hierarchy already noted in Figure 3.1, but the effects are made considerably more complicated by the initial spatial configuration of the lattice.

Third, a species' rank in the competitive hierarchy is not a sufficient basis for predicting how the initial spatial configuration of neighbors will affect its survival. This can be seen by replacing the species names with their ranks in the five grazing treatments. One can then ask whether the rules are constant for a species ranked, say, third, regardless of which species actually occupied that position. Survival rules translated into ranks are shown in Table 3.4; if anything, they are less consistent than the rules based on the identity of species. However, our results do not totally escape generalization on the basis of ranks. As might be expected, species 1 (i.e., that with the highest competitive ability) survived any arrangement of its competitors. In four of the five matrices, species 2 also survived any arrangement of its competitors. For species 3, the identity rather than rank of the first and second species was important; location at an edge was not essential to survival. For species 4, location at an edge was always required if the species was to survive, often with a particular immediate neighbor but not necessarily species 3.

3.6   Concluding Comments

Here we briefly revisit the four main issues of this chapter. First, plant ecologists have developed various ways of manipulating local densities of individuals to investigate competition in the field. Second, these field experiments tell us that interspecific competition is widespread; positive interactions also occur, but have been little studied. Even more poorly understood are +/– relationships. Such interactions might seem unlikely between plants, but Schwinning and Parsons (1996a, 1996b) recently examined the interaction between the grass *Lolium perenne* and white clover (*Trifolium repens*) and concluded that the spatial and temporal dynamics are similar to those seen between predator (in this case, the grass) and prey (the clover). Evidently there are more kinds of interactions between grassland plants than might at first be imagined.

Third, the few studies that have looked at several species in a community allow one to make some preliminary comments about competition matrices. The matrices tend to be transitive, but species' ranks alter with treatments such as grazing and nutrient addition. This is true of studies measuring invasion as well as those simply measuring short-term performance, and it is a property that is important for modeling grassland communities as spatio-temporal processes. Even though a species' competitive rank is contingent upon a range of environmental factors, it seems possible that relative competitive abilities might be predicted from traits, given a sufficient knowledge of both the trade-offs between traits and the fitness consequences of different trait values in different environments. In grasses, tiller size seems to be a promising trait for use in predictions.

Finally, our exploration of the spatial dynamic consequences of different transitive competition matrices leads us to suggest that the spatial configuration of competing species imposes an extra level of complexity on the system so that knowledge of species' competitive ranks is not sufficient to predict the outcome, at least over the medium term.

Acknowledgments This work was supported in part by the Natural Environmental Research Council (UK). We are grateful to the Northmoor Trust and the staff of Little Wittenham Nature Reserve for providing field facilities. We thank O. Miramontes and S.H. Roxburgh for comments on an early version of the manuscript. Figure 3.1 was reprinted with permission of the British Ecological Society. Table 3.2 was reproduced with permission of the Ecological Society of America.

4

Spatio-temporal Patterns in Grassland Communities

Tomáš Herben, Heinjo J. During, and Richard Law

4.1 Introduction

This chapter turns from the neighborhood processes of plants (discussed in Chapters 2 and 3) to the spatial patterns of plant communities. It provides some empirical background on what plant ecologists have learned about spatial patterns and their dynamics. In doing this, we examine criteria for distinguishing between patterns generated internally by biotic processes within communities and patterns generated externally by structural features of the abiotic environment. Although the focus of the chapter is primarily empirical, we describe several conceptual models that plant ecologists have considered while thinking about the dynamics of spatial pattern.

Spatial patterns play a pivotal part in plant community dynamics. When we look at these patterns, we see the outcome of a complex series of past events – including the biotic processes of birth, death, and movement, and some element of chance – together with the structure of the abiotic environment. At the same time, we see a baseline on which future events depend; how spatial patterns unfold through time depends on their previous states. Through these dynamics, large-scale patterns emerge that cannot be predicted from local processes, because these processes are coupled to the global pattern of the community. This coupling will become a major theme in later chapters of the book.

4.2 Spatio-temporal Patterns in Plant Communities

Ecologists have long recognized the importance of spatial pattern in terrestrial plant communities and have devoted considerable effort to documenting the patterns that occur (Blackman 1935; for reviews see Pielou 1968; Greig-Smith 1983; Kershaw and Looney 1985). The locations of plants in such communities can be thought of as patterns in a two-dimensional space.

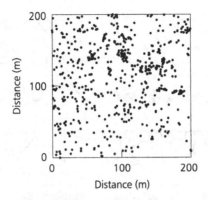

Figure 4.1 Locations of longleaf pine trees in an area 200 m × 200 m. *Source*: Cressie (1991).

A pattern can be recorded as a map of points (Figure 4.1) if there is plenty of space between individuals at ground level, as in woodlands and desert communities. Alternatively, a pattern can be recorded as a map of clumps of finite area (Figure 4.2) if conspecifics cannot be readily separated from one another and the edges of clumps are distinct (Pielou 1968). If, as is often the case, clumps do not form clear boundaries, spatial location can be measured to the level of discrete cells in a lattice (e.g., Thórhallsdóttir 1990a; Herben *et al.* 1993), as shown in Figure 4.3. When dealing with multispecies plant communities, individuals or clumps can be indexed by species and if necessary by other qualities as well.

Much of the early research on spatial pattern was designed to establish whether plants of a single species are independently distributed in space or, if not, whether the departure from randomness is toward overdispersion (regularity) or underdispersion (clumping). In formal terms, this work tests the null hypothesis that spatial pattern is completely random, that is, generated by a homogeneous spatial Poisson process (see Chapter 5). This research showed that patterns are often highly nonrandom (Greig-Smith 1983; Kershaw and Looney 1985). In grassland communities, departures from randomness are typically toward clumping (see, for instance, the spatial pattern of the grass *Nardus stricta* in Figure 4.3). Such clumping is not surprising in view of the importance of clonal growth among the grasses and other taxa that dominate grassland communities, because one property of clonal growth is that offspring tend to be placed close to the parent plant. However, clustering is by no means inevitable; in other kinds of communities, such as those in arid environments, overdispersion can sometimes be detected. In these arid environments, plants that occur close to one

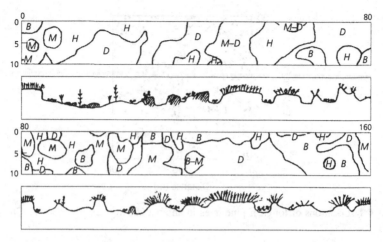

Figure 4.2 Patchy spatial structure of a 160 cm × 10 cm plot in a Breckland grassland. Abbreviations: B = Building phase; M = Mature phase; D = Degenerate phase; H = Hollow phase. *Source*: Watt (1947).

another may run an increased risk of death due to competition or allelopathy (Phillips and MacMahon 1981).

Although nonrandomness is unsurprising, and deemed uninteresting in some quarters, it has a crucial bearing on the dynamics of plant communities. Processes in plant communities take place in small neighborhoods (see Chapter 2); in addition to random differences in neighborhoods from one plant to another, the nonrandom spatial patterns often observed in the field lead us to expect systematic differences in neighborhoods. Such spatial structure is very likely to affect the births and deaths that lie at the heart of community processes, and to ignore it would be to miss a critical coupling in the dynamics. Nonetheless, surprisingly few researchers have studied the dynamics of spatial pattern in plant communities. Plant ecologists interested in temporal dynamics have until recently turned more to animal ecology for inspiration (for a review, see Harper 1977), and the innate mobility of many animals means that spatial pattern is a less obvious feature of these communities.

A notable early exception was the work of Watt (1960). Starting in the 1930s, Watt tracked the fine-scale spatial structure of grassland communities in the Breckland in England every year for several decades, generating the kind of data needed to gain insight into spatio-temporal processes in plant communities. However, few researchers followed his lead; a survey of permanent plots in Britain (Hill and Radford 1986) did not report a single study (apart from forestry studies) containing spatial patterns of multispecies communities through time.

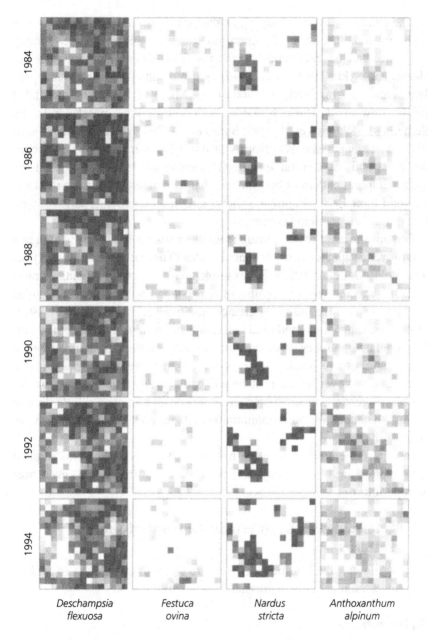

Figure 4.3 A spatio-temporal process in a montane grassland community at the Krkonoše, Czech Republic. The plot (50 cm × 50 cm) is divided into 15 × 15 cells and is dominated by four grass species. The depth of shading in each cell is proportional to abundance; white indicates absence of the species. *Source*: Law *et al.* (1997).

The situation is now changing as data on spatio-temporal processes in grassland and other kinds of communities are becoming available. Among these are data on calcareous grasslands from Öland, Sweden (van der Maarcl and Sykes 1993; Sykes et al. 1994); results from several independent studies of grasslands in Britain (e.g., Mitchley 1988; Thórhallsdóttir 1990a; Law et al. 1993; McLellan 1995), limestone grasslands in the Netherlands (Willems et al. 1993; Sykes et al. 1994; Mitchley and Willems 1995), savanna grasslands in North Carolina (Sykes et al. 1994), and mountain grasslands of Central Europe (Herben et al. 1993, 1995); and the spatio-temporal pattern of bryophytes, that is, mosses and liverworts (During and Lloret 1996). Figure 4.3 shows a representative time series of spatial pattern in a grassland community.

Perhaps the most striking result from these studies has been the high rate of turnover of species at small spatial scales (Thórhallsdóttir 1990a; Sykes et al. 1994; McLellan 1995). For example, in the Krkonoše mountain grasslands, no species showed significant temporal autocorrelations at a given location in space for time lags greater than five years (Herben et al. 1993, 1995); in savanna grasslands and grasslands in Sweden and the Netherlands the cumulative number of species in microsites more than doubled over a time interval of five years. The bryophyte communities also showed strong microscale turnover, whereas at larger spatial scales little change was observed. In general, macroscopic structures change relatively little despite the fast dynamics of these communities (Pärtel and Zobel 1995; van der Maarel 1996).

In sum, the empirical research has shown that spatial patterns in plant communities are often highly nonrandom and far from fixed, even over short periods of time.

4.3 Externally versus Internally Generated Spatial Patterns

Here we turn to the question of what drives the nonrandom and rapidly changing spatial patterns in grassland communities.

On boundaries

The first important issue is whether nonrandom spatial patterns are generated by processes operating within the community or whether the pattern is imposed on the community from outside. Externally generated patterns are not central to the perspective of this book (the dynamics of imposed patterns are far less intricate than those of emergent patterns), but internally generated patterns certainly are. The distinction between the two is crucial and immediately raises the question of how to define internal and external in a conceptually consistent way.

In some cases the distinction appears clear-cut. Some patterns are generated by factors that, for practical purposes, are independent of the vegetation, such as the depth of soil on rock outcrops (Burgman 1987; Ohsawa and Yamane 1988). Such factors are usually deemed external: they affect the vegetation but are not affected by it. Even here it is arguable that the time period could be extended sufficiently for the pattern-generating mechanism to be affected by the vegetation dynamics, as soil development is ultimately influenced by the plant community.

In other cases, appropriate boundaries (or scales) in space and time are less easily decided and it is more difficult to distinguish between external and internal. Consider, for instance, the effects of large vertebrate grazers on grassland vegetation. In savanna or prairie ecosystems (e.g., Collins and Glenn 1988), such grazers form an obvious part of the ecosystem and are responsible for much of the pattern in the vegetation. At the same time, the changes that occur in the spatial pattern of the vegetation over the course of time feed back to the behavior of the grazers. Here the coupling is mutual – grazer behavior affects spatial patterns of vegetation, which affect grazer behavior – and it makes sense to think of the grazers as internal to the system. In comparison, the fine-grained structure of small, dense tussocks of *Nardus stricta* intermingled with loose turfs of *Deschampsia flexuosa* in Czech mountain grasslands (Herben *et al.* 1993; see Figure 4.3) has hardly any effect on the behavior of the grazers, even though the vegetation pattern itself may strongly depend on their action. Here the feedback loop is broken, and the grazers may be thought of as external to the system as far as the vegetation dynamics are concerned. Detailed knowledge of a system is needed to determine the appropriate boundaries, and such knowledge is often lacking.

Covariation of pattern in community and environment

One might hope to settle the issue in practice by looking at the extent to which spatial pattern in a plant community correlates with heterogeneity in the physical and chemical environment. Obviously, pattern in the environment does exist and is, at least in some cases, correlated with the pattern in vegetation (Stark 1994). However, plants change their environment in various ways, and the presence of environmental heterogeneity does not preclude the possibility that the heterogeneities might have been

generated by the plants themselves. A good example of how pattern in
the environment depends on vegetation is the horizontal light variation in
forests (Pearcy *et al.* 1994) or in grasslands (Silvertown and Smith 1989;
Tang and Washitani 1995).

Another example is the fine-scale correlation of vegetation and nutrient availability (Jackson and Caldwell 1993; Robertson and Gross 1994).
Such heterogeneity is expected to change through time, at least at small
scales, due to continuous uptake by plants and microorganisms and input
by weathering and decomposition (Stark 1994). Here again, environmental
heterogeneity can be said to be internally generated by plants themselves.
A nice example is given by Schlesinger *et al.* (1996), who found 35–76%
of the variation in soil nitrogen in grasslands of the Chihuahuan Desert
of New Mexico to be within 20 cm of individual plants. This is likely to
be the result of local accumulations of soil nitrogen under *Bouteloua eri-
opoda*, a perennial bunchgrass. Although relatively little is known about
the spatial and temporal scales of below-ground heterogeneity, the available
data show below-ground patterns with scales similar to those of vegetation
above ground and short-term feedbacks between vegetation and soil.

Correlations between plant pattern and environmental pattern are
widespread, but we have to go beyond these correlations to understand the
causal pathways involved. Experiments that manipulate the plants and test
for changes in the environment are needed to check for the existence of
feedbacks. In some cases these experiments are straightforward; for example, removing a plant has an immediate effect on the light environment in
its neighborhood. In other cases, such as plant–soil relationships, appropriate manipulative experiments are more difficult to design and perform.
However, until such feedbacks are ruled out, we cannot conclude that the
spatial pattern we observe in a community is imposed on it externally by
the environment.

4.4   Concepts in Spatio-temporal Processes in
       Plant Communities

Plant ecologists have been more concerned with documenting spatial patterns and how they change through time than with studying and understanding their dynamics in formal models. With a few notable exceptions
(e.g., Pacala *et al.* 1996), this applies as much to recent research on spatio-
temporal patterns as it does to earlier work on "snapshots" of spatial patterns. Nonetheless, certain kinds of pattern dynamics have led to informal
conceptual models in the minds of ecologists, and at least one of these, the

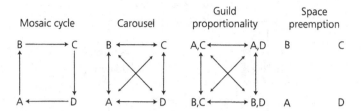

Figure 4.4 Schematic depiction of four models of plant community dynamics. A, B, C, and D are species and arrows indicate possible transitions between states. In the guild proportionality model, A and B are in the same guild (as are C and D) and do not occur together.

mosaic cycle, has become quite influential in ecological thinking. Here, we outline four conceptual models of vegetation dynamics; Figure 4.4 summarizes their main features. The data we focus on come mainly from grasslands because these communities have a relatively simple structure and because, perhaps for this very reason, most studies on pattern and process have been conducted in grasslands.

Mosaic cycles and related patterns

The mosaic cycle stems from work on a variety of grassland, heathland, and woodland communities; the data were first synthesized by Watt (1947) in a famous paper on pattern and process in the plant community. Watt noted, for example, the patchy structure of a grassland in the Breckland (see Figure 4.2). The grass *Festuca ovina* plays a prominent part in this community: when a seedling grows and develops into a young vigorous plant, soil accumulates around it and forms a hummock. This building phase is followed by a mature phase, during which the original plant becomes separated into small fragments, which later become colonized by lichens, leading to the degenerative phase as the hummock becomes eroded. Eventually the cycle returns to its starting point, the hollow phase. These and other data led Watt to picture plant communities as spatial mosaics of patches. The sequence of events at a single spatial location is cyclic, going through the building phase to maturity to degeneration to the hollow phase, as shown in Figure 4.5. Watt (1947) did not himself refer to this as a mosaic cycle, but the term has come to be associated with cyclic turnover of the state of local spatial patches (Remmert 1991; van der Maarel 1996).

A key notion in the mosaic cycle is environmental change due to the presence of particular plant species. This change produces conditions that are favorable to a set of species different from those resident at the site. In many cases, the change is closely linked to the life cycle of one dominant

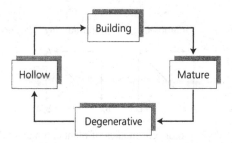

Figure 4.5 Cyclic sequence of patch states occurring at a single location corresponding to the patch states in a Breckland grassland (see Figure 4.2).

species, whose morphology thus forces a certain spatio-temporal pattern on the community. This is well illustrated by Watt's Breckland grassland, where the dominant species is *Festuca ovina*. The bunch grasses in Spanish "steppe" areas are another example; these grasses strongly affect light and moisture conditions in their immediate surroundings, leading to a characteristic zonation of bryophyte and lichen communities around them (Martinez-Sanchez *et al.* 1995). Although the individual grass tufts are long-lived, the dominant pattern of widely separated "bunches" gradually shifts due to tuft mortality and re-establishment elsewhere.

The carousel model

The carousel model is an outcome of research on species-rich limestone grasslands on the Swedish island of Öland carried out by van der Maarel and his colleagues. In these grasslands a large number of plant species are mixed at a fine spatial scale; in 0.01 m^2 quadrats the mean number of species present was approximately 16 in 1986 (van der Maarel and Sykes 1993). What is particularly interesting about the dynamics of this system is the rapid turnover of species at a fine spatial scale over subsequent years. Tracking the species composition of the same quadrats over time, they found that the cumulative number of species was 24 five years later, yet through the period the number actually present remained at about the 1986 level (Table 4.1). This led van der Maarel and Sykes (1993) to suggest that each site can be colonized by most of the species present in the community.

To interpret their observations, van der Maarel and Sykes (1993) introduced the metaphor of a carousel (or merry-go-round), the idea being that species move around the spatial region occupied by the community and sooner or later reach every location in the space. It is implicit in their model that there is no particular order in which species appear; in other words, the current state of any microsite is essentially independent of its previous state.

Table 4.1 Total and cumulative number of species in 0.01 m^2 and 0.25 m^2 quadrats in an alvar grassland on the island of Öland, Sweden. All values are means of 40 observations. *Source*: van der Maarel and Sykes (1993).

Quadrat size	1986	1987	1988	1989	1990	1991
Mean number of species per quadrat						
0.01 m^2	16.3	13.9	14.9	14.5	12.1	14.9
0.25 m^2	26.1	25.8	26.8	25.9	27.1	27.3
Cumulative number of species in quadrat since 1986						
0.01 m^2	16.3	19.0	20.6	21.7	22.8	24.1
0.25 m^2	26.1	29.3	30.5	31.5	32.6	33.6

The carousel model thus differs from the mosaic cycle in that there is no definite order in which species replace one another at a given location.

Rapid turnover of species of the kind envisaged in the carousel model appears to be a common phenomenon in grassland systems (for a review, see van der Maarel 1996). A similar process at a much smaller scale is evident in ephemeral bryophytes. They occur in the soil diaspore bank (in an inactive state) in large numbers, but appear above ground only for a short time after some kind of small-scale disturbance (e.g., frost heaving in spots with locally reduced cover of their larger plant competitors, worm casts, ant hills). The plants live just long enough to produce new diaspores, which are little dispersed in space and thus remain in the local diaspore bank. The fact that such diaspores have been found nearly everywhere in the soil of grasslands, forests, and other communities (During and Ter Horst 1983; During *et al.* 1988) suggests that in the long run each microsite in the communities will have been such a gap; the actual aboveground pattern of these species then reflects shifting patterns in vegetation structure and animal activities.

Because the carousel model is a simplified view of the community dynamics, it should not be expected to give a detailed description of processes in species-rich communities. Further analyses on the data set obtained by van der Maarel and Sykes (1993) indeed showed that some microsites tended to be consistently species-rich, whereas others were consistently species-poor (Wilson *et al.* 1995a; van der Maarel *et al.* 1995). This may be indicative of some fine-scale niche differentiation, or perhaps just a high persistence of species with slower dynamics. It is unclear to what extent the turnover of species is really generated within the community by processes such as limited module life span and competition. Indeed, spatio-temporal patterns of communities with structure imposed externally by random local disturbances may be very similar to those of communities with structure generated internally.

Guild proportionality

The concept of guild proportionality was put forward by Wilson and Roxburgh (1994) to explain combinations of species co-occurrence in a lawn in New Zealand. Using a point sampling system, they were able to identify groups of species that tended to exclude one another, as opposed to species from different groups that did not do so. For example, at locations where two species were present, it was more likely that one would be a grass and one a forb (herbaceous dicotyledon) than that both would be grasses or both forbs. They interpreted this as being due to a system of several guilds of species, each of which locally saturates to a fixed low number of species from a larger species pool. Only two guilds were identified in the community studied by Wilson and Roxburgh (1994), perhaps because of the low statistical power of the technique used. By referring to the groups as "guilds," they were implying that the nonrandom co-occurrence of species within guilds is due to competitive exclusion. Guild proportionality prevents the system from attaining all combinatorially possible species sets locally, even though the species may have fast dynamics at a small spatial scale. Like the carousel model, no constraint is imposed on the order in which species replace one another locally, but in contrast to it, there is an additional constraint that some combinations of species are more likely than others; in the extreme, some combinations could be absent altogether.

How general a feature of grasslands guild proportionality is remains to be established; moreover, techniques for establishing the guild structure of a community are themselves a matter of debate (Goldberg 1995; van der Maarel *et al.* 1995; Wilson *et al.* 1995a). In our opinion, the evidence for the existence of such a rigid guild structure within grassland communities is rather weak. Still the guild proportionality model of community structure is a feasible notion and cannot be dismissed on the basis of the evidence currently available.

Space preemption

Sometimes "ownership" of a site confers a competitive advantage on the owner and leads to continued occupation of the site by a single (set of) species. An example of this is found in certain peat bogs that show a pronounced mosaic of elevated sites (hummocks) and depressions (hollows), each with a specific set of species. Analysis of soil cores shows that such occupation may last for several centuries at least, and there are demonstrated cases of it lasting for several millennia (Casparie 1969, 1972). As in the mosaic cycle, the species occupying the site induce environmental change there; but in contrast to the mosaic cycle, the environmental change favors the species already present. The presence of a plant species thus feeds back, producing conditions that support the species already occupying the site.

As far as we know, plant ecologists have not yet given a name to this process; we refer to it as space preemption. Founder events in communities with space preemption have major effects on subsequent spatial patterns and maintain a remarkably constant spatial pattern for components of the community. Space preemption is quite different from all of the earlier models because there is little turnover in species composition at each location in space and the species' spatial distributions are determined primarily by random effects at the beginning of the habitat colonization.

It is not known how common space preemption is; even in the case of peat bogs, some early research suggested a mosaic cycle as an appropriate model (Watt 1947), although this was shown to be incorrect by later observations. In most communities, founder effects are probably weakened by plant mortality and subsequent establishment of shoots or seedlings of other species.

When are these models spatial?

At their simplest, the models discussed here are concerned only with events at a single site. Depending on the model, the sequence of states may be random or cyclic, or there may be no change in state at all. In the context of this chapter, it is important to know the extent to which events at one site interact with those at neighboring sites. Such a coupling may then generate large-scale spatio-temporal structure in the community.

The importance of interactions between neighboring sites depends on the specific processes that drive the dynamics. In the case of the mosaic cycle, Watt (1947) made it quite explicit that spatial coupling of events in neighboring patches played a major role in the dynamics. Some degree of neighbor dependence is also known from the gap-phase dynamics of forests; data from the Barro Colorado tropical forest (see Chapter 13) show that, where a gap occurs, additional gaps are likely to form in its immediate neighborhood.

In the carousel and guild proportionality models, no specific assumption is made about dependence of dynamics at one site on the state of its neighborhood. If species taking part do not produce daughter plants on stolons or rhizomes [e.g., the bryophytes of During *et al.* (1988) or annual species of the calcareous grassland species of van der Maarel and Sykes (1993); see also Wilson *et al.* (1995a)], the spatial coupling is likely to be rather weak. Clonal growth, however, often constrains the species that appear at the site to those that occur in neighboring sites and strengthens the spatial coupling (Law *et al.* 1993; Herben *et al.* 1995). In the space preemption model, it is hard to envisage any major spatial coupling; the local state at a given microsite is determined by initial conditions at that microsite only.

In some cases, spatial coupling is imposed by an external, but spatially homogeneous, factor: the very clear wave-like pattern in coastal heaths in westernmost France, for instance, seems to be caused by the continuous effect of strong westerly, salt-laden winds from the sea, causing gradual mortality of branches on one side with continuing growth on the other side. A similar process of wave-like pattern formation due to clonal growth of plants was described by Watt in his classic 1947 paper (see discussion of dwarf Callunetum). Similar patterns may be generated by moisture retention of dryland vegetation growing on a sloping terrain. This vegetation tends to form horizontal stripes that slowly move up the slope, apparently due to the capture of runoff water from the bare stripes in between by specialized pioneer species growing at the upper rim of the stripes (Cornet *et al.* 1988).

The data thus show that spatial dependence is a common property of those communities to which the mosaic, carousel, and guild proportionality models apply, although it is not necessarily implied by these models. A possible classification of some field systems into classes with strong and weak spatial coupling is suggested in Table 4.2; it should be understood that, with knowledge as it stands at present, this can be no more than a tentative classification.

4.5 Ergodic and Non-ergodic Communities

The obvious way to distinguish communities according to the models discussed in Section 4.4 is to look at the sequence in which species appear and disappear at some spatial location over the course of time. The carousel model, at its simplest, envisages that the location will be visited by every species in the community as time progresses and that there will be no particular order in which the species occur. This is in contrast to the mosaic cycle, in which species appear and disappear at the location in a particular

Table 4.2 Classification of several field systems into different underlying models

Underlying model	Weak spatial coupling	Strong spatial coupling
Mosaic cycle	Spanish "steppe" (Martinez-Sanchez *et al.* 1995)	Regeneration waves (Watt 1960)
Carousel	Soil bryophytes (During *et al.* 1988) Alvar grassland (van der Maarel and Sykes 1993)	Perennial grasslands (van der Maarel and Sykes 1993)
Guild proportionality	–	Mown grassland (Wilson and Roxburgh 1994)
Space preemption	Hummock and hollow of peat bogs (Casparie 1969)	–

sequence that is repeated over and over again. A community in which space preemption operates is quite different because the species composition is frozen in time; the fact that species differ from one location to another is not reflected at a single location followed through time. Guild proportionality is different again, as the states a site can take are restricted; states involving combinations of incompatible species are less frequently found than those of compatible species.

These distinctions lead to an issue of what, if any, plant community can be said to satisfy an assumption of ergodicity (see Box 4.1 for an introduction to the notion of ergodicity). The assumption of ergodicity is that sample averages equal ensemble averages; roughly speaking, this means that the average long-term state of a plant community at a single location is the same as its average state across different locations at a single time.

There are two reasons for raising the issue of ergodicity. The first is to clarify an important distinction between some of the conceptual models above. A community operating under the rules of the carousel model is ergodic, at least in the simplest form of the model. Given a long enough period of time, each location will be visited by each species in proportion to its abundance at different locations in space. This is in contrast to a community to which space preemption applies. Here, the community cannot be ergodic; the time average at a single location will be nonzero only for the single species that reached the location first, whereas the ensemble average will have nonzero values for all the species in the community. A community with a mosaic cycle is ergodic but has an additional property that the states follow one another in a specified sequence. In the presence of guild proportionality, ergodicity would again apply, but the tendency for certain

Box 4.1 Ergodicity

Ergodicity (from the Greek: *ergon*, work; *hodos*, path) is a general mathematical notion. A process is said to be ergodic if and only if its sample average is equal to its ensemble average. But this is somewhat cryptic and requires explanation.

Think of a sample $f(u)$, which is a function of some variable u. We might, for instance, be thinking of population size through time, in which case the sample f is population size and the argument u is time. Arguments other than time are equally possible, for example, population size could be given as a function of spatial location at a single point in time. The sample average S is given by

$$S = \lim_{U \to \infty} \frac{1}{U} \int_0^U g(f(u))\, du \ .$$

Notice that we think in terms of a function $g(f)$ of the sample rather than of the sample itself, as this makes the notion much more general; the function $g(f)$ can be thought of as a filter through which the sample is observed. At its simplest, we could have $g(f) = f$; in the case of population size as a function of time, S is then the population size averaged over time. Another possibility is $g(f) = (f - \overline{f})^2$, in which case S would be the variance of population size averaged over time. Taking the limit $U \to \infty$ gives the average of the sample as the variable u becomes large.

To understand what is meant by the ensemble average, one needs to think of the probability $m(f)$ that the sample f takes each state, for instance the probability that population size takes values $0, 1, 2, \ldots$, and so on. In the context of a stochastic process, one may think of many realizations of the sample that together form an ensemble; associated with this ensemble is a probability $m(f)$ for each state at some point in time. The function $m(f)$ is said to be the density of the measure of f, and the ensemble average E is given by

$$E = \int g(f)\, m(f)\, df \ .$$

Thus, for the process to be ergodic, (1) the sample average and ensemble average must exist, and (2) the relation $S = E$ must be satisfied.

Because it is a rather general notion, ergodicity can mean different things in different contexts. The context here is a spatio-temporal process of a plant community. The object of study is a multispecies time series, so the sample $f(u)$ is some measure of the population size – such as number or biomass – of each species at a particular location in space and the argument u is time. There is no need to think of a filter through which to observe the sample, so we take $g(f) = f$, and the sample average is simply the long-term average over time of the population sizes. The ensemble

continued

Box 4.1 *continued*

average can be thought of in terms of the same stochastic process at many independent and equivalent locations; at some point in time, one can then envisage a probability associated with each state of the community over this ensemble. The ergodicity assumption is that the sample average equals the ensemble average.

species not to occur together would be reflected by correspondingly low values for the probability of these states.

The second reason for raising ergodicity is methodological. Plant ecologists do not have long time series available and are unlikely to have them for many years to come. Under these circumstances, it is natural to ask what, if anything, could be learned about the average state in the long term from the state at many locations at a single time. The answer is that information from a single point in time can be used, but the ergodic assumption must be satisfied. Hara *et al.* (1995), for instance, were explicit in making this assumption when they estimated parameters in a model of tree growth in a multispecies community.

There are some caveats to keep in mind when applying the notion of ergodicity to plant communities. When making use of information across space to construct an ensemble average, the locations must be sufficiently remote from one another to be treated as independent. This criterion puts a lower bound on the size of the spatial region from which locations can be drawn to construct the ensemble average. At the same time the locations must be equivalent, otherwise they cannot be treated as replicates of the same stochastic process. This second criterion puts an upper bound on the spatial region; clearly, if we include locations in quite different environments, the stochastic process is going to be different. An analogous issue arises when following a community at a single location over time: if the time period is too long, changes in the environment are bound to occur. This is an intrinsic difficulty in using observations from permanent plots, since ecological conditions are never constant in time (van den Bergh 1979; Silvertown 1980). The matter is often complicated by the fact that the critical signals in the environment responsible for dominance shifts are not known and therefore no correction can be made for them (Stampfli 1995; Rosén 1995; Walker *et al.* 1994). In view of the trade-off between the criteria of independence and equivalence, how to choose an appropriate spatial and temporal scale in plant ecology is ultimately a matter of judgment.

4.6   Concluding Comments

Grassland communities often have fast dynamics at small spatial scales and show pronounced nonrandom spatio-temporal patterns. The extent to which these dynamics can be said to be internally generated depends in part on where system boundaries are drawn. At present, information on which to establish appropriate system boundaries is in short supply (grids with cells of only a few square centimeters, study durations of a few years), and the mechanistic understanding of the system being studied is often insufficient. On the other hand, there is nothing that strongly contradicts the notion of structure being generated internally, and there are various properties of birth and death processes that lead naturally to development of nonrandom spatial pattern.

The literature on plant ecology contains several informal models of how plant community dynamics may proceed in space and time; we have identified in particular the mosaic cycle, the carousel model, guild proportionality, and space preemption. These models have contrasting consequences for local turnover of species and implications for the assumption of ergodicity. However, understanding of the spatial component of plant community dynamics is as yet very limited and stands to be greatly enhanced by the current growth in mathematical knowledge of spatio-temporal processes of the kind described in the chapters that follow.

Acknowledgments Figure 4.1 was reprinted by permission of John Wiley & Sons Inc. from *Statistics for Spatial Data* by N.A.C. Cressie, Copyright © 1991, John Wiley & Sons. Figure 4.2 was reprinted by permission of the British Ecological Society.

5

Statistical Modeling and Analysis of Spatial Patterns

David R. Cox, Valerie Isham, and Paul Northrop

5.1 Introduction

This chapter deals with statistical methods for the analysis and interpretation of empirical data. It covers three main subjects: essentially descriptive statistical methods, the development of stochastic models, and more formal statistical methods interrelating the models with empirical data. Spatial and spatio-temporal data arising in ecology can be of many different types. We aim to give some general principles with outline illustrative examples. In Sections 5.3 and 5.4, we concentrate largely on observations in the form of sets of points in space and time (for example, animals or plants of a particular species).

Systems and data under study may be classified in numerous ways (e.g., Figures 4.1 to 4.3). For example, they may be based on quadrat counts – that is, sums or integrals over finite areas – or on essentially continuous observation in space, or sampled at a discrete grid of points in space or along line transects. Similarly, data may be observed in continuous time or aggregated over time intervals, or they may be sampled at discrete time points, such as yearly intervals. The analysis may concentrate on a single feature of interest or, more commonly, on the interrelationships between different features, for instance, on the interdependence of different species or on the dependence of one or more features on explanatory variables. Data may be observed at one time point and several or many spatial positions, or may be replicated in time as well. For simplicity of exposition, however, in Section 5.2 we deal briefly with data replicated only in time (see Section 5.2, Descriptive time series analysis).

One very important aspect that we do not discuss is study design: for general discussions of experimental design, see Cox (1958); for discussions of sampling methods, see Thompson (1992) and Thompson and Seber (1996). Study design includes the design of sampling schemes and the size, number, and siting of quadrats, for example. If experimentation is possible, statistical principles of the design of experiments will apply. In any case, key issues such as which features to measure to achieve interpretable conclusions have statistical aspects.

5.2 Descriptive Analysis

We begin with an outline of some essentially descriptive methods, that is, techniques that are not highly specific to a detailed stochastic model. The distinction between these and more formal methods is not clear-cut, but on the whole the relatively descriptive methods tend to be most useful in the preliminary inspection of data. These methods include checks of data quality, especially the isolation of portions of data of suspect quality; checks for the absence of major unanticipated features; and the presentation of conclusions in a simple form. These methods may also be useful in indicating the broad type of model likely to be appropriate for more detailed analysis.

We deal with data referring to two-dimensional space and to time. Thus, to some extent the methods are an amalgam of those for spatial processes (Matérn 1960; Greig-Smith 1964; Ripley 1981; Diggle 1983; Cressie 1991; Renshaw 1991) and the much more extensively developed methods of time series analysis (see, e.g., Priestley 1981, and Diggle 1990).

It may be tempting to regard time as a third dimension and to apply methods for data in three-dimensional space. While this is formally possible, in particular with methods in the frequency domain (i.e., based on some form of spectral analysis), the directional character of time is lost; thus for most purposes we preserve the special character of the time dimension.

Data configurations

The types of data that may arise can be classified in various ways depending especially on the spatial and temporal arrangement of the data points. We confine ourselves to data recorded at discrete, equally spaced time points, for example, from annual observations. Then we may have

- a small number of spatial points recorded repeatedly over a longish time period;
- a rich configuration of spatial data recorded at a small number of time points with the spatial configuration remaining fixed;
- a combination in which there is substantial replication in both time and space;
- the probably rarer possibility that the spatial areas observed at different time points are different, so that each time point can be regarded as generating an independent replicate of the spatial processes involved.

Descriptive time series analysis

For ease of description, we suppose that we have a fairly small number of time series, with each series consisting of a single variable and the different series corresponding to the same variable at different places. The following broad guidelines are helpful for preliminary analysis, although not all are relevant in every application.

- Plot the series in a form such that the series in different places are comparable.
- At any stage of the analysis it may be useful to transform the value y_t at time t, for example, to log y_t.
- Examine for defective data, such as those induced by instrument failure, false zero values, and extreme fluctuations.
- Examine for trends by dividing each series into sections and calculating from each section summary statistics such as means, standard deviations, and, especially for qualitative data, marginal frequencies.
- With long series it may be possible and useful to distinguish between trend, long-term systematic structure, and long-range stochastic dependence or scaling. In the last of these (Cox 1984), stochastic components of variation of longer and longer wavelength enter in such a way that statistical summaries (such as standard deviations) of temporal averages of sections of the data vary with section duration in a log–log relation. The slope of such a relation (i.e., the index of the power law) is different from that for a series in which departures from total randomness of interest concern only values close together in time.
- In the subsequent analysis for shorter-term structure, stability of conclusions can be examined by making the analyses separately within each section – for example, within each half of the data. It may also be necessary first to eliminate trends.
- Examine periodic variation of known wavelength (e.g., seasonal variation in monthly or quarterly data) by forming two-way tables (e.g., month by year) and calculating marginal means.
- Examine local structure by lagged scatter plots of y_{t+h} versus y_t for small values of the lag h, looking for correlation, nonlinearity, and outliers.

If the data are concerned with point events occurring in time, the same broad principles apply. There is a distinction between methods based directly on counting numbers of occurrences in suitably defined time periods and methods based on the distribution of, and interrelationships between, intervals between successive points. The latter are especially suitable for examining local structure.

With data from a modest number of sites, the above scheme, including the comparison of related plots at different sites, needs to be supplemented by a study of the relation between local structure at different sites. If $y_{i;t}$ denotes the observation at site i at time t, then plots of $y_{i;t}$ versus $y_{j;t+h}$ for $h = 0, \pm 1, \ldots$ and $i \neq j$ are a natural starting point connecting the corresponding features to the geographical relation between sites i and j.

Similar methods apply if, as will usually be the case, there are several or indeed many different features measured at each time and each place instead of a single variable $y_{i;t}$. Especially for the study of local structure, approaches not based on specific subject-matter arguments are not likely to be fruitful when many features are involved. Examination of relations between all possible pairs of variables may be feasible, but approaches beyond that soon become impracticable or, even if practicable, are insufficiently sensitive. Again, purely explanatory variables may often be present (see, e.g., Thomson *et al.* 1996); adjustment for such variables through some form of regression analysis may be important, calling for at least a purely empirically based formal model. Derived variables may be used to reduce dimensionality. Simplification of structure typically amounts to finding meaningful conditional independencies among a large set of variables.

More formal time series methods

The approaches sketched above are quite strongly related to more formal methods of time series analysis. These may be roughly classified as

- linear and quadratic methods in the time domain, centering on autocorrelations, autoregressive and moving average models, and generalizations, including Markov chain models for qualitative data;
- quadratic methods in the frequency domain, that is, spectral methods including periodogram analysis: while broadly equivalent to the corresponding time domain methods, these methods have a number of technical statistical advantages, such as simpler assessment of sampling errors and greater stability for missing and irregular data patterns;
- nonlinear models, taking these in a very broad sense, but essentially defined here as systems whose properties cannot be captured via linear and quadratic statistics.

There are a considerable number of books on the first two topics, ranging from the introductory account by Diggle (1990) to the careful discussion of the theoretical basis by Brockwell and Davis (1991). The third topic is much more recent and is best appreciated via the review paper and discussion by Tong (1995).

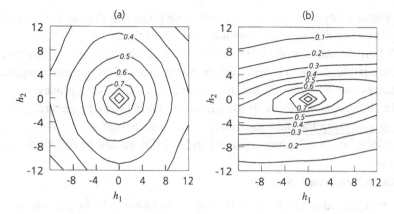

Figure 5.1 Illustration of spatially (a) isotropic and (b) anisotropic correlation structure.

Descriptive methods for spatial data

We now turn to systems in which rather extensive spatial data are available, possibly only at a limited number of time points. To some extent we can directly adapt the simple ideas presented in our discussion of descriptive time series analysis. Thus we may divide a large area into a smallish number of subareas and then compute summary statistics within each subarea to examine for trend and stability. For these and other analyses, establishing a suitable spatial coordinate system can be important. This system may be based on geographical, hydrological, or meteorological gradients and in some cases may best be curvilinear. In some instances the N–S, E–W map coordinates may be adequate.

The most difficult aspect of using these methods is probably the examination of short-range structure, especially if the data points are irregularly separated in space.

If the data values fall on a spatial lattice – that is, are of the form $y_{i,j:t}$ for $i, j = 0, \pm1, \ldots$, where i, j refer to a biologically relevant coordinate system – then correlations between sites neighboring in the direction of the coordinate axes, or correlations that can reasonably be assumed to be spatially isotropic, are fairly directly detected from relevant scatter plots. For correlations that are not isotropic, more elaborate methods, for example in the frequency domain (Priestley 1981; Ripley 1981), may be needed. Figure 5.1 gives examples of isotropic and anisotropic spatial correlation structures using empirical data sets. The analogue of the temporal autocorrelation coefficient is the correlation between points (h_1, h_2) apart; for an isotropic process, this will be a function only of $|h| = \sqrt{h_1^2 + h_2^2}$. With a relatively small number of time points, it may be sufficient to conduct the above analyses separately at each time point and then to examine the stability or evolution of important features over time.

Where the data consist of point events, the analogue of interval methods in time is the study of inter-point distances, particularly nearest-neighbor distances (see Section 5.3, Simple properties of the Poisson process).

In all these analyses, with sufficiently extensive data the general principle of analyzing the data separately through meaningful subdivisions and examining the conclusions for trend and stability should be followed.

Data rich in space and time

The first step in descriptive analysis is again likely to be inspection, as far as is feasible, of all the data for broad features, including portions of data of questionable quality. Separate temporal and spatial analyses may then be done, possibly at a sample of spatial and time points, respectively. In some contexts it may be interesting to examine the relation between the temporal and spatial autocorrelations. Under the so-called frozen-field hypothesis (Taylor 1938), these autocorrelations are of the same shape related by a scaling constant, a speed. That is, the correlation at a fixed time between values a distance h apart in space is the same as the correlation at a fixed point in space between values h/v apart in time, where v is a constant with the dimensions of a speed. Departures from such a relation may have a substantive interpretation.

The most difficult informal analysis is likely to be that pointing toward a spatio-temporal model. In terms of empirical regression equations and associated graphical methods, two somewhat different approaches can be used:

- regress $y_{i,j;t}$ on relevant features at time $t - 1$, for example, $y_{i,j;t-1}$, $y_{i,j\pm1;t-1}$, $y_{i\pm1,j;t-1}$, and so on;
- include in the regression features at time t itself, for example, $y_{i,j\pm1;t}$, $y_{i\pm1,j;t}$.

The former has a more direct specification in terms of a potential causal process that might have generated the data. It is, however, likely to lead to substantial unrepresented correlation structure, which the second approach may be able to absorb, albeit at the cost of direct interpretability.

5.3   Stochastic Models

Stochastic processes play an important part in modeling many biological and physical systems. In general, such models seek to represent only the main observable features of the system, often replacing complicated deterministic dynamics with simple random processes or with a simpler deterministic system augmented by the inclusion of random "noise." Typically, they have rather few unknown adjustable parameters, and these represent interpretable elements of the system. Although there are important simple models of some general applicability, one should normally start from a particular biological situation and construct a model appropriate for it. The following are among the roles such models play:

- Gaining general understanding of an underlying process. For example, in studying the way in which competing populations interact, one might construct alternative models for the interaction mechanisms and aim to acquire evidence to support one in preference to another by comparing model properties with those of empirical data.

- Estimating behavior for situations similar to those that applied when available data were collected (interpolation) or for situations where rather different circumstances apply (prediction). For example, for temporally evolving processes, data are often collected over a rather short time period, on the basis of which a model is fitted. It may then be of particular interest to investigate the longer-term behavior of the fitted model, perhaps to see whether two species stay in coexistence or whether one gradually dominates while the other becomes extinct.

- Answering "what if" questions. Models are often used to determine what treatment and control strategies would be optimal if the model held and to examine sensitivity to perturbations from the model. For example, if a satisfactory model is available for the spread of a particular infectious disease in the absence of any preventive treatment, then alternative treatment strategies can be compared by using the model, thus reducing the need for costly and time-consuming experiments.

In this section, however, we concentrate on some simple stochastic models. These models are developed for the mathematical tractability of their properties and for their important role in providing both a framework within which the development of models motivated by specific subject matter can take place and a structured hierarchy of component parts from which those models can be constructed. We start by discussing purely spatial processes and then proceed to some remarks about their temporal evolution.

Spatial patterns arising as realizations of ecological processes can be of many different types. Suppose that the feature of interest is the value of a single (real-valued) function $Y(x)$ as it varies over the space. If the function varies continuously so that the pattern is generated by its level contours, a suitable model (possibly after appropriate transformation of Y) might be a Gaussian process characterized by second-order properties (Ripley 1988).

Such processes are particularly widely used in geophysics, where the *variogram*, which shows the dependence of the variance, var$(Y(x') - Y(x))$, on the spatial separation $x' - x$, plays an important role and the method of *kriging* (i.e., generalized least squares) is used to interpolate the surface between observed spatial locations. An extensive discussion of the analysis of geostatistical data is given by Cressie (1991).

Alternatively, if the function takes only a discrete set of values, then the pattern will naturally focus on subregions or "patches" that have the same value; examples include maps of land use or vegetation type. However, if the function takes only two values (say, 0,1) and the level-1 patches shrink to spatial points, then we obtain a qualitatively different type of pattern made up of point events; examples include the locations of trees (Figure 4.1) or of diagnoses of leukemia made over some specified time period. If the point events have a distinguishing feature, then a type or "mark" (which could be discrete or continuous, univariate or multivariate) may be attached; for example, in Carey (1996) the points and marks are respectively the locations of medfly captures and the number captured at that location. In another recent application, Wälder and Stoyan (1996) discuss a point process of tree locations with marks specifying tree sizes and use the variogram in a spatial analysis of these marks. In the rest of this section, discussion will be confined to processes of point events and some examples of other processes, including patch processes, that have an underlying point process structure. Some relevant references are Cox and Isham (1980), Daley and Vere-Jones (1988), Snyder and Miller (1991), and Karr (1991).

Poisson processes

The idea of a homogeneous Poisson process is that the point events of interest occur completely *at random*. In other words, no part of the space is more likely to contain point events than any other (homogeneity), and the presence or absence of an event at a particular location is not influenced in any way by the presence or absence of any other events elsewhere (the property of independent increments). The events occur singly. Thus, such a process provides a base against which to compare other processes where mechanisms of clustering or inhibition of events are thought to occur either because of some feature (inhomogeneity) of the underlying physical space, or because of interactions or other dependencies between the events themselves.

The formal definition of a general (spatial) Poisson process is as follows. Let $N(A)$ represent the number of point events (abbreviated to *points* in the following discussion) in a spatial region A, where A is an arbitrary subregion of some fixed region Ω of the plane. Also, let δx represent an extremely small region of area $|\delta x|$ centered on a location x. Then N is a Poisson process if

1. the probability of there being one point in δx is $\lambda(x)|\delta x| + o(|\delta x|)$;
2. the probability of there being no point in δx is $1 - \lambda(x)|\delta x| + o(|\delta x|)$;
3. for all non-overlapping spatial regions A and B, the numbers of points $N(A)$ and $N(B)$ are independent random variables.

The term $o(|\delta x|)$ has a precise mathematical meaning and for practical purposes can be treated as a quantity that is negligible in comparison with $|\delta x|$. Conditions (1) and (2) taken together imply that the chance of more than one event in δx is $o(|\delta x|)$ and is thus negligible as $|\delta x|$ shrinks to zero.

It follows from these assumptions that $N(A)$ has a Poisson *distribution* with mean $\Lambda(A) = \int_A \lambda(x) \, dx$. The function $\lambda(x)$ is the *rate* or *intensity* of the process. Also, if it is known that $N(A) = n$, then the n points are independently distributed over A with probability density function $\lambda(x)/\Lambda(A)$.

If $\lambda(x) = \lambda$, then the process is *homogeneous*, and because $\Lambda(A) = \lambda|A|$, the n points are *uniformly* distributed over A. This property provides a simple means of constructing a realization of a spatial Poisson process by using sequences of uniform random digits to generate Cartesian coordinates of the point. (A typical example of such a realization consisting of 40 points in the unit square is given in Figure 5.2a.) It is important to understand that the uniform assumption for individual points does not translate into a process in which the points are at all evenly spaced. In general, a realization of a homogeneous Poisson process exhibits considerable irregularity with loose (and even sometimes quite tight) clusters of points and sizable empty spaces, and it is against this appearance that evidence for clustering or inhibitory mechanisms must be judged.

Various operations applied independently to the points of a Poisson process leave the underlying structure unchanged. For example, if the points are translated by identically distributed random displacements, the resulting process remains Poisson. Similarly, if the process is *thinned* by deleting each point with probability p, the final process is again Poisson, with its rate function multiplied by $1 - p$; and if two independent Poisson processes are *superposed*, that is, their points are pooled, the superposition is a Poisson process whose rate function is the sum of the individual rates.

These invariance results mean that the Poisson process occurs as the limiting process when operations of these sorts are applied to rather arbitrary point processes under suitable limiting conditions. For example, the structure of the arbitrary process can be removed if its points are subject to sufficiently varied displacements or are deleted with a probability p that

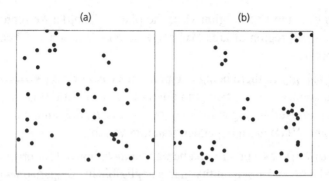

Figure 5.2 Realizations from (a) a homogeneous Poisson process (40 points) and (b) a Neyman–Scott process with 10 parents, each with 4 offspring dispersed according to a circular normal distribution.

approaches 1. Also, a type of Central Limit theorem applies, by which the superposition of a large number of processes will be approximately Poisson as long as there is little dependence between them and no one process dominates the rest. Formal statements of such results and further references can be found in Chapter 4 of Cox and Isham (1980). However, the practical importance of these results is that, for the class of point processes, the Poisson process plays a role corresponding to that of the normal distribution within the class of probability distributions, and many empirical point processes are approximately Poisson because they arise from these sorts of operations applied to processes that may originally have been far from Poisson.

Simple properties of the Poisson process

If N is a Poisson process in a fixed spatial region, then for any subregion A, the number $N(A)$ of points in A has a Poisson distribution, with equal mean and variance and therefore with a unit index of dispersion (the ratio of variance to mean). If there are repeated observations of counts on A, then the sample index of dispersion can be used to test the hypothesis that the counts arise in a Poisson process. Alternatively, if data for only a single time point are available but it is reasonable to assume that the process is spatially homogeneous, then the spatial region can be subdivided into a number of subregions, say, $A_1, \ldots, A_n$, all with the same area as A, to give replicate counts that can be used to calculate the sample index. If the process exhibits clustering relative to a Poisson process, then more of the replicates will have large numbers of points than would be expected under the Poisson hypothesis, and more will have relatively few points, leading to a variance that exceeds the mean and an index of dispersion that is greater than 1. The process is then said to be *overdispersed*. Similarly, an *underdispersed* process has index values that are less than 1, corresponding to processes with replicates in which the points are rather more evenly spaced than is the case for a Poisson process. However, it is worth emphasizing that the fact that a point process has a unit index of dispersion does not necessarily indicate that it is a Poisson process.

For a general spatial point process with point events that occur singly, just as for the Poisson process above, the limiting probability of a point (at x) per unit area as that area shrinks to zero is the *intensity* $\lambda(x)$ of the process. Often, there will also be interest in the *conditional intensity*, that is, the limiting conditional probability of this point per unit area, given a point at some origin 0 outside δx. For the Poisson process, the independent increment property means that the conditional and unconditional intensities are the same, but for a clustered process the conditional intensity typically will be greater than the unconditional intensity when $|x|$ is suitably small, and vice versa for a process with inhibition. This leads to a means of testing the null hypothesis that data originate in a homogeneous Poisson process by estimating the function $K(r)$ – defined to be the expected number of points at most a distance r from the origin given a point at the origin – relative to the overall rate of the process. Under the null hypothesis, $K(r) = \pi r^2$, so that it is usual to plot $\sqrt{K(r)/\pi}$ against r; a Poisson process then corresponds to a straight line with unit slope. Only if there is strong evidence against this hypothesis is it reasonable to begin to investigate particular mechanisms for clustering or regularity of points. Details on the estimation of K and its use in hypothesis testing are given in Ripley (1981) and Cressie (1991).

Two related quantities are the *nearest-neighbor distances*; these are the distances either from an arbitrary location or from an arbitrary point of the process to the nearest point of the process. For a homogeneous Poisson process, both of these distances have the same distribution function (again because of the independent increment property), which is $1 - \exp(-\lambda\pi d^2)$ where the distance d satisfies $d > 0$. The corresponding density function, $f_D(d)$, is shown in Figure 5.3. Based on this result, various statistics can be devised to test a homogeneous Poisson hypothesis about an observed process, although care is needed in allowing for possible edge effects due to the observation of the process in a bounded region. If the region is sufficiently large, perhaps the simplest strategy is to measure the nearest-neighbor distances only from those locations or point events that are more than some critical distance (say, r_0) from the edge of the region, thus creating a boundary region of width r_0 (for details see Ripley 1981 and

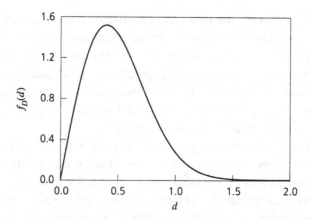

Figure 5.3 Nearest-neighbor distance density function for a homogeneous unit intensity Poisson process.

Cressie 1991). Distances are measured *to* points in the boundary region, but distances *from* locations and points in this region are not included in the statistics.

Poisson-based models

We have seen that clustering or regularity is normally regarded relative to the homogeneous Poisson process. The appropriate type of model will depend on the sort of mechanism involved. For example, for the clustering of plants or of cases of disease, it may be sensible to use a nonhomogeneous Poisson model whose variable rate function reflects the fertility of the soil or density of the population.

This rate function could itself be a *random* process if additional variability between realizations is required. Such a process is called a *doubly stochastic Poisson* (or Cox) process. Any single realization will be that of a Poisson process, so that observations on several realizations will be needed to distinguish the doubly stochastic Poisson process from the normal Poisson process. The doubly stochastic model might be appropriate if, for example, each realization were a microscope slide of a section of tissue from a different individual. For such a process any count $N(A)$ will be overdispersed, that is, the variance will exceed the mean.

On the other hand, the clustering of plants or cases of disease might be due to a parent–offspring mechanism (either literally in the case of plants or representing an infectious mechanism for diseases). In such instances, it might be appropriate to weaken the independence of point events by starting from a Poisson process of "parent" events and adding a cluster of "offspring" around each of these. In the simplest case, the *Neyman–Scott Poisson cluster process* (illustrated in Figure 5.2b), the clusters around distinct parents are independent and identically distributed, with random numbers of offspring per parent, and offspring are independently and identically located relative to their parents. A related cluster process is used by Pfeifer *et al.* (1996) in spatial models for the ecology of tidal flats.

If a model for an underdispersed process is required, one possibility is to start from a Poisson process and make the realizations more regular by filling gaps and/or removing clusters. Various simple mechanisms are mathematically tractable: for example, with some specified probability, one or both of a pair of points within a critical distance could be deleted.

A third cause of clustering or regularity could be the actual attraction or inhibition of the points themselves. For example, some species of animals or birds form cooperative groupings, whereas others maintain exclusive territories. Such processes are probably most naturally modeled by Markov point processes, described in the next section.

Markov random field models

The intuitive idea behind the definition of a Markov random field is that the presence or absence of a point of the process at a particular location x is much more likely to be influenced by other points that are relatively near to x than by points that are far away. In particular, the effects of these faraway events will only be indirect, acting through their effects on those locations in between them and x. For example, the points of the process might be the locations of especially large plants of a particular species. Competition for nutrients and light tends to make such plants spatially well separated, so that the presence of a large plant at y would militate against another such plant at x if x and y are close together. On the other hand, if x and y are farther apart, it is the locations of large plants in between x and y that are most important and, *given these*, x will be unaffected by a large plant at y. Such ideas are formalized mathematically in the notion of a Markov random field.

To define a Markov random field on a bounded region Ω of the plane, we must first define the notion of regions forming neighborhoods. This notion can be quite general, but for most biological and physical processes a natural definition is to say that two locations are *neighbors* if they are within some critical distance, r_0 say. Now suppose that for any arbitrary set Δ, $\partial\Delta$ represents the boundary region surrounding (and outside) Δ of width r_0. Then, a point process on Ω is a Markov random field if, given the points on the boundary region $\partial\Delta$, the points on Δ are independent of those outside $\Delta \cup \partial \Delta$. Equivalently, given the process on the region outside Δ, the process on Δ depends only on the points on $\partial \Delta$.

Next, we must define the idea of sets of mutually neighboring locations. Specifically, a set of locations $x = \{x_1, \ldots, x_n\}$ is a *clique* if every element of x is a neighbor of all other elements of x. Then, for a Markov random field, the joint probability density for exactly n points in Ω at locations $x = \{x_1, \ldots, x_n\}$ has the form

$$f(x) = \lambda^n e^{-\lambda |\Omega|} \prod \phi_c(x) \,, \tag{5.1}$$

where $\phi_c(x)$ only involves $\{x_i : x_i \in c\}$ and the product is over all subsets c of x that are cliques. In this density, $\lambda^n e^{-\lambda |\Omega|}$ is the corresponding joint probability density for a homogeneous Poisson process, and $\prod \phi_c(x)$ gives the contributions from each clique and determines the clustering or inhibition relative to a Poisson process. Thus the contribution of neighboring (interacting) groups of points can be modeled explicitly.

For example, suppose that in a particular process a single point at x_i contributes an amount β to the density $f(x)$, regardless of the location x_i. Similarly, a pair of points at locations (x_i, x_j) contributes an additional amount γ if x_i and x_j are neighbors [so that (x_i, x_j) is a clique]. Suppose, further, that points at cliques of more than two neighboring locations do not make any additional contributions to the density over and above those already made by the single points and their constituent pairs. For this process, $\prod \phi_c(x) = \alpha \beta^n \gamma^{s(x)}$, where $s(x)$ is the number of neighbor pairs in x (i.e., pairs of points within the critical distance r_0). Such a process is called a *homogeneous, pairwise-interaction process*. Note that for the function $f(x)$ to represent a valid joint probability density, either $\gamma = 1$ (the Poisson process) or there must be inhibition between the points (i.e., $\gamma < 1$). The special case $\gamma = 0$ is a *hard-core* process in which each point is a distance of at least r_0 from any other. Otherwise, if $\gamma > 1$, the function $f(x)$ does not represent a probability density since it cannot be normalized over the bounded region Ω to give a total probability of 1.

An important result, useful for generating realizations of Markov random fields, is that such fields can be obtained as equilibrium processes of particular spatial birth–death processes, in which the birth and death rates at location z immediately after time t depend on the state of the process at time t only in terms of the process on ∂z. See Isham (1981) for an introductory review of Markov random fields.

The above description of Markov random fields concentrates on point processes, which can be regarded as having a binary variable attached to each location of a continuous space. It is worth noting that Markov random fields can also be defined on discrete spaces and are often used as models of processes on regular lattices (Preston 1974; Isham 1981). In this case, the variables defined at each lattice point need not be binary and may take continuous or discrete values. Such variables could represent counts of plants or crop yields in a regular array of quadrats. The essential point is that, in a Markov random field, each site in the underlying space has a set of neighbors (in the case of a lattice, it will usually be the four or eight nearest lattice vertices), and the random variable at that site, given the values at its neighbors, is conditionally independent of the variables at other sites. As in the point process case described above, this condition leads to a convenient product form for the joint distribution of the process.

Point-process-based models

Point processes can be used as a basis for constructing models of other types of processes. Perhaps the simplest example is to suppose that independent random sets (e.g., disks with random radii) are located in an appropriate way at each point. Then a spatial process can be defined as the number of these sets that overlap each spatial position. Such a mechanism would be one way of constructing a "patch" process having a discrete set of (integer) values. More generally, if each random set has a random mark (which might be discrete or continuous-valued) then the value of the spatial process at a particular position could be the sum of all the marks attached to the sets overlapping the position. Similarly, to construct a process whose realizations are continuously varying "surfaces," one possibility is to attach a simple spatial function to each point event. For example, one might take a spherically symmetric "Gaussian" form centered on the point, decaying in a quadratic exponential manner from a randomly distributed maximum. A variety of such processes have been used successfully to represent rain cells in precipitation models (Le Cam 1961; Waymire *et al.* 1984; Wheater *et al.* 1996).

Spatio-temporal models

There are many ways a spatial process might evolve over time. In particular, if the process evolves on a regular spatial grid (as, for example, is the case with many managed crops, where the spread of disease is of interest), then the methods and results of probabilistic cellular automata are especially important (see Chapters 6, 7, and 11). Here, we discuss only the evolution of a spatial Poisson process, but similar evolutionary mechanisms can be considered for the other Poisson-based models described in this chapter.

The simplest basic spatio-temporal model is a Poisson process consisting of point events in space *and* time. For such a process, the earlier definition of a Poisson process applies, but for a region of three-dimensional space rather than a plane. This gives point events that have a temporal position as well as a spatial one (locations and dates of disease onset, for example). In general, the intensity of this process is a function $\lambda(x, t)$ of time t as well as position x; in some special cases this might be a simple combination of functions of space and time separately, such as a product $\lambda_1(x)\lambda_2(t)$.

It is often likely to be physically appropriate for these space–time events to have a temporal lifetime (e.g., the duration of infectiousness in the case of a disease). This lifetime could be regarded as an additional random coordinate of the point event and modeled by generating points in an appropriate four-dimensional space. However, it is usually more sensible to keep the spatial and temporal aspects of the process distinct and to model them separately. For example, by taking the lifetimes to be exponentially distributed variables, independently between distinct point events and independent of their spatio-temporal position, the model retains strong Markov properties.

Additionally, in some specific processes, the point events do not remain at fixed spatial locations but move around the space during their lifetimes. For each event, such movement might be in a random direction and at a random speed that remain fixed during the lifetime, or one or both of these might vary in time. A Brownian diffusion is one fairly simple possibility. Of course, the more variables that are involved to specify the spatio-temporal process, the more scope there is for complicated dependencies between these variables. While Poisson-based models in which virtually all additional variables are mutually independent are usually reasonably tractable, allowing appropriate dependencies is often problematic for formal analysis, not least because there seldom are good empirical grounds for assuming a particular form of joint dependence.

5.4   Model Fitting

Fitting statistical models to spatio-temporal data is complicated by the fact that the random variables corresponding to realized data values are generally correlated, to a greater or lesser extent depending on their separation in space and time. We suppose that we have a model (or models) thought to be appropriate for describing an observed spatio-temporal process from which we have obtained random samples of data. We view the data as a collection of realized values generated from the model under consideration. Typically, data sets consist of a single realization of the process.

Such models generally have one or more unknown adjustable constants or *parameters* to be estimated, and in this section we describe some methods of parameter estimation. Parameters may or may not have an empirical interpretation. For a given problem it may be possible to propose *ad hoc* parameter estimates. Indeed, for more complex problems it may be that an element of subjectivity is necessary because more formal methods are difficult to implement. Often, we hope to use a method that in some sense produces the "best" estimates. Generally, we hope that our estimator is *unbiased*, so that its expectation equals the quantity being estimated, that is, it is correct on average. One or more parameters may be estimable directly from the data independently of the other parameters of the model. For example, the velocity of rain events can be estimated using spatio-temporal correlation structure (Northrop 1996). Consideration of uncertainty in the parameter estimates and goodness of fit of the model is useful in choosing the best model and method of fitting.

If the data are specified as $x_1, \ldots, x_n$, we suppose that the model gives the probability or probability density of these values as $f(x_1, \ldots, x_n; \theta)$, where $\theta = (\theta_1, \ldots, \theta_p)$ is a set of unknown parameters.

Maximum likelihood estimation

The basis for most formal methods of analysis in light of a given model is the likelihood function, $L(\theta)$, defined as the probability or probability density of the observed values regarded as a function of the unknown parameters; that is, $L(\theta) = f(x_1, \ldots, x_n; \theta)$ for a given set of data $x_1, \ldots, x_n$. Provided this function can be computed from the model in a reasonably accessible form, various procedures are available for point and interval estimation of parameters and testing of hypotheses about parameters.

Intuitively, $L(\theta)$ indicates the ability of a particular value of θ to explain the data, and this provides some motivation for considering in particular the maximum likelihood estimate $\hat{\theta}$, the value that maximizes $L(\theta)$ for the data under analysis. Except for anomalous problems – for example, in which the likelihood approaches its maximum only as one or more components of θ tend to infinity – the estimate $\hat{\theta}$ has good properties. For large sample sizes it is asymptotically unbiased, has minimum variance among unbiased estimators, and is normally distributed.

Numerical algorithms for finding maxima of relatively complicated functions are widely available. If the component observations $x_1, \ldots, x_n$ are independent (although this will rarely be the case in the kinds of model considered here), $L(\theta)$ will be a product of contributions, best dealt with by

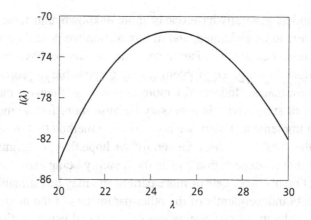

Figure 5.4 A typical log-likelihood function for the Poisson model.

working with the log-likelihood, $l(\theta) = \log L(\theta)$. Even in problems with
dependent observations, $l(\theta)$ is nearly always to be preferred to $L(\theta)$.

As a simple illustration of the method of maximum likelihood, consider
counts of plants in n quadrats of area A. If plants occur with intensity λ
over the area of interest and, for example, do not tend to cluster together, it
might be appropriate to fit a Poisson model. If we let X_i denote the random
number of plants in the ith quadrat, then under our model X_i has a Poisson
distribution of mean λA. The probability distribution of X_i is

$$P(X_i = x) = \frac{1}{x!}(\lambda A)^x e^{-\lambda A}, \quad x = 0, 1, \ldots \tag{5.2}$$

so that, because under the model the X_i are mutually independent, the log-
likelihood can be written as

$$l(\lambda) = \ln(\lambda A) \sum_{i=1}^{n} x_i - n\lambda A - \sum_{i=1}^{n} \ln x_i! . \tag{5.3}$$

The log-likelihood is maximized by setting its first derivative to zero, giving
a maximum likelihood estimator of $\hat{\lambda} = \bar{X}/A$, where $\bar{X} = n^{-1}\sum_{i=1}^{n} X_i$.
Figure 5.4 shows the form of this log-likelihood for data with a mean
of 24.9. Alternatively, in cases where it is impractical to obtain either a
complete mapped pattern of locations or quadrat counts, λ can be estimated
using a sample of nearest-neighbor distances from randomly chosen locations. Care should be taken that the sampled locations are sufficiently separated for their nearest-neighbor distances to be considered independent.
If we let D_i be the distance from an arbitrary location in the area to the

nearest plant, then it can be shown that πD_i^2 has an exponential distribution with mean $1/\lambda$ (see Section 5.3, Simple properties of the Poisson process). A maximum likelihood estimator of λ is thus $\hat{\lambda} = m/(\pi \sum_{i=1}^{m} D_i^2)$. This method can be extended to consider the kth nearest-neighbor distances. Increasing k may give a more precise estimate of λ by reducing the sampling variance of $\hat{\lambda}$, but may lead to increasing bias due to edge effects and problems in ensuring independence.

The generalized method of moments

We use the term *moment property* to refer to any property of the data for which a theoretical expectation can be calculated under the model proposed. The generalized method of moments (Hansen 1982) is an *ad hoc* fitting procedure in which a weighted sum of the squared errors between n $(n \geq p)$ selected moment properties of the data and their theoretical expectations is minimized. We refer to this sum of squares as the *objective function*. The number of model expressions used can be greater than p (the number of parameters) if desired. The sets of weights used in the objective function should reflect the sampling variability of the individual properties involved. The principal difficulty in using this method is the subjective choice of fitting properties to include in the objective function, since different choices of fitting properties give rise to different estimated parameter sets.

In simple cases it may be possible to obtain algebraic expressions for the estimates of $\theta_1, \ldots, \theta_p$, although in more complex problems numerical methods will have to be used.

We return to the Poisson example of the previous subsection (Maximum likelihood estimation). A method-of-moments estimate of λ can be obtained by choosing the sample mean $\bar{x}$ as the relevant moment. The estimate has theoretical value λA and on equating observed and theoretical values we recover, in this special case, the maximum likelihood estimate. In this case, the method-of-moments estimate is the same as the maximum likelihood estimate. This situation is not uncommon in simple problems.

In cases where it is not available directly, the sampling distribution of $\hat{\theta}$ can be examined using simulation. We repeatedly generate samples from the fitted model and obtain a value of $\hat{\theta}$ for each of these sets of simulated data. Assuming that the true underlying process is the same as that of the model, we have a random sample from the multivariate sampling distribution of $\hat{\theta}$. We can then investigate whether $\hat{\theta}$ is unbiased and calculate the variability of the components of $\hat{\theta}$. In addition, we can investigate the degree of association between the components of $\hat{\theta}$.

Spectral and simulation methods

The complex dependencies generally present in spatio-temporal data (and hence in any "realistic" spatio-temporal model) can lead to difficulties in formulating a likelihood in a useful form. The use of traditional maximum likelihood estimation is thus impracticable. Difficulties in deriving expressions for important theoretical properties may limit the usefulness of the generalized method of moments.

One solution is to work in the frequency domain, calculating the Fourier transform of the data (see, e.g., Priestley 1981) and treating the resulting Fourier coefficients as data (rather than using the raw data themselves). An approximate likelihood function can be derived since the sample Fourier coefficients are asymptotically (i.e., for long temporal runs of data over a large area) independent and normally distributed. The log-likelihood involves the estimated periodogram of the data, which can be smoothed to improve computational efficiency. This method is referred to as *Whittle's method* in the field of time series analysis (see, e.g., Dzhaparidze and Yaglom 1983).

This method is valuable for relatively complicated models for which the spectral density can be calculated in fairly simple form. It has the advantage over the generalized method of moments that it uses all the available data, albeit only in terms of the second-order spectral density, rather than just a few summary statistics. However, this may itself cause computational difficulties. A subjective choice of fitting properties is not required, although the reliance on second-order properties may mean that other important properties – for example, the presence or absence of a particular species – may be poorly reproduced. Chandler (1996, 1997) describes a spectral approach for fitting models to rain gauge (i.e., purely temporal) and rainfall radar (i.e., spatio-temporal) data, respectively.

An alternative method of inference for models that have intractable distribution theory but are straightforward to simulate is described by Diggle and Gratton (1984). Estimated log-likelihoods are constructed for a range of values of the parameter(s) to be estimated using kernel density estimation on simulations of the assumed model. The position of the maximum of this estimated log-likelihood is used as an approximate maximum likelihood estimate.

Image analysis and fitting Markov random field models

Data in the form of images are widely available for many spatial ecological variables, for example, satellite images of tree density over an area of a forest or of the variation in color of wheat in a field. Such images generally consist of a rectangular array of measurements over square pixels, where each measurement is subject to recording error. Interest often focuses on the problem of *image restoration*, that is, making inferences about the true (i.e., measured without error) image T. This true image will generally be contaminated by random *noise*, which will often be correlated between neighboring pixels. Additionally, the image may also be *blurred*. Blurring can be modeled by assuming that a blurring function has smoothed the image by replacing each pixel value with a linear combination of itself and surrounding values.

We consider a Bayesian approach to this problem. We assume a *prior* model, $\pi(T; \theta)$, with parameter vector $\theta = (\theta_1, \ldots, \theta_p)$, for T. For example, this could be a Markov random field model. We also assume a model, $f(Y|T; \phi)$, for the distortion of this true image to the observed data Y. The parameter ϕ can often be set using physical considerations – for example, from knowledge of atmospheric distortion. It is common for θ to be set to a value that has given satisfactory results in the past. Using Bayes's theorem it can be shown that, given the data, the *posterior* distribution of T, $\pi(T|Y; \theta, \phi)$, is given by

$$\pi(T|Y; \theta, \phi) = \frac{f(Y|T; \phi)\, \pi(T; \theta)}{\sum_T f(Y|T; \phi)\, \pi(T; \theta)}. \tag{5.4}$$

The normalizing constant in the denominator involves a sum over all possible true images and is generally difficult to evaluate. Therefore, it is usual to note that $\pi(T|Y; \theta, \phi) \propto f(Y|T; \phi)\pi(T; \theta)$ and work with this non-normalized posterior distribution. We may wish to find the posterior mode – that is, the T that maximizes $f(Y|T; \phi)\pi(T; \theta)$ and is thus the most likely true image given the data – or, better still, find the posterior mean. Iterative methods of maximizing and simulating from $f(Y|T; \phi)\pi(T; \theta)$ are discussed by Ripley (1988, Chapter 5).

Note that the parameters of the prior model for T have not yet been estimated using the data Y. The spatial dependencies arising in, for example, Markov random fields often make computation of the full likelihood difficult. Alternative approaches that seek to simplify complicated likelihoods or utilize simulation (increasingly viable due to advances in computing power) are then necessary.

As an illustration of the latter, we shift the emphasis of the Bayesian analysis above. We wish now to make inferences about the models we assume to have produced T, and consequently Y, rather than attempt to restore T from Y. In particular, we are interested in estimating θ, the parameter vector of the model for the underlying spatial process. In other

words, the emphasis shifts from reconstructing a "true process" from a
noisy observation of it, to estimating the biological process that is considered to have generated the data. In dealing with this latter, more ambitious
task, it may be that the assumptions about $\pi(T; \theta)$ that are adequate for
image reconstruction need more critical study. In particular, the model may
be adequate for the local structure but inadequate for longer-term relations.

We assume a prior distribution $g(\theta, \phi)$ for θ and ϕ and, as above, models
$\pi(T; \theta)$ and $f(Y|T; \phi)$ for the production of T and the distortion of T to
Y, respectively. The posterior distribution $g(\theta, \phi|Y)$ of θ and ϕ is given by

$$g(\theta, \phi|Y) = \frac{f(Y|T, \phi)\,\pi(T; \theta)\,g(\theta, \phi)}{\int_\theta \int_\phi f(Y|T, \phi)\,\pi(T; \theta)\,g(\theta, \phi)\,d\phi d\theta}. \tag{5.5}$$

Again, we may wish to use the posterior mode or posterior mean to estimate
θ and ϕ. In some circumstances it may be appropriate to assume that ϕ is
known from physical considerations, leaving θ to be estimated.

In many cases this estimation is still a nontrivial problem due to dependence between the components of θ, but simulation from $g(\theta|Y)$ is possible
using the *Gibbs sampler* (Geman and Geman 1984). As before, we note
that $g(\theta|Y) \propto f(Y|T, \phi)\,\pi(T; \theta)\,g(\theta, \phi)$ to avoid evaluating the normalizing constant. We start with an initial parameter vector $\theta^{(0)}$ and, during
the nth iteration, discard the current value of θ_i and replace it with a value
drawn from $g(\theta_i|\theta_{\{j:j\neq i\}}, Y)$ sequentially for $i = 1, \ldots, p$. If the Gibbs
sampler is run for a "large number" of iterations (so that dependence on
$\theta^{(0)}$ is minimal), the parameter vector $\theta^{(n)}$ obtained after n iterations can be
considered to have been sampled directly from $g(\theta|Y)$. Thus, after a suitably long "burn-in" period, a random sample from $g(\theta|Y)$ can be obtained
by using every kth $\theta^{(n)}$, where k is large enough for successive variates to
be assumed to be independent. The posterior distribution can thus be studied using this sample from it. Implementation of the Gibbs sampler and
related algorithms is discussed in Gilks *et al.* (1996).

Although a Markov random field may be a realistic model for the *local* dependence structure of a spatial field, it may not be adequate for the
analysis and interpretation of spatial data. For example, finite realizations
from a Markov random field defined on a discrete set of "colors" may be
dominated by one color (Ripley 1988, p. 80).

Other likelihood-based methods of estimating the parameters of Markov
random field models aim to simplify their complicated likelihoods. The
coding method (Besag 1974) splits the variates into two distinct sets so that
those in the first set, the "dependents," are conditionally mutually independent given those in the second set, the "conditioners." This is generally

relatively easy to achieve for lattice data. Conditional maximum likelihood estimates of the unknown parameters can then be obtained using the conditional likelihood of the dependents given the conditioners. The roles of the sets can be reversed to give a second set of parameter estimates, which can then be combined with the first in a suitable manner. Different models can be compared using likelihood ratio tests if the same coding scheme has been used for both models. This method is wasteful of data, particularly if the spatial dependence structure results in small coding sets. A method that should make more efficient use of the data is *pseudo-likelihood* estimation, introduced by Besag (1975), in which the product of the univariate conditional distributions of each variate given the rest is maximized to produce parameter estimates.

Assessment of fit

The previous discussion concentrates on the estimation of parameters for a given model. There are the complementary problems of testing the adequacy of a given model and the comparison of and possible choice between different models.

Broadly speaking, there are two approaches for assessing model adequacy. The first is to consider one or more features of the system not used in fitting and to check whether the observed values agree with those implied by the fitted model. A suitable significance test determines whether any discrepancies are reasonably accounted as random fluctuations plausible under the model. The other possibility is to determine a set of quantities, called (generalized) residuals, that under the model should show no systematic structure. Any clear structure determined graphically or otherwise is evidence of departure from the model and usually a clue to the kind of modification to the model that is needed.

It is worth noting that simulation is available for evaluating those properties of the fitted model that were not available explicitly for use in parameter estimation. For example, the distribution of extreme values may be important. In the case of a purely temporal process, we may wish to investigate the times at which the level of the variable of interest rises above (an upcrossing) and falls below (a downcrossing) a threshold and study how time spent above the threshold between these level crossings depends on the value chosen for the threshold. Extending this concept to space is equivalent to studying the distribution of the area of *islands* created by imposing a threshold on the surface of the variable of interest over two-dimensional space.

If two or more models are available, usually the most cautious procedure is to see whether one, both, or neither model is reasonably adequate.

Automatic procedures for model selection have at best a very limited role in intermediate stages of analysis and usually should be avoided altogether. In fitting a complex model, problems may be experienced concerning parameter identifiability. In particular, different parameter estimates may give a similar fit to the data, raising the question of which estimate should be used. For example, the distinction between a small number of large clusters of points and a large number of small clusters may not be clear-cut. Empirical considerations may indicate which estimate is more plausible.

Models are inevitably idealized and, especially with large amounts of data, it may make sense to use a model for which lack of fit has been detected, provided the lack of fit is reported and its possible impact on the conclusions is considered, at least in outline.

5.5 Concluding Comments

In this chapter we have given a brief survey of statistical approaches to the analysis and interpretation of empirical data from spatio-temporal processes. The data may be of various types involving substantial replication in time and/or space. We have covered three main areas: essentially descriptive statistical methods, the development of stochastic models, and some more formal statistical methods.

Descriptive analyses are especially valuable in checking for both data quality and the absence of major unanticipated features, for indicating the broad class of model likely to be appropriate for more detailed analysis, and for presenting conclusions in a simple form.

Stochastic models play an important role in gaining a fuller understanding of complex biological or physical systems by capturing the main observable features using a relatively small number of interpretable parameters. They are particularly useful in determining optimal treatment or control strategies. Simple, mathematically tractable classes of stochastic model can be used as building blocks in the construction of subject-specific models.

Fitting statistical models to empirical spatio-temporal data is complicated by the complex correlation structures that are usually present. Nevertheless, standard methods may be appropriate and their implementation is considerably eased by modern computer-intensive methods of numerical optimization.

We have attempted to cover a very large field and have given little in the way of technical detail or illustrative examples, both of which are essential for detailed understanding. For this reason and for further reading, however, we have provided many pointers to the relevant literature.

Part B

When the Mean-field Approximation Breaks Down

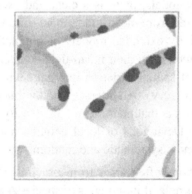

Introduction to Part B

Part B of the book turns from the field to models of ecological processes in spatially structured environments. We hope eventually for a seamless transition from field-based rules of interaction among neighboring plants and animals to computer simulations. The results given in Chapter 3 illustrate how far plant ecologists have gotten in the field, and Chapter 6 shows what theorists can do by listening carefully to ecologists. But at the present state of the art, most models, while motivated by ecological and evolutionary phenomena, are based on assumptions about spatial processes in nature. From study of these models, there is a clear and exciting message: new phenomena, unexpected from mean-field models, are very often evident.

Why should spatially extended models of population and community dynamics differ so much from their mean-field counterparts? A major reason is the existence of spatial variation in local environments. As Part A emphasizes, organisms very often interact with their neighbors, and it is the density of these neighbors that matters, not the density averaged over some large spatial region. Deviations of local neighborhoods from the global average are of two kinds: systematic and random.

- Systematic deviations often arise from previous interactions between neighbors. For example, if individuals of species A and B interact antagonistically and reproduce locally, fewer individuals of A will be found around those of B than expected from their global densities (and vice versa). Such deviations are described by local correlations, and dealing with them becomes a major theme later in the book (Chapters 13, 14, and 18 to 21).
- Random deviations are due to the finite number of neighbors. These local fluctuations play an important role when organisms respond non-linearly to different local environments because the individuals' response, averaged across environments, is not equivalent to their response to the average environment. In such cases, chance fluctuations do not cancel out, and they may have macroscopic consequences (Chapters 7, 19, and 21).

The separation of systematic and random deviations does depend to some extent on the approach used; yet both types of deviation are often present, and the distinction is a useful aid to understanding the dynamics. The aim of Part B is to illustrate and emphasize the phenomena that can result from these local correlations and fluctuations.

To construct models of spatio-temporal processes, researchers have to make some basic choices about space, time, and state variables. Each of these quantities may be continuous or discrete, giving eight combinations, underpinned by different kinds of mathematics. Some approaches are more commonly used than others, and the chapters in Part B illustrate two of the more widely adopted approaches. Most popular among biologists are cellular automaton (CA) models. These are usually stochastic and update discrete states of cells (typically representing individuals) on a discrete spatial lattice according to the state of cells in some neighborhood in discrete time (Chapters 6 to 8). The popularity of CA models is well deserved: as they are rule based, biologists can readily turn them into algorithms for numerical simulation of individuals interacting in some spatial region, as discussed in Chapter 6.

In another widely used approach, models take state to be a continuous variable, envisaged as a locally well-mixed density or concentration, with dynamics that are continuous in space and time. These models are written as partial differential equations (PDEs), most commonly reaction–diffusion equations, with terms describing (1) interactions ("reactions") that depend on locally well-mixed densities of individuals and (2) diffusion across space by random movements of individuals. Biologists often find reaction–diffusion equations more difficult than CA models to understand and turn into computer algorithms, but reaction–diffusion equations do have the potential to give important insights into the development of large-scale spatial structure, as Chapters 9 and 10 illustrate. How these two modeling paradigms are related is discussed in detail in Chapter 9.

CA and PDE models are two points in a broader spectrum of alternative approaches to spatial modeling. The relative merits of different approaches depend on the biological problem posed, and Chapter 11 illustrates how the modeling framework needs to be changed as the scale moves from the individual to the local population to the global population.

In their study of spatially extended models, theorists are motivated by a wide variety of biological phenomena, as is evident from the chapters in Part B. In Chapter 6, Wissel illustrates the insights that CA models can give into effects of space on dynamics of three ecological communities. The first deals with rabies in populations of foxes; using the CA model and given some simple rules for infection, a wavelike pattern of spread develops over space with a good match to field data. The second is a shrub community in an arid environment; a CA model shows the importance of rainfall events, external to the community, in driving the long-term community dynamics. The third community is a boreal forest; the CA model describes the spread of fires through the forest and gives a close resemblance to the size and shape of fires observed in the field.

Chapter 7 introduces a framework for analysis of prebiotic evolution based on a set of replicators that share a common metabolism. The replicators are bound to a surface and depend on local diffusion of metabolites from other replicators within a small neighborhood to synthesize the monomers they need to replicate themselves. Comparing results of a CA model for this system with those of its mean-field counterpart, Czárán and Szathmáry show that the community of replicators depends on the absence of local correlations for survival. All replicators must be present in the same locality for successful metabolism, and a replicator that becomes rare has an advantage over those that are common. Such a system illustrates the importance of local fluctuations and has the potential to produce a stable community of replicators, notwithstanding their inherent tendency toward competition.

In Chapter 8, Nowak and Sigmund describe evolutionary games that are played with neighbors on a spatial grid. They show that interactions between neighbors have major effects on the relative success of individuals exhibiting cooperative and selfish behavior. In particular, clusters of cooperative individuals can develop and persist, even without repeated interactions between pairs of individuals. This is not possible in the absence of spatial structure, and the introduction of space thus increases the range of conditions under which cooperation is to be expected.

Chapter 9 explores discrepancies between CA and PDE models that are intended to describe the same reaction–diffusion system. Drawing examples from the interaction of polymers in early evolution, Cronhjort demonstrates the various pitfalls awaiting the unattentive modeler. He explains how different assumptions that are deeply ingrained in the two types of spatial models sometimes can result in incompatible predictions of spatiotemporal dynamics. Processes used for illustration are interesting in their own right: rotating spirals, their resistance to parasites (chemical species that accept catalysis from a member of the cycle but do not catalyze the self-replication of any other member), and self-generating clusters that can split, chase, and collapse are fascinating instances of non-mean-field behavior.

Chapter 10 elaborates on one of the biological themes broached in Chapter 9, spatial self-structuring in hypercycle models. Hypercycles are small cyclic reaction networks where each chemical species catalyzes the self-replication of the next species in the cycle. Boerlijst uses PDE models of hypercyclic reaction–diffusion systems while accounting for the discreteness of individuals by using a cut-off at low densities. He discusses how spatio-temporal patterns generated in these systems can greatly change the course of evolution. Specifically, the formation of spirals and spots can make hypercycles resistant to parasites. Such dynamics are best explained by analyses of selective pressures arising at the level of emergent spatial patterns.

Chapter 11 discusses how diffusive coupling between patches can stabilize small predator–prey cycles at the cost of large ones. Jansen and de Roos also consider the effective decoupling of fluctuations in different patches; this decoupling comes about due to demographic stochasticity as well as to the existence of a deterministic but chaotic spatial attractor. The resulting spatio-temporal processes lead to effective stabilization of overall population densities, and the message is that spatially extended predator–prey cycles are damped down relative to their mean-field counterparts.

It is evident from the examples given in Part B and elsewhere in the book that the dynamics of systems that incorporate space can be much richer than those based on mean-field approximations. Biologists and mathematicians are only just starting to chart this territory and to consider what the implications are for dynamics of living systems.

6

Grid-based Models as Tools for Ecological Research

Christian Wissel

6.1 Introduction

Ecosystems typically are spatially heterogeneous. Most textbooks on ecology state that spatial structures play an essential role in the dynamics and functioning of ecosystems, for example, in their biodiversity. However, only a limited number of ecological field studies have investigated the spatial distribution of ecological elements (e.g., individuals of different species) and the ways this distribution changes over time. Ecological data are often collected over short time periods (3–5 years) at a few representative locations. But the questions of interest to ecologists often relate to longer time scales and larger areas. Ecological models can help to answer these questions by processing these short-term, small-scale data and deducing their consequences for longer periods of time and for larger areas.

Usually, however, not all the information needed for a model is available from hard data. Weak data, estimations, and plausible assumptions have to be added to achieve a closed problem. To keep the model realistic, it should be founded on good ecological knowledge and data and not be based solely on assumptions. For this reason, close collaboration between the modeler and field ecologists is necessary if the models are to produce ecologically relevant results.

At this point, it is necessary to remember that a multitude of possible objectives can be addressed by models. Models can be divided into two categories: those that aim at general *understanding* of ecological processes, and those that aim at *predicting* future events. Classic ecological models that use mathematical equations, especially differential equations, primarily belong to the first category. Much of their structure is driven by the need to keep the mathematics feasible. For this reason, they are seldom suitable for dealing with specific ecological situations. The use of mathematics can lead to a strong distortion of the ecological information used as the basis for an adequate description of a specific ecological situation. In most cases, mathematical models give a highly idealized description of reality. But by omitting specific details by which ecological systems differ, they open the possibility of obtaining some general results. Mathematical models can provide a theoretical basis for many ecological problems and may provide some general insights into ecological processes.

The general starting point of any modeling is determining the problem the model should solve. If a model is intended to provide some understanding of or prediction for a rather specific ecological situation, one may use a "rule-based" simulation approach. In this method the model is formulated not in mathematical terms but directly in a computer programming language. With the help of statements, rules for the change of ecological objects (e.g., individuals, populations, parts of ecosystems) are formulated. This process is often made easier by the one-to-one correspondence of ecological processes and such rules. The structure of the model can be adapted to the specific ecological question and the available information; it is not dictated by mathematical feasibility. In this way, communication between the modeler and the field ecologists is greatly facilitated.

Because powerful computers are now available, in principle it is possible to include many details in the model. One should resist this temptation or the resulting model will be overly complex. Such models contain so many details that a complete presentation in a paper is impossible because of space limitations in journals or elsewhere. Only rough descriptions of such models are normally given: they seldom provide good understanding of the ecological processes, and they cannot be checked. For these reasons, simulation models should only contain those key factors that are important for the question being investigated.

6.2 Grid-based Simulation Models

There is a powerful rule-based simulation approach for spatio-temporal problems in ecology. Traditionally this is called a cellular automaton model (Allouche and Reder 1984; Wolfram 1986; Hogeweg 1988; Ermentrout and Edelstein-Keshet 1992; Silvertown et al. 1992; Halley et al. 1994). Today, however, the term "grid-based" model is preferred, because it is more comprehensive; the cellular automaton model usually has restrictions that hinder modeling of near ecological reality.

The principal procedure of grid-based modeling is very simple (see Box 6.1). The art of such modeling consists of filling the model with biological life (Green 1989; Bradburg et al. 1990; Clark 1990; Moloney

---

Box 6.1 General procedure for grid-based modeling

Subdivide the modeled area of an ecosystem into a grid of squares and
choose the following:

- Size of grid squares; size of system (number of grid squares)
- Possible ecological states of grid squares
- Length of time step
- Rules for changes of the ecological states during a time step

---

et al. 1991, 1992; Urban *et al.* 1991; Czárán and Bartha 1992; Jeltsch
et al. 1992, 1996, 1997; Wissel 1992; Jeltsch and Wissel 1994; Wu and
Levin 1994; Wiegand *et al.* 1995; Hendry *et al.* 1996). The area of an ecological system is subdivided into a grid of cells (although in some cases
one has to deal with models in a three-dimensional space). Using a grid
of square cells is convenient in ecology; spatial data are often collected
on such grids, and most ecological problems have a lower limit of spatial
resolution given by the size of an individual or by the scale at which the
processes of interest act.

The first step of modeling is to select the size of the squares. The size
is determined by the question being addressed by the model, the processes
that must be modeled to answer the question, and the information available
as input to the model. The size of the whole modeled system – that is, the
number of squares – must also be chosen so as to guarantee that all spatial
patterns important for the question and processes can evolve.

The second step is to define the different ecological states possible for
these squares. There are many ways to describe ecological situations;
for instance, the ecological states of a square may be described quantitatively using discrete or continuous numbers (e.g., number of individuals or
amount of biomass) or qualitatively (e.g., presence or absence of a species).
The number and type of possible ecological state descriptors must be chosen according to the question addressed by modeling, the processes to be
included in the model, and the information available. The spatial state of the
whole system is described by the ecological state assigned to each square
and the location of all cells.

The third step is to choose a proper time step for describing the changes
of the ecological states. Because reproduction and other activities frequently are related to the seasons of the year, a time step of one year is
often appropriate. The change in ecological state of each square during the
time step must be described by a set of rules. Once again, a great variety

of rules is possible. They may be quantitative or qualitative (e.g., by using if–then statements). Quite often there is some uncertainty concerning these rules. In this case, one can work with probabilities for the transition from one state to another using a random number generator, available in most programming languages.

All the steps of this grid-based modeling approach – namely, choosing the size of a square, the possible ecological states of a square, the time step, and rules for changes in the ecological state – must be adapted to the problem to be answered by the model, to the processes to be modeled for solving this problem, and to the available ecological information used as model input. For this reason, all the steps are strongly interlinked. Almost no mathematics is needed for this type of modeling. Thus, the modeler's mathematical skill is less important than his or her ability to describe the states of the cells and the rules of their changes based on his or her ecological understanding of the problem and the available ecological information. Constructing a model should be done in cooperation with field ecologists who hold the information needed as input for the model. In this cooperation, the language of rules instead of mathematics is a great advantage because it is a language that biologists can understand.

Because grid-based modeling is very flexible and ideally is adapted to the specific question under consideration, the concrete realizations of the model are quite different in different cases. Therefore, in the following sections some examples are given that demonstrate the universality of this method and illustrate various forms this type of model can take.

6.3 Spread and Control of Rabies

In the past few decades rabies has spread through Europe from east to west. The spreading front has the spatial form of a damped oscillation. There are strong indications that the spreading front moves without changing its form, implying that the number of rabies cases in a location shows the same damped oscillation (Sayers *et al.* 1985) as a function of time (see Figure 6.1). One simply has to multiply the time by the spreading velocity to obtain the corresponding spatial dependence. The first problem a model should solve is to explain the development of this unusual form of a spreading front.

In Central Europe, the red fox (*Vulpes vulpes*) is the main agent for the spread of rabies. Its main biological properties that are important in connection with rabies are as follows. The red fox lives in territorial clans of a few adults and some cubs. Rabies is a viral disease transferred by bites.

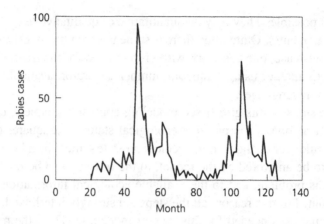

Figure 6.1 Rabies cases recorded in an area near Metz (France) versus time. *Source*: Sayers *et al.* (1985).

If a fox is infected, it enters an infectious state and dies after some weeks. Because of the close contact between the animals, once one member has rabies the entire clan becomes infected. On average four subadult foxes leave their home territory in October or November and may migrate distances of up to 30 km. In this way, they may recolonize territories left empty because of fox deaths from rabies. If the migrating subadult foxes start from an infectious territory, they may transfer rabies over long distances.

These few facts are sufficient to construct a preliminary model (Jeltsch *et al.* 1997). This preliminary model is kept extremely simple to gain an understanding of the principal reasons for the peculiar form of the spreading front (see Box 6.1). The territories are represented by 5 km² squares, which fits to the typical size of a clan territory. Possible states of a territory are healthy, infectious, or empty. The transition rules are as follows. When a territory is in the infectious state, all foxes therein are infectious and will die. Consequently, the territory is empty after the next time step, which is chosen to be one year because reproduction and long-distance migration of subadults occur once annually. Because of the high birth rate of foxes, an empty territory is assumed to be occupied and in the healthy state after one time step. A healthy territory may become infectious in one time step. Because the corresponding processes are not known in detail, one can give only a probability that a healthy territory becomes infectious. Also, because the incubation period and the time to extinction of an infectious clan are variable, a healthy territory may become infectious and then empty within one year.

We begin by showing the equivalence of a simple grid-based model and one based on equations. As we are interested in the spread of rabies in one direction, we can work with a one-dimensional model. Therefore, we use a one-dimensional grid (i.e., stripes), whose location is described by the coordinate x. Let $H(x, t)$ be the fraction of healthy territories in stripe x at time t. Correspondingly, let $I(x, t)$ and $E(x, t)$ be the fraction of infectious and empty territories in stripe x at time t, respectively. The probability that a healthy territory in stripe x is infected in year t is chosen to be

$$T(x, t) = 1 - \exp(-S(x, t)) , \tag{6.1}$$

with

$$S(x, t) = a \sum_u \exp(-bu^2) \, I(x - u, t) . \tag{6.2}$$

Thus, the infectious territories in a stripe $(x - u)$ can infect a healthy territory at x over a distance u with a probability that declines with u like a normal distribution. The probability that an infectious territory becomes empty in one year is p; the probability that it remains infectious is $1 - p$.

With these definitions, we can write

$$I(x, t + 1) = T(x, t) \, H(x, t) \, (1 - p) , \tag{6.3}$$

$$E(x, t + 1) = I(x, t) + T(x, t) \, H(x, t) p , \tag{6.4}$$

$$H(x, t + 1) = E(x, t) + (1 - T(x, t)) \, H(x, t) . \tag{6.5}$$

The last line means that all empty territories are occupied and healthy in the next year and that a fraction $(1 - T(x, t))$ of the healthy territories remain uninfected. The other lines are obtained in the same way. Because the sum of the fractions $H + I + E = 1$, one has to iterate only two of the three equations numerically.

A typical result is shown in Figure 6.2. The spreading front does indeed take the form of a damped oscillation. It moves with a constant velocity in direction x. Thus, at a location on the x-axis, the fraction of infectious territories $I(x, t)$ has exactly the same form as a function of time t. This result is very robust; that is, it is valid for a large range of parameter values of a, b, and p. The oscillation is created by the fact that the large number of infectious territories at the very front are empty in the next time step and consequently can contribute to the spread of rabies only after the following time step. This effect can be demonstrated in the model by extending the

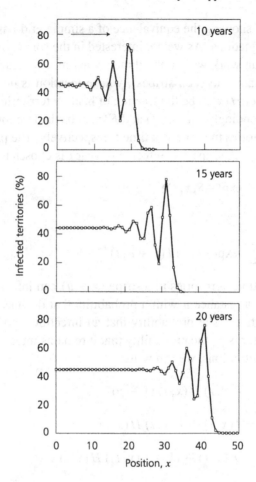

Figure 6.2 Spreading front of rabies from numerical simulation. Graphs show the percentage of infected territories versus position x after 10, 15, and 20 years. Parameter values: $a = 17$, $b = 0.05$, and $p = 0.75$.

time that an empty territory requires to become occupied again. In this case, the wavelength is enlarged as expected.

Instead of solving the above equations, one can simulate the same processes using a two-dimensional grid. Each square (territory) can be in one of three states: healthy (H), infectious (I), or empty (E). The transition rules in one time step are given in Box 6.2. The correspondence to the equations is obvious. The probabilities are realized by a random number generator. The results agree completely with those in Figure 6.2 if the number of infectious squares in each stripe perpendicular to the spreading direction x are summed and divided by the total number of squares in a stripe.

Box 6.2 Transition rules for the simple rabies model

Possible states of the grid squares are healthy (H), infectious (I), and empty (E). Transitions between these states appear with the following probabilities:

States	*Transition probability*
$H \to I$	$T(1 - p)$
$H \to E$	Tp
$H \to H$	$(1 - T)$
$E \to H$	1
$I \to E$	1

where
$$T = \sum_I 1 - \exp(-ae^{-bu^2}) \; .$$

The sum runs over all infectious squares with u being the distance to the square under consideration.

This simple model, which can be formulated using equations or a grid-based approach, is sufficient to deduce the form of the spreading front. But if one wishes to investigate more of the biological details, such as the different modes of infection, one must extend the model. Expansion is necessary, for instance, if one wishes to investigate the effect of different control strategies in the model. Below, a more detailed model is outlined that incorporates rules for methods of control such as barriers (for more details see Jeltsch *et al.* 1997).

For investigating the effect of different control strategies, the time step is chosen to be two months, because this matches the time period of infection and because the migration of subadults also takes place during two months (October and November). Thus the transition rules of Box 6.2 are now for a time step of two months with the following changes. In each time step, there is a probability that an infectious territory infects a neighboring territory through contacts across the border. In addition, there is a long-range transfer of rabies by migrating young foxes.

The main facts known about the migration are roughly modeled in the following way. Four subadults leave each occupied territory in October or November. Their direction of movement is chosen at random. Each subadult moves from square to square; there is a 50% probability that it will continue in the original direction and a 50% probability that it will deviate from this direction by one square to either side. For each territory

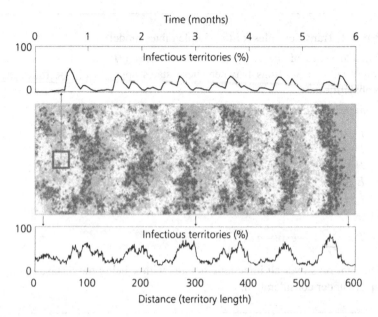

Figure 6.3 Large-scale distribution pattern of rabies in a region of 600 × 300 territories after 120 time steps (20 years) for the spatial model. Direction of spread is from left to right. Black indicates infectious territories; gray, healthy territories; and white, empty territories. (Top) Total number of infectious territories in the window versus time. (Bottom) One-dimensional projection perpendicular to the direction of spread, that is, number of infectious territories in a stripe versus position of the stripe (unit: territory length).

entered by a fox, a probability exists that the fox will stay there. This probability is greater for empty territories than for occupied ones, and it increases linearly with the number of territories entered. If a subadult starts from an infectious territory, it infects with a certain probability the healthy territories it passes through and the territory where it stays. If a subadult starts from a healthy territory it may recolonize the territory where it stays if the territory is empty. This rule is a substitute for the transition from E to H in Box 6.2.

The damped oscillation of the spreading front is again found for a large range of parameter values that are not unreasonable from a biological perspective. A typical pattern is shown in Figure 6.3. Thus this model is at least as good as the first one, but it allows more insight into biological mechanisms. Comparing the projection perpendicular to the spreading direction in the lower part of Figure 6.3 with the levels of infection in Figures 6.1 and 6.2 shows that the form of the spreading front of this improved model fits the empirical data better than that of the spreading front of the simple model. The damped oscillation is found only on a larger scale; on a smaller

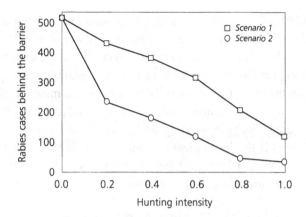

Figure 6.4 Effect of a combination of two barriers in control of rabies. Number of rabies cases behind the barriers versus hunting intensity (percentage of clans killed per year) over a period of 10 years. In barrier A (width of 5 territories) a percentage of clans are killed just before the migration period. In barrier B (width of 10 territories) 70% of the foxes are vaccinated. Scenario 1: barrier A in front of barrier B. Scenario 2: barrier A inside barrier B.

scale (see window in Figure 6.3) a permanent oscillation is found. The damped oscillation for the spreading front is found only if both neighbor infection and long-range infection by subadults are included in the model.

This improved model has been used to investigate the effect of a barrier on the spread of rabies. The barrier is generated by a vaccination program, by reducing the number of occupied territories through hunting, or by both vaccination and hunting. Vaccination and hunting each have the effect of reducing the probability of infection. Reasonable values for all model parameters can be found from different sources. In the model, the success of a barrier is measured by the number of rabies cases that appear beyond it in the protected area. Clearly, this success depends on the width of the barrier and the degree of vaccination or the extent of hunting. The model shows that when vaccinations are used in the barrier, the immunization rate must exceed 30% to have any noticeable effect; however, immunization rates higher than 80% do not improve the effect. In reality, the vaccination method currently applied gives an immunization rate of about 70%. The combined effect of two barriers is shown in Figure 6.4. Hunting takes place in barrier A (width of 5 territories). A vaccination program with a 70% immunization rate is instituted in barrier B (width of 10 territories). The result of this strategy is shown as scenario 1 in Figure 6.4. The effect of the double barrier can be improved considerably by shifting the hunting barrier into the vaccination barrier; in other words, a migrating fox first encounters a barrier (width of 5 territories) with vaccination and hunting and then a barrier (width of 5 territories) with vaccination only (scenario 2). Although the effort in both scenarios is the same, the second is more effective.

This model provides a basis for investigating various applied problems. Depending on the questions under consideration, some minor modifications may be necessary. For instance, it has been found that a vaccination applied in the whole area results in a strong decline in the number of rabies cases after a few years if the immunization rate exceeds 60%. In that case, endemic rabies stays at a low level for several years and may disappear by chance, as has occurred in eastern Germany.

6.4 Dynamics of a Dwarf Shrub Community

The model presented in Section 6.3 is an example of a grid-based model where the ecological states and transition rules are rather simple. This simplicity is due to the fact that more detailed biological information is not available and is not necessary for answering the large-scale questions. In this section an example is shown where many biological details are well known and can be included directly in the model (Wiegand *et al.* 1995, 1997; Wiegand and Milton 1996).

In semiarid Karoo, South Africa, there is a region near the town of Prince Albert that is dominated by a dwarf shrub community. On large plains, individuals of different shrub species are spaced out, with large areas of bare ground between them. The community is dominated by five shrub species. Three species (*Brownanthus ciliatus*, *Ruschia spinosa*, and *Galenia fructicosa*) are so-called colonizers. They establish themselves in gaps containing sufficient bare ground. Two species (*Pteronia pallens* and *Osteospermum sinuatum*) are successor plants that establish themselves in the shade of colonizers and take over the site if the colonizer dies. These five species are the subject of the model in this section. An essential external factor is rainfall: this region's annual mean rainfall is 167 mm with strong fluctuations over the course of time.

Detailed data for all five species are available (Wiegand *et al.* 1995), including information on seed production, germination, and dispersal, and mortality and competition. The question of interest is how these data and factors concerning small temporal and spatial scales combine over the whole system. What community dynamics can be expected in the long run? What are the main driving mechanisms of these dynamics? The first step in modeling is to cull those data that are essential for answering these questions. The data then must be transformed into rules in a computer program.

Box 6.3 Structure of the model for a dwarf shrub community

States of the grid squares are characterized by the following:

- Bare ground or shrub species present
- Age and size of shrub
- Number and species of seeds/seedlings

Rules for changes of state are used corresponding to the following processes:

- Growth of individuals*
- Seed production*
- Germination*
- Survival of seedling*
- Seed dispersal
  Establishment in safe sites (gaps with bare ground for colonizer plants or shade of an adult plant for successor plants)
  Hierarchical competition
  Death (dependent on age)

* These processes depend on yearly rainfall, which is determined in a submodel.

Because of the multitude of factors and rules in the model, only a few examples can be explained here. For more details about the model, see Wiegand *et al.* (1995).

The size of a grid square is chosen to be 0.5 m × 0.5 m; each grid square can be occupied by one adult individual at most. In this way it is guaranteed that the shrubs will be spread out in the model. The ecological states (see Box 6.3) of the squares are characterized by bare ground or by the species present. The age and size of the adult shrub have to be assigned, as do the species and the number of seeds or seedlings in a square. For each of the five species, the processes given in Box 6.3 are described by a set of rules, some of which are as follows. Growth of individuals depends on age and rainfall. In Figure 6.5, the number of seeds produced annually by an adult plant of *O. sinuatum* is shown versus the rainfall in the growing season. A threshold at about 20 mm exists below which no seed production occurs. The regression line in Figure 6.5 is used as a quantitative rule for a single shrub. Similarly, a minimum amount of rainfall is necessary in the subsequent germination period if the seeds produced are to germinate. The resulting seedlings survive the critical period after germination only if the species-specific rainfall condition is fulfilled. Equivalent rules exist for the other species. To apply these rules, a submodel that determines the rainfall

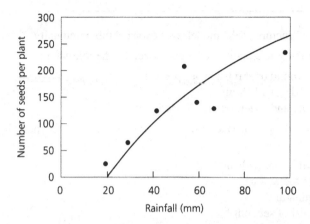

Figure 6.5 Number of seeds produced by a single plant of the shrub *Osteospermum sinuatum* per year versus rainfall in the growing season. Points are data from field investigation. The curve is the regression used in the model.

for each year is needed. Rainfall data, which exist for the past 97 years, can be used for this purpose.

Once the number of potential surviving seedlings produced by each adult plant has been determined, where the seedlings become established must be specified. Data on typical seed distribution for these species gathered from field studies can be used for this purpose. In the model, the seeds produced by an individual in a square are distributed according to these distribution data, giving the number of seeds and the resulting seedlings for each square (the third state characteristic in Box 6.3). To establish themselves the seeds must fall into a safe site. For a plant requiring shade, a safe site is a square occupied by a colonizer that offers shade. For a colonizer, a safe site is an empty square that is surrounded by a species-specific number of empty squares. This empty area prevents competition from adult plants. Only one of the seedlings present can take over the square. This self-thinning is in accordance with the rule that only one adult per square is possible. If there are seedlings of different species in a square, a clear hierarchy determines which species takes over the site (Wiegand *et al.* 1995). Colonizer *B. ciliatus* wins over *G. fructicosa*, which in turn wins over *R. spinosa*. Of the plants requiring shade, *O. sinuatum* outcompetes *P. pallens*. A seedling of a shade species can wait up to nine years until its host colonizer plant dies before taking over the site. Finally, each adult dies when its maximal life span is reached. More details of the rules can be found in Wiegand *et al.* (1995).

Using these rules, simulations on a grid of 77 × 53 squares (about 1000 m^2) are investigated. A typical result is shown in Figure 6.6. Results

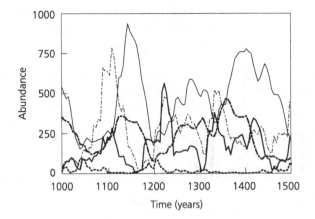

Figure 6.6 Typical community dynamics in a dwarf shrub community. Abundances (number of adult plants in the modeled area) of the five species vary over time due to the model rules and the distribution of rainfall.

with the same characteristics are found if, instead of the rainfall data, a model is used that produces rainfall values that correspond to the typical real rainfall in this region. The typical community dynamics show no equilibrium. Instead, strong fluctuations appear, resulting in strong variations of the relative abundances of the species. But all five species coexist. Large recruitment events are followed by a collapse after a time determined by the life span of the corresponding species. During some periods (1260–1300 years) the system is relatively constant. An ecologist who observes the system only during such periods would get a misleading impression of the system dynamics, as the typical time scale for the dynamics of the system is several decades.

The most noticeable property of the system's dynamics is the recruitment events. To gain a better understanding of them, Figure 6.7a shows the effects of variations in rainfall on the species *R. spinosa*. The rainfall during the season of each year is transformed into the number of seeds per plant using the rule shown in Figure 6.5. The white columns indicate years where there is enough rainfall during the seed production phase, but not enough during the germination period. The gray columns indicate years where there is enough rainfall during the seed production and germination phases, but where rainfall during the subsequent phase is too low to enable survival of the seedlings. Only the years indicated with black columns have enough rainfall in all three phases to permit seed production and germination, and seedling survival.

In Figure 6.7b, the black columns, indicating years suitable for recruitment, are shown together with the total abundance of *R. spinosa*.

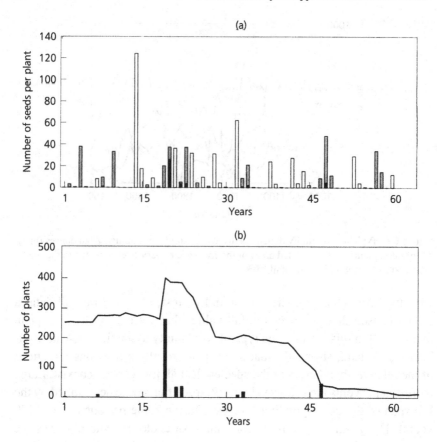

Figure 6.7 Relation of recruitment events of the shrub *Ruschia spinosa* to rainfall. (a) Seed production per plant versus time based on empirical rainfall data; white columns indicate no germination; gray, germination but no seedling survival; black, germination and seedling survival. (b) Black columns are as above; curve shows the number of adult plants of *R. spinosa* versus time.

In year 19, a strong increase in abundance is found, as expected from the recruitment of seedlings (black column). But in year 47, no increase is found despite seedling recruitment. Obviously, rainfall is not the only factor responsible for recruitment events. A more detailed analysis of year 47 shows that in this year the density of shrubs in the model is rather high, leaving little open space with bare ground between the shrubs. *R. spinosa* is a colonizer plant that requires much space for establishment, which is not available in this year. In general, one finds that the recruitment events are caused by the accidental coincidence of adequate rainfall in three different seasons and the existence of sufficient space for new recruitment. Thus, three external, temporal conditions and an internal, spatial one must all be fulfilled.

On the farms in this region, areas occupied by the shrub community are used for sheep grazing. Many farms are overgrazed and show deteriorated vegetation. Thus, another purpose of this model is to help develop a sustainable grazing strategy. Grazing can be modeled as a reduction in the growth of shrubs and the production of seeds to simulate feeding on the blossoms. As expected from the results in Figure 6.6, the community's reaction to good or bad grazing strategies often can be seen only decades later. By testing alternative strategies in the model, a sustainable grazing strategy may be found that preserves the typical structure and dynamics of the community. In any case, a suitable grazing strategy must ensure that inflorescences are not eaten by sheep in years of potential recruitment.

6.5   A Generic Forest Fire Model

In Mediterranean and boreal forests, fire is a regular factor that determines the composition and spatial structure of vegetation. Good data on the influence of fire on natural forests are available, especially for Canada. After fire in an area, a succession of the vegetation (i.e., a distinct sequence of different plant communities) appears. The occurrence of another fire sets the vegetation back to the first stage of the succession. Not all parts of a forest burn in every fire. Thus, forests are composed of a mosaic of patches at different stages of succession. This factor is the principal cause of boreal forest diversity. Therefore, understanding the mechanisms that determine the size and form of burned areas is a matter of considerable interest. Are catastrophic large fires anthropogenic, or do they occur under natural conditions?

A grid-based fire model (Ratz 1995, 1996) was developed to explore these issues. To obtain a general understanding of the mechanisms involved, the model design was kept simple. The size of the modeled forest area is 200 × 200 squares, with one square corresponding to 4 ha of forest. To exclude boundary effects, only the interior 100 × 100 squares are evaluated in the model. The ecological state of a square is described by its age since the last fire. This age is related to a successional stage of the vegetation, which is not considered in more detail in the model.

The rules for change of the states during a time step of 10 years are shown in Box 6.4. The age (measured in units of 10 years) increases in one-unit steps until it reaches its maximal value of 50 units. As there is usually a fire before this maximum is reached, its precise value is unimportant. If

---

Box 6.4 Rules of the forest fire model

The states of grid squares are characterized by the age a since the last fire. The states change in time steps of 10 years. If no fire occurs, a increases by one unit each time step. No increase in a occurs beyond a maximum of 500 years. A fire sets a to zero. Five randomly positioned lightning strikes occur during each time step. A square burns with the probability $f = i + ca^2$ if it is hit by lightning or borders a burning square.

---

a square burns, its age is set back to zero. Fire is normally caused by lightning or other unpredictable factors. Therefore the occurrence of lightning is described by a probability set to five strikes per time step. If a square is hit by lightning, it burns with the flammability probability f. With flammability $f = 0.5$, for the 200×200 grid there are 1.6×10^{-6} fires per year per ha, which corresponds to empirical values. The position of a lightning strike is chosen at random. The four squares adjoining a burning square also burn with the probability f. In this way a fire spreads over the modeled grid. Because the flammability probability is $f < 1$, the fire stops somewhere by chance, thus burned areas of different size and form are created. Over the course of time these areas overlap and produce a mosaic with patches of different ages, as shown in Figure 6.8. The pattern in the figure resembles the empirical data of mosaics in boreal forests.

Two alternative ways of depicting flammability are under discussion in the literature: constant flammability independent of the age of the site, and flammability that increases with age. The second case seems reasonable, as the amount of flammable material should increase with age (i.e., with time since the last fire). In the model, the flammability of a square of age a is chosen to be

$$f = i + ca^2 . \tag{6.6}$$

This function for the increase of f with a gives the best agreement with field data (see discussion below), but other functions give similar results. With an appropriate choice of parameters i and c, both alternatives can be realized in the model. A detailed analysis of the mean intervals between fires indicates that an increase in flammability with age is more realistic.

Despite the simplicity of the model, a more detailed comparison with a real boreal system is useful. Data from Alberta, Canada (Eberhart and Woodard 1987), give a detailed description of the fine structure of individual fires. The following measures are based on the fact that the area (a_b)

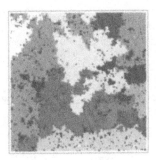

Figure 6.8 A typical mosaic of even-aged patches generated by a model of fire events (the darker the patch, the older the vegetation).

actually burned differs from the area (a_o) enclosed by the outer perimeter (p_o) of the fire because there are unburned "islands" within the fire. The following fine-structure measures are used:

- Fraction of undisturbed area $1 - a_b/a_o$
- Shape index $p_o/(4\pi a_o)^{1/2}$
- Edge index $p_w/(4\pi a_o)^{1/2}$, where p_w is the whole perimeter including the edges of inner islands
- Median of the areas of islands
- Number of islands per 100 ha

The shape and edge indices are the quotient of the respective perimeter and the perimeter of a circle with an area equal to that enclosed by the outer perimeter. These five measures are determined for five different fire size classes. The corresponding 25 empirical values are shown in the left-hand column of Figure 6.9. The same procedure was carried out for the model simulations; the results are shown in the right-hand column of Figure 6.9. In the model run, the size of a grid square was not fixed. One of the 25 fine-structure measures of the model is fitted to the corresponding empirical values, which gives a size of 4 ha per grid square. Although the model is very simple and does not include any ecological detail, the agreement between model and empirical data is surprisingly good. This holds true for a large set of the parameters i and c. Only extremely low and extremely high flammabilities give poor agreement; however, these flammabilities surely are unrealistic, as they produce results where almost the entire area is in a very early stage of succession (extremely low flammability) or in a very late one (extremely high flammability).

In addition to the spatial form of the fires, the sizes of burned areas are of interest. Over the course of time, fires of very different sizes appear in the

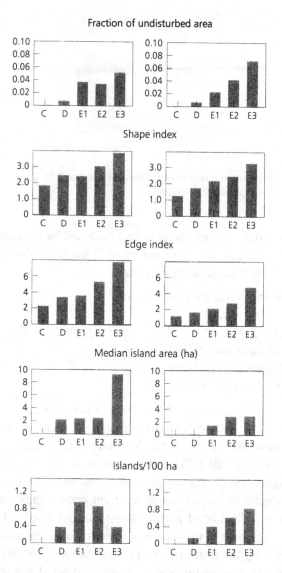

Figure 6.9 Comparison of (left) fine-structure measures of real fires (*Source*: Eberhart and Woodard 1987) with (right) model fires for different fire size classes. Parameter values: $i = 0.3$ and $c = 0.00012$. Fire size classes: C = 20–40 ha; D = 41–200 ha; E1 = 201–400 ha; E2 = 401–2000 ha; E3 = 2001–20 000 ha.

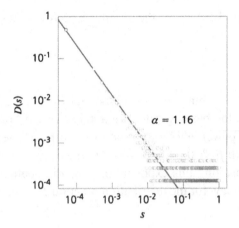

Figure 6.10 Fire size distribution. Typical frequency distribution of fire size $D(s)$ versus fire size s (fraction of area burned) in double logarithmic scale for $i = 0.2$ and $c = 0.0001$. The straight line shows the least squares fit of D to s, using Equation (6.7).

model. From these fires a frequency distribution is determined, as shown in Figure 6.10. The straight line in a double logarithmic scale means that the distribution function $D(s)$ of the fire size s follows a power law

$$D(s) \propto s^{-\alpha} , \qquad (6.7)$$

with $\alpha = 1.16$. This power law holds true for a large set of parameter values of i and c. Again, only the unrealistic cases of extremely high and extremely low flammability show different behavior.

A power law of this kind is typical for so-called self-organized criticality. The classic example of this behavior is a sandpile where grains of sand are steadily poured onto the top. The sandpile develops to the state of a critical steepness, at which point avalanches of very different sizes occur. Their size distribution shows a power law like that of the distribution function above. The analogy between fires and avalanches is as follows: the fire system develops to a critical frequency and spatial distribution of the different age classes, that is, different successional stages with different flammabilities. At this critical state, the power law of the size distribution $D(s)$ of fires can be found. The slope of the sandpile is responsible for the occurrence of avalanches; the age (and consequently the flammability) is responsible for the appearance of fires. Having seen the properties of self-organized criticality in the model, it would be interesting to examine the field data to see if the same power law exists there. Indeed, data exist that can best be fitted by a power law.

In self-organized criticality, the duration t of avalanches is related to the size s by a power law:

$$t \propto s^\beta .$$ (6.8)

In the model, the burning time of individual fires can be defined as the number of times the burning rules are applied until a fire is extinguished (see Box 6.4). A power law for the relation of burning time t to fire size s is indeed found with $\beta = 0.54$, reconfirming the existence of self-organized criticality. By inserting this exponent into the distribution function $D(s)$, one finds a power law for the distribution function $D(t)$ of the burning time with

$$D(t) \propto t^{-\gamma} ,$$ (6.9)

with

$$\gamma = \alpha/\beta .$$ (6.10)

The good agreement of the model to available field data suggests that fires in real boreal forests show self-organized criticality. This criticality seems to be a generic property, as no ecological details of the forest and the burning process have been used in the model. An immediate consequence of the power law of self-organized criticality is that extremely large fires need not be caused by human activity alone, but can appear under natural conditions, albeit with low frequency.

6.6   Concluding Comments

Grid-based modeling has been shown to work *without mathematics*. Instead, the model is directly formulated in a computer program using *rules*. In this way the biological information can be translated into the model rules without being distorted for mathematical needs. As there is a vast number of possibilities for the formulation of these rules using programming languages, this rule-based approach can be adapted to a specific problem and to the available information. The shrub community model is an example of a rather detailed description, whereas the fire model is rather abstract. The example of rabies shows that different models of the same system should be different for different questions, which can be achieved using the grid-based approach. Thus, grid-based modeling is a flexible tool with universal application.

The example of the shrub community demonstrates an instance where grid-based modeling seems to be the only possibility for modeling a particular concrete case study. It is difficult to imagine how this system could be described with the help of mathematical equations. In such a study, close cooperation between modelers and field ecologists is especially important. The use of rules instead of mathematics facilitates communication considerably in instances of this kind. Yet the fire model shows that general problems also can be addressed by grid-based modeling. The fire model does not rely on any specific ecological details, thus its results should be applicable to a wide range of fire ecosystems.

The models described in this chapter assume that the same rules apply to all squares of the grid. This is equivalent to assuming that the underlying environmental setting is spatially homogeneous. The spatial patterns that appear in a model of this type are created by the biological interactions inside the system in a self-organizing manner. However, in real situations, the environmental setting may show its own spatial pattern. In this case, this environmental pattern interferes with the self-organized biological pattern, producing new results that strongly depend on the scales of both patterns. In ecological modeling, little research has been done in this direction. In principle it is clearly possible to include environmental heterogeneity in a grid-based approach. The rules for each square have to be modified according to the state of the environment: a heterogeneous landscape is reflected in heterogeneous rules.

There are no recipes for designing a grid-based model. The basic method is very simple, but the way it is applied differs from model to model, as demonstrated by the examples given here. The art of modeling consists of using ecological reasoning and the available information to describe the ecological states of grid squares and the rules of their changes. Models always must be adapted to the problem to be solved, to the processes to be modeled, and to the available information.

7

Coexistence of Replicators in Prebiotic Evolution

Tamás Czárán and Eörs Szathmáry

7.1 Introduction

The role of spatial population structure in promoting cooperation and mutualism has recently received much interest in a number of theoretical ecological and evolutionary studies (e.g., Nowak and May 1992; Hammerstein and Hoekstra 1995; Killingback and Doebeli 1996). It has also been emphasized in the context of the origin of life for essentially the same reason: it is a means to establish coexistence of potentially competing template replicators, thus allowing increased capacity of information storage and transmission by the population as a whole (see Maynard Smith and Szathmáry 1995 for review). There are three known, detailed model approaches:

- Structured deme-type models (Wilson 1980; Michod 1983; Szathmáry 1992)
- Replication–diffusion systems as modeled by cellular automata (Boerlijst and Hogeweg 1991b)
- Group selection of replicators encapsulated in compartments (Szathmáry 1986; Szathmáry and Demeter 1987; Maynard Smith and Szathmáry 1993)

The motivation for these studies originates with a seminal paper by Eigen (1971) arguing (1) that primitive genomes must have been segmented (consisting of physically unlinked genes); (2) that these unlinked genes must have had the tendency to compete with one another; and (3) that, consequently, some mechanism ensuring their coexistence was needed. Eigen saw the hypercycle (Figure 7.1), a system of cyclically interacting molecular mutualists, as fulfilling this role. However, in a spatially homogeneous setting, the hypercycle is vulnerable to parasitism: a cheating replicator that does not give catalytic aid to any member of the cycle can kill the cycle off, provided it receives more catalytic help from the cycle than the

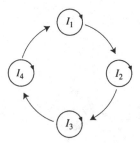

Figure 7.1 Cooperation of replicators ($I_1, \ldots, I_4$) in a hypercycle. Circular arrows indicate replication; curved arrows, direct catalytic help in replication.

member that it competes with in the first place. It has become apparent that compartmentation is an efficient means to separate bad genes from good ones (Maynard Smith 1979; Eigen *et al.* 1981). The stochastic corrector model was presented as an alternative to hypercycles as a means of integrating genetic information dispersed in unlinked replicators. The model demonstrates that, once compartments exist, their internal organization need not be hypercyclic: a so-called metabolic coupling of replicators is sufficient (Figure 7.2), whereby genes contribute to the good of the compartment by catalyzing its metabolism at various points (Szathmáry and Demeter 1987). The stochastic corrector model assumes that there is an optimal template composition of compartments that gives the highest protocell division rate (for a review, see Szathmáry and Maynard Smith 1995). Variation between compartments is generated by the stochastic effects during template reassortment at cell division as well as during template replication. Natural selection between the compartments acts on this variation.

Structured deme-type models essentially lead to the same conclusion: interactions confined to immediate neighbors promote coexistence. Michod (1983) has demonstrated the hypercycle's resistance to parasites in this setting. The viability of the hypercyclic and metabolic systems has also been demonstrated in such a context (Szathmáry 1992).

The cellular automaton approach applied to the problem of information integration (Boerlijst and Hogeweg 1991b; see also Chapters 9 and 10) is rather different from both the stochastic corrector and the structured deme framework. It is basically a discretized reaction–diffusion system: replication and diffusion of templates is imagined to take place on an adsorbing surface, without compartmentation. A hypercycle can be resistant to parasites in such a reaction–diffusion system provided there are more than four replicators. The reason for this provision is that spiral waves emerge as spatial manifestations of the intransitive circle of mutualistic interaction

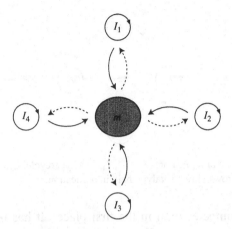

Figure 7.2 Cooperation of replicators $(I_1, \ldots, I_4)$ through a common metabolic system (m). Circular arrows indicate replication; curved dashed arrows, supply of monomers for replication; curved solid arrows, catalytic help for monomer production through metabolism.

assumed in the hypercycle model. Without the spirals (e.g., with fewer replicators) this particular system collapses if a parasite invades. Cronhjort and Blomberg (1995) have conducted numerical studies of the partial differential equation model of the same reaction–diffusion system and have found that the section of parameter space allowing for parasite resistance is smaller in this model than in the cellular automaton of Boerlijst and Hogeweg (1991b). The spirals are spontaneously emerging self-organized "units of selection" in the reaction–diffusion approach; the role they play in this approach is in some respects similar to that played by compartments in the other two models. Spiral waves as units of selection are of course much less definite than the compartments of the structured deme and the stochastic corrector models: whether a replicator molecule belongs to a certain spiral is not always easy to determine.

The fundamental difference between the cellular automaton model presented in this chapter and that of Boerlijst and Hogeweg (1991b) is the use of a stochastic corrector mechanism: the dynamic link among the replicator types is realized through a common metabolism, not through direct, intransitive hypercyclic coupling. Moreover, unlike in the cellular automaton approximations to reaction–diffusion systems, the discrete, "corpuscular" appearance of the replicator macromolecules is not a methodological compromise in our model but an essential feature of the object to be investigated: populations of macromolecules can hardly be imagined as fluids consisting of an infinite number of sizeless mass points. We show that discrete individuality can be, and in our model is, of profound dynamic importance (see also Durrett and Levin 1994b).

Using the cellular automaton model of the metabolic system, we show that

- metabolic coupling can lead to coexistence of replicators despite an inherent competitive tendency;
- parasites cannot easily kill the whole system;
- complexity can increase by natural selection;
- by varying the critical neighborhood size and the diffusion rate, one can approximate the behavior of other models.

7.2   Metabolic Replication: A Cellular Automaton Model

The dynamics of the nonspatial version of the metabolic system are as follows:

$$\frac{dx_i}{dt} = x_i \left[k_i m(x) - \Phi(x) \right] , \tag{7.1}$$

where x_i stands for the concentrations of template I_i, x is the vector of these concentrations, and k_i is the specific replication constant of template I_i. The term $m(x)$ is a multiplicative function of the concentrations of all the templates, and $\Phi(x)$ is an outflow term representing a selection constraint (constant total concentration). This formulation is formally identical to that given by Eigen and Schuster (1978) for a "minimum model of primitive translation." As they correctly noted, the fact that replication of any template is impossible without the presence of all the other templates does not prohibit the system from undergoing competitive exclusion: $m(x)$ is the same in all equations, hence the system essentially behaves as a collection of Malthusian competitors whose dynamics are influenced by a common time-dependent factor.

It is assumed that the replicators I_i have two functions: as templates they are necessary for their own replication (autocatalysis), and as "ribozymes" (RNAs able to act as enzymes) they contribute to metabolism, producing monomers. It is not our aim to model metabolism in detail; therefore, for the purposes of the present chapter, the model given in Equation (7.1) is sufficient.

Space, time, and the representation of metabolism

We now assume that replication takes place on the surface of a mineral (possibly pyrite) substrate. The replicator molecules themselves are of a finite size, therefore the number of replicators bound to a unit area of the substrate is constrained. We consider a two-dimensional square lattice of binding sites as the scene of the replication–diffusion process; each of the sites can harbor at most a single macromolecule. The lattice is toroidal (the opposite edges of the grid are merged in both dimensions) to avoid edge effects.

At $t = 0$, half the sites are occupied by n different types of macro-molecules (we call the number n of replicator types the *community size*). The replicator types are equally abundant in the initial pattern, and individual molecules are randomly assigned to sites. The other half of the sites are initially empty. Time is discrete; replication, decay, and diffusion take place in each generation of the simulation.

The effect of monomer-producing metabolism is implicit in the model; it acts directly on the replication process through a *local metabolic function*. This function is local in the sense that its arguments are the copy numbers $f(i)$ of replicator types i ($i = 1, \ldots, n$) within certain localities (neighborhoods) of the lattice. In accordance with the assumption that the presence of a complete set of replicators is necessary for metabolism to produce monomers for replication, the within-neighborhood copy numbers $f(i)$ must act multiplicatively in the metabolic function. A simple option for a specific form of the metabolic function $M(f_s)$ at a site occupied by a replicator s is the geometric mean of the copy numbers $f_s(i)$ within the metabolic neighborhood of s; that is,

$$M(f_s) = \left[\prod_{i=1}^{n} f_s(i) \right]^{1/n} \tag{7.2}$$

The metabolic neighborhood is either the Moore neighborhood (i.e., the 3×3 sites centered on s, whereby the size of the neighborhood is $A = 3 \times 3 = 9$), or a larger ($A = 5 \times 5 = 25$; $A = 7 \times 7 = 49$; $A = 9 \times 9 = 81$; etc.) square neighborhood around the focal replicator molecule s (Figure 7.3a). Notice that $M(f_s)$ is zero if any of the replicator types is missing from the metabolic neighborhood of s, and that the larger and more uniform the copy numbers of the different replicator types within the metabolic neighborhood, the more efficient the metabolism at the given locality. By choosing Equation (7.2) as the metabolic function, we assume that (1) con-specific replicators within the same neighborhood help replication, and (2) the focal replicator supports its own replication. Assumption (1) can be interpreted as metabolism being somewhat faster locally in the presence of more catalysts. The actual effect should be rather weak and should vanish

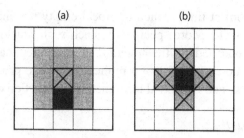

Figure 7.3 Types of neighborhood applied in the simulations. (a) Metabolic neighborhoods of the replicator in the focal site (gray site with ×) beside an empty (black) site. Gray sites = 3 × 3 neighborhood; gray + white sites = 5 × 5 neighborhood. (b) The replication neighborhood (gray sites with ×) of an empty (black) site.

as the copy number increases. This feature is properly reflected in the metabolic function given in Equation (7.2): if a replicator type is already present in a replication neighborhood, then a second, third, etc., copy of it does not add too much to the focal template's replication chances. Assumption (2) implies that a metabolite diffuses out of the neighborhood in which it was produced on a time scale longer than that of the catalyzed reactions of metabolism. Because the "habitat" of the reaction–diffusion system is an absorptive mineral surface, this is again an appropriate assumption. The size of the metabolically effective neighborhood is an implicit measure of metabolite and monomer diffusivity: larger neighborhoods represent faster diffusion of the intermediate metabolites and the monomers.

Values of the metabolic function given in Equation (7.2) obviously are equal for replicator molecules with identical metabolic neighborhoods, which also means that in any proper mean-field approximation to $M(f_s)$ – that is, in any of the possible corresponding models omitting spatial heterogeneity – m should in fact be independent of s, but dependent on the overall replicator abundance vector x. In this sense, $m(x)$ in Equation (7.1) is the simplest possible mean-field approximation to Equation (7.2). Note that by dropping both assumptions (1) and (2) above, the metabolic function in the mean-field model [Equation (7.1)] would be different, and, consequently, Equation (7.1) would lead to coexistence of the replicators.

Replication, decay, and diffusion of replicators

The *replication neighborhood* of an empty site consists of its four orthogonal nearest neighbors (see Figure 7.3b). If the replication neighborhood contains a macromolecule capable of replication (i.e., one with a nonzero

metabolic function) at time t, then the focal empty site may become occupied at time $t + 1$. More specifically, each replicator molecule in the
replication neighborhood has the *potential* to place a replica of itself in the
empty site. The potential $C(s)$ of a replicator s is the product of a type-
specific constant k_s and the actual local metabolic function of s:

$$C(s) = k_s M(f_s) . \tag{7.3}$$

During the simulations, the constants k_s were evenly spaced on the interval
[2.0, 6.0] (e.g., for a community of size $n = 3$, the replication constants
were $k_1 = 2.0, k_2 = 4.0, k_3 = 6.0$). The empty site itself also has a constant
potential C_e of remaining empty – we set $C_e = 2.0$ in all simulations. The
chance P_s that the molecule s will replicate in the empty site is proportional
to its replication potential:

$$P_s = \frac{C(s)}{C_e + \sum_{l=1}^{4} C(l)} , \tag{7.4}$$

where l are the macromolecules in the replication neighborhood. The probability that the empty site remains empty by time $t + 1$ is

$$P_e = \frac{C_e}{C_e + \sum_{l=1}^{4} C(l)} . \tag{7.5}$$

Spontaneous decay is a simple probability event in the model; the chance
that any replicator will disappear from the lattice is a constant p_d, independent of the type of macromolecule. We set $p_d = 0.2$ in all runs of the
model. The complete extinction of any one replicator type from the lattice
is, of course, fatal to all the other types as well, as this prevents all of them
from replicating and there is no process to counter the decay.

The diffusive movements of the macromolecules are explicitly modeled by the algorithm of Toffoli and Margolus (1987), which preserves
the number and the frequency distribution of the molecules, resulting in
a Brownian-like trajectory for each molecule on the lattice. We divide the
grid into 2 × 2 subgrids first. One elementary diffusion step is then completed by (1) rotating each subgrid independently 90° clockwise or coun-
terclockwise with equal probability and (2) shifting the grid frame one site
in any one of the four diagonal directions with equal probability. The intensity of macromolecule diffusion can be set by specifying the number
of such elementary diffusion steps per generation. Note that the diffusiv-
ities of the macromolecules and those of the small-molecule intermediary

metabolites and monomers can be controlled independently by different parameters: macromolecule mobility depends on the number of diffusion steps per generation, whereas metabolite and monomer diffusivity are implicit in replication neighborhood size.

The updating procedure can be either synchronous or asynchronous, representing discrete or continuous time scales for replication, respectively. Because this is known to affect coexistence and spatial pattern in certain cellular automaton models (Huberman and Glance 1993), we tried both updating procedures. In the synchronously updated version, neighborhoods are taken from the grid state at t to calculate the actual state of the cells at $t + 1$. The program systematically scans all cells, looking first at whether each cell is empty. If a cell is not empty, the replicator there decays with the probability characteristic of its type. If it is empty, the replicators in the replication neighborhood (if there are any) have a chance to place a copy of themselves there, according to the replication algorithm described above. Diffusion takes place after decay and replication.

For asynchronous updating we use a random-sequence procedure: one randomly chosen site is updated at a time. To ensure that on average each site is updated once per generation, each generation comprises a number of updates equal to the number of lattice sites. The decay-replication procedure is the same as in the synchronously updated version, but the grid from which the neighborhoods are taken is updated after each decay-replication event and elementary diffusion steps are evenly spread over the time interval of the generation. The results of the randomly updated version of the program are not qualitatively different from those of the synchronously updated model; in fact, they have also proved to be extremely similar in quantitative terms in almost all runs as well. Therefore, we present results only for synchronous updating.

We have completed a few (usually three) replicate runs for each parameter combination; due to the large lattice size (300×300), replicate runs are quite similar in all cases. The parameters and the actual values applied in the simulations are summarized in Table 7.1.

7.3   The Phenomenology of Coexistence

There are two outcomes for the model: the process either ends up in the absorbing state (global extinction) or it reaches a quasi-stationary abundance distribution inside the positive orthant of the state space (persistence). There is no third possibility. One of the most important results of the simulations is the somewhat surprising finding that the model is capable of

Table 7.1 Parameters of a model for metabolic replication on a surface (see text for details).

Parameter	Description	Value
C_e	Potential of empty cells to remain empty	2.00
p_d	Decay probability (per generation) of replicators	0.20
k_i	Replication constant of replicator type i	2.00–8.00
n	Number of replicator types (community size)	3–9
A	Metabolic neighborhood size	3×3 to 9×9
d	Number of diffusion steps per generation	1–300

producing coexistence in a substantial part of its parameter space, even though the mean-field model [Equation (7.1)] predicts the competitive exclusion of all the inferior replicator types, and thus the subsequent collapse of the entire system to the single type with the largest k_i whatever parameters we choose. Notice that adding a diffusion term would not help to save Equation (7.1), as the metabolic function would still be the same for all replicators at any given location; that is, the dominance order of the replicator types would remain the same everywhere. Local density differences can induce local differences in the speed of the extinction process through the metabolic function m, but extinction remains inevitable everywhere in the spatially inhomogeneous partial differential equation case as well. This means that extending the metabolic replicator system in space is in itself not sufficient (albeit necessary) for coexistence.

From the viewpoint of template coexistence, the most relevant parameters of the model are (1) community size n (Table 7.1), (2) metabolic neighborhood size A, and (3) diffusion rate d. Figure 7.4 shows the effects of changing these parameters within reasonable limits, displaying the time series of replicator numbers for a single run of the program with each parameter set. We now take a closer look at the effects of each of these parameters in turn.

Diffusion and coexistence

The diffusive movement of the replicators seems to play a central role in replicator coexistence. It is a general observation that increasing the number of diffusion steps (d) per generation is beneficial for the persistence of the replicator system (Table 7.2). The reason is straightforward, bearing in mind that the more diverse the replicator set in a metabolic neighborhood, the larger the value the metabolic function M will take at that locality, and thus the higher the probability for a nearby empty site to be occupied in the next generation. Spatial aggregation is definitely disadvantageous for all

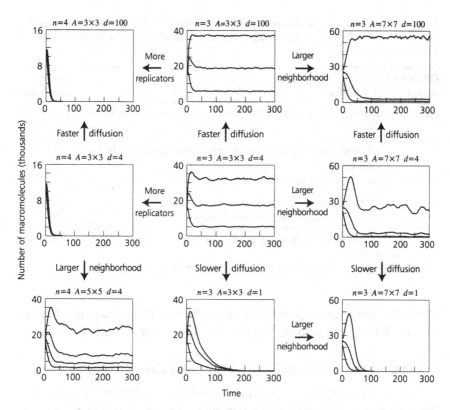

Figure 7.4 Time series of simulation results with different community sizes (n), metabolic neighborhood sizes (A), and diffusion rates (d). Parameters: $k_1 = 2.0$, $k_2 = 4.0$, and $k_3 = 6.0$ if n is 3; $k_1 = 2.0$, $k_2 = 10/3$, $k_3 = 14/3$, and $k_4 = 6.0$ if n is 4; other parameters as in Table 7.1.

replicator types, because a metabolic neighborhood with too many identical replicator molecules is likely to lack one or more of the other types of replicator. In this case, the metabolic system is locally incomplete: monomers are not produced, therefore replication ceases at regions of monospecific or nearly monospecific aggregation. Replication itself induces aggregation, since mother and daughter molecules are always immediate neighbors after the replication event. To avoid this aggregation, a certain level of spatial mixing is necessary, and this mixing is provided by diffusion.

Nevertheless, increased diffusion might also accelerate the extinction process of very sparse patterns (those with too many empty sites) by dispersing potentially cooperative replicators far from each other. This effect does not really act against coexistence, however, because it does not change the outcome of the dynamics in qualitative terms – extinction could not be

Table 7.2 Summary of the effects of parameter changes on persistence. Abbreviations: n = number of replicator types; A = metabolic neighborhood size; d = diffusion rate; ↑= increase; ↓= decrease.

Parameter change	Direct effect	Effect on persistence
n ↑	↓ Chance that a metabolic neighborhood contains a complete set of replicators	↓
A ↑	↑ Chance that a metabolic neighborhood contains a complete set of replicators	↑
	↓ Advantage of the rare	↓
d ↑	↑ Spatial mixing (high replicator density)	↑ [a]
	↑ Spatial mixing (low replicator density)	↓ [b]

[a] The system approaches one in which group selection occurs on neighborhood configurations (Wilson 1980).
[b] Replicators are dispersed, driving the system more rapidly to extinction.

avoided in very sparse patterns anyway. What diffusion actually does to such a dilute system is to shorten the time it takes for extinction to occur.

Community size, metabolic neighborhood size, and coexistence

The effects of community size (n) and metabolic neighborhood size (A) on the persistence of the metabolic replicator system are closely dependent on one another (Table 7.2). The key to understanding this cross-dependence is the direct effect these two parameters have on the metabolic function M. It is obvious that if community size n exceeds $A - 1$, none of the replicators can ever replicate because no metabolically sufficient set of macromolecules can fit into the metabolic neighborhood, thus the system dies out. In such a case, increasing A (that is, assuming higher diffusivity for the intermediary metabolites and monomers) can make the system persistent. Alternatively, persistence can also be achieved by decreasing the community size, which in fact always promotes coexistence. This is a property that may not be welcome in a model designed to explain how community size might have *increased* during prebiotic evolution. Note, however, that by *not* assuming in this model that more replicators can catalyze a more sophisticated metabolic pathway (i.e., that metabolic efficiency is an increasing function of community size), we are making it particularly difficult for community size to increase. What we wish to show is that even in such a worst-case metabolic system it is possible to preserve, or even increase, the number of coexistent replicator types n.

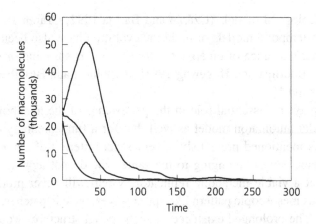

Figure 7.5 Time series of simulation results for parameters $n = 3$; $A = 9 \times 9$; $d = 4$; $k_1 = 2.0$; $k_2 = 4.0$; $k_3 = 6.0$; other parameters as in Table 7.1.

Although increasing metabolic neighborhood size is beneficial (due to the increased chance of metabolic complementation), increasing this neighborhood also has an adverse effect on replicator coexistence, as the middle-right graph of Figure 7.4 and the still larger metabolic neighborhood size of Figure 7.5 demonstrate. The simplest phenomenological explanation for the harmful effect is that if the metabolic neighborhoods are large, the replicators "feel" the global averages of the macromolecule densities instead of the strictly local ones, and the dynamics of the system can be approximated with the nonpersistent mean-field model given in Equation (7.1). To take the most extreme case as an example, the metabolic functions are obviously identical for all individual macromolecules if the metabolic neighborhood is the whole lattice.

Because this model does not assume that increasing community size has a beneficial effect on the efficiency of metabolism, the size of viable communities is limited: too many replicator types need large metabolic neighborhoods to contain their metabolism, but large neighborhoods themselves act against coexistence. We find that communities of sizes greater than nine tend to become extinct within the range [2.0, 6.0] of replication constants k and with the decay constants p_d set to 0.2.

7.4 Spatial Pattern and the "Advantage of the Rare" Effect

For most spatially explicit dynamic systems it is simply the emergence of spatially inhomogeneous patterns that makes them persistent – in fact, it is usually for this reason that space is a dynamically important consideration in many ecological models (Czárán and Bartha 1992; Czárán 1997). Recent spatio-temporal models of prebiotic evolution have also been built on the dynamic relevance of emergent patterns such as spiraling or standing waves (see Boerlijst and Hogeweg 1991b; Cronhjort and Blomberg 1995; Chapters 9 and 10).

Space plays an essential role in the persistence of the metabolic replicator cellular automaton model as well, but for a fundamentally different reason. As mentioned previously, a very strict criterion of persistence is that macromolecules belonging to the same type do not aggregate. This implies that a viable metabolic replicator system will never produce any conspicuous mesoscopic pattern like spiral waves, mosaic patches, or steep gradients. The prolonged existence of such spatial structures would obviously be deleterious, because they represent a high level of aggregation – the system either gets rid of them quickly or goes extinct altogether.

Given that a uniform spatial distribution of the macromolecules is a prerequisite for coexistence, one can ask what makes the macromolecules coexist. In what sense is the model different from the poor mean-field approximation, which is also built on the assumption of overall spatial uniformity?

Coexistence is very closely tied to the fact that, not unrealistically, we assume macromolecule "individuals" to be discrete objects in the cellular automaton model, with a finite spatial extension. This assumption is implied by constraining the maximum number of replicator molecules to one per site, which in turn implies that it is distinctly possible that a sufficiently small metabolic neighborhood does not contain even one of the replicator types. A macromolecule s with a small replication constant k_s has an obvious disadvantage in terms of replication potential for occupying an empty site with a copy of itself. But it also has an advantage when it becomes rare: a rare replicator type is more likely to be complemented by the common types in the same metabolic neighborhood than vice versa. Replication ceases wherever the rare macromolecule type disappears. The "advantage of the rare" effect may be sufficient to maintain coexistence by favoring the inferior type, even if it is otherwise seriously handicapped by its low replication constant.

The rare replicator type has an additional advantage if the metabolic neighborhood is of size 3×3 (Moore neighborhood). In this case, it has a good chance of finding itself adjacent to an empty site with a common replicator type on the opposite side of the gap. Both molecules fall outside each other's metabolic neighborhood, but whereas the rare type may easily obtain metabolic help from another copy of the common type, the common type most probably lacks a rare molecule type in its Moore neighborhood. In such a case, the common replicator is excluded from competition with the other three macromolecules, thus increasing the chance that the rare type will replicate.

For the "advantage of the rare" mechanisms to operate, the system must be both discrete and spatial (see Durrett and Levin 1994b). Discreteness ensures that metabolic neighborhoods can be incomplete – in a spatial, but continuous system like the reaction–diffusion version of Equation (7.1), any finite area of space carries a positive (albeit potentially very small) concentration of all the replicator types, which maintains the relative competitive advantage of the common everywhere until the system collapses (see Section 7.3). Variation in the composition of metabolic neighborhoods is realized in space; reducing this variation by any means reduces the chance of coexistence, even if the discrete individuality of the macromolecules is maintained. This situation would be best approximated by running the simulation model with the metabolic neighborhood size set equal to lattice size, as explained in Section 7.3 (see subsection on Community size, metabolic neighborhood size, and coexistence).

Diffusion cannot homogenize metabolic neighborhoods below the spatial resolution level of one site, therefore the advantage of the rare does not vanish as the rate of diffusion increases. On the contrary, an increased diffusion rate promotes coexistence (see Section 7.3, Diffusion and coexistence). With a sufficiently large diffusion rate, the system approaches the following dynamics: local replication is followed by random reassortment of groups, which in turn is followed by local replication, and so on. This is practically a "trait group" model *sensu* Wilson (1980), for which template coexistence has been demonstrated analytically for both the hypercyclic and the metabolic systems (Szathmáry 1992). Neighborhood interaction represents a kind of temporary compartmentation that helps maintain the complete set of metabolically active replicators.

7.5 Resistance to Parasites and the Evolution of Community Size

Defining a parasite of the metabolic replicator system is straightforward: a parasitic replicator is one that receives metabolic help from the cooperative members of the system but does not itself contribute to metabolism in any way. In biochemical terms, a parasite uses the monomers produced by the cooperative replicators but does not catalyze any elementary reaction of monomer production. If the replicators are free to mutate and thus to

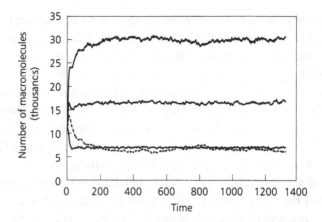

Figure 7.6 The effect of a parasitic replicator with a large replication constant ($k_p = 8.0$) appearing in a persistent three-replicator system. The time series of the parasite is shown as a dashed line. Notice that the parasite is suppressed by the cooperative replicators but remains persistent. Parameters: $A = 3 \times 3$; $d = 100$; $C_e = 2.0$; $p_d = 0.2$; $k_1 = 2.0$; $k_2 = 4.0$; $k_3 = 6.0$.

produce parasitic (as well as other, possibly cooperative) variants, the long-term viability of the metabolic replicator system critically depends on the system's resistance to such parasites.

Parasites are introduced into the model as an extra replicator type with a large replication constant k_p. The parasite is the fastest replicator in every simulation with a parasite present: $k_p > k_i$ ($i = 1, \ldots, n$). The updating rule is such that cooperative molecules do not need the presence of the parasite within their metabolic neighborhood to replicate, but all the cooperative types must be present for the parasite to replicate. More specifically, the local parasite copy number was omitted from the product of the metabolic function $M(f_s)$ in Equation (7.2).

The main conclusions of simulations with parasites are that (1) even fast parasites usually cannot kill a system that otherwise is viable, and (2) once established, parasites are not excluded, but (for small enough metabolic neighborhoods) their abundance is efficiently controlled by the cooperative replicators. A characteristic time series of a small community with a small metabolic neighborhood is given in Figure 7.6. Notice that the parasite, although having an obvious prior advantage in terms of its replication constant k_p, is the least abundant of all replicator types for most of the time series. The explanation for its low abundance is the reverse of the "advantage of the rare" effect: the parasite has little chance to replicate wherever it is abundant. The greater the density of the parasite, the smaller the probability that a parasite molecule has a metabolically complete set of cooperative replicators within its metabolic neighborhood. This effect becomes weaker with larger metabolic neighborhoods; in this case, the parasite achieves high copy numbers within the lattice but still does not drive the system to extinction (see Figures 7.7a and 7.7b).

Because parasites cannot do much harm to the cooperative system and the system cannot completely eliminate its parasites, the evolution of community size can proceed through the "domestication" of replicators that began as parasites. There is a possibility that a positive conversion by mutation will occur: the system can incorporate the parasitic sequence into the metabolic machinery and make it work for the common good. The beneficial effect on metabolism might be facultative at first, giving only some local replication advantage to neighborhoods containing the converted parasite, but later it may become an essential part of the metabolism. Once the interaction is obligate, the altered system is one member larger and metabolically more effective than the original. This mechanism allows for the development of an increasingly complex biochemical system, which is a process tightly coupled to the increase of coexistent and fully functional genetic information. Figure 7.7 shows some stages of this process in the evolution from a three-member to a four-member metabolic system. A comparison of Figures 7.7b and 7.7c shows that the conversion of a true parasite into an obligate mutualist (cooperator) does not affect the original cooperators; it is in the "interest" of the parasite to become a cooperator in the sense that by doing so it can increase its quasi-equilibrium abundance by about 20%. This applies even to the case where metabolic efficiency is not dependent on community size; in a similar model assuming a positive correlation between metabolic efficiency and community size, the conversion of the parasite would be beneficial to each of the replicators through increased metabolic support. A dynamic model of the evolution of community size based on these considerations is under development.

The model also has a valid interpretation in purely ecological terms, quite apart from the current focal context of "prebiotic ecology." The type of interaction we assume among the replicator populations is cooperation mediated by a common metabolism. By replacing "metabolism" with any "resource" produced by the joint efforts of the cooperators and needed for their survival and/or reproduction, the system becomes a model of obligate ecological mutualism between populations that also compete for the resource they produce together. Although this describes a rather specific type of ecological interaction, there are cases that match the assumptions

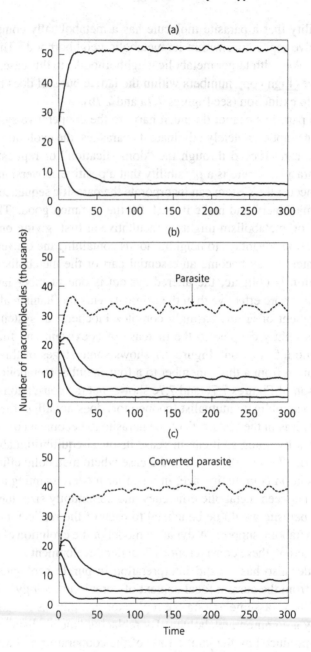

Figure 7.7 A possible scenario for the evolution of community size. (a) Persistent three-replicator system with parameters $A = 5 \times 5$, $d = 100$, $C_e = 2.0$, $p_d = 0.2$, $k_1 = 2.0$, $k_2 = 4.0$, $k_3 = 6.0$. (b) Same system after introduction of a parasite with $k_p = 8.0$. (c) Same system with the parasite converted into a cooperative member (i.e., just as essential as the other three members) of the system.

in many respects, for example, the symbiosis of fungi and algae in some lichens.

Ecological reinterpretation of the model leads to another scenario of the conversion of a resource–parasite interaction to obligate mutualism (see Michalakis *et al.* 1992, and references therein) through the "domestication" mechanism described above, if the interaction between the host and the parasite is local and the interacting populations consist of discrete individuals.

7.6   Toward a Dynamical Theory of Surface Metabolism

The finding that there is coexistence without any mesoscopic emergent pattern is robust and counterintuitive. It is due to the inherent discreteness (i.e., the discrete nature of the replicator molecule populations) and spatial explicitness of the model – aspects that grasp essential features of the living world in general and of macromolecular replicator systems in particular. An inferior (that is, slowly replicating) molecule type does not die out since there is an advantage to being rare in the system: a rare template has a much better chance of finding itself in a metabolically sufficient neighborhood than a common template. This effect stems from the joint effect of discrete individuality and spatiality: neither is sufficient without the other.

Surface dynamics seem to become more and more important for understanding the origin of life in general. Wächtershäuser (1988) points out that chemical evolution leading to increasingly complicated networks is likely to have taken place on surfaces, especially on those of pyrite. Chemical dynamics on a surface essentially occur in two dimensions. This fact has important thermodynamic and kinetic consequences. For example, an appropriate surface can act as a catalyst for the reactions in question. Water is liberated from the surface following condensation reactions, leading to larger molecules. The loss of water molecules renders the reaction favorable due to the increased entropy of the system as a whole. Surface dynamics of replicators with unlimited heredity (for which the number of possible types greatly exceeds the number of existing individuals, as it applies to nucleic acid genomes of sufficient lengths) are a natural outgrowth of these "primordial pizza" dynamics (see Maynard Smith and Szathmáry 1995). In this chapter, we have shown that the spatial distribution of discrete individuals on a surface crucially changes the outcome of selection, even for non-hypercyclic, but metabolically cooperative, systems.

A possible line for further research is to investigate a dynamical model of surface metabolism (M. Cronhjort, personal communication). Replacing m with an explicit model of metabolism seems to be essential: one should not forget that in the course of evolution, metabolism has always been an autocatalytic network of small intermediates. It is the reactions of such a network, comprising small organic molecules, that the enzymatic templates (ribozymes) used to catalyze. Metabolism in turn produced the monomers for replication. (Ribozymes were, presumably, replaced by protein enzymes rather early in evolution.) Modeling such a system entails considerable complications: for example, the small molecules (intermediates and monomers) necessarily have a much higher diffusion rate than the template macromolecules, and the number of variables necessarily increases with the number of reactions in the network. Coevolution of the network with the templates (see Wächtershäuser 1992), or, put more broadly, of metabolism with heredity, is a tremendously exciting problem likely to lead to some novel types of models.

Acknowledgments This work was supported by grants from the Hungarian Scientific Research Fund (OTKA T 012793 and T 019524).

8

Games on Grids

Martin A. Nowak and Karl Sigmund

8.1 Introduction

The theory of games and the theory of cellular automata seem at first glance to be totally unrelated (despite being created at about the same time – some 50 years ago – and boasting the same father, John von Neumann). In a recent and rather surprising development, the two disciplines have been brought together. The resulting *spatial evolutionary game theory*, which was first used to shed light on the emergence of cooperation, has grown rapidly during the past five years and has proved useful in other biological and economic contexts. In this survey chapter, we concentrate on the game-theoretical aspect, as the other contexts are well covered in other chapters of this volume. Thus, we stress mainly the effects spatial structures have on frequency-dependent selection.

Let us start with an arbitrary 2×2 game (Box 8.1), that is, a game between two players each having two strategies: (1) to cooperate (denoted by C) and (2) to defect (denoted by D). A player using C receives a payoff R (the reward) if the co-player uses C, and S (the sucker's payoff) if the co-player uses D. A player using D obtains the payoff T (the temptation) against a C-player, and P (the punishment) against a D-player. We interpret C as helping in a common enterprise and D as withholding help; therefore, we assume that the payoff R for two C-players is larger than the payoff P for two D-players. Because the evolution of cooperation is only interesting if it is threatened by unilateral defection – that is, if a unilateral D-move is good for the defector and bad for the cooperator – we assume $T > R$ and $R > S$. What is *a priori* less clear is the ranking of P and S. The ranking $P > S$, gives the well-known Prisoner's Dilemma game, where D is the dominant strategy. This game has been used since the early 1970s (Trivers 1971) as a paradigm for the evolution of cooperation (see also Axelrod and Hamilton 1981; Sigmund 1995). However, one can argue that $P < S$ is also an interesting situation: it reflects the dilemma of two players who each

---

Box 8.1 Payoff matrices for symmetric 2 × 2 games

	If the co-player plays C	If the co-player plays D
If I play C, I receive	Reward R	Sucker's payoff S
If I play D, I receive	Temptation T	Punishment P

A payoff matrix like the one above describes a Prisoner's Dilemma if its matrix elements obey the following inequalities:

$$T > R > P > S .$$

The Chicken game (also known as the Snowdrift game or the Hawk–Dove game) is defined by

$$T > R > S > P .$$

The following two payoff matrices provide specific examples:

$$\begin{pmatrix} 3 & 0 \\ 4 & 1 \end{pmatrix} \qquad \begin{pmatrix} 3 & 1 \\ 4 & 0 \end{pmatrix}$$

  Prisoner's Dilemma   Chicken game

---

prefer to play C even if the other plays D. This has been described as the Snowdrift game by Sugden (1986): even if one player refuses to help, the other would be prepared to dig a path through the snowdrift for both, rather than sit tight and freeze. The ranking $P < S$ leads to $T > R > S > P$, and hence to the game called Chicken by classical game theorists, and Hawk–Dove by evolutionary biologists.

In the usual setting of evolutionary game theory, one assumes a well-mixed population of players matched randomly against each other and multiplying at a rate that increases in proportion to their payoff. In the Chicken game, this leads to a mixed population of C- and D-players. In the Prisoner's Dilemma game, the C-players are doomed to extinction. If the probability that players meet for another round (and recognize each other) is sufficiently high, however, then cooperation can be an evolutionarily viable outcome in the resulting *repeated* Prisoner's Dilemma game. In particular, there exist populations using cooperative strategies that cannot be invaded by minorities of players who always defect. This point was made most forcefully in Axelrod's book *The Evolution of Cooperation* (1984).

In Chapter 8 of his book, Axelrod describes an iterated Prisoner's Dilemma game played on a spatial grid. He considers a territory subdivided into square cells, with each cell occupied by one player. Each player plays against his or her four neighbors in a repeated Prisoner's Dilemma game. The total scores are computed. If a player has neighbors who are more successful, he or she switches to the strategy that obtained the highest score (if there is a tie among the most successful neighbors, one is picked at random). Thus, Axelrod views neighbors as role models whose behavior can be imitated. But one can obviously also interpret the updating differently, as the formation of a new generation, with each cell being taken over by an offspring of the previous owner or of one of the neighbors, depending on who did best in the previous generation – a kind of colonization. Axelrod shows that it is at least as easy for a strategy to protect itself from a takeover in such a territorial structure as it is if the co-players are chosen randomly in the population (i.e., without regard to spatial structure).

Most important, Axelrod shows how a single invader playing the strategy of always defecting (*AD*) can spread in a population of *Tit For Tat* (*TFT*) players (who cooperate in the first round and then do whatever their co-player did in the previous round). This yields fascinating snowflake-like patterns of defectors bypassing islands of cooperators (Figure 8.1).

8.2 One-round Games

In Axelrod's investigations, territoriality was primarily seen as a device for ensuring continuity of the interaction: after defection, the defector cannot run away, but must face an eventual retribution.

Interestingly, cooperators can persist even if there is *no* retribution, that is, even if the game is never repeated. Persistence was first demonstrated by Nowak and May (1992; see also Sigmund 1992; Nowak *et al.* 1995a). Like Axelrod, Nowak and May considered a large lattice with each cell occupied by one player. The players engage in *one* round of the Prisoner's Dilemma game against each of their neighbors. [This could be the four neighbors to the north, south, east, and west, or the eight nearest neighbors corresponding to a chess king's move. Somewhat in the tradition of Axelrod (in whose tournaments each strategy played against itself as well as against other strategies), each player on the lattice could also interact with him- or herself. The main results of the spatial model are independent of these details.] The players are either cooperators or defectors, that is, they play either *C* or *D* in each of the one-shot Prisoner's Dilemma games against their neighbors. Afterward, the next generation is formed: each cell is taken over by a copy of the highest-scoring strategy within the neighborhood (consisting of the previous owner of the site and the nearest neighbors). This simultaneous updating leads to a deterministic transition rule and defines a cellular automaton with the interesting property that the transition rule depends, not only on the states of the nearest neighbors, but

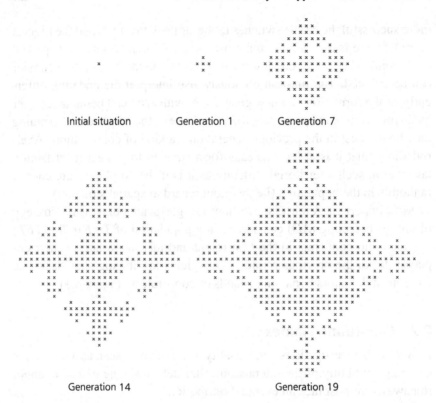

Initial situation   Generation 1   Generation 7

Generation 14   Generation 19

Figure 8.1 Axelrod's snowflakes. Crosses denote defectors, all of whom descend from a single defector who invaded a lattice populated with *Tit For Tat* players. *Source*: Axelrod (1984).

also on those of *their* nearest neighbors (i.e., of a 5×5 square, if the nearest neighbors are the eight cells of a chess king's move).

Let us consider, for instance, a configuration where the left half of the lattice is occupied by cooperators and the right half, by defectors (Figure 8.2). A cooperator sitting at the edge interacts with five cooperators and three defectors, and hence obtains $5R + 3S$ as the payoff, whereas a defector can exploit at most three cooperators and hence obtains the payoff $3T + 5P$. For many parameter values, cooperators earn more than defectors, and hence advance their front by one step to the right per generation. Similarly, if a 2×2 block of cooperators is surrounded by defectors (Figure 8.3), then a cooperator is boosted by the interaction with three other cooperators and obtains $3R + 5S$ points, whereas a defector, as an "outsider," can exploit at most two cooperators and earns $2T + 6P$ points. Again, it is possible for cooperators to spread. Of course, lone defectors will always do well. But by prospering, they surround themselves with defectors and diminish their

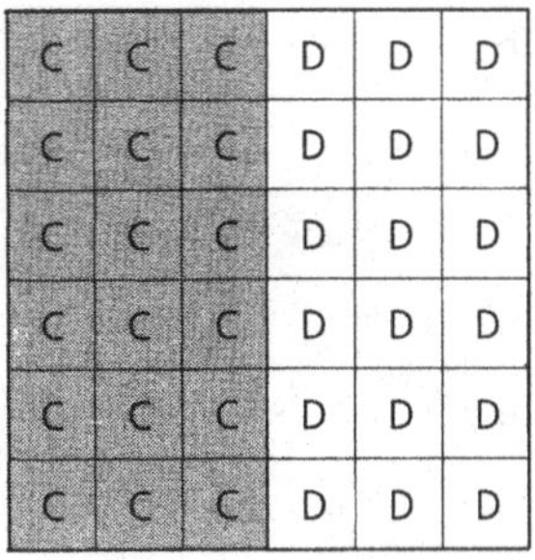

C	C	C	D	D	D
C	C	C	D	D	D
C	C	C	D	D	D
C	C	C	D	D	D
C	C	C	D	D	D
C	C	C	D	D	D

Figure 8.2 Phalanx of cooperation. Cooperators sit on the left half of the plane, defectors on the right half. If $5R + 3S > 3T + 5P$, the front of cooperators advances to the right.

own return. For some parameter values, a square of cooperators can expand at the corners but shrink along the sides: this process yields highly intricate patterns of growth, an endlessly milling spatio-temporal chaos with metastatic tentacles flailing at each other.

Nowak and May's cellular automata can display most of the gadgetry of Conway's Game of Life – periodic blinkers, for instance, or gliders consisting of teams of cooperators moving through a sea of defectors. If one starts with a symmetric initial condition, the emerging configuration retains this symmetry and can lead to wonderful kaleidoscopic patterns reminiscent of Persian carpets and Andalusian tiles.

The basic message of the Nowak–May model is that, for a substantial subset of the parameter space (i.e., the payoff values R, S, T, and P) and for most initial conditions, defectors and cooperators can coexist forever, either in static irregular patterns or in dynamic patterns with chaotic fluctuations along predictable long-term averages. In stark contrast to the case of well-mixed populations without spatial structure, where cooperators are bound to die out, cooperators frequently persist if the population has a territorial structure. In retrospect, this could have been guessed from Axelrod's simulations leading to the growing snowflake of invading defectors and persisting islands of cooperators. Indeed, the probability of a further round in the repeated game was chosen by Axelrod to be as low as one-third, so that the average "repeated" game consists of only 1.5 rounds!

Most of the results of Nowak and May (1992, 1993) consist of numerical explorations of the model's behavior for different payoff values. They consider the case $P = S$, which is the limiting case between the Prisoner's Dilemma game and the Chicken (or Snowdrift) game, and normalize such that this common value is 0 and the value of R is 1 (such an affine linear normalization, which is standard practice in game theory, obviously does not

D	D	D	D	D	D
D	D	D	D	D	D
D	D	C	C	D	D
D	D	C	C	D	D
D	D	D	D	D	D
D	D	D	D	D	D

Figure 8.3 Island of cooperation. A 2 × 2 island of cooperators in a sea of defectors.

affect the structure and dynamics of the game). Thus the only remaining parameter is T, the temptation to defect. Starting with a random initial configuration, some values of T lead to an overall structure of defectors interspersed with tiny regions of cooperators (Figure 8.4). For other values of T, the lines of defectors become connected. For still larger values, chaotic regimes appear: the board is covered with defectors, but many small clusters of cooperators persist. These clusters continue to grow as long as they do not come too close to each other. Surprisingly, the average frequency of cooperators converges to about 32%, independent of the initial condition, as long as T has a value between 1.8 and 2.0 (Figure 8.5). Other neighborhood structures and different lattices – for instance, hexagonal lattices – lead to similar results.

A new twist was introduced by Huberman and Glance (1993), who stressed that if the cells on the lattice were updated, not *synchronously* as is usual in cellular automata, but *asynchronously* – picking up an individual cell at random and replacing the player sitting there with the highest-scoring player within the neighborhood – then the evolution leads to the all-out victory of defectors, rather than to the survival of a certain percentage of cooperators. Huberman and Glance claimed that because synchronous updating requires a global clock, it leads to a rather implausible model. They concluded that spatial games offer little hope for explaining the evolution of cooperation in biological communities. A similar argument was made by Mukherji *et al.* (1995).

It quickly became clear, however, that this effect of asynchrony was only valid for a rather limited part of the parameter space. Substantial but distinct regions of the parameter space lead to the persistence of cooperators with both synchronous and asynchronous updating.

In more detail, Nowak *et al.* (1994a, 1994b) added several elements of *stochasticity* to the spatial model. Asynchronous updating of randomly

(a) (b) (c)

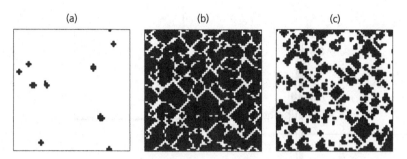

Figure 8.4 The signature of temptation. Different regimes for different values of T: (a) defectors interspersed with small islands of cooperators; (b) lines of defectors on a background of cooperators; (c) spatio-temporal chaos of defectors and cooperators. *Source*: Lindgren and Nordahl (1994).

chosen sites is just one such factor. Another biologically plausible assumption is that the scores obtained by the cells within the neighborhood do not fully determine which player will take over in the next generation (the one with the highest total payoff), but only specify the *probability* that a player will take over. [A somewhat related account of the role of stochasticity upon the updating rule can be found in Mukherji *et al.* (1995).] More generally, if A_i denotes the score of the player at site i, and s_i is 0 if site i is occupied by a defector and 1 if occupied by a cooperator, then the probability that site j will be occupied by a cooperator in the next generation is given by $(\sum A_i^m)^{-1} \sum A_i^m s_i$, where the sum is extended over all neighboring sites i of site j, and where m is a positive real number characterizing the degree of stochasticity of the takeover mechanism. If $m = 1$, the chances of colonizing the site are proportional to the score. The limiting case $m \to \infty$ gives the original deterministic rule: the site goes with certainty to whoever achieved the highest score. For $m = 0$, on the other hand, we have random drift: the score no longer plays a role, all players in the neighborhood have an equal chance of taking over. Figure 8.6 shows the outcome for the simulations with $0 \le m \le \infty$ and $1 < T < 2$ (and, as before, $R = 1$ and $S = P = 0$). It is interesting to compare the effect of synchronous and asynchronous updating in this wide setting of parameter values. As noted by Huberman and Glance (1993), the coexistence of C and D for large values of T holds (for $m \to \infty$) only in the synchronous case. But with probabilistic updating (i.e., $m = 1$), it actually holds for *more* T values in the asynchronous case. It must be stressed, however, that for $m = 1$ the region where cooperation persists is rather small. The advantage of belonging to a patch of cooperators cannot be efficiently "exported" when $m = 1$, a fact that has also been noted by Wilson *et al.* (1992) in a related model.

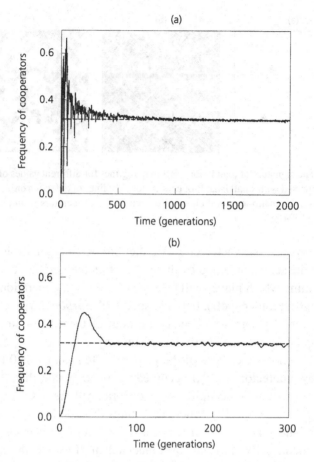

Figure 8.5 Average cooperation. For many initial conditions, the frequency of cooperators converges to 0.3178... for $1.8 < T < 2.0$ (players interact with their eight neighbors and with themselves); (a) starts with a single defector, (b) starts with a random mixture of cooperators and defectors. *Source*: Nowak and May (1993).

One can randomize the game still further by introducing random neighborhood grids, or random dispersal. These changes lead to no essential alterations of the picture.

Furthermore, Nowak *et al.* (1994b) have established that cooperation can easily be maintained if there exists a certain probability that cells in the neighborhood of individuals with low payoff remain empty. In this case, defectors tend to subvert their own livelihood. Cooperators can then invade the barren patches. This holds even for the unfavorable case $m = 1$, and for very large temptation values T.

Several other results are less intuitive. For certain parameter values it may happen that cooperators vanish despite consistently having a higher

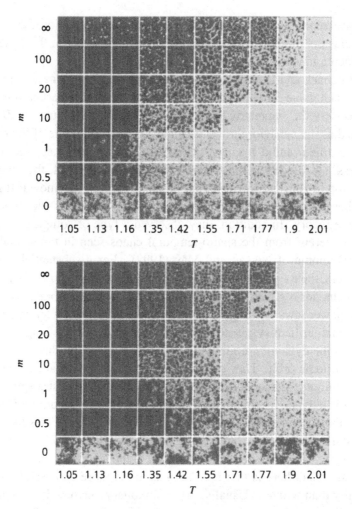

Figure 8.6 Synchronous versus asynchronous updating. Black indicates cooperation; gray, defection. The squares give the outcome of spatial simulations for different values of m and T (see text). The upper figure is based on synchronous updating; the lower figure, on asynchronous updating. *Source*: Nowak *et al.* (1994a).

average payoff than the defectors. The reason is that the few defectors close to cooperators have a higher score and can spread. The low payoff of defectors surrounded by defectors and the high payoff of cooperators surrounded by cooperators do not matter; what counts is the interface where cooperators and defectors are in close proximity.

An even more surprising result of the stochastic game with $m = 1$ is that populations of cooperators can become extinct even though isolated cooperator cells can have more than one offspring on average (i.e., although their basic reproductive ratio is greater than 1). This extinction occurs when larger clusters of cooperators cannot grow significantly. In this case, random fluctuations may wipe them out in one blow.

Killingback and Doebeli (1996) extended this type of analysis to cover the Chicken game (or Hawk–Dove game). As mentioned earlier, this game has precisely the same structure as the Snowdrift game, but now *D* is interpreted as the behavioral rule "escalate a conflict until one of the contestants is harmed" and *C* means "stick to displaying, that is, keep the conflict harmless." Killingback and Doebeli show that, in general, the long-term proportion of Hawks (or "defectors," in the context of the Snowdrift game) is smaller than the equilibrium proportion predicted by classical evolutionary game theory. In addition, they observed a type of complex dynamics that is different from the spatio-temporal chaos seen in the spatial Prisoner's Dilemma of Nowak and May (1992). For a substantial range of parameters, their system organizes itself into a critical state in which its dynamic behavior is governed by long-range spatial and temporal correlations and by power laws (Killingback and Doebeli 1998). An interesting property of such critical systems is the existence of extremely long transients on which the dynamics are initially very complicated but then relax into a simple periodic orbit. They conjectured that suitable spatial extensions of any evolutionary game with a mixed evolutionarily stable strategy will exhibit critical dynamics for appropriate parameters. Killingback and Doebeli (1996) also extended the game to include conditional strategies like *Retaliator* (start by displaying, but escalate the conflict if the opponent escalates) and *Bully* (escalate the conflict, but retreat if the adversary also escalates). The basic result here is that *Retaliator* is much more successful with territoriality than without. Usually, the evolutionary outcome is a population of *Retaliators*, occasionally interspersed with a few *Doves*. This is quite different from the evolutionary outcome without territoriality.

Another one-round game with more than two strategies is the Rock–Scissors–Paper game (a cyclic arrangement of three strategies, each dominating its successor). A spatial version of this game was briefly studied by Nowak *et al.* (1994b), who showed that it may lead to spiral waves with the three strategies endlessly pursuing each other.

Feldman and Nagel (1993) investigated a variant of the spatial Prisoner's Dilemma where the players are not updated in every round but can accumulate payoff; against this, they must pay a certain fee (the same for all players) to stay in the game. If their savings are eaten up, they are replaced by a wealthier neighbor. The authors observed widespread local coordination on the lattice.

8.3 Repeated Games

The main message so far is that neighborhood structure seems to offer a promising way out of the Prisoner's Dilemma toward the emergence of co-operation. There are many alternative explanations of the prevalence of cooperation, but, arguably, none require less sophistication on the part of the individual agents than those with spatial structure. The latter need no foresight, no memory, and no family structure. Viscosity suffices.

As soon as one considers more highly developed agents who are long-lived enough to interact repeatedly, have enough memory to keep track of their past, and are smart enough to adapt to their co-players' moves, co-operation becomes much easier to sustain, even in the well-mixed case. A retaliatory strategy like *TFT* or *Grim* (the strategy of cooperating until the adversary defects for the first time, and from then on relentlessly defecting) can invade a population of *AD* players provided its initial frequency exceeds a certain low threshold. Further evolution can lead to populations where cooperation is more robustly established and is proof against occasional mistakes. One frequent outcome is *Pavlov*, the strategy that starts with a cooperative move and then cooperates if and only if in the previous round both players used the same move [both playing C or both playing D, see Nowak and Sigmund (1993)]. *Pavlov* is a strategy based on the win-stay, lose-shift principle: a player repeats his or her previous move if the payoff was large (T or R) and tries the other move if the payoff was small (S or P).

It turns out that in models incorporating repeated interactions *and* territoriality, the probability for a cooperative outcome is high indeed. This property was firmly established by the extensive computer simulations of Lindgren and Nordahl (1994). What was less predictable was which strategies would dominate the cooperative regime.

Lindgren and Nordahl (1994) normalized the payoff values to $R = 1$ and $S = 0$, with $1 < T < 2$ and $0 < P < 1$ for the Prisoner's Dilemma. They investigated the infinitely repeated Prisoner's Dilemma game, where the probability for a further round is equal to 1. Like Nowak and Sigmund (1993), they also assumed that players occasionally make mistakes; in their simulations, one move in a hundred was mis-implemented. This means that the initial move is no longer relevant: it only affects the beginning of the interaction, which plays no role in the long run because, in the limit, the overall payoff is now the mean of the payoff per round.

Lindgren and Nordahl (1994) started by investigating strategies with memory spans of 0. This corresponds to unconditional C or D strategies for the one-shot Prisoner's Dilemma. Using a four-cell neighborhood, they

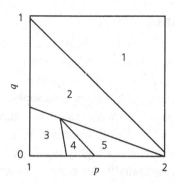

Figure 8.7 Lest we forget. Asymptotic behavior of memory-0 strategies: (1) leads to a homogeneous state of defectors, (2) to small domains of cooperators (3) to percolation networks of defectors in a background of cooperators, (4) to spatio-temporal chaos, and (5) to rigid lines of cooperators; here, $p = \frac{T-S}{R-S}$ and $q = \frac{P-S}{R-S}$. *Source*: Lindgren and Nordahl (1994).

showed that the evolutionary outcome could be as encapsulated in Figure 8.7. For $T + P > 2$, defectors take over. The remaining parameter values lead to a coexistence of defectors and cooperators either in a stable form (islands of cooperators or percolation networks of defectors) or in the form of spatio-temporal chaos.

Next, Lindgren and Nordahl considered strategies that take into account the *co-player's* previous move – *TFT* is one such strategy, *Anti-TFTat* (which does the opposite of what the opponent did in the previous round and, not surprisingly, never establishes a firm foothold) is another. The remaining two strategies are the unconditional *AD* and *Always Cooperate* (*AC*). It turns out that only a tiny corner of the parameter space – far less than one-thousandth of it – leads to the emergence of a homogeneous *TFT* population (indeed, paired *TFT* players suffer severely from errors, as they engage in long runs of alternating defections). A much larger part of the parameter space leads to the dominance of the *AC* strategy. Other parameter regions lead to spatio-temporal chaos or spiral waves, still others to frozen configurations of *AD* and *TFT*.

By enhancing the sophistication of the players so that they can base their next moves on both their co-player's previous move and *their own* previous move (i.e., on the full outcome of the previous round), one obtains 16 strategies that can interact in rather complex ways (see also Nowak *et al.* 1995b). More than half the parameter space leads to a cooperative regime dominated by the *Pavlov* strategy (Figure 8.8). The remaining parameter values lead either to an outcome dominated by a parasitic strategy like *AD* or *Bully* (the strategy of always defecting except after having been punished in the previous round), or to complex behavior with high diversity.

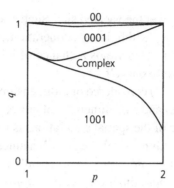

Figure 8.8 *Pavlov*'s domain. Asymptotic behavior of strategies based on the outcome of the last round. In the region marked 1001, *Pavlov* prevails. In the regions marked 00 and 0001, a cooperative regime is established; p and q are as in Figure 8.7. *Source*: Lindgren and Nordahl (1994).

One can, of course, consider strategies with even longer memories, based, for instance, on the outcome of the previous *two* rounds. In fact, Lindgren (1991) has developed a genetic algorithm where mutations providing for ever-larger memories can be introduced and eventually selected. If there is no upper bound to the memory length, the evolutionary simulation can lead to extremely diverse patterns consisting of many strategies coexisting in highly entangled spatio-temporal arrangements. Many of the resulting strategies share with *Pavlov* the property that any error leads to a short period of mutual punishment after which, as of a common accord, the players simultaneously return to cooperation. Lindgren and Nordahl (1994) stress that asynchronous updating has little effect on the simulations (although it excludes spiral waves). They argue convincingly that one cannot speak of *the* Prisoner's Dilemma game: for different parameter values one can obtain very different strategic interactions.

At present, there seems little hope for obtaining a full analytic understanding of these phenomena. As with cellular automata in general, most of the results are based on computer simulations and are at best a very crude intuition of some local effects. An interesting attempt at an analytic treatment was made by Nakamaru *et al.* (1997), who considered the interplay of *TFT* with *AD*, as in Axelrod's book (Axelrod 1984). In contrast to Axelrod and the other authors mentioned so far, they assume that the fitness gained by each player (i.e., the total payoff obtained against all neighbors) translates, not into a *transmission rate* (the probability of invading a neighboring cell), but into a *mortality rate*. This assumption yields a variant of asynchronous updating where the probability that a cell is updated depends on the success of the player sitting in that cell, whereas each neighbor has the same chance of taking over the vacated site. Such a setup offers some scope for spiteful behavior. By withholding cooperation, one decreases the survival rate of a neighbor. This yields a chance of inheriting the neighbor's site – although at a cost to oneself.

Nakamaru *et al.* (1997) considered one-dimensional lattices (where each cell has two neighbors) and two-dimensional lattices (with eight neighbors per cell). The outcome of the spatial simulations is very different from both that of the complete mixing model (i.e., the dynamics without spatial structure) and that of a first, crude approximation called the *mean-field approximation*. But another approximation, the *pair approximation* (Chapters 13 and 18), based on neglecting correlations except among nearest neighbors, frequently leads to analytical results in good agreement with the numerical simulations. In particular, both computer simulation and pair approximation show that *TFT* can invade a population of defectors if the probability of a further round is sufficiently high. Interestingly, the model of Nakamura *et al.* (1997) never leads to the coexistence of cooperators and defectors, which is a frequent outcome in the Nowak–May (1992) model.

Another analytic approach has been investigated by Eshel *et al.* (unpublished). These authors consider players on a one-dimensional lattice. Each player interacts with the k nearest neighbors to the left and to the right. There are two available strategies. After each generation, a player either sticks to his or her own strategy or switches to that of one of the n players to the left or to the right, with probabilities proportional to the neighbors' scores. However, a player can switch strategies only if at least one of the two immediate neighbors uses the alternative strategy. This model is based on an interaction neighborhood and a (possibly different) propagation neighborhood. Eshel *et al.* (unpublished) give conditions on the stability and invasibility of strategies that, for large values of k and n, only depend on the value of k/n. This helps to interpret the concept of population viscosity in terms of inclusive fitness (see Hamilton 1964). In particular, cooperation wins in the Prisoner's Dilemma game if each player observes a large number of players before imitating one.

A similar distinction between one-dimensional propagation neighborhoods and interaction neighborhoods (arranged on the circle) is made by Hoffman and Waring (1996). They studied all deterministic memory-1 strategies for the repeated Prisoner's Dilemma using numerical simulations and found that localization of learning always fosters cooperative behavior, but that the effect of localization of the interaction is not clear-cut and depends on details of how the competing strategies are clustered.

Other papers dealing with the spatial version of the repeated Prisoner's Dilemma include Mar and St. Denis (1994, and unpublished), Grim (1995), and Kis (unpublished). Mar and St. Denis (unpublished) actually studied a more general variant, where each player can opt for a larger or smaller degree of cooperation. Kirchkamp (unpublished) presents particularly extensive simulations exploring a wide range of learning rules. In particular, he also allows for stochastic timing, not just of the updating, but of the interaction as well.

8.4 Extensions and Related Work

The spatial games of Nowak and May (1992, 1993) emerged at the same time as some models in theoretical ecology and prebiotic evolution, all of which stress the same message: territoriality favors diversity. This point is discussed further in Chapters 9 and 10 in this volume. Durrett and Levin (1994b) provide a general survey of spatial aspects.

Herz (1994) analytically classifies the behavior of 2×2 games in terms of their payoff structure. The spatial versions of some of these games lead to an uphill march in an abstract fitness landscape, whereas other games become bogged down due to a phenomenon called "frustration" in the physics of Ising-type models. The updating rule used by Herz is not evolutionary in the proper sense (neither the transmission rate nor the mortality of a player is affected by the score). However, it is based on a *Pavlov*-like principle that can be interpreted as a learning mechanism: if the player obtains a total score against his or her neighbors that is above a certain threshold, the player uses the same strategy in the following round; if the score is below that threshold, he or she switches to the alternative strategy. The simplicity of this updating rule provides the basis for a theoretical analysis of the resulting cellular automaton.

Another, very different approach to the spatial modeling of the Prisoner's Dilemma game can be found in the one-dimensional reaction–diffusion models of Hutson and Vickers (1995) and Ferrière and Michod (1995), which lead to traveling waves of *TFT* players swamping the inveterate defectors.

The spatial, iterated Prisoner's Dilemma has mainly been used to study the effects of neighborhood on the evolution of cooperation between members of the same species. Recently, Doebeli and Knowlton (1998) have suggested an extension in order to study mutualistic interactions between members of different species. In their model, two different species occupy two different, superimposed lattices. Interspecific interactions occur between individuals at corresponding sites in these lattices, and the payoffs from these interactions determine the outcome of competition within each lattice. Doebeli and Knowlton show that in this setting, spatial structure is even more essential for the evolution of mutualism than in one-lattice games. (In fact, they considered versions of the iterated Prisoner's Dilemma in which investment could vary continuously and always evolved to 0 in the non-spatial games.)

Finally, we mention that lattice models of artificial societies have a long tradition, as can be seen from the work of Schelling (1971) and Sakoda (1971), who used them to simulate migration and segregation behavior based on neighborhood rules. Some recent, related investigations have been done by Hegselmann (1996), for instance, who analyzed a *solidarity game*, which is an extension of the Prisoner's Dilemma, and Epstein and Axtell (1996), who studied highly elaborate artificial societies.

8.5 Concluding Comments

A wealth of computer simulations of artificial populations now exist showing that cooperation can be stably sustained in societies of simple automata. The introduction of spatial structure has shown that cooperation becomes even more likely if interactions are restricted to neighbors. The venerable rule of cooperating with neighbors is certainly not new. But placed alongside other spatial models, it offers wide perspectives for the emergence and stability of cooperative societies. In particular, it shows that even if the interaction between two individuals is not repeated, cooperation can be sustained in the long run.

This positive message of cyber-sociology should not obscure another, darker aspect. In territorial structures, it is easily possible to distinguish between us (me and my neighbors) and them (the foreigners). In fact, most of the challenges experienced by territorial bands of hominids might have been caused by other bands. Cooperation within groups was crucial for dealing with conflicts between groups. It is quite likely that the fiercest of these conflicts were always fights for territory.

9

The Interplay between Reaction and Diffusion

Mikael B. Cronhjort

9.1 Introduction

A large variety of spatio-temporal patterns can be observed in so-called reaction–diffusion systems: these systems model several different species that can interact with ("react to") each other and diffuse in space. Such patterns, which depend on assumptions regarding the reaction and diffusion mechanisms, may be stationary or may change with time. The emerging patterns are often of great importance in both determining and understanding the properties of spatio-temporal dynamics.

Reaction–diffusion systems are applied to many different problems in biology, chemistry, and physics. Depending on the situation, different simulation techniques can be used. In general, there are two different modeling frameworks for reaction–diffusion systems: partial differential equations (PDEs) and cellular automata (CA). In PDE models, populations of each species are described by concentrations that vary in space and time. In CA models, populations are represented by units or individuals that move on a grid and interact with neighboring units. While in some cases PDE models may best capture the dynamics of real reaction–diffusion systems, in other cases CA models are preferable.

- In ecological predator–prey systems, fluctuations in space and in time play a vital role in the dynamics of the two populations. Consequently, simulation models for these systems should allow for fluctuations and describe the position and the life span of individuals: a CA model would be a good choice for a large individual-based simulation.
- An extensively studied chemical reaction–diffusion system is the Belousov–Zhabotinsky reaction, which can give rise to intricate patterns (Winfree 1974; Ross *et al.* 1988). These patterns are often quite regular, and PDEs describe their dynamics well.

■ In physics, reaction–diffusion systems are studied in the context of the temperature of a hardening epoxy material, for example, or in the context of the Ginzburg–Landau equation (Frisch and Rica 1992), which is related to superconductivity. Again, PDEs often are a suitable tool for describing such dynamics.

CA and PDE models therefore represent two competing modeling paradigms, both of which capture, in a simplified way and to a certain extent, important aspects of real reaction–diffusion systems. In some cases it is difficult to decide which alternative is to be preferred. Both modeling frameworks contain simplifications that result in loss of details that may be important for the dynamics of real systems. While often it would be ideal to use a highly detailed, stochastic, individual-based model, such simulations of large (often three-dimensional) spatial systems require enormous computing power. When emerging patterns are large compared with their constituent units (particles, molecules, individuals), simulating large systems is inevitable. Owing to limited computer resources, the model must then be highly simplified. Yet, CA and PDE models often can be derived as simplifications or limits of an underlying detailed, stochastic, individual-based model. In PDE models, which can be regarded as large-scale descriptions, populations of individuals are described by their concentrations and small-scale correlations are neglected. CA models, on the other hand, operate at a molecular level and focus on local correlations; however, they do not handle long-range correlations and large-scale patterns of concentrations as well as PDEs. For many real systems, both CA and PDE models can capture dynamical essentials. In some cases, however, simulations based on CA and PDE models may give widely different results, even though each model is designed to describe a particular real reaction–diffusion system as well as possible (within the limitations of each framework). Therefore, from a practical point of view it is important to learn when to use which model: to do so, one ought to be familiar with the strengths and weaknesses of each modeling framework. In many situations, however, it is still difficult to tell which model is "best," and there may even be cases where neither model provides a satisfactory description of a particular real system.

In this chapter we investigate merits and shortcomings of CA and PDE models by studying systems recently considered in speculations on the origin of life (Boerlijst and Hogeweg 1991a, 1991b; Hogeweg 1994; Cronhjort and Blomberg 1995, 1997a, 1997b; Chacón and Nuño 1995). These studies concentrated on systems in which spatial organization provides a compartmentalization of the system. Once spatial pattern formation has divided a system of molecules into compartments, separated from each other by "firewalls," infection of one compartment by a harmful molecule remains confined. In extant organisms, compartmentalization is provided by cell membranes and a complex machinery that allows the transport of only certain molecules across membranes. At a primitive stage, before such machinery was developed, a similar function might have been provided by pattern formation. This hypothesis can be tested by simulations and experiments, with the simulations identifying promising candidate systems for experimental verification.

9.2 The Models: Cellular Automata versus Partial Differential Equations

Before we discuss modeling details, we must look at those reaction–diffusion systems that our models are supposed to describe. For the sake of concreteness, we do this in the context of hypothetical explanations for the origin of life.

Reaction–diffusion systems for the origin of life

The first living organism was probably preceded by a set of macromolecules in a chemical "soup." We assume that these macromolecules were already able to replicate by forming copies of existing macromolecules. Replication of each macromolecule is supposed to be governed by catalytic support from other macromolecules. Thus the rate of replication of a certain macromolecule depends on the presence or absence of catalytic molecules in its immediate neighborhood. This scenario is analogous to the "RNA world" proposed by Gilbert (1986), where the same type of molecules occur as self-replicating units and as catalytic supporters for replication. Such macromolecules are the key ingredients in the models for the origin of life considered here. The real prebiotic soup also comprised smaller molecules such as metabolites, monomers, and waste products. Some of the models discussed in this chapter include such smaller molecules. For an analogy between chemical systems, such as that discussed here, and ecological systems, see Box 9.1.

The macromolecular species and their catalytic couplings constitute a catalytic network. One of the most frequently discussed catalytic networks is the hypercycle, as presented by Eigen and Schuster (1979). The hypercycle contains a number of RNA-like polymer species that catalyze their replication cyclically; that is, the first species catalyzes the replication of the second, which catalyzes the replication of the third, and so on. The hypercycle is closed by the last species, which catalyzes the replication of the first one. Another special case, and the simplest possible "network," is

Box 9.1 Analogy between chemical and ecological systems

Although this chapter draws its illustrations from chemical systems for the origin of life, a similar formalism can be applied to ecological systems. Different macromolecules in a chemical system correspond to individuals of different species in an ecological system. Chemical/ecological analogues also exist for the other concepts:

- Metabolites/food
- Catalytic support/mutualistic interaction
- Reactions/encounters
- Replication/reproduction
- Decay/death
- Compartmentalization/habitat fragmentation

Chemical and ecological spatial systems can be either two- or three-dimensional. Chemical reactions often take place in a gas or a solution (three dimensions), but many chemical molecules stick to surfaces (two dimensions). For most terrestrial biological systems, the surface of the Earth (two dimensions) is the natural arena, but many bacteria and aquatic organisms move in three dimensions.

a single macromolecular species that catalyzes its own replication; we refer to this as an auto-catalytic species.

Evolutionary stability, however, is problematic for systems with catalyzed self-replication. This is most easily understood in the mean-field approximation. In this approximation, the spatial system is assumed to be fully homogeneous, that is, the concentrations of species do not vary in space but only in time. Hypercycles and other catalytic networks are vulnerable to so-called parasites. A parasite is a species that receives catalytic support for its replication from some species in the network but itself engages in no catalytic activity. Such parasites are likely to arise by mutations of the macromolecules included in the network. The parasite and the original species compete for the same resources, for example, in the form of activated monomers, which are the building blocks of all polymer species. If a parasite has a higher replication rate than the original species from which it has mutated, or if it decays more slowly, it will outcompete and replace the original species, at least locally. In homogeneous states, as assumed by the mean-field approximation, the parasite may easily spread over the entire system and globally replace its ancestor. Such an occurrence may eventually lead to the destruction of the entire catalytic network. With a compartmentalization of the systems, however, the harmful effects caused by the parasite can be limited to only one compartment.

Box 9.2 Cellular automaton models

In CA models, species appear as states of points on a grid (Wolfram 1986). The state of each grid point may represent either an empty space or an individual of any of the considered species. At most one individual can occupy a grid point at any time. The change in the state of any grid point is governed by the updating rules of the CA and depends on the state of the considered grid point as well as on the states of its neighbors. CA models can be updated synchronously or asynchronously. In synchronous models, at each time step all points are simultaneously updated according to the CA rules. In asynchronous models, at each basic step a single point of the grid is selected at random and changed according to the CA rules. The rules, which are often probabilistic, correspond to events like reproduction, diffusion, and death. Diffusion, in fact, can be realized in many ways. Most straightforward is to move a selected individual in a random direction, either (1) by moving it to an empty point or (2) by swapping states with another occupied grid point. Diffusion can also be simulated (3) by random rotations of 2 × 2 regions of points: after rotating all such regions on the grid, the 2 × 2 partition is shifted one grid point diagonally, and further rotation-shift steps may follow. With the last two mechanisms, movement is not restricted to empty points.

Cellular automata

In this chapter we consider several different CA models tailored to describe different phenomena. In all of them, units on the grid represent individuals of different macromolecular species. For a brief general introduction to CA models, see Box 9.2. Rules in the CA model represent growth, decay, and diffusion of units. For example, during each time step a state corresponding to a molecule can be reproduced in a neighboring empty grid point, it can decay and be replaced by the empty state, and it can move by diffusion. Because neighboring molecules catalyze replication, we assume that the probability for growth at one grid point depends on the states of neighboring grid points. In two-dimensional simulations, the considered neighborhood consists of the eight grid points surrounding a certain grid point (Moore neighborhood), but sometimes results have also been compared with those based on four nearest neighbors (von Neumann neighborhood) to check for robustness of conclusions. In three-dimensional simulations, neighborhoods consist of only the six nearest neighbors.

Most simulations described below use periodic boundary conditions. For modeling spirals and scroll rings, reflecting boundary conditions have sometimes been used. Some simulations are performed with synchronous

Box 9.3 Partial differential equation models

In PDE models, populations of individuals of each species are represented by concentrations – that is, by continuously varying variables. In these models, space and time are also continuous. The dynamics of concentrations u_i in species i are given by PDEs

$$\frac{\partial u_i}{\partial t} = f_i(u_1, u_2, \ldots, u_N) + D_i \Delta u_i \, ,$$

where $f_i(u_1, \ldots, u_N)$ are functions that describe the reaction part and $D_i \Delta u_i$ describe the diffusion part of the reaction–diffusion system; N is the number of different species; and D_i are diffusion coefficients. The diffusion operator Δ is a second derivative with respect to spatial coordinates; for two dimensions x and y we have $\Delta = \frac{\partial^2}{\partial x^2} + \frac{\partial^2}{\partial y^2}$. Such equations are often solved numerically, after discretization of space and time. There are so-called explicit and implicit methods for the numerical calculation (see Crank 1975). Each method has its drawbacks: in explicit methods the time step must be small to avoid numerical instabilities, whereas implicit methods in each time step rely on solving a system of coupled equations for $u_i, \ldots, u_N$. We use an explicit method, where the PDEs are approximated by a system of first-order finite difference equations. The basic approximations are

$$\frac{\partial u_i}{\partial t}(x, y, t) \approx \frac{1}{\Delta t} \left[u_i(x, y, t + \Delta t) - u_i(x, y, t) \right] \, ,$$

$$\Delta u_i(x, y, t) \approx \frac{1}{(\Delta x)^2} [u_i(x + \Delta x, y, t) + u_i(x - \Delta x, y, t)$$
$$- 2u_i(x, y, t) + u_i(x, y + \Delta y, t) + u_i(x, y - \Delta y, t)$$
$$- 2u_i(x, y, t)] \, ,$$

where x and y are coordinates of two-dimensional space, t is time, Δt is the discretization step for time, and Δx and Δy are those for space.

updating, others are based on asynchronous updating (see Box 9.2). Synchronous models were used mainly for studying the dynamics of spirals and scroll rings; three-dimensional simulations, in particular, demand a lot of computing power, and synchronous models are easily run on massively parallel computers. For example, the three-dimensional simulations described below were performed on a Connection Machine CM200. In synchronously updated models, diffusion mechanism (3) of Box 9.2 is used. To examine the significance of the diffusion mechanism, especially for simulations of chasing, we have used asynchronous updating, since this offers more possibilities for varying diffusion mechanisms. With asynchronous updating, we can restrict movement to only empty grid points (so-called excluded-volume diffusion) giving a kind of "cage effect," where high concentrations hinder diffusion. It is also possible to give different species different rates of diffusion. Of course, asynchronous models can also be without a "cage effect" and with all species having the same rate of diffusion. We can therefore see that implementing diffusion asynchronously can, in some cases, reveal qualitative features that cannot be obtained in synchronous models. For the systems treated in this chapter, however, there do not seem to be any essential, qualitative differences between the two ways of updating or the three mechanisms of diffusion.

The basic CA models presented here do not consider monomers. Instead, all CA models have an implicit growth limitation, simply because the number of available sites is limited. An additional, explicit growth limitation can be introduced by a factor that decreases the probability for replication when the total number of macromolecules increases. Alternative explicit growth limitations exist: one is to introduce resource dependences, such as in the form of activated monomers needed for assembling macromolecules, as suggested by Blomberg *et al.* (1981). There is also an implicit cutoff level in CA models, a lower limit for how small concentrations can be. This level corresponds to a configuration comprising a single unit of a certain species in a given region.

Partial differential equations

For a brief general introduction to PDE models see Box 9.3. For the reaction–diffusion systems considered in the origin-of-life example, catalytic networks are described by the equations

$$
\frac{\partial X_i}{\partial t} = \overbrace{M \sum_{j=1}^{N} k_{ij} X_j X_i}^{(i)} - \overbrace{g_X X_i}^{(ii)} + \overbrace{D_X \Delta X_i}^{(iii)} \, , \quad i = 1, \ldots, N \, , \tag{9.1}
$$

$$
\frac{\partial M}{\partial t} = \underbrace{k_M}_{(iv)} - \underbrace{g_M M}_{(v)} \underbrace{- LM \sum_{i,j=1}^{N} k_{ij} X_j X_i}_{(vi)} + \underbrace{D_M \Delta M}_{(vii)} \, , \tag{9.2}
$$

where X_i denotes the concentration of polymers of type i, M is the concentration of monomers, and N is the number of different polymer species. Terms on the right-hand sides of the equations correspond to different dynamic processes in the system.

The first term on the right-hand side of Equation (9.1), labeled (i), corresponds to catalyzed replication of polymer X_i: the growth rate is proportional to the concentrations of activated monomers M, of templates X_i, and of each polymer X_j, which catalyze the replication of X_i by rate constants k_{ij}. Different choices for the matrix K correspond to different catalytic networks. For example, for a hypercycle network with six members we could have

$$
k_{ij} = \begin{pmatrix}
0 & 0 & 0 & 0 & 0 & 1 \\
1 & 0 & 0 & 0 & 0 & 0 \\
0 & 1 & 0 & 0 & 0 & 0 \\
0 & 0 & 1 & 0 & 0 & 0 \\
0 & 0 & 0 & 1 & 0 & 0 \\
0 & 0 & 0 & 0 & 1 & 0
\end{pmatrix} . \tag{9.3}
$$

For an auto-catalytic species there is only a single rate constant, for example, $k_{11} = 1$. Noncatalytic growth that would result in a linear term in Equation (9.1) is neglected. Term (ii) corresponds to decay of polymers; g_X is a decay rate constant. Term (iii) corresponds to diffusion; D_X is the diffusion coefficient.

Equation (9.2) describes the activated monomers that limit polymer growth. In contrast to CA models, PDE models need some type of explicit growth limitation, which prevents the system from growing without bounds. The monomers, with concentration M, are assumed to be produced at a constant rate k_M, as described by term (iv). They decay to an inactive form with a decay rate constant g_M, as described by term (v). Term (vi), which contains a double sum, corresponds to consumption of monomers due to replication of polymers. The number of monomers needed to produce a polymer is denoted by L. Term (vii) corresponds to diffusion of monomers. As monomers are much smaller than polymers, the diffusion coefficient D_M for the monomers is different from (usually greater than) the diffusion coefficient D_X for the polymers. For more details on how the second derivative Δ arises in the diffusion terms (iii) and (vii), see Chapter 22.

PDE models are strictly deterministic, which often makes them convenient to use. For instance, determinism implies that a single simulation is sufficient to evaluate the consequences of each initial condition. This is not the case for CA models: due to the fluctuations arising from probabilistic updating rules, even when starting from the same initial condition different simulations usually give different results. For CA models we must thus collect several simulations to assess the genericity of results. The determinism of PDE models corresponds to the determinism we experience in everyday (macroscopic) life, for example, the reliable experience that apples fall down to the ground. Here, averaging over an exceedingly large number of particles ensures determinism. By comparison, CA models are random and often more akin to the microscopic world, where fluctuations are important: a smoke particle may fly in any direction (assuming that the air is still), and where it will finally settle is essentially unpredictable. Of course, many real processes lie somewhere between the falling apple and the flying smoke particle. These "in-between" processes are the trickiest to simulate, and it is difficult to tell which modeling framework best captures them.

The "pure" PDE models described above allow for arbitrarily low concentrations. As these concentrations cannot occur in finite, individual-based systems, it is often desirable to add a cutoff rule to the PDE, which can be done in many different ways.

The most straightforward formulation of a cutoff rule is after each time step to set concentrations equal to zero at all grid points where concentrations are below a given cutoff value. The cutoff value often is chosen to correspond to the presence of a single individual on a grid point. This simple formulation, however, violates mass conservation and may produce strange and spurious effects, especially if the chosen cutoff values are too high, as described, for example, by Cronhjort and Blomberg (1995).

A better alternative, although further away from the PDE concept, is to introduce a probabilistic cutoff rule: concentration values below the cutoff value are either decreased to zero or increased to the cutoff value. The probability for setting a concentration to the cutoff value equals the ratio of current concentration to cutoff value. This second cutoff rule on average conserves mass – that is, on average individuals do not appear or disappear due to the cutoff.

9.3 Spiral and Scroll Ring Patterns

In this section and the next, I present simulation results for the models described in Section 9.2. The main issue here is to describe situations where the two different types of model give qualitatively different results, despite the fact that each model has been constructed to simulate the same system. To illustrate discrepancies, we investigate two particular catalytic networks: in this section we consider the hypercycle, and in the next section, a single auto-catalytic polymer species in a soup of monomers. In general, patterns obtained from different models look similar at first glance. When parasites are introduced, however, systems may respond in completely different ways, revealing important discrepancies between the models. As already mentioned, the parasites considered would be fatal under mean-field conditions: all parasites grow faster or decay more slowly than the original species with which they compete. In the mean-field approximation, therefore, these parasites would outcompete the original species.

CA models in two dimensions

Boerlijst and Hogeweg (1991a, 1991b) have shown that, in a two-dimensional CA model, hypercycles with five or more members can spontaneously give rise to rotating spiral patterns. Each such spiral consists of several "arms," one for each species of the hypercycle; arms follow each other in space in the same sequence as species catalytically support each other in the hypercycle network. In each arm, one species dominates, while others may be present at lower concentrations. Since replication of each species in the hypercycle is catalyzed by the species in the preceding arm of the spiral, each arm grows at one of its two edges and decays at the other. This process of growth and decay results in the rotating motion of each spiral.

Parasites introduced into such a spatial spiral system begin to grow where they find themselves in a region that provides catalytic support for their replication; elsewhere, they may be unable to grow. If the hypercycle network contains six or more members, parasites cannot grow across the spiral arms toward the center of the spiral (M. Cronhjort, personal observations). In such spirals, parasites are instead transported outward by the rotating motion of the spiral. When parasites reach the periphery of a spiral, where the infected spiral arms meet with spiral arms from neighboring spirals, the parasites decay. Again, this is because the parasites are unable to grow against the outward motion of the newly encountered spiral arms, hence they cannot advance into other spirals either. To "kill" a spiral (i.e., to infect its entire area), parasites must occur close to the spiral center, where the spiral arms are generated. In the CA model, spirals with six or more spiral arms are therefore resistant to parasites occurring outside spiral centers. The number of species needed to make a spiral resistant to parasites may depend on the details of the CA model and on parameter values. Boerlijst and Hogeweg (1991a) report local extinction and resistance to parasites for seven or more species, but I have also found it for six species (M. Cronhjort, personal observations). The mechanism for resistance to parasites depends on local extinction: because the species that catalyzes the replication of the parasite is not present in every spiral arm, all arms without the catalyst

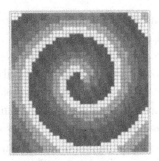

Figure 9.1 A spiral with five members in a PDE model. The shade of gray at each point indicates which species is present at highest concentration.

form barriers to the parasite's advancement. In a system consisting of many spirals, this corresponds to compartmentalization.

Contrary to the formation of spirals, however, the detailed properties of local extinction are model dependent. Specifically, in the CA models we have used hypercycles with six or more members exhibit local extinction, yet in other CA models seven members may be needed.

PDE models in two dimensions

In PDE models with five or more hypercycle members, spiral patterns readily emerge from most initial states [see Eigen and Schuster (1979), Section VIII.7c, for an explanation of why five or more members are critical]. These spirals are similar to those obtained in CA models. Some particular initial states, however, may give rise to homogeneously oscillating states. This is in contrast to CA models, where local fluctuations always break the symmetry of homogeneous states, resulting in spontaneous spiral formation. A spiral obtained from a PDE model is illustrated in Figure 9.1. Initial states based on randomized initial concentrations generally give rise to patterns comprising several spirals. In pure PDE models without a cutoff rule, when a parasite is introduced into a system of spirals, the spirals cannot resist the parasite: parasites are able to spread inward in spirals and can also invade neighboring spirals. This result is expected, since at all grid points of the PDE model all species are present at some concentration. The higher the number of species, the lower their local concentrations become during the dips induced at each point by the spiral rotation; nevertheless, concentrations always remain larger than zero. Pure PDE models therefore display no local extinction. This means that parasites can advance from one grid point to the next and take over each of them as if they were parts of homogeneously oscillating systems, for which the mean-field results apply.

To achieve local extinction in a PDE model, one must add an explicit cutoff rule to the PDE. Combined with a suitable cutoff rule, PDE models can then reproduce the resistance to parasites that is observed in CA models. In general, any of the cutoff techniques described above can be used, but for high cutoff values one might prefer a mass-conserving technique, because otherwise the cutoff may affect the concentrations and the rotation of spirals too much. The cutoff mechanism leads to the complete absence of the particular species catalyzing a parasite from certain regions of the spiral; in these regions the parasite decays. This process prevents the parasite from advancing inward, and the spirals consequently become resistant to parasites. Unfortunately, this means that in PDE models the cutoff value is an important free parameter. Its value determines how many hypercycle members are needed to achieve local extinction. If there are many members in the hypercycle, then the spirals have many arms and a low cutoff is sufficient for local extinction to occur. With fewer members (in excess of five), local extinction results only for higher cutoff values. If the number of species is five or less, the cutoff rule cannot cause local extinction in spirals. This can be compared with the results for CA models: one may say that CA models possess an implicit cutoff rule, which can cause local extinction only if the number of hypercycle members is six or more.

Scroll rings in three dimensions

The implicit cutoff of CA models affects results of simulations in three dimensions even more than in two dimensions. Here, the basic patterns are not spirals but scroll rings, which are, however, related to spirals. Scroll rings consist of scroll waves (which possess spiral cross sections) rotating around a so-called singular filament; these filaments are in general circular, as illustrated in Figure 9.2. In contrast to spirals in two dimensions, however, scroll rings are unstable: the singular filaments of scroll rings contract and finally disappear (Cronhjort and Nyberg 1996). In a PDE model without a cutoff rule, when a scroll ring collapses it gives rise to a transient state of concentric spherical waves propagating outward; this state soon decays into a final state of homogeneous oscillation.

The collapse of a contracting scroll ring again highlights the discrepancies between CA and PDE models. Models that display local extinction (i.e., CA models with at least six hypercycle members and PDE models with a cutoff rule) cannot oscillate homogeneously, since local extinction implies global extinction for a homogeneous state. For such models, the collapse of a scroll ring is followed by sudden complete decay of all

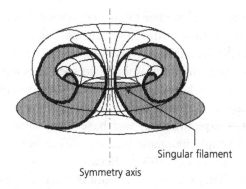

Figure 9.2 Sketch of a scroll ring. A sector of the ring is removed to display its interior. Cross sections perpendicular to the singular filament (not shown in detail here) have a structure similar to that of spirals in two dimensions.

polymer concentrations instead of leading to a homogeneously oscillating state. Starting simulations from randomly generated initial states reveals discrepancies between CA and PDE models similar to those found for systems approaching homogeneous states. For CA models, the initial states have randomly selected species at each grid point; for PDE models, all species have randomly chosen concentrations at each grid point. In three-dimensional PDE models without a cutoff rule, scroll ring patterns are formed from the initial states. In three-dimensional CA models, scroll rings are formed only if there are five hypercycle members. For six or more members, all polymer concentrations soon decay and patterns never arise.

It is worth noting that in three dimensions, a CA model with five members gives a homogeneously oscillating state. In comparison, in two dimensions spirals form spontaneously due to local fluctuations. My interpretation of these findings is that the contraction of scroll rings, which is very strong for small rings, prevents their spontaneous formation, even though local fluctuations continuously perturb the homogeneous state.

9.4 Cluster Dynamics

Spirals and scroll rings are not the only self-organized patterns that can distort mean-field predictions: we now discuss the formation, movement, and regeneration of cluster structures.

Formation of clusters

Consider a single auto-catalytic polymer species, referred to here as the main species. In addition to this main species, reactions may involve parasites and monomers. Here, the monomers play a crucial role: with a suitable explicit growth limitation based on monomer concentrations, both PDE and CA models of this reaction–diffusion system can exhibit cluster formation. Clusters are spatially localized aggregations of polymers; in other words, inside clusters the concentration of the main species is high. Clusters are surrounded by regions where the polymer has low concentrations but the monomer is present at high concentrations. Similar clusters ("spots") have been demonstrated experimentally and have been obtained in numerical simulations for a simple inorganic reaction–diffusion system by Pearson (1993) and Lee *et al.* (1993, 1994). Because clusters are formed due to monomer–polymer interactions, they result in a pattern that is fundamentally different from spirals and scroll rings. For spirals, pattern formation leads to the spatial segregation of different polymer species. Clusters, on the other hand, arise from polymer species growing in spatial aggregation. Cluster formation can be obtained not only for auto-catalytic species, but also for a large number of other catalytic networks, such as hypercycles. Where clusters contain several polymer species, species inside the cluster may display additional patterns such as spirals. Such additional patterns are not treated in this chapter; for simplicity, we only consider clusters composed of a single, auto-catalytic species.

In PDE models, clusters can arise if monomers have a very high diffusion rate. This seems to be a reasonable assumption, since a monomer is much smaller than a polymer. In CA models, on the other hand, an explicit growth limitation must be applied for cluster formation (in addition to the implicit growth limitation of CA models, which implies that growth becomes difficult when there are only a few empty grid points). This explicit growth limitation must not be "too local": it should be a function of the "regional" concentration of polymers (i.e., of the frequency of occupied grid points in some region surrounding a given grid point). The region that is considered needs to be sufficiently large, at least larger than a cluster, so that all points within the cluster and the points surrounding it experience the same growth limitation; in the CA models described here, the chosen region is the entire grid. With this choice, the CA and the PDE models are not completely equivalent. In the CA model, the spatial dynamics of monomers are not modeled. Rather, assuming infinitely fast monomer diffusion, their spatial dynamics are replaced by a global growth limitation on polymers.

We have used two different formulations of this explicit global growth limitation, both giving similar patterns and both resulting in chasing as described below. In the first formulation, the probability of replication is proportional to a factor $(1 - c_{tot}/c_0)$, where the total concentration c_{tot} is the fraction of grid points occupied by any polymer, and c_0 is a fixed reference concentration indicating the maximal total concentration ($0 < c_0 \leq 1$). The size of the clusters depends on the value of c_0. In Figure 9.3, $c_0 = 0.2$. In the second formulation, monomers are used. At each time step, there is a certain number of monomers, which are available for replication on all grid points – that is, the monomers do not belong to any particular point. The probability of replication is proportional to the number of monomers. For each replication the number of monomers is reduced by a given amount, and at each time step the monomers are restored toward an equilibrium value. Regardless of which global growth limitation is used, each polymer has a certain probability of decay.

Parasites here have the same effects as in the hypercycle models: they do not provide any catalytic support, but they receive catalytic support for their replication from the main species. When parasites are introduced into clustered patterns of PDE and CA models, entirely different results are obtained. In CA models, a cluster can avoid being destroyed by a parasite by growing away from the parasite in a process called chasing (Cronhjort and Blomberg 1997b). In PDE models, however, each cluster that is infected by a harmful parasite is "killed," while a system of several clusters may resist a parasite if a suitable cutoff rule is applied. Without a cutoff, all clusters in the system are "killed" by a parasite once it has been introduced into one of them. Essentially, the cutoff can prevent polymers from diffusing between clusters: clusters that are "killed" by parasites may then be replaced through division of uninfected clusters, as demonstrated by Cronhjort and Blomberg (1997a).

Because CA and PDE models respond so differently to parasites in clusters, we investigate the two mechanisms of chasing and regeneration and show how each of them depends on the modeling framework.

Chasing of clusters

In a CA model, when a parasite is introduced into a cluster, the parasite begins to grow in number. As it grows, it outcompetes the main species in the cluster and the cluster begins to decline. However, the cluster also starts to grow at those edges where the parasite is not present, since the explicit global growth limitation (based on very fast monomer diffusion) allows the cluster to respond to the loss caused by the parasite. After a while the parasite forms a layer on one side of the cluster (Figure 9.3). As the main species decays at the parasitized edge and grows at the parasite-free edge, the cluster moves. The parasite cannot encircle the main species

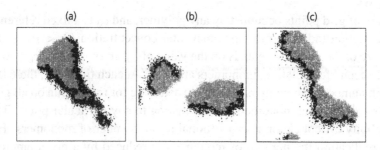

Figure 9.3 An auto-catalytic species (gray) coexists with a parasite (black). (a) A parasite with three times more catalytic support than the auto-catalytic species gives a long, narrow cluster. (b) More equal parameter values give a more regular cluster shape. Collisions can split the cluster. (c) When the decay rate of the main species is high (here it is five times the decay rate of the parasite), the cluster is wide and short. Since the explicit growth limitation in the displayed CA models is global, in the long run only a single cluster will persist.

unless it has far better replication or decay parameters: even with replication constants differing up to a factor of about three in favor of the parasite, or decay probabilities differing up to a factor of at least five, the parasite cannot encircle the main species. As the main species in the cluster can grow faster along the parasite-free edge than the parasite can grow along the parasitized edge, the cluster moves instead of being killed and the parasite thus "chases" the main species. The shape of the moving cluster varies depending on the parameters (see Figure 9.3). Without the parasite, the cluster is essentially circular. In addition to chasing, the parasite may also split a cluster and then destroy one of its parts. Although chasing is most clearly demonstrated by combining diffusion mechanism (1) of Box 9.2 with asynchronous updating, it also occurs for other choices.

These results are quite different from those obtained for PDE models. As clusters have an essentially homogeneous internal structure, mean-field results apply in PDE models when a parasite infects a cluster: the parasite spreads throughout the cluster and "kills" it. Adding a cutoff rule does not change this outcome. In CA models, chasing may save a cluster from a parasite because polymer positions have strong local correlations in space; such correlations are completely neglected in PDE models. Small-scale correlations are important for two reasons. First, in CA models a polymer's probability of replication depends on its immediate neighbors: there must be both an empty grid point and a polymer providing catalytic support. The main species at the parasite-free edge of an infected cluster has many neighbors that provide catalytic support as well as many empty grid points into which it can replicate. Therefore, the main species can grow faster at this edge than the parasite can grow along the infected edge, where there are fewer empty sites and competition with the main species is strong. Second, in CA models auto-catalysis is more beneficial than obtaining catalytic support from another species, even if catalytic parameters are the same. At replication, an "offspring" is placed adjacent to the catalyzing "parent." This implies that there is a high chance that offspring and parent will meet again and mutually catalyze their replication in the future. These two consequences of small-scale spatial correlations mean that effective reaction dynamics in CA models of cluster formation are quite different from those in PDE models: PDE models are based on the absence of such small-scale, or local, correlations.

Regeneration of clusters

In the CA models described above, we assumed an explicit global growth limitation corresponding to infinite monomer diffusion and we obtained one cluster. If the monomer diffusion is very high, but not infinite, several clusters may coexist, as has been demonstrated in a PDE model by Cronhjort and Blomberg (1997a). Inside clusters the concentration of polymers is high and, because monomers are consumed by the replication of the polymers, the concentration of monomers is low. In between clusters, the concentration of polymers is low and monomers are present in high concentrations. Monomers flow into clusters due to gradients in their concentrations. For instance, from an initial state containing only a single cluster, the cluster grows outward due to monomers flowing in from the surroundings. But as the cluster grows, polymer concentrations in the interior begin to decay as monomer concentrations there are lowered too much. This results in a division of the cluster. The same procedure is repeated until clusters are evenly distributed over the entire system and all clusters are more or less the same size. Thus in PDE models clusters can readily divide, whereas in the CA models described above they cannot, since the splitting mechanism requires large-scale heterogeneities of monomer concentrations.

In pure PDE models without a cutoff rule, all clusters are connected to each other by regions of low polymer concentrations. If a parasite is introduced into a cluster it "kills" that cluster; it then spreads to adjacent clusters, which in turn are killed. Cluster division, induced by the decay of infected clusters, cannot save the system from the parasite: clusters that arise from divisions will be infected by the parasite before the division is completed.

If a suitable cutoff rule is added to PDE models, clusters can be separated from each other by regions in which the polymer concentrations are

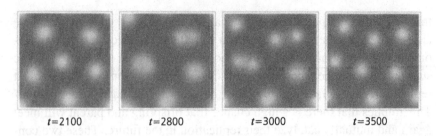

$t=2100$ $t=2800$ $t=3000$ $t=3500$

Figure 9.4 A sequence of pictures showing regenerating clusters in a PDE model combined with a cutoff rule. Shades of gray correspond to the levels of concentration of the auto-catalytic species, with white indicating high concentrations. Here, the cutoff value is 0.01. For comparison, the concentration of polymers in the interior of the clusters is approximately 0.3. In the first picture ($t = 2100$), the system is shown before a parasite is introduced; this is a steady state. A parasite inserted into the central cluster soon "kills" that cluster, but it also spreads to a neighboring cluster (above and left of the original one). In the next picture ($t = 2800$), the two infected clusters have already been "killed." The parasite does not spread further because, due to the cutoff, clusters are almost completely separated from each other by empty (black) points. When the infected clusters have decayed, division takes place ($t = 3000$). In the last picture ($t = 3500$), the division is completed and a new steady state is soon established.

essentially zero; this is illustrated in Figure 9.4. Now it is more difficult for a parasite to spread from one cluster to the next. To infect neighboring clusters, the parasite needs some exchange of polymers between clusters before the parasite has killed the primarily infected cluster. When a cluster has been "killed," the parasite also decays, and soon monomer concentrations increase. Then the uninfected neighboring clusters begin to grow: some may divide and replace the killed cluster with a new, uninfected one.

So far, systems with many coexisting clusters have only been demonstrated in PDE models. The reason is that coexistence and division are features regulated by the rapid diffusion of monomers. Preliminary simulations indicate that similar states may be obtained in interactive CA–PDE models, where the polymers are modeled by a CA part and the monomers, by a PDE part. Such models, however, are complicated and difficult to evaluate. Owing to cluster formation, these models provide some resistance to parasites, but there is also chasing of infected clusters. Such chasing may spoil the resistance to parasites, because the chased cluster moves and therefore can spread the parasite to uninfected clusters. Further studies of this CA–PDE model and of simpler CA models for cluster regeneration are warranted.

We must make a clear distinction between discrepancies that are expected to disappear if the CA and PDE models are made more comparable and discrepancies that are expected to remain. Coexistence of several clusters and splitting of clusters are features that are expected to be found in both types of model, whereas chasing due to small-scale spatial correlations is expected to be found only in CA models.

9.5 Concluding Comments

In this chapter we have studied simulations of reaction–diffusion systems in two different modeling frameworks, CA and PDE models. Corresponding models designed to describe the same reaction–diffusion system within the two frameworks often give similar patterns. One cannot, however, take for granted that these patterns, although similar, will respond identically when, for example, parasites are introduced. There are situations where even small differences between the corresponding models have qualitative consequences.

In this chapter, I have flagged four mechanistic discrepancies between CA and PDE models. These are critical for understanding the phenomenological discrepancies between both frameworks as observed in the examples.

Inherent cutoff concentrations. CA models have implicit cutoff concentrations corresponding to the smallest possible nonzero concentration at which a single individual can be present in a given region. This implicit cutoff is not easily modified. It allows for local extinction in CA models (with at least six hypercycle members), implying, for instance, resistance to parasites, and decay of homogeneous states in three-dimensional systems. PDE models, on the other hand, allow for arbitrarily low local concentrations. Because each species is present at each point in PDE models, each species "encounters" all other species, even if their concentrations are very low. For situations where the cutoff effect of the CA is desired, the PDE models can be combined with a suitable explicit cutoff rule. One must then decide on a reasonable cutoff value, which can be difficult and may be critical, for example, for establishing resistance to parasites.

Inherent growth limitations. Similar to the implicit cutoff concentration, there is also a maximal concentration in CA models, amounting to an implicit growth limitation. Growth stops when the lattice is full. For many systems, this is a convenient feature, but the implicit growth limitation does not allow for competition between growth in different regions. Hence, for example, cluster formation does not occur unless an extra growth limitation is applied. In contrast, PDE models always require the specification of an explicit growth limitation. For both types of model, however, the critical question is how to underpin a particular growth limitation by mechanistic considerations.

Small-scale spatial correlations. In CA models, local correlations are important, since each grid point contains at most one individual that interacts with a small number of neighbors. The significance of such small-scale correlations is illustrated by the difference between how clusters react to parasites in CA and PDE models. We have seen that chasing, which occurs in CA models, depends on local correlations between catalysts, which are completely neglected in PDE models (with or without cutoff rules). In PDE models, all species have continuous concentrations and, therefore, all species are present at each point.

Stochastic versus deterministic dynamics. PDE models are deterministic, whereas CA models contain numerous random events. For this reason, CA models are more suitable for microscopic simulations or for the study of other systems where numbers are low. On the other hand, PDE models are geared to macroscopic simulations, where only local spatial averages matter. Usually it is more convenient to use a deterministic model because each initial condition needs to be tested only once. In CA models, many stochastic realizations are needed before one can draw conclusions concerning dynamical essentials.

We have also seen that the fact that a certain model provides a good representation of one feature of a reaction–diffusion system does not automatically imply that other features are also well captured. Therefore, caution is necessary when drawing conclusions from models and simulations: it certainly is not possible to decide once and for all that a certain modeling framework is "better" than a competing one. Ultimately, it is the features of the reaction–diffusion system under study (e.g., rapid local stirring or inherent discreteness of concentrations) and the characteristic spatial and temporal scales that determine which modeling framework is most appropriate for answering a specific question.

10

Spirals and Spots:
Novel Evolutionary Phenomena through
Spatial Self-structuring

Maarten C. Boerlijst

10.1 Introduction

The concept of a hypercycle was introduced in the early 1970s (Eigen 1971) as a model for cyclic helping of self-replicating entities. Figure 10.1a shows a schematic diagram of a hypercycle: each member of the cycle replicates itself and supports the replication of the next member. Eigen and Schuster (1979, 1982; Eigen 1992) suggested a role for hypercycles of ribonucleic acid molecules in prebiotic evolution. They showed that in prebiotic evolution there exists a so-called information threshold: the length of molecules is restricted by the accuracy of replication. In a hypercycle, each separate molecule species is constrained by the maximum string length, but the species can combine their information and thus cross the information threshold.

An important objection to the hypercycle theory has been raised by Maynard Smith (1979; see also Bresch *et al.* 1980): because there is no selection for the giving of catalytic support to the replication of another molecule, this property cannot be maintained. Giving catalytic support is an "altruistic" property; that is, it does not increase the number of copies of the molecule itself, but increases those of another, competing species.

As a result, a hypercycle is vulnerable to invasion by so-called parasites. Figure 10.1b shows a hypercycle with a parasite. The parasite is capable of self-replication on its own; in addition, it receives catalytic support from species 1 but does not give catalytic support to any other molecule. If species 1 gives more support to the parasite than to species 2 [i.e., $\kappa_{par} > \kappa_2$ in Equation (10.5) below], the parasite will be selected in favor of species 2 and the entire hypercycle will be lost. There seems to be a large class of

(a)   (b)

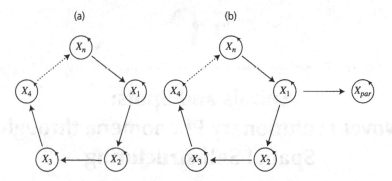

Figure 10.1 Schematic diagram of a hypercycle. The hypercycle consists of n self-replicating molecule species X_i. (a) Each species provides catalytic support for the subsequent species in the cycle. (b) Parasitic species X_{par} coupled to the hypercycle.

parasites that are fatal to hypercycles; that is, hypercycles are evolutionarily unstable.

Recently, a twist to the parasite story has become apparent (Boerlijst and Hogeweg 1991a, 1991b, 1995b; May 1991; Boerlijst 1994), namely, the effects of spatial pattern formation can render the hypercycle resistant to otherwise deadly parasites. In spatial models hypercycles structure themselves into spiral waves. Within a spiral each species is located behind its catalyst, and because it locally outcompetes the catalyst the spirals rotate in time. In such a system, a parasite typically can "kill" a spiral it has infected at the center, but it fails to invade other spirals. It has been shown that this phenomenon applies both in discrete stochastic cellular automaton (CA) models (Boerlijst and Hogeweg 1991a, 1991b) and in continuous deterministic partial differential equation (PDE) models with spatial heterogeneity (Boerlijst and Hogeweg 1995b). However, it has been suggested that parasite resistance cannot be obtained in homogeneous PDE models (Cronhjort and Blomberg 1995; Cronhjort 1995; see also Chapter 9).

In this chapter I demonstrate parasite resistance of spirals in a homogeneous PDE model and discuss the similarities and differences between the CA and PDE models. The PDE considered here is closely related to a model by Pearson (Pearson 1993). In this model [and in the corresponding chemical system (Lee *et al.* 1994)] so-called self-replicating spots and many other fascinating patterns are observed. I show that hypercycles can also generate self-replicating spots that spontaneously divide and decay, thus creating a kind of protocell. Furthermore, the spots also provide resistance to parasites. The two mechanisms for parasite resistance are compared.

10.2   A Spatial Hypercycle Model

In the model presented here, different polymer strains X_i compete for the building blocks of replication, the charged mononucleotides M.

Chemical kinetics

We include the following processes.

- Self-replication:

$$X_i + M \xrightarrow{\rho_i} 2X_i \, , \tag{10.1}$$

  in which ρ_i is the non-catalyzed replication rate of polymer type i.
- Catalyzed replication:

$$X_i + X_{i-1} + M \xrightarrow{\kappa_i} 2X_i + X_{i-1} \, , \tag{10.2}$$

  in which κ_i stands for the catalyzed replication rate of polymer type i, catalyzed by the preceding member of the hypercycle (in the case of $i = 1$, this is the last member of the cycle).
- Decay:

$$X_i \xrightarrow{\delta_i} \, , \quad M \xrightarrow{\delta_m} \, , \tag{10.3}$$

  in which δ_i and δ_m denote the decay rate of polymer i and monomers, respectively.
- Influx of monomers:

$$\xrightarrow{I} M \, , \tag{10.4}$$

  in which I stands for the influx rate of monomers (the rate at which they become charged).

Note that the monomers M are scaled in units needed to replicate exactly one polymer. I chose $I = \delta_m$ so that the carrying capacity of monomers in the absence of polymers is scaled to 1.

Concentration dynamics

We study the following reaction–diffusion-type PDE model:

$$\frac{\partial X_i}{\partial t} = \kappa_i M X_i X_{i-1} + \rho_i M X_i - \delta_i X_i + D_X \Delta X_i \, ,$$
$$i = 1, \ldots, n \, , \tag{10.5}$$

$$\frac{\partial M}{\partial t} = \delta_m(1 - M) - \sum_i (\kappa_i M X_i X_{i-1} + \rho_i M X_i) + D_M \Delta M \ . \quad (10.6)$$

All processes (10.1) to (10.4) are included in this model (with the convention $X_0 = X_n$). A Laplacian operator Δ is added for the diffusion (see Chapter 9; see also Chapter 22). The numerical simulations are forward Euler integrations of the finite-difference equations resulting from discretization of the diffusion operator. The spatial mesh consists of a rectangular grid of 150×150 elements up to 400×400 elements with toroidal boundary conditions. The time step was varied from $t = 0.1$ to $t = 0.01$. The default parameter settings used were $\kappa_i = 5$, $\delta_i = 0.08$, $\delta_m = 0.02$, $D_X = 0.1$, and $D_M = 0.2$. The non-catalytic replication rate ρ_i was varied, with $\rho_i = 0.1$ for the spiral waves and $\rho_i = 0$ for the self-replicating spots.

10.3 Spirals and Spots

First, I introduce the spatial patterns under study. The patterns in Figures 10.2a to 10.2d were started from a small initial patch of five polymer species that constitute a hypercycle.

With self-replication, $\rho_i = 0.1$ (Figures 10.2a and 10.2b), the patch develops into spiral waves. Figure 10.2a shows the polymer distribution. Within a spiral each species is located behind its catalytic supporter, and the spirals rotate in time. Figure 10.2b, the monomer distribution, shows that the spirals in fact have a circular core (Mikhailov *et al.* 1994); in the middle of the spirals there is very little catalysis, which is reflected in a high monomer concentration. The monomer concentration is lowest in the spiral arms on the boundaries between polymer species, where catalysis is strongest. The pattern is very stable; it repeats itself after every full rotation of the spirals.

Without self-replication, $\rho_i = 0$ (Figures 10.2c and 10.2d), the patch develops into self-replicating spots. Figures 10.2c and 10.2d show the polymer and monomer distributions, respectively. For these parameters, the "virgin" concentration of monomers (i.e., the non-catalytic equilibrium $M = 1$) is simply too small to allow for global polymer persistence. However, on a spatial domain, a "spot" of polymers can locally persist through the influx of monomers from around the spot. When such a spot increases in size, the middle of the spot runs out of monomers and the spot divides. Spots can also disappear when they are perturbed, for example, when other spots come too close. The pattern is chaotic with a positive Lyapunov exponent (see Pearson 1993).

(a)   (b)

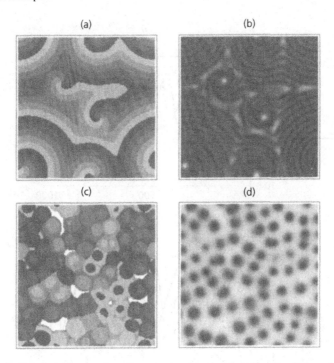

(c)   (d)

Figure 10.2 Spirals and spots in a five-member hypercycle. The simulation starts with the "virgin" monomer state $X = 0$ and $M = 1$, except for a small 10×10 patch in which $M = 0.5$ and a random polymer species $X_i = 0.5$. The situation is shown at $t = 10\,000$. (a, b) Spiral waves for $\rho_i = 0.1$. In (a) the shade of gray distinguishes the dominant polymer species at each location, with X_1 being the darkest. In (b) the monomer distribution is shown; darker shades indicate smaller M values. (c, d) Self-replicating spots for $\rho_i = 0$. In (c) the polymer distribution is shown as in (a); white indicates empty grid points. In (d) monomer distribution is shown as in (b). For other parameters, see accompanying text; for actual concentrations, see Figures 10.3a and 10.3b.

10.4 Local versus Global Extinction

In Figure 10.3a, a time plot of a grid point in the spiral waves of Figure 10.2a is shown (the grid point is located on the periphery of a spiral). This time plot is compared with the dynamics of the nonspatial variant of the PDE in Figure 10.3b [an ordinary differential equation (ODE) is obtained by simply omitting the Laplacian operators in Equations (10.5) and (10.6)]. Both the PDE and ODE show limit-cycle behavior, but the limit cycle in the spiral waves has a smaller amplitude and higher frequency. This damping of the oscillations can be understood by the notion that in a spiral wave the oscillations are driven by the spatial waves instead of by the local limit cycle. In Figure 10.3c, the ODE and PDE are compared for

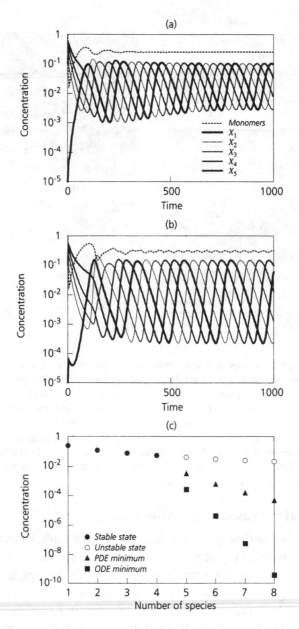

Figure 10.3 Comparison of spatial versus nonspatial dynamics. (a) Time series of species densities at a grid point in the spiral waves shown in Figure 10.2a. (b) Time series of the (nonspatial) ODE analogue of the PDE shown in (a). (c) Bifurcation diagram for an increasing number of species in the hypercycle. For $n > 4$ the steady-state dynamics switch to limit cycles in the ODE and to spiral waves in the PDE. The minimum concentrations for these cycles are plotted.

Figure 10.4 Polymer distribution in a 15-species hypercycle. For interpretation of shades of gray see Figure 10.2a.

hypercycles with different numbers of members. For four or fewer members, the ODE has a stable steady state, which in the PDE leads to a stable homogeneous solution. For five or more members, the ODE shows limit-cycle behavior and the amplitude of the cycle quickly increases with the number of species in the hypercycle. In the PDE the amplitude of the oscillations also increases, but much less dramatically than in the ODE. Of course, it is not realistic to assume that population sizes can become infinitely small. Currently there is considerable debate on how to introduce extinction explicitly (see Section 10.6). Here, we simply consider (local) populations that are smaller than 0.00001 to be extinct and we explicitly set them to zero. In the ODE, this implies that hypercycles of six or more members become extinct. In the PDE, hypercycles of nine or more members show local extinction of species, but globally the spirals persist because the extinct species are locally reintroduced by the next wave. Due to the spiral waves, there seems to be no upper limit to the number of species that can coexist. A hypercycle of 15 members is shown in Figure 10.4. Interestingly, in this case the core of the spiral becomes essentially linear.

We have not yet discussed the local dynamics within the self-replicating spots. In this case it is not possible to compare the PDE directly with the ODE, as the ODE has no polymer persistence for these parameter settings. Within a spot the oscillations are driven by an ODE-type local limit cycle in the middle of the spot [comparable to so-called target patterns, see Tyson and Keener (1988)]. Therefore, as in the ODE, the oscillations within spots quickly increase in amplitude as the size of the hypercycle increases. With an extinction threshold set at 0.00001, only hypercycles with five or fewer members can form self-replicating spots. Hypercycles of six or more members become extinct.

(a)   (b)   (c)

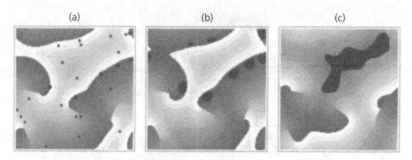

Figure 10.5 Parasite invasion of spiral waves. Starting with the polymer distribution shown in Figure 10.4, 25 random grid points are infected with $X_{par} = 0.5$. Grid points dominated by parasites are indicated in black. The parasite increases over the time period shown, but eventually goes to extinction (not shown) and the spiral pattern is restored. (a) $t = 10$; (b) $t = 50$; (c) $t = 300$.

10.5 Resistance to Parasites

This section describes the consequences of introducing a "deadly" parasite into the spirals and spots shown in Figures 10.4 and 10.2d, respectively. Replication of the parasite is catalyzed by polymer species X_1, with $\kappa_{par} = 6$ (and thus $\kappa_{par} > \kappa_2$). All other parameter values are identical to those of the other polymer species. The parasite is introduced at concentration $X_{par} = 0.5$ at 25 randomly chosen grid points.

Resistance of spirals

A parasite invasion of the spirals of Figure 10.4 is shown in Figures 10.5a to 10.5c. In Figure 10.5a, at $t = 10$ after infection, the parasites have increased in the vicinity of polymer species X_1, their catalytic supporter. Parasites that do not invade near species X_1 quickly become extinct because they do not receive catalytic support. In Figure 10.5b, at $t = 50$, the total parasite count has increased, but the parasite regions follow the "normal" direction of growth within the spirals, which is toward the periphery. In Figure 10.5c, at $t = 300$, the parasite regions have reached the boundary of the spiral domains. Here they run out of catalytic support, since species X_1 is locally outcompeted from all sides. As a result, at approximately $t = 470$ (not shown), the parasite is completely wiped out and the original situation is restored.

The mechanism of parasite resistance of spiral waves has been described in detail (Boerlijst and Hogeweg 1991a). The ODE-type competition between the parasite and species X_2 for catalysis of species X_1 essentially helps the parasite in the wrong direction, namely, toward the boundary of

(a)   (b)   (c)

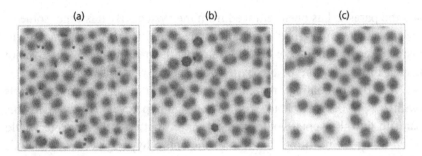

Figure 10.6 Parasite invasion of self-replicating spots. Starting with the monomer distribution shown in Figure 10.2d, 25 random grid points are infected with $X_{par} = 0.5$. Grid points that have $X_{par} \geq 0.01$ are indicated in black. The parasite "kills" all spots that it infects, but thereafter the parasite has no support and becomes extinct. (a) $t = 10$; (b) $t = 100$; (c) $t = 600$.

the spirals, where all species run out of catalytic support. To reach into the core of the spirals, the parasites have to compete against the catalytic waves of the other species in the hypercycle. Consequently, all properties of the species in the hypercycle become important to the outcome of the competition. Also, the number of species in the hypercycle plays a crucial role, as it determines the delay of the next wave of species X_1. In the case of the parasite defined above, the hypercycle must consist of at least seven members to be resistant. For shorter hypercycles the parasite is able to persist until the next wave of species X_1. On average it can make progress toward the core of spirals; once it is there, it kills off all other species and removes the spiral region.

In Figure 10.5a, by chance no parasite has invaded a spiral core. But, if this does happen, the spiral whose core is infected will be completely wiped out by the parasite. Thereafter, the parasite region may in turn be completely removed by the other spirals or a small parasitic region (a "cyst") may persist, depending on the number of spirals that were infected (for details see Boerlijst and Hogeweg 1991a).

Resistance of spots

Resistance of spots to parasites is quite straightforward. In Figures 10.6a to 10.6c, the spots of Figure 10.2d are infected with parasites. In Figure 10.6a, at $t = 10$ after infection, the parasites have increased within spots where species X_1 is abundant. In Figure 10.6b, at $t = 100$, some of the initially infected spots have already disappeared, others are completely taken over by the parasite. After taking over a spot, the parasites run out of catalytic support; in Figure 10.6c, at $t = 600$, all infected spots have disappeared and the parasite has been removed from the system. By this time, some of the spots that were "killed" have already been refilled by movement and division of nearby uninfected spots. Note that the parasite in this case cannot persist in the absence of the hypercycle, because it lacks self-replication ($\rho_{par} = 0$). In contrast, the parasite that was introduced into the spirals (with $\rho_{par} = 0.1$) is able to exist without the hypercycle. Introducing this parasite into the spots would be fatal, as in that case the parasite would reach all spots, resulting in a loss of the entire hypercycle. We conclude that parasite resistance is stronger in spirals than in spots.

10.6 Concluding Comments

Spirals and spots are emergent patterns that can spontaneously develop in a two-dimensional medium with hypercycles. In this chapter, I have shown that both patterns can provide resistance to parasites. The key property that causes this resistance is similar in both cases, namely, both the spirals and the spots create an environment of effectively isolated subpopulations. However, the two patterns are found in quite different parameter regimes. Spirals are only formed in hypercycles with five or more species, whereas spots can only exist in hypercycles with up to five members (depending on the local extinction level). Furthermore, self-replicating spots only exist in small parameter ranges. Spiral waves are the default pattern in (sufficiently large) hypercycles; they are formed for most parameter combinations that allow for polymer persistence. Only on the boundary of polymer persistence will other patterns be formed. In fact, there is a whole "zoo" of possible patterns, with even more patterns than observed in the Pearson system (Pearson 1993). An exciting open question is how the system will evolve within this parameter region.

A requirement for resistance to parasites is that there is no "gene flow" between different spots or spirals. In the spots this is always the case, whereas in the spirals there must be a minimum number of members in the hypercycle to achieve this property. Even a one-member hypercycle has resistant spots (Böddeker and McCaskill, unpublished), whereas for spiral resistance at least seven members are necessary. The space between the spots simply does not contain enough monomers to allow for polymer persistence. In the spirals, local extinction of polymers only takes place if the local limit cycle is large enough. Therefore, resistance to parasites is much easier to achieve in spots than in spirals. However, spirals can resist much stronger parasites, namely, parasites that can persist in the absence of the hypercycle. Furthermore, in the spirals there is selection for an increase of catalysis caused by selection at the level of the spirals (Boerlijst and Hogeweg 1991b); polymers with increased catalysis constitute faster rotating spirals, and such spirals can increase their domain. In the spots, selection for an increase of catalysis has yet to be demonstrated. One could imagine, for example, selection for rapid division of spots.

In this chapter a strictly deterministic PDE model with continuous variables is used. I have shown that the mechanism of parasite resistance due to spatial pattern formation does not depend on stochasticity or discreteness of variables (although the local extinction threshold does introduce a certain aspect of discreteness). PDEs are an attractive formalism because they are closely linked to well-studied ODEs and they give hope for mathematical analysis. There are, however, a few problems with using the PDE formalism for studying evolving systems. First, local extinction is impossible in a strict PDE; there will always be diffusion of unrealistically small proportions from populated areas. We "solve" this problem by simply introducing a threshold population size below which variables are set to zero. This hardly affects the dynamics of the spirals and spots, because both patterns are driven by their centers, and local extinction does not occur in the center of spirals and spots. Another, more accurate treatment of local extinction is to describe small populations with a stochastic discrete model (Böddeker and McCaskill, unpublished). Second, in a PDE there is no notion of an individual. This property makes it difficult to study invasion dynamics of mutants, which are often dominated by discreteness and random drift. Furthermore, a separate equation has to be added to the PDE for each new mutant type, making them impractical, especially in cases where ecological and evolutionary time scales cannot be separated, causing many mutant types to be present at the same time (Savill 1997). Third, PDE approximations often break down near bifurcation points. In the case of four-member hypercycles with spiral wave parameters the local dynamics are at a branching point from a stable steady state to a limit cycle. The PDE shows a very slow and stiff convergence to a global steady state. However, adding any kind of stochasticity shows a very different type of attractor, namely, so-called breaking spiral waves or spiral turbulence (Boerlijst and Hogeweg 1991b). It is difficult (but not impossible) to describe this type of spatial pattern in a PDE (Panfilov and Hogeweg 1993).

The notion that spatial self-structuring can affect evolution obviously is not restricted to hypercycles. A number of authors (Boerlijst *et al.* 1993, Savill *et al.* 1997, Keeling and Rand 1995, Rand *et al.* 1995) have shown that the same principles apply in parasitoid–host and predator–prey systems, for example. In these systems pattern formation of turbulence or spiral waves can cause selection for "prudent" predators. Another exciting application is selection for competition strength in coral reefs (Johnson and Seinen, unpublished). In this case, the emergence of patchy patterns evokes selection for intermediate competition strength of corals. Self-structuring patterns can provide a valuable source of information for creating new possibilities for life. Once a self-structuring pattern is present, it can "trap" the dynamics so that they become subordinate to the dynamics of the newly emerged pattern. Considering self-structuring patterns merely as "constraints" (as do Maynard Smith *et al.* 1985, for example) underestimates the creative and guiding power of self-structuring phenomena in the evolution of life.

11

The Role of Space in Reducing Predator–Prey Cycles

Vincent A.A. Jansen and André M. de Roos

11.1 Introduction

Throughout the history of ecology, the interaction between predators and their prey has received attention from ecologists. Some of the longest and most well-known data series in ecology are from predator and prey populations, and predator–prey models are among the oldest in the field. Despite these efforts, a discrepancy exists between the behavior of most models and that of natural predator and prey populations. Most predator–prey models predict lasting periodic oscillations in population densities. Oscillations have been observed in nature, but they do not seem nearly as common or as pronounced as models predict.

Several explanations have been suggested for the qualitative difference between the behavior of the models and reality. Most of these involve additional mechanisms, such as invulnerable life stages of the prey or optimal foraging behavior of the predator. However, experimental studies show that the equations used in classical models do indeed give a reasonable qualitative description of the behavior of predator and prey populations (Gause *et al.* 1936; Maly 1969; Harrison 1995). All these studies compare model predictions with laboratory-scale experiments in which the densities often exhibit oscillations that can lead to population extinction (Huffaker 1958; Gause 1969; Luckinbill 1974). The models thus seem to describe the interaction between predators and their prey in laboratory experiments reasonably well, but fail to capture the properties of natural populations. Because there is no reason to believe that laboratory and natural populations are fundamentally different, the spatial scale at which the predator–prey system exists must play a crucial role in preventing population oscillations in the field.

The spatial scale at which observations of ecological systems are carried out strongly influences their outcome. Hence, explanations for these observations should also take into account the spatial extent of the mechanisms considered. Modeling population dynamics should intuitively start at the level of the individual, at which the organisms reproduce, die, and interact, thus giving rise to population dynamics. At the individual level, it is appropriate to use a model that describes the whereabouts of all individuals and the interactions between them. However, describing the dynamics of the local or the global population in this manner is a daunting task. To unravel the dynamics at larger spatial scales, it would be preferable, or even necessary, to abstract individual-based descriptions of dynamics in terms of variables that are measurable at the scale of the local population. Subsequently, descriptions of the dynamics of the global population can be cast in terms of coupled local populations. The key issue is to find a non-phenomenological procedure to bridge the spatial scales – an issue that pervades many chapters in this book.

In this chapter, we describe models for three different spatial scales: those of the individual, the local population, and the global population. We first discuss some results from individual-based models. We then present a simple two-patch model for predator–prey interaction, followed by a generalization of this model to an arbitrarily large number of patches. Finally, we discuss how these results may be related to each other.

11.2 Individual-based Predator–Prey Models

For a spatially explicit description of interacting predator and prey individuals, one needs to keep track of all individuals and their positions. It will come as no surprise that these models are computationally complicated and intensive. Finding simplified models for such spatial, individual-based systems is an art in itself. Here, we try to make inferences directly from an individual-based, stochastic simulation model and argue that the emerging spatial dynamics reflect a system of coupled local populations where the local populations have a characteristic spatial extent. The dynamics of the (local) populations as observed at the characteristic spatial scale are close to the dynamics of the nonspatial, or homogeneously mixed, analogue of the spatial model. This naturally reduces the study of the global population dynamics to (1) finding the characteristic spatial scale of the local populations and (2) studying the dynamics of the coupled set of local populations.

The results we discuss are from "discrete-entity simulations" (de Roos *et al.* 1991; Wilson *et al.* 1993, 1995b; McCauley *et al.* 1993). These simulations describe a large lattice (128 × 128) on which predators and prey live. Every grid point can either be empty or occupied by at most one predator or one prey at a time. Hence, the whereabouts of every individual are known at every point in time. Predators and prey can move around independently in various ways. For instance, all individuals can be randomly redistributed over the entire lattice at every time step, resulting in homogeneous movement. Another possibility is diffusive movement in which the individuals randomly walk through the grid for a variable number of steps. The third option is a stationary population that can only occupy new space by growing into it. If a destination site is already occupied by an individual of the same species, movement does not take place. Once movement has taken place, a sequence of interactions follows: Prey individuals reproduce with a certain probability into a neighboring site. If no free site is available, the offspring is aborted. Similarly, predators die with a certain probability. If, as a result of individual movement, a prey and a predator end up at the same site, the predator eats the prey and subsequently reproduces with a certain conversion efficiency. The offspring is placed on the grid in the neighborhood of the parent. (For details of this simulation procedure see McCauley *et al.* 1993.)

Within this framework, the rules at the individual level can be varied to reflect different types of behavior for the predator and prey. We discuss some results from rules that have effects comparable to logistic prey growth and a type II functional response, given by the Holling disk equation (Holling 1965). This choice enables us to compare the behavior of this model with that of the classic nonspatial predator–prey model of Rosenzweig and MacArthur (1963), which is based on similar assumptions. In the simulations, prey growth is locally density dependent, because a prey offspring is aborted if the neighboring site where it attempts to establish itself is occupied. The type II functional response is mimicked by preventing the predators from eating for a number of time steps after they have consumed a prey. This results in a handling time for the predators (de Roos *et al.* 1991).

The simulation model behaves very much like the Rosenzweig–MacArthur ordinary differential equation (ODE) model *when the prey is stationary and the predators move homogeneously*. Both models have two qualitatively different types of dynamics: either the densities approach a stable equilibrium or they exhibit stable oscillations. Decreasing the predator death rate causes a transition from stable to oscillatory dynamics. Quantitatively, the transition occurs at similar values for the predator death rate in both the simulation and the ODE models (McCauley *et al.* 1993).

In contrast, when the predators move diffusively and the prey remain stationary the simulation model behaves very differently: a decrease in the predator death rate has very little effect on the observed dynamics, and for all parameter combinations the densities deviate little from constant densities (de Roos *et al.* 1991; McCauley *et al.* 1993).

Why the difference? One possible explanation is that the classic ODE model does not accurately represent the interactions among the individuals when they move diffusively. For example, even though the rules for predator behavior in the simulations lead to a type II functional response at the population level *when movement is homogeneous*, diffusive movement might lead to a type III functional response when measured at the population level. It could be argued that the latter causes the observed stabilization of dynamics. However, as the observed functional responses in the simulation model with homogeneous or diffusive predator movement are very much like those in the classic model (de Roos *et al.* 1991), this explanation seems unlikely. A similar conclusion holds for the prey growth function. Hence, the explanation is probably not that the ingredients of the classic model are wrong, but that the spatial character of the interactions in the simulation model qualitatively changes the dynamics.

Spatial interactions can only influence the population dynamics if the population densities are not spatially homogeneous. Indeed, under diffusive predator movement the populations in the simulation model are clustered. Such clusters imply that, from an individual's point of view, the environment looks relatively homogeneous as long as it is perceived through a spatial window of about the size of the cluster. This locally experienced density might bear no relation to the overall densities as measured on large spatial scales. However, the dynamics at such local scales are similar to those predicted by the nonspatial model, where individuals mix homogeneously (de Roos *et al.* 1991). Figure 11.1 shows the coefficient of variation (the standard deviation divided by the average) of the time dynamics in prey density versus the size of the spatial window within which these dynamics are observed. For very small window sizes the coefficients of variations for diffusive movement and homogeneous movement are comparable; for larger window sizes the curves deviate. Also, the (temporal) autocorrelation function of these dynamics in prey density shows a convergence to the autocorrelation predicted by the nonspatial model when the observations on prey density are carried out at smaller spatial scales (de Roos *et al.* 1991).

The window size where the curves start to deviate can be designated as the characteristic spatial scale of the system. Measured at spatial scales much larger than this characteristic scale, the dynamics are the cumulative

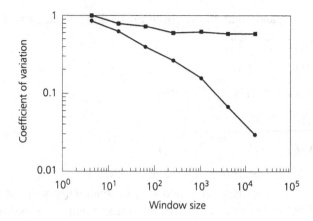

Figure 11.1 Variability in prey density in terms of the coefficient of variation of the discrete-entity simulations. Squares indicate homogeneous predator movement; circles indicate diffusive predator movement. *Source*: de Roos *et al.* (1991).

result of processes in only loosely coupled regions. At much smaller spatial scales, the dynamics are increasingly influenced by stochastic fluctuations due to the finite number of individuals present. At the characteristic spatial scale the system thus behaves in a more deterministic manner than at any other scale (Rand and Wilson 1995; see Chapter 12).

11.3 A Deterministic Model of Two Coupled Local Populations

We make the step from a description at the level of the individual to the level of the local population using the observation that at the characteristic spatial scale determinism is maximal and the populations behave as if they are well mixed. This justifies a description of local populations in terms of their mean densities only. To describe the populations beyond the characteristic spatial scale, we couple a number of local populations diffusively by assuming that every individual has a fixed probability of leaving its local population and migrating to another.

We investigate how fluctuations in the densities in the local populations affect the dynamics of the global population. Ideally, one would derive the equations that describe the local populations from the rules that govern the simulation model. Because it is still an open question how this should be done, we make life simple by boldly assuming that the local populations behave like standard predator–prey models. The simplest predator–prey model is the Lotka–Volterra model, and we start our investigations with the simplest spatial extension of this model, the two-patch Lotka–Volterra

model (Comins and Blatt 1974). In terms of dimensionless variables, the
model can be specified as

$$\dot{N}_1 = rN_1 - N_1P_1 \,,$$

$$\dot{P}_1 = N_1P_1 - \mu P_1 + \frac{m}{2}(P_2 - P_1) \,,$$

$$\dot{N}_2 = rN_2 - N_2P_2 \,,$$ (11.1)

$$\dot{P}_2 = N_2P_2 - \mu P_2 + \frac{m}{2}(P_1 - P_2) \,,$$

where the dot means differentiation with respect to time. It is assumed
that the local prey density, N_i, increases exponentially with rate r in the
absence of predators and that there is a linear contact rate between local
prey and predators. The densities are scaled such that all prey mass that is
eaten reappears as predator mass. In the absence of prey the local predator
density, P_i, decreases exponentially with rate μ. The two populations are
coupled through migrating predators, which move to a neighboring patch at
rate $\frac{m}{2}$. (The migration rate is divided by two because in the more general
situation of a chain of connected patches in which every patch has two
neighbors the migrants leave a patch at rate m. Because this system consists
of only two patches, half the emigrants will be reflected from the system's
boundary and the effective migration rate is $\frac{m}{2}$.) The prey are assumed to
be stationary as in the discrete-entity simulations.

As long as the densities of the predator and the prey are the same in both
patches (i.e., $P_1 = P_2$, $N_1 = N_2$, the system is completely homogeneous),
the net effect of migration is zero and the model is identical to the nonspatial Lotka–Volterra model. The dynamics of the nonspatial model are well
known: a family of closed orbits surrounds a neutrally stable equilibrium,
and the densities oscillate eternally with an amplitude that depends on the
initial conditions.

For Lotka–Volterra models in a spatially continuous domain the densities eventually become homogeneous (Murray 1975). The same holds
for the two-patch Lotka–Volterra model: all differences in density between the patches disappear asymptotically for any positive initial condition. [This can be shown using the Lyapunov function $V_1(t) + V_2(t)$, where
$V_i(t) = N_i + P_i - \mu \ln N_i - r \ln P_i$.] One therefore might not expect the dynamics of the spatial Lotka–Volterra model to yield any insights that cannot
be gained from the nonspatial Lotka–Volterra model. Surprisingly, this is
not the case. Figure 11.2 shows a solution of Equations (11.1) starting with
small differences in densities between the patches. Although the differences between the prey and predator densities in the two patches eventually

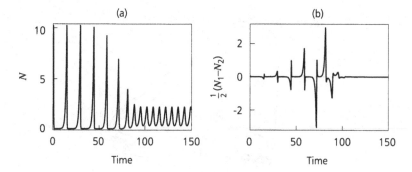

Figure 11.2 A solution of Equations (11.1) for $r = \mu = 1$. (a) Average prey density (N). (b) Differences between prey densities in the two patches (ψ_N). The predator densities exhibit similar fluctuations (not shown).

disappear, they initially increase (see Figure 11.2b). As the differences increase, the amplitude of fluctuations in the average densities are reduced.

To understand the behavior of the model given by Equations (11.1), it helps to consider the geometry of the four-dimensional state space. The part of the state space in which there are no differences in densities between the patches – that is, $N_1 = N_2$ and $P_1 = P_2$ – is represented by an invariant two-dimensional plane. On this diagonal plane the dynamics are given by the (nonspatial) Lotka–Volterra model: closed orbits surround an equilibrium. The orbit shown in Figure 11.3 starts near a closed orbit in this plane, moves away from it, and then returns to another closed orbit in the diagonal plane on which the amplitude of the fluctuation is much smaller. Not all closed orbits in the diagonal plane have similar stability properties: some attract while others repel.

To establish which of the closed orbits attract and which repel, we analyze their (local) stability. To this end we introduce variables for the average prey and predator densities, $N = \frac{1}{2}(N_1 + N_2)$, $P = \frac{1}{2}(P_1 + P_2)$, and a vector that contains the differences between the patches,

$$\psi = \begin{pmatrix} \psi_N \\ \psi_P \end{pmatrix} = \begin{pmatrix} \frac{1}{2}(N_1 - N_2) \\ \frac{1}{2}(P_1 - P_2) \end{pmatrix} . \tag{11.2}$$

In new variables, the system reads as follows:

$$\begin{aligned} \dot{N} &= rN - NP - \psi_N \psi_P \\ \dot{P} &= NP - \mu P + \psi_N \psi_P \\ \dot{\psi}_N &= (r - P)\psi_N - N\psi_P \\ \dot{\psi}_P &= P\psi_N + (N - \mu - m)\psi_P . \end{aligned} \tag{11.3}$$

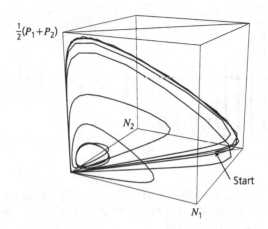

Figure 11.3 The solution visualized in Figure 11.2 in a projection in the N_1, N_2, $\frac{1}{2}(P_1 + P_2)$ space. The orbit starts at the arrow.

The dynamics of Equations (11.1) close to the diagonal plane are given by a linearization in the neighborhood of $|\psi| = 0$:

$$\dot{N} = rN - NP$$
$$\dot{P} = NP - \mu P \ , \tag{11.4}$$

and

$$\dot{\psi} = \begin{pmatrix} r - P & -N \\ P & N - \mu + \lambda m \end{pmatrix} \psi \ , \tag{11.5}$$

where the variable λ is introduced for later use and here takes the value -1. Notice that the system for the averages, Equations (11.4), is identical to the nonspatial Lotka–Volterra model. The linearized equations for the differences, Equation (11.5), are driven by the Jacobian of Equations (11.4) with an additional diagonal term representing migration.

The closed orbits in the diagonal plane are represented by solutions of Equations (11.4) and (11.5) with $\psi = 0$. To see whether a closed orbit attracts or repels, we take an initial value for N and P and integrate Equations (11.4). Because Equations (11.4) are decoupled from Equation (11.5), we can take $\psi \neq 0$ without changing the solutions of Equations (11.4); in other words, in the linearization we can introduce differences between the patches while keeping the mean densities behaving as if there are no differences. A closed orbit repels if $|\psi|$ increases in the long run. A more formal, but essentially similar method of establishing the stability of the closed orbits is determining the dominant Floquet multiplier of Equation (11.5) (see Box 11.1).

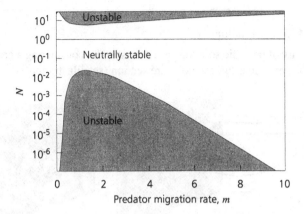

Figure 11.4 Minimum and maximum prey densities over the closed orbit of Equations (11.1) with a multiplier at −1 for different values of the predator migration rate. Closed orbits with a larger amplitude than the closed orbit with multiplier at −1 are unstable and repel; closed orbits with a smaller amplitude are neutrally stable and attract orbits over which the densities in the patches differ. Therefore, this graph also represents the maximum and minimum prey densities in the two-patch Lotka–Volterra system given by Equations (11.1) after transients have died out. Parameter values: $r = \mu = 1$.

The closed orbits of the Lotka–Volterra model have two multipliers equal to 1 and hence are neutrally stable. Therefore, the closed orbits on the diagonal plane, described by Equations (11.4), also have two multipliers equal to 1 and two others, determined by Equation (11.5), that represent the spatial interactions. Because the dynamics on the diagonal are identical to the nonspatial dynamics, the spatial interactions can destabilize these orbits but can never stabilize them.

Figure 11.3 suggests that closed orbits on which the densities oscillate with a large amplitude can be unstable, while smaller closed orbits attract. Figure 11.4 shows that this is indeed the case for a large range of migration rates. The large closed orbits are unstable because the absolute value of one of the multipliers exceeds 1 (it is real, negative, and smaller than −1). For small closed orbits, the absolute value of the multipliers determined by Equation (11.5) is smaller than 1. These closed orbits attract orbits from outside the diagonal plane. (They are neutrally stable because they also have two multipliers at 1.) Between the neutrally stable and the unstable closed orbits lies a single closed orbit with a multiplier at −1. Figure 11.4 shows the minimum and maximum prey values attained at this closed orbit.

The two-patch Lotka–Volterra model can thus behave very differently from its nonspatial counterpart. In the nonspatial model fluctuations of all amplitudes are possible. In contrast, in the spatial model homogeneous

Box 11.1 Floquet multipliers

The stability of periodic solutions is established by considering a cross section to the periodic orbit, the Poincaré section (see illustration).

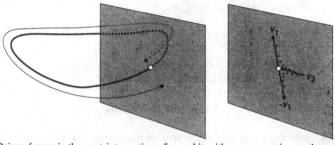

The Poincaré map is the next intersection of an orbit with a cross section to the periodic solution, the Poincaré section (left). The periodic orbit is a fixed point of the Poincaré map; it is unstable if nearby orbits move away from it (right). v_1 and v_2 are respectively the unstable and stable eigenvectors of the linearized Poincaré map $B(T)$.

Starting from the intersection point of the orbit with this plane, the system will obviously return to this same point in the state space every T time units, where T is the period of the solution. The periodic solution is stable if orbits that are close to it in the state space converge to it over time. Whether this occurs or not can be studied by analyzing a related discrete time system, the so-called Poincaré map, which maps the intersection point of a particular orbit with the Poincaré section to the intersection point after the next turn-around. The periodic solution is stable if the fixed point of the Poincaré map (which corresponds to the intersection point of the Poincaré section and the periodic orbit) is stable.

The stability of the discrete system can be determined from its linearization, but with a twist: to arrive at a computational procedure we linearize the underlying continuous time system in the neighborhood of the periodic orbit. The calculation is the same as for linearizing around an equilibrium, that is, we temporarily fix a point $x(t)$ on the periodic orbit and linearize the right-hand side of the differential equation around this point to arrive at

$$\frac{dy}{dt} = F(t)\, y \,, \tag{a}$$

where $y(t)$ denotes a small perturbation relative to $x(t)$, and $F(t)$ denotes the local matrix of partial derivatives of the right-hand side of our original differential equation for x, evaluated at the point $x(t)$. The only difference from the usual linearization around an equilibrium is that F now depends on time in a T-periodic fashion due to the movement of x around the periodic orbit, so that $F(t + T) = F(t)$. Periodic linear systems of differential equations such as (a) do not allow an explicit solution. But that does not matter as we only need numerical solutions from $t = 0$ to $t = T$ for the special initial conditions   *continued*

Box 11.1 *continued*

$$y(0) = \begin{pmatrix} 1 \\ 0 \\ \vdots \\ 0 \end{pmatrix}, \ldots, y(0) = \begin{pmatrix} 0 \\ \vdots \\ 0 \\ 1 \end{pmatrix}. \tag{b}$$

These are easily calculated. We glue the resulting columns $y(t)$ behind each other into a matrix $B(t)$. $B(t)$ indicates how the neighborhood of $x(0)$ is transformed into the neighborhood of $x(t)$ by the dynamical system acting over a time t. For the special value $t = T$

$$y(T) = B(T)y(0), \tag{c}$$

we get a map of the neighborhood of $x(0)$ onto itself. The eigenvalues m_i of $B(T)$ are the so-called Floquet multipliers of the linear system. In the present context they are also referred to as the Floquet multipliers associated with the periodic orbit. For such special periodic systems $B(T)$ necessarily has one eigenvalue $m_1 = 1$ corresponding to an eigenvector pointing from $x(0)$ in the direction of the orbit; the other Floquet multipliers m_i, $i = 2, \ldots$, are equal to the eigenvalues of the linearized Poincaré map (see, e.g., Hartman 1964). If all $|m_i| < 1$ for $i = 2, \ldots$, the periodic orbit is asymptotically stable. If at least one $|m_i| > 1$, it is unstable.

---

large amplitude solutions can be diffusively unstable; consequently, in the long run fluctuations with a large amplitude will not be observed.

11.4 Larger Spatial Domains

The possibilities for oscillations in a two-patch Lotka–Volterra system are restricted because closed orbits with large amplitude can be diffusively unstable. The coupling between the two patches can bound the oscillations of the entire predator–prey system. Often, the spatial domain will be much larger than twice the characteristic spatial scale. Are the oscillations in densities of larger systems also reduced? If so, how do results from two-patch systems relate to systems with more patches? To answer these questions, we investigate the behavior of the equivalent of Equations (11.1) with n patches:

$$\dot{N}_j = rN_j - N_j P_j ,$$

$$\dot{P}_j = N_j P_j - \mu P_j + m \sum_{i=1}^{n} c_{ij} P_i . \tag{11.6}$$

Again, it is assumed that prey is stationary, hence the change in prey density in patch j, N_j, depends only on the local prey and predator densities. The predator density in patch j, P_j, changes through reproduction after prey consumption, through predator death (which depends only on local densities), and through migration of predators. The rate of predator emigration from patch j is given by mc_{jj}. The predator density in patch j changes through immigration from patch i to j at rate mc_{ij} (note that the use of indices is contrary to the convention in the deterministic literature). The matrix $C = (c_{ij})$ depends on the spatial organization and the size of the patches. Although our analysis below holds more generally, we only consider a linear chain of n equal-sized, identical patches. In this case,

$$
C = \begin{pmatrix}
-\frac{1}{2} & \frac{1}{2} & 0 & \cdots & & & 0 \\
\frac{1}{2} & -1 & \frac{1}{2} & \ddots & & & \vdots \\
0 & \frac{1}{2} & -1 & \frac{1}{2} & \ddots & & \\
\vdots & \ddots & & & \ddots & & \vdots \\
& & \ddots & & & & 0 \\
\vdots & & & \ddots & \frac{1}{2} & -1 & \frac{1}{2} \\
0 & \cdots & & \cdots & 0 & \frac{1}{2} & -\frac{1}{2}
\end{pmatrix}.
\tag{11.7}
$$

The parameter m hence represents the maximum emigration rate from a single patch.

We now analyze the stability of spatially homogeneous solutions, that is, solutions for which the densities in all patches are equal. Because we assume that all patches have the same size and their local dynamics are identical, the migration terms cancel when no spatial differences in the densities exist, and the dynamics of Equations (11.6) again reduce to those of the nonspatial Lotka–Volterra system. As with the two-patch Lotka–Volterra model, the model has a two dimensional diagonal subspace in which closed orbits surround a neutrally stable equilibrium. Using the same method as for the two-patch model, it can be shown that solutions of Equations (11.6) with positive initial conditions converge to the diagonal: differences in densities disappear asymptotically.

To find out whether the possible range of fluctuations is reduced in the multi-patch model as in the two-patch model, the stability of the closed

---

Box 11.2 Local stability analysis in multi-patch models

In a system of connected, identical patches, in the absence of migration the local dynamics in all patches can generally be described by

$$\dot{x} = f(x) \tag{a}$$

[x is a vector containing the k different species' densities, $f(x) : \mathbb{R}^k \to \mathbb{R}^k$ is a vector-valued function]. A homogeneous solution in such a multi-patch system would mean that the densities in all patches are the same and thus net migration is zero. Let $s(t)$ denote the time-course of all densities in a patch in the homogeneous solution, $s(t)$ necessarily is a solution of Equation (a). In Appendix 11.A it is shown that such a homogeneous solution is stable if, for all i, $\psi = 0$ is an asymptotically stable equilibrium for

$$\dot{\psi} = [Df(s(t)) + \lambda_i M] \psi , \tag{b}$$

and unstable if $\psi = 0$ is an unstable equilibrium for at least one i. [Here λ_i is an eigenvalue of the matrix C, which describes how the patches are connected, M is a diagonal matrix with the species' migration rates on its diagonal, and $Df(s(t))$ is the Jacobian of f, evaluated at $s(t)$.]

This offers a generally applicable method to uncouple the local dynamics and reduce the complexity of spatial models.

---

orbits on the diagonal plane needs to be assessed. Although at first a stability analysis of a $2 \times n$ dimensional system might seem forbidding, it turns out to be simple: the problem can be reduced to n decoupled two-dimensional systems similar to Equation (11.5) (see Box 11.2). This enables us to relate the stability of the multi-patch model to that of the two-patch model by a scaling of the migration rate. The only information that is needed about the spatial structure are the eigenvalues of the connectivity matrix C. We now demonstrate the effect of the size of the spatial domain on the oscillations in the predator–prey model. The eigenvalues of the matrix C, given above, equal $\lambda_i = -1 + \cos \frac{i}{n}\pi$, where $i = 1, \ldots, n$. For a chain consisting of a single patch, there are no spatial interactions and hence no restrictions on the possible fluctuations. For a chain of two patches, there is one eigenvalue of C different from zero with value -1 (the value of λ used in the previous subsection). The maximum possible fluctuation can be read from Figure 11.4. Next consider a chain of three patches. The relevant eigenvalues are $-\frac{1}{2}$ and $-\frac{3}{2}$. To establish the size of possible fluctuations for a given m, we have to read the graph in Figure 11.4 at $\frac{1}{2}m$ and $\frac{3}{2}m$. Because closed orbits that are unstable for at least one of these two values will not be observed, only the most restrictive value matters. In

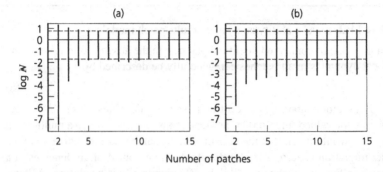

Figure 11.5 Maximum size of the neutrally stable closed orbits in a linear chain of patches versus chain length. This also represents the maximum observable oscillation after transients have died out. The dashed lines correspond to the extrema from Figure 11.4. Parameters: $r = \mu = 1$; in (a) $m = 10$, in (b) $m = 0.2$.

this way we can construct a diagram of the maximum possible fluctuations versus the length of the chain (Figure 11.5). With increasing chain length, the number of eigenvalues increases and it becomes more likely that a value lies within the range of migration rates for which the oscillations are maximally reduced in Figure 11.4. The range of predator migration rates for which the fluctuations will be reduced therefore increases with the size of the spatial domain (the number of patches). For migration rates larger than that for which the curve in Figure 11.4 attains its minimum, the amplitude of the maximum observable fluctuations converges toward the minimum for a two-patch system, as can be seen in Figure 11.5a. For smaller migration rates the observable fluctuations converge with increasing domain size toward either the minimum value possible or a larger value (Figure 11.5b). Because for a linear chain of patches the eigenvalues of C lie between 0 and -2, this value is the maximum fluctuation possible for twice the migration rate in the two-patch case.

11.5 The Spatial Rosenzweig–MacArthur Model

The solutions of the multi-patch Lotka–Volterra model become spatially homogeneous, which prevents statistical stabilization as it occurs in the discrete-entity simulations. This is not a general property of multi-patch models. The Lotka–Volterra models are useful for obtaining some detailed insights into spatial predator–prey systems, but the well-mixed counterparts to the discrete-entity simulations are not of the Lotka–Volterra type but of the more general Rosenzweig–MacArthur type. Therefore, we briefly discuss our present understanding of the spatial Rosenzweig–MacArthur model, starting with a two-patch version (Jansen 1994, 1995).

When the equilibrium of the nonspatial Rosenzweig–MacArthur model is unstable a stable limit cycle exists. The two-patch version of this model has a spatially homogeneous limit cycle that can be unstable for intermediate predator migration rates. This spatially homogeneous limit cycle is stable when the predator migration rate in a two-patch model is low. Beyond a critical value of predator migration, the limit cycle is unstable, as are the larger neutrally closed orbits of the two-patch Lotka–Volterra model. In contrast to the Lotka–Volterra model, the differences between the patches in the spatial Rosenzweig–MacArthur model persist, giving rise to qualitatively different dynamics. With predator migration slightly larger than the critical value, the dynamics are intermittent: most of the time they exhibit what appears to be a regular oscillation that is interrupted from time to time by "bursts" of irregular behavior (Bergé *et al.* 1984). In the two-patch Rosenzweig–MacArthur model the nearly regular oscillation appears when orbits dwell in the neighborhood of the unstable spatially homogeneous limit cycle. The bursts of irregular behavior occur when orbits leave the vicinity of the unstable homogeneous limit cycle in a fashion similar to that shown in Figure 11.2. For a short time the densities in the patches differ greatly and the amplitude of the oscillation is reduced. Then the differences in density decrease again and, slowly, the amplitude of the oscillation increases as the orbit reapproaches the unstable homogeneous limit cycle. For migration rates just over the critical value, this sequence of events is repeated in a chaotic fashion (type III intermittency), for larger migration rates it is repeated regularly and the dynamics are quasi-periodic (Jansen 1994).

In larger spatial domains the stability of the multi-patch Rosenzweig–MacArthur model can be calculated using the method described in Box 11.2. In the two-patch model the homogeneous limit cycle can be unstable for a relatively small range of intermediate predator migration rates. This range increases roughly with the square of the number of patches. In larger spatial domains the homogeneous limit cycle is stable only for very low or extremely high values of the predator migration rate. The homogeneous limit cycle loses its stability similarly to the two-patch model: for predator migration rates exceeding a critical value the dynamics become intermittent and a spatio-temporal pattern develops. The pattern is formed by groups of neighboring patches that oscillate almost in synchrony and are separated by regions in which the oscillations are reduced (Figure 11.6).

This behavior is comparable to the behavior of coupled map lattices (spatio-temporal intermittency, Kaneko 1993) and forms a transition between an ordered pattern and fully developed spatio-temporal chaos. The

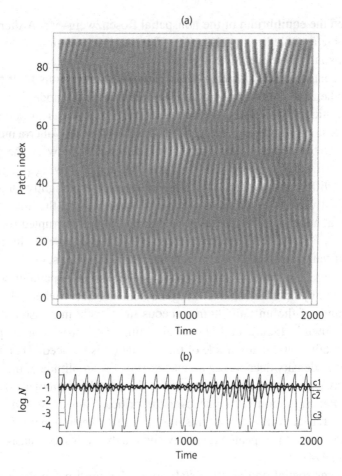

Figure 11.6 Solutions of a 90-patch Rosenzweig–MacArthur model, given by the equations $\dot{N}_j = N_j(1 - N_j/2) - bP_jN_j/(1 + N_j)$, $\dot{P}_j = bP_jN_j/(1 + N_j) - P_j + m\sum_{i=1}^{n} c_{ij}P_i$. The patches are arranged in a linear chain with reflecting boundaries; the c_{ij} are given in the text; $b = 9.96$. For these parameters the homogeneous limit cycle is unstable for $0.2736 < m < 1832$; in this simulation $m = 0.28$. (a) Prey densities in space and time. The darker the shading, the higher the prey densities. (b) Prey densities versus time, averaged over all patches (c1), in patch 40 (c2), and in the nonspatial model (c3).

phases of separated groups of synchronized patches are not related to each other. Therefore, the oscillations in the spatially averaged densities are reduced for two reasons: (1) because of the reduction in amplitude of the oscillations in the transition regions between groups of synchronized patches, and (2) because of the averaging combined with the phase differences between the synchronized groups of patches. In large spatial domains the total densities become nearly constant, again by statistical stabilization.

11.6 Concluding Comments

We now discuss the relation between the results of the discrete-entity simulations and the simple multi-patch models. We claim that in both models the occurrence of large amplitude oscillations is prevented by statistical stabilization. However, the discrete-entity and multi-patch models do differ in some important aspects. First, discrete-entity simulations account for all spatial scales, from the individual up to the system size, whereas the multi-patch models only distinguish scales larger than the local-population scale. Second, demographic stochasticity is an inherent part of models that explicitly account for individual organisms, such as the discrete-entity simulations. In contrast, all processes in the multi-patch models are fully deterministic and phrased in terms of densities. Third, in the discrete-entity simulations movement is accounted for at the individual level and is fundamental for both the interactions among individuals (e.g., predation events) and the dispersal of individuals between different parts of the system. The multi-patch models only account for the dispersal of individuals between different parts of the system as a diffusion process, while mixing at the scale of the local population, which forms the basis of individual interactions, is assumed to be rapid and homogeneous.

We have indicated how the limited mobility in discrete-entity simulations naturally leads to a subdivision of the spatial scales into two categories: those that are larger than the characteristic spatial scale and those that are smaller than it. Therefore, the spatial structure in the discrete-entity simulations is somewhat akin to the structure artificially imposed by the multi-patch models. In addition, the prey growth rates and functional responses in regions smaller than the characteristic scale are similar to those in the simulations with homogeneous movement, which in large systems should be similar to Rosenzweig–MacArthur-type differential equations.

This view is based on the models we have studied so far and certainly needs to be substantiated more rigorously. However, if it is true, only two differences remain between the discrete-entity simulations and the multi-patch models: (1) both the local dynamics and individual movement in the discrete-entity simulations are strongly influenced by demographic stochasticity, and (2) the purely diffusive movement of individuals in the discrete-entity simulations need not necessarily lead to purely diffusive movement between local populations.

The question of whether the stabilizing mechanism in the two model frameworks is the same cannot be answered unambiguously with the information currently available. In our opinion, the key issue is to resolve the role of demographic stochasticity in the discrete-entity simulations. A speculative answer would be that the mechanisms are the same in the sense that both can be considered statistical stabilization, but that the causes of the effective uncoupling of the populations over the characteristic scale are of a different nature. In the multi-patch model, the uncoupling is due to deterministic forces that cause symmetry breaking and to the chaotic nature of the attractor. In the discrete-entity simulations, it is primarily a result of the independence of the individual stochastic events. These two mechanisms are not mutually exclusive but complement and reinforce each other. For parameter values for which symmetry breaking does not occur, the effect of demographic stochasticity may still be reinforced by the deterministic trend. This makes statistical stabilization a robust phenomenon that can explain many of the differences in dynamic behavior between spatial and nonspatial predator–prey systems.

Acknowledgments We thank Howard Wilson, John Prendergast, Jason Matthiopoulos, and John Lawton for comments on the manuscript.

Appendix 11.A   Stability Analysis of a Multi-patch System

Consider a system of n patches in which k species live. The density of the jth species in patch i is described by $x_{i,j}$. The densities of all k species are given by the vector $x_i = (x_{i,1}, \ldots, x_{i,k})^T$. To keep track of the densities of all species in all patches, we describe the state of the whole system with a $k \times n$ matrix $X = (x_1, \ldots, x_n)$ that has the densities of a particular species as rows and the densities of the k species in a particular patch as columns. We assume that, from the perspectives of the interacting species, all patches are identical environments and that the local dynamics are defined by Equation (11.10). The global dynamics of the spatial system are given by a combination of local interactions and dispersal:

$$\dot{X} = F(X) + MXC , \tag{11.8}$$

where $F(X) = (f(x_1), \ldots, f(x_n))$. The matrix M is a $k \times k$ diagonal matrix that has the migration rates of the species on its diagonal. The matrix C is an $n \times n$ matrix that describes how patches are connected in the system.

We perform a local stability analysis for solutions of Equation (11.8) that are spatially homogeneous, that is, for which for all i; $x_i(t) = s(t)$, where $s(t)$ is a solution of Equation (a) in Box 11.2. Such flat solutions, which are denoted by $S(t) = (s(t), \ldots, s(t))$, exist when C has a left eigenvector $(1, \ldots, 1)$ with eigenvalue 0.

We transform $X(t) - S(t)$ into $\Psi(t)$ using the linear transformation $\Psi = (X - S)A$, where A is an $n \times n$ matrix that is invertible, hence $X - S = \Psi A^{-1}$. The time derivative of Ψ is given by

$$\dot{\Psi} = (\dot{X} - \dot{S})A$$
$$= (F(X) - F(S))A + M(X - S)CA$$
$$= (F(X) - F(S))A + M\Psi A^{-1}CA . \tag{11.9}$$

Matrix A is chosen such that $A^{-1}CA = \Lambda$ is diagonal. This can be done if all eigenvectors of C are different by choosing $A = (w_1, \ldots, w_n)$, where w_i is a right eigenvector of C – that is, $Cw_i = \lambda_i w_i$ – and by choosing $A^{-1} = (v_1, \ldots, v_n)^T$, where v_i is a left eigenvector of C – that is, $v_i C = \lambda_i v_i$. The matrix Λ has the eigenvalues λ_i of C on its diagonal. We choose $v_1 = (1, \ldots, 1)$, which is a left eigenvector of C with $\lambda_1 = 0$. Next we linearize the system around the homogeneous solution S:

$$\dot{\Psi} = (F(X) - F(S))A + M\Psi\Lambda$$
$$= Df(s)(X - S)A + M\Psi\Lambda + h.o.t.$$
$$\approx Df(s)\Psi + M\Psi\Lambda . \tag{11.10}$$

Thus A transforms Equation (11.8) in the neighborhood of the spatially homogeneous solution in a system of n decoupled subsystems, given by Equation (b) in Box 11.2. Hence, a spatially homogeneous solution of Equation (11.8) is asymptotically stable when $\psi_i = 0$ is an asymptotically stable solution of Equation (b) in Box 11.2 for all i. Note that since the subsystem for $i = 1$ is the linearization of Equation (a) in Box 11.2 around $s(t)$, a homogeneous solution S can only be asymptotically stable if $s(t)$ is an asymptotically stable solution of Equation (a) in Box 11.2.

In general there are no methods to determine the stability of Equation (b) in Box 11.2 directly from Equation (a). However, the method described here can greatly reduce the numerical effort, because it permits extrapolating results from one spatial system (the simplest being a two-patch system) to more complicated systems.

Part C

Simplifying Spatial Complexity: Examples

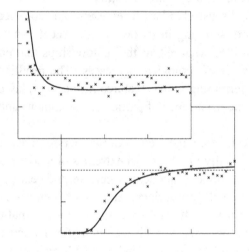

Introduction to Part C

Because individuals react only to their local environment, ecological interactions are intrinsically spatial. It is the local environment that affects light absorption, nutrient or food intake, and predation risk, thereby indirectly impinging on growth, births, deaths, and movements. Part A showed how, in real life, the local environment is influenced more by near neighbors than by neighbors at greater distances. Various examples presented in Part B showed that these local ecological interactions can have a dramatic influence on population dynamics. Clearly, mean-field approximations can tell only a small part of the ecological story.

Each example discussed in Part B represents an ecological or evolutionary problem worth studying in its own right. Yet it is natural to want to go further and ask to what extent the relationships that emerge apply to more general classes of ecological processes. The resulting research program aims at determining which features in the interplay of mechanisms are essential for the occurrence of particular phenomena and which are coincidental.

This agenda can be approached from two perspectives, an intuitive and a formal one. (Actually these are just extremes of a continuum of research strategies in which both components figure in different proportions.) Intuitive approaches seek appropriate metaphors drawn from our physical or geometrical imagination. If used unaided by more formal tools, these approaches have two drawbacks: their unsystematic character severely limits their scope in complicated settings, and the resulting insights are not always trustworthy. Yet, results based on intuitive reasoning are easy to communicate and have considerable immediate appeal because they engender a feeling of "real understanding."

Currently available formal approaches cover only a rather small subset of the ecological mechanisms and phenomena of interest. In our opinion, this should not be so much a reason for experimentalists or theorists to seek other pastures as a reason to try to extend the reach of formal methods. In ecology we are unavoidably confronted with complicated patterns of interaction, and formal tools are all we have for tackling such problems in a systematic way. Part C links the surprising range of phenomena originating from the spatially localized nature of ecological interactions exemplified in Part B to the formal approaches for analysis of spatial interactions explained in Part D. This is done by means of concrete examples showing how the tools can be brought to bear on ecological questions. Part C also illustrates additional relevant tools that are not yet well embedded in a systematic framework.

Chapter 12 shows that, from an appropriate perspective, ecological spatial complexity can be less than it appears at first sight. The chapter starts with the question of how one recognizes natural spatial scales generated by ecological interactions against the backdrop of a homogeneous physical environment. Wilson and Keeling answer this question using as examples both a static pattern generated by a genetic competition mechanism and a dynamic one generated by a predator–prey interaction, obtained by running a grid-based artificial ecology. After identifying the appropriate spatial scales, the authors show how a substantial degree of low-dimensional determinism can be found by considering spatial averages at an intermediate scale, above the scale at which the inherent demographic fluctuations dominate and below the scale at which spatial averaging reduces all spatio-temporal patterns to a homogeneous blur. The chapter's overall message is that a fair amount of formalism may be needed to extract clear signals from a spatially explicit ecological model, but that the quest for reduced descriptions of heterogeneous systems is far from hopeless.

The remaining chapters in Part C fall into one of three categories. Chapters 13 and 14 deal with small-scale patterns; Chapter 15, with intermediate-scale patterns; and Chapters 16 and 17, with large-scale patterns. Small-scale patterns are necessarily stochastic, since they involve small numbers of individuals, whereas large-scale patterns tend to be more deterministic.

Part C provides a first glance at so-called pair approximations and moment closures, recently developed formal tools for dealing with small-scale spatial heterogeneity. In grid-based models, the behavior of each individual is influenced only by a few neighboring individuals. If those neighbors are equivalent and act additively on births, deaths, and movements, mean rates for these three types of events depend only on the average number of neighbors. If we refer to individuals as singlets and to pairs of adjacent individuals as doublets, the change in the average density of singlets depends only on the average density of doublets. Unfortunately, the rate of change of the latter depends on the average density of triplets. Pair approximations and moment closures try to capture triplet densities as functions of doublet and singlet densities, for example, by assuming conditional independence: triplets behave as if they were formed from doublets by random assemblage. Equations describing the change in singlet and doublet densities are also called moment equations. In the general case, doublet densities, or second moments, depend on the distance between paired individuals. The resulting functions are also called correlation densities, and the notion of moment dynamics is used interchangeably with that of correlation dynamics. In current practice, the term pair approximation is reserved for modeling in discrete space. In principle, however, pair approximations are just special types of moment closures that derive their simplicity from restricting attention to pair correlations at nearest-neighbor distances (a restriction that is only meaningful in discrete space). Examples in Chapters 13 and 14 show how pair approximations and moment equations can successfully describe ecological change under small-scale spatial heterogeneity. A systematic treatment of pair-approximation techniques and moment closure methods can be found in the first four chapters of Part D.

Chapter 13 demonstrates the utility of pair approximations for understanding grid-based models. Iwasa presents three successful applications, ranging from forest dynamics to bacterial competition. The first example studies plants that can reproduce both vegetatively and by seed, with a linear trade-off between the two modes. Mean-field results predict that equilibrium plant density is independent of relative investment in either mode, whereas spatially explicit simulations and pair-approximation results both predict a single maximum at intermediate levels of relative investment. The second example illustrates bistability in the spatially explicit dynamics of forest gaps; the bistability is predicted by pair approximation but is absent from mean-field equations. The model was fitted quantitatively to field data from Barro Colorado Island, Panama, using pair approximation. A third example studies competition between colicin-producing and colicin-sensitive strains of *Escherichia coli*. Whereas mean-field equations predict either bistability or success of the sensitive strain, pair-approximation results largely replace bistability with success of the colicin-producing strain, in accordance with laboratory experiments.

Chapter 14 introduces moment equations for continuous space and presents applications to two spatial Lotka–Volterra models as examples. As in Chapter 13, the quantitative match between results obtained from spatial simulations and those from a moment closure based on a conditional independence hypothesis turns out to be surprisingly good. In the first model, Law and Dieckmann pit two species against each other, one being the better competitor and the other the better disperser. Mean-field results predict survival of the better competitor, whereas moment equations correctly forecast development of a spatial structure that causes the poorer disperser to suffer more from intraspecific competition, resulting in a reversal of the competitive outcome. The second example investigates how spatial scales of dispersal and competition affect equilibrium mean densities in a single species with local logistic density regulation. Mean-field results can only be trusted when both scales are large. For smaller, yet similar, scales, clumping causes reduced equilibrium densities. When competition neighborhoods are small relative to the scale of dispersal, more regular distributions of individuals result, allowing for increased equilibrium densities.

Chapter 15 investigates the evolution of transmission rates in a grid-based host–parasite system. In the mean-field approximation, evolution maximizes the basic reproduction ratio of parasites, which is proportional to the transmissibility and lifetime of the parasites. The fact that the basic reproduction ratio should be larger than 1 sets a lower bound to viable transmissibilities. In spatial systems accounting for the discreteness of individuals, an upper bound also exists: parasites that are too virulent quickly kill all locally reachable hosts and therefore die out. In addition to these ecological considerations, Keeling discusses evolutionary implications. In a system with spatial structure, adaptation of transmission rates stops near the upper critical value, whereas in mean-field models rates continue to increase (unless a trade-off between transmission and survival is imposed). This chapter is featured in Part C because it illustrates how to simplify complexity at intermediate spatial scales: based on a deterministic caricature that describes the frequency dynamics of spatial aggregates that differ in their numbers of hosts and parasites, the essential behavior of the full spatial model is well recovered. Although, at present, the reduction technique employed is specific to the host–parasite model investigated here, it appears to hold wider promise. One additional message of this chapter is that evolution may proceed considerably slower in a spatial setting than under mean-field conditions.

Chapters 16 and 17 analyze the dynamics of invasion waves, a class of large-scale spatial phenomena that are particularly well understood. Invasion waves are especially relevant for evolutionary considerations, as adaptive innovations often sweep through spatial populations in a wavelike fashion.

Chapter 16 investigates the expansion of epidemics in agricultural crops, ranging from the microscale, where foci with diameters of a few meters develop from single infections, to the continental scale, where an epidemic rages over a continent in one or a few growing seasons. The economic importance of epidemics has resulted in detailed quantitative experimentation and modeling. Zadoks has participated in this research from its early stages and from this perspective describes its history and results. Initial approaches used deterministic numerical simulations. Because basic reproduction ratios of fungal pathogens are very high in agricultural epidemics, their life-history characteristics cannot be neglected; consequently, reaction–diffusion models are inadequate. Early simulation models were based on delay-differential equations for pathogen density coupled with integral equations for the spatial redistribution of spores. A second generation of models taking an analytical approach were phrased in terms of integral equations in space and time; the corresponding mathematical framework is summarized in Chapter 23, including simple recipes for extracting spatial expansion rates. When supplied with real life-history data, these integral-equation models provide surprisingly accurate predictions of the speed of focus expansion. Third-generation models take better account of spatial inhomogeneities and the stochasticity inherent in long-range spore dispersal; they again rely mainly on numerical simulations.

Chapter 17 provides a link between spatial game theory, described in Chapter 8, and reaction–diffusion models, treated in Chapter 22. Ferrière and Michod show how reaction–diffusion equations can be constructed for evolutionary games and review the mathematics available for predicting outcomes of competitive spatial processes. In such settings, strategies can overcome disadvantages of rarity by forming clumps. Supported by such "base camps," rare strategists can start to conquer space in a wavelike fashion. The chapter investigates invasion waves for the competition between *Tit For Tat* (*TFT*) and *Always Defect* (*AD*), two strategies in the iterated Prisoner's Dilemma. In particular, the effects of mobility rates and types of memory (determining how past actions of other players are remembered) on competitive outcomes are discussed. In a spatially inhomogeneous setting, the mobility rates strongly influence the rate at which fresh meetings occur. For the *TFT* strategy to spread, *TFT* players need to encounter a sufficient number of their own kind in the front of an invasion wave and should not be suckered too often by *AD* players newly moving into *TFT* clumps.

By showcasing "methods at work" to simplify spatial complexity, Part C should provide a gentle transition to the systematic treatments of techniques offered in Part D.

12

Spatial Scales and Low-dimensional Deterministic Dynamics

Howard B. Wilson and Matthew J. Keeling

12.1 Introduction

Biological populations are composed of individuals whose movements are usually limited in space. Consequently, even in a uniform habitat, ecological systems will (and do) show heterogeneity and patchiness at a broad range of temporal and spatial scales. In recent years, ecologists have become increasingly aware of how important an influence spatial heterogeneity has on the dynamics of populations (e.g., Fahrig and Paloheimo 1988; Gilpin and Hanski 1991; McCauley *et al.* 1993) and persistence of species (e.g., Hastings and Wolin 1989; Reeve 1990; Taylor 1990; Hassell *et al.* 1991a; Comins *et al.* 1994; Kareiva and Wennergren 1995). For this reason, there is a strong need to understand and quantify the dynamics and patterns in spatially extended systems.

In this chapter, a number of methodologies or tools for such analyses are introduced. We begin by addressing the question of how to choose the appropriate scale for monitoring a spatial system. Results obtained from both models and field studies are influenced by the scale at which observations and measurements are made (e.g., Sugihara *et al.* 1990). For many ecological systems, sampling on too small a scale produces data that are noisy and dominated by stochastic effects. Alternatively, sampling on too large a scale averages out the interesting effects. This is because at large scales the dynamics in distant parts of the area sampled are effectively independent. By measuring the temporal variance of the population density as a function of the area sampled, we show that there exists a different scaling behavior above and below the scale at which different parts of the system begin to act independently. This procedure provides us with a clearly defined intermediate scale at which to observe dynamics that maximize the ratio of "deterministic" information to stochastic fluctuations. This scale is an inherent characteristic of the ecosystem and the appropriate scale of measurement. It is related to scales proposed by a number of authors (e.g., Carlile *et al.* 1989; de Roos *et al.* 1991; Levin 1992). This technique also indicates the aggregation (or, more precisely, the expected deviation from the mean) at every scale. Using the intermediate scale identified above, we show for a particular example that at this scale the dynamics are essentially low-dimensional and deterministic. We do so using classical techniques for the approximate reconstruction of the state space of relatively low-dimensional, deterministic dynamical systems. This combination of techniques provides a characterization of the features of the complex spatial patterns under consideration in terms of a low-dimensional vector. This vector provides a natural approximate parameterization of the local states of the ecosystem, allowing for very effective data compression. These techniques also facilitate the development of new approaches to analyzing ecosystem data for monitoring and management purposes, for example, detecting long-term structural change or drift in stochastic systems with complex dynamics.

In the next section we introduce two spatial models by which we exemplify our methodology. In Section 12.3, the spatial scale methodology is derived and applied to these models. Finally, in Section 12.4, the tools for reconstructing deterministic state spaces are introduced and applied.

12.2 Two Models from Evolutionary Ecology

In this section, two types of model are introduced: a deterministic coupled map lattice and an individual-based stochastic model. These models display a variety of dynamics ranging from a fixed spatial configuration to complex chaotic behavior.

Gene competition

The model considered in this section is situated on a large two-dimensional square lattice of sites. At each site the local dynamics are described by a map having the states of the site and its four nearest neighbors as its arguments [models of this type are customarily called coupled map lattices; see Kaneko (1992)]. The boundary conditions are toroidal, or wraparound, so that opposite edges of the lattice are joined. This map represents competition between two genotypes, labeled A and a. Only p_α, the proportion of loci in the population at site α that are of type A, is monitored. Each site on the lattice is assumed to support a very large constant population size, so that the local change in gene frequency can be represented by a deterministic map.

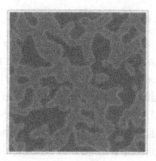

Figure 12.1 An example of a final state spatial pattern of the genetic model. The lattice size is 200×200 and $W = 0.1$ so that homozygotes are only weakly favored. This produces large patches with smooth boundaries. Dark gray indicates sites with all A; medium gray, sites with all a; and light gray, sites with a mixture of a and A.

The dynamics progress as follows: in each iteration, individuals within a site and the local neighborhood mate randomly, with all pairs equally fertile, to produce the offspring at the focal site. Individuals survive according to their fitness. The fitness of a homozygote (AA or aa) is larger than the fitness of a heterozygote (Aa). Let this difference in fitnesses be W, so that

$$W = \frac{\text{Fitness of homozygote} - \text{Fitness of heterozygote}}{\text{Fitness of heterozygote}}. \tag{12.1}$$

The dynamics of p_α are then given by the map

$$p'_\alpha = \frac{W p_\alpha \bar{p}_\alpha + \frac{1}{2} p_\alpha + \frac{1}{2} \bar{p}_\alpha}{2 W p_\alpha \bar{p}_\alpha - W p_\alpha - W \bar{p}_\alpha + W + 1}, \tag{12.2}$$

where $\bar{p}_\alpha$ is the average value of p at α and the four neighboring sites. Essentially, this represents the number of type A alleles multiplied by the fitness of type A divided by the total fitness. If the homozygote genotypes are fitter ($W > 0$), standard theory tells us that either one form (AA) or the other (aa) will dominate. In the spatial model, however, uniform patches of genotypes develop whose scale is related to W. When W is large, small patches tend to develop; when W is small, large patches develop.

From an initial condition where the value at each site is a uniformly distributed random number between zero and one, spatial structure slowly emerges, and after around 200 iterations (generations) a fixed stable pattern has developed (see Figure 12.1). The underlying mechanisms of this process can be compared to a variety of problems where the states 0 and 1 are attracting and there is limited coupling to the neighborhood.

Predator–prey–resource dynamics

In this model, physical space is again represented by a two-dimensional $\ell \times \ell$ lattice of sites. Each site can be in any one of a number of discrete states. The neighborhood of a site (defined below) determines a probability distribution on a finite set of local events. Combined in an appropriate manner, these probabilities determine a probability distribution for the ecosystem state at time $t + 1$, given the state at time t.

Each site can be in one of five states: resource, predator, predator plus resource, prey, or empty. The behavior of the two entities in the predator plus resource site is assumed to be independent. This assumption allows us to specify the event probabilities in terms of events affecting each entity separately. The boundary conditions are periodic, as if the lattice were situated on a torus.

It is simplest to explain our model by focusing on how the individuals behave. This completely defines the event set and the event probabilities for a given neighborhood state. All sites are updated synchronously. The order of site updating is random, and once a site changes from one state to another it is ignored by any further updating in the time step. In this description, sites adjacent to α refer to the four nearest sites $\mathcal{N}(\alpha) = \{\alpha + \beta : |\beta_1| + |\beta_2| \leq 1, \beta \in \mathbb{Z}^2, \beta \neq (0, 0)\}$. The rules are as follows:

- A resource, if there are empty sites in its neighborhood, produces with probability R_b one offspring, which is deposited into one of the (randomly chosen) empty neighboring sites.
- A prey, if there are resource sites in its neighborhood, moves into a (randomly chosen) neighboring resource site and depletes the resource. Thus the adjacent resource site becomes a prey and the original site becomes empty. If no resources are available in adjacent sites, a prey moves randomly into one of the adjacent sites. There is a certain probability P_b of a prey giving birth into a randomly chosen adjacent empty site. This probability is zero if a prey has not eaten for a certain amount of time t_{P_b}. A prey dies if it has not eaten for a certain amount of time t_{P_d} (where $t_{P_d} > t_{P_b}$).
- An empty site does nothing.
- A predator's neighborhood is larger than the neighborhoods of the prey and resource species. Its immediate neighborhood comprises the eight nearest sites: $\mathcal{N}(\alpha) = \{\alpha + \beta : |\beta_1| \leq 1, |\beta_2| \leq 1, \beta \in \mathbb{Z}^2, \beta \neq (0, 0)\}$. If there are prey sites in its neighborhood, a predator moves into and depletes the prey from one of the (randomly chosen) neighboring prey sites. As outlined for the prey, there is a certain probability Q_b of the

Table 12.1 Parameter values for the predator–prey–resource model. All times are in time steps of the model.

R_b	P_b	t_{P_b}	t_{P_d}	Q_b	t_{Q_b}	t_{Q_d}
0.4	0.2	2	8	0.1	5	8

predator giving birth into a neighboring empty site. This probability is zero if a predator has not eaten for a certain amount of time t_{Q_b}. A predator dies if it has not eaten for a certain amount of time, t_{Q_d} ($t_{Q_d} > t_{Q_b}$). If there is no neighboring prey, a predator has the ability to hunt, in that it can sense prey much farther away than it can move in one time step. A predator can "see" prey anywhere in a 48-site neighborhood, that is, the neighborhood is $\mathcal{N}(\alpha) = \{\alpha + \beta : |\beta_1| \leq 3, |\beta_2| \leq 3, \beta \in \mathbb{Z}^2, \beta \neq (0,0)\}$. A predator is restricted in a manner similar to that of its prey species in that it can only move one site in one time step. A predator thus moves to the site in its eight-site neighborhood that is nearest (in the Euclidean sense) to the located prey. If two or more sites are equally near, then one of the nearest sites is chosen at random.

The model is meant to simulate a predator–prey–resource community to probe biological and mathematical mechanisms that might emerge from a spatially extended system. It is not meant to be biologically realistic. For instance, all the species are moving and reproducing on similar spatial and time scales, which may not always be true. All the simulations in this chapter have the parameter values given in Table 12.1.

12.3 Identifying Spatial Scales

We investigate the effects of scale in our models simply by increasing or decreasing the square window of size L through which we observe the dynamics (where L is the total number of sites in the window and $L \leq \ell \times \ell$). When the boundary conditions are cyclical, all sites in the model are dynamically equivalent. In this case, the window can then be positioned anywhere on the lattice. For other boundary conditions the position of the observation window may make a difference. A time series x_t is generated for each window size by recording the total population size within the window at each time step. The genetic model is a little different, however, because a fixed stable pattern develops. In this case, a "time series" is generated by fixing the window in time and moving it in space until the full spatial pattern has been sampled.

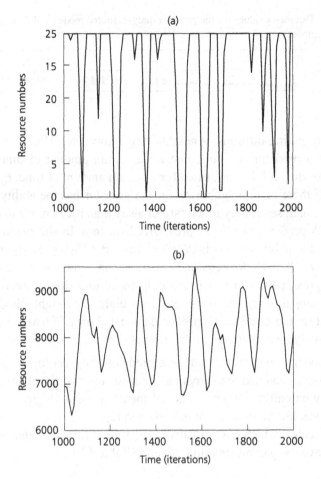

Figure 12.2 Two time series of the number of resource sites in (a) a 5 × 5 window and (b) a 100 × 100 window. The lattice size is 150 × 150.

Randomness dominates model behavior at the scale of single sites, either due to the stochastic nature of the processes (in the predator–prey–resource model) or due to the random initial conditions (in the genetic model). When the window is small, observations are made on a similar scale and the behavior of the time series is thus dominated by stochastic fluctuations. At intermediate scales, the dynamics are dominated by deterministic behavior (Figure 12.2 shows typical time series for the predator–prey–resource model from windows of different sizes). On large lattices, distant parts of the lattice begin to act independently or become uncoupled (this assumption may hold for many biological systems; a further discussion of this issue can be found in Chapter 11). Hence, when the window size L is large, there is a decoupling of the phases of the oscillations. As L is increased further an averaging effect occurs, and the Central Limit theorem predicts that the size of these fluctuations in average density decreases at a rate proportional to $L^{-1/2}$. The dynamics of the spatial averages of an infinite system will therefore tend to be perceived as fully static. This fact can be used to identify a spatial scale L_s above which distant parts of the lattice act independently.

Consider a time series x_t of length T from a window of size L, where $\sqrt{L}$ is not small compared with any spatial correlation length and T is large compared with any correlation time. This time series can be thought of as the summation of time series from a conglomeration of k independent patches from within the window of size L. Each of these independent patches is of size L_s and there are $k = (L/L_s)$ patches, each with variance V_s. A scale L_s is sought at which all the patches of size L_s have essentially identical statistics and the correlation between two patches i and j is exponentially small in the separation between them.

The variance of the sum of two variables is equal to the sum of the two variances plus the correlation between the two. In this case, there are k patches, all of size L_s and variance V_s. Thus

$$\text{Var}(L) = \sum_{i=1}^{k} V_s + \sum_{\substack{i,j=1 \\ i \neq j}}^{k} \text{Cov}_{i,j} \ . \tag{12.3}$$

When two variables are independent, the correlation between them is zero. When the observation window is larger than the scale L_s, the patches of size L_s become decoupled and essentially independent (so that the correlation between two time series from patches i and j is zero). In fact, there is a very small correlation c between the patches (for a more detailed exposition, see Rand and Wilson 1995) because the length T of the time series is finite and the dynamics in each patch are on the same attractor (when this attractor is chaotic then $c \to 0$ as $T \to \infty$). Hence

$$\text{Var}(L) = \frac{L}{L_s} V_s + \left(\frac{L}{L_s}\right)^2 c$$

$$\Rightarrow \quad \frac{\text{Var}(L)}{L} = \frac{V_s}{L_s} + \frac{L}{L_s^2} c \ . \tag{12.4}$$

The important point here is that when $L \gg L_s$, the variance scales quite differently with the window size L than when $L < L_s$. The scaling of $\text{Var}(L)/L = f_L$ above L_s only comes from the correlations due to patches

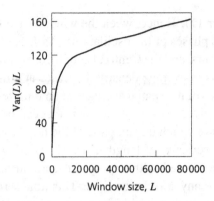

Figure 12.3 The variance of the resource time series as viewed through a window of size L. The lattice size is 300×300.

being on the same attractor. This scaling is linear in L with gradient c/L_s^2, which is very small. For sufficiently large observation times ($c \rightarrow 0$), $\mathrm{Var}(L)/L$ should become approximately constant for L above L_s.

In Figure 12.3 we plot $f_L = \mathrm{Var}(L)/L$ against L for the predator–prey–resource model. This plot, called a *fluctuation diagram*, shows a clear change in the scaling behavior at approximately $L = 10\,000$; this is the scale L_s.

Recall that for the genetic model a fixed stable pattern develops. In this case, the window is moved around the lattice until the full pattern has been sampled. $\mathrm{Var}(L)$ is then the variance in the total number of loci of type A in a window of size L, where the variance is calculated over the whole pattern at one point in time rather than for one fixed point in space at various times, as in the predator–prey–resource example. Alternatively, one can fix the window in space but calculate the variance over many different simulations, each starting with a different random initial condition. The result is seen in Figure 12.4. Above length scale L_s (≈ 1400), f_L is approximately constant, as predicted by our theory. In addition, however, there is a maximum in f_L when $L \approx 700$. The shape of the curve f_L is informative about the aggregation present at various scales.

Consider a lattice where each site is an independent random variable. The fluctuation diagram is then a constant value equal to the variance in the random variable. If the curve f_L increases with L, the behavior for larger windows lies farther from the mean than we would expect from the behavior of the smaller, supposedly (but not actually) independent windows. Hence, an increasing f_L curve reflects greater aggregation. Similarly, if the curve

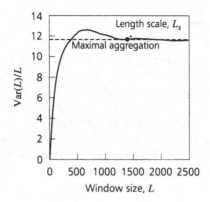

Figure 12.4 The fluctuation diagram for the genetic model. The length scale (L_s) of approximately 1400 sites is clearly identifiable, as $\mathrm{Var}(L)/L$ is approximately constant above this scale.

decreases, the behavior for larger windows lies closer to the mean than expected. This indicates the existence of a restoring mechanism.

Interpreting the fluctuation diagram in terms of aggregation and disaggregation can be made clearer by studying a simple example. Consider a lattice filled with zeros and ones in a checkerboard pattern, with each square consisting of $n \times n$ sites. Recall that $f_L = \mathrm{Var}(L)/L$ and that $\mathrm{Var}(L)$ is the variance in the sum of all the sites in the window averaged through time. In this case, it is assumed that either the window moves around the checkerboard or the checkerboard pattern moves, so that the window samples the whole pattern. Any small window will usually be filled with a single value, so that the sum within a window of size L will be either L or 0. The mean value will be $L/2$, the variance $\mathrm{Var}(L) \approx L^2/4$, and $f_L \approx L/4$, so that each new larger window indicates that the system is more aggregated than would have been expected from smaller windows. However, there comes a point $L = n^2$ above which every window of size L is guaranteed to contain a mixture of zeros and ones and hence f_L begins to decrease. The decrease continues until $L = 4n^2$, at which point each window at any time contains equal numbers of ones and zeros, so that $\mathrm{Var}(L) = 0$ and $f_L = 0$. Therefore, the maximum value of f_L is at the scale of the checkerboard square – the most aggregated spatial scale.

Returning to the results from the genetic model in Figure 12.4, the maximum at $L \approx 700$ informs us that this is the average size of a uniform genetic patch.

Box 12.1 Embedding

Takens (1981) and Packard *et al.* (1980) have demonstrated that it is possible to reconstruct a multidimensional state space from a scalar time series. This reconstructed space is topologically equivalent to the original and thus preserves important geometrical invariants such as the dimension of the attractor and the Lyapunov exponents of a trajectory (for reviews see, e.g., Eckmann and Ruelle 1985; Casdagli *et al.* 1991; Sauer *et al.* 1991). Although there are many choices of possible coordinate systems – for example, successive time derivatives – the easiest and most widely used are delay coordinates, $x_t = (x_t, \dots, x_{t-(E-1)\tau})$, together with the next state map:

$$x_t = F(x_{t-\tau}, \dots, x_{t-(E-1)\tau}, x_{t-E\tau}, x_{t-\tau}, \dots, x_{t-E\tau}) \,,$$

where E is an integer known as the *embedding dimension* and τ is the so-called *delay time*. The basic idea behind this methodology is that the past (and future) states of the system hold information about the present, unobserved, state variables.

Unfortunately, Takens' theorem does not tell us what the "best" choices are for E and τ, and often the "best" choice will depend on the particular analysis being performed. However, Takens has shown that choosing E greater than or equal to $2K + 1$, where K is the dimension of the original attractor (which of course may not be known *a priori*), should lead to a faithful representation of the structure of the attractor. In principle, the chosen embedding dimension should be large (Grassberger *et al.* 1991), but practical considerations (e.g., limited amounts of time series data) mean that smaller embedding dimensions usually give better results. The delay time, τ, should not be so small that successive points are highly correlated, otherwise successive points will not contain significantly more information than the previous point. However, the time between points should not be too large, otherwise important information about the local attractor structure will be missed. One should choose a value as an appropriate balance between these two considerations. Alternatively, the choice for τ is often motivated by the underlying biology of the system being studied or is set by the way the data have been collected. For example, it may be appropriate to set τ to be one generation of the organism being studied, or the data may be collected seasonally or annually, which defines a minimum value for τ.

In practice, one should use a range of values for both E and τ and check the robustness of the results (e.g., see Sugihara *et al.* 1990; Sole and Bascompte 1995).

12.4 Dynamics, Determinism, and Dimensionality

Locally, the predator–prey–resource model is stochastic, since the rules governing the transition from one state to another are probabilistic. We wish to show that, this stochasticity notwithstanding, the overall population dynamics can be well approximated by deterministic dynamics. By this we mean that there exists some time-independent functional relationship, F, between the present, x_t, and the past, $x_{t-1}, x_{t-2}, \ldots,$

$$x_t = F(x_{t-\tau}, \ldots, x_{t-(E-1)\tau}, x_{t-E\tau}) + \xi_t , \tag{12.5}$$

where E is an integer known as the *embedding dimension*, τ is the so-called *delay time* (see Box 12.1), again an integer, and ξ_t is a stochastic component whose magnitude is small so that $\xi_t \ll |x_t - x_{t-1}|$. This last condition is important because when it holds, the dynamics are dominated by the deterministic component. In this case practical techniques exist to detect the determinism. No attempt is made to determine the function F, only to demonstrate that such a function must exist.

By definition, determinism means that for a given initial state the system will evolve in a fixed and predictable manner. If the system returns to this state, it will evolve in the same manner as before. To test this in the predator–prey–resource model, time-delayed vectors, X_t, are formed from the time series (see Box 12.1). Choosing some state at time t, X_t, we find a previous state, $X_{t-t'}$, very close (in the Euclidean sense) to our original state, that is, $|X_t - X_{t-t'}|$ is small. In a deterministic system, the evolution of these two states is very similar. In the case of the predator–prey–resource model, the deterministic part of the population dynamics is found to be chaotic. This means that nearby states diverge from each other exponentially fast at a rate determined by the largest Lyapunov exponent (see Box 12.2). The dashed curve in Figure 12.5 shows the typical divergence of two nearby states; the exponential divergence is clear. This divergence is interpreted as further evidence of determinism, as a stochastic system is unlikely to produce such a pattern (stochastic systems may diverge at a linear rate but usually not at an exponential rate). The average of the logarithm of the divergence calculated over all vectors in the time series is shown as the continuous curve in Figure 12.5. This is a straight line, indicating exponential divergence (the best-fit straight line has a correlation coefficient close to 1). Exponential divergence is only expected at small scales in the phase space, below the scale at which nonlinearities are important (see Box 12.2). This system is highly nonlinear and so this scale, although it will vary at

Box 12.2 Lyapunov exponents and chaos

If the initial state of a system is slightly perturbed, the exponential rate at which the perturbation increases (or decreases) with time is called a characteristic exponent. This exponential rate is calculated by a linearization of the system. The Lyapunov exponents of an orbit in an attractor are found by linearizing the system around that orbit; that is,

$$\lambda_i = \lim_{t \to \infty} \frac{1}{t} \ln(|i\text{th eigenvalue of } A(t, t_0)|) \, ,$$

where $A(t, t_0)$ is the linear system that maps points at time t_0 to points at time t. For a discrete-time nonlinear dynamical system, $A(t, t_0)$ must be found using the product of the Jacobian matrices (the matrices of partial derivates at points on the orbit) along the orbit.

The Lyapunov exponents therefore reflect the average rate at which perturbations increase or decrease around the orbit. There are as many Lyapunov exponents as there dimensions of the state space, and each exponent corresponds to whether the perturbation will increase or decrease in a particular direction (for a review, see Eckmann and Ruelle 1985). If the largest (dominant) Lyapunov exponent is positive, then nearby orbits will tend to diverge exponentially fast. This property of chaotic orbits is called "sensitive dependence on initial conditions," since systems starting in practically indistinguishable initial conditions will soon behave very differently – predictability is rapidly lost. This local exponential expansion is clearly incompatible with motion restricted to an attractor, unless a global folding process brings widely separated trajectories close again and again. This folding process is due to the nonlinearities in the system.

For a system with a stable equilibrium, the largest Lyapunov exponent is negative, indicating that perturbations near the equilibrium decrease in time. A limit cycle in a continuous time dynamical system has one Lyapunov exponent equal to zero, corresponding to motion along the periodic orbit; all other Lyapunov exponents are negative, reflecting convergence onto the attractor. Chaotic attractors in continuous time have at least one positive exponent, indicating divergence of nearby trajectories; one exponent equal to zero, corresponding to motion along an orbit; and at least one negative exponent, corresponding to trajectories converging onto the attractor.

There are a number of methods for calculating Lyapunov exponents (e.g., Wolf *et al.* 1985; Sano and Sawada 1985; Eckmann *et al.* 1986; Nychka *et al.* 1992). In this chapter a method originally proposed by Wolf *et al.* (1985) is used. This method involves taking two nearby orbits in the phase space and calculating their rate of convergence or divergence. Given an embedded time series (see Box 12.1), we choose an initial point X_0 and

continued

Box 12.2 *continued*

locate the nearest point Y_0 (in the Euclidean sense) in the remainder of the
time series and denote the distance between the two points by $d(0)$. At a
later time T, the distance between the successors X_1 and Y_1 of these points
is $d'(1)$. T should be small enough that only the small-scale attractor structure is examined (the trajectories should not have gone through a folding
region, which would lead to an underestimation of the exponent). A new
data point Y_1' is then chosen so that its separation from X_1 is small, and so
that it approximately preserves the direction in embedded space between
X_1 and Y_1. The new distance between the points is $d(1)$. This procedure
is repeated throughout the time series to obtain the dominant Lyapunov
exponent, λ:

$$\lambda = \frac{1}{MT} \sum_{k=1}^{M} \ln \frac{d'(k)}{d(k-1)} \,,$$

where M is the total number of replacement steps. (Although natural logarithms have been used here, it should be noted that, historically, Lyapunov
exponents have usually been expressed as logarithms to base 2.)

---

different points on the attractor, is relatively small. Hence, the figure only
shows the divergence for a relatively short time.

After allowing initial transients to die away, the dynamics from the
predator–prey–resource model settle onto an attractor. The spatial structure of the states in the attractor is complex, nonhomogeneous, and time
dependent. Resources are patchily distributed, with prey associated with
edges of patches and predators associated with prey. We wish to address
the question of the attractor's dimensionality. If the dimensionality of the
attractor is small, in principle it may be possible to achieve a large amount
of data compression by using this low-dimensional parameterization. It also
provides further evidence for determinism, because a stochastic system has
infinite dimensions as the noise tends to fill out the embedded state space.
Thus, low dimensionality is evidence of a nearly deterministic system.

At any one time, an embedded vector X_t exists that represents the state
of the system at that time and is a point on the attractor. Given these embedded vectors, we wish to calculate the dimension of the linear space that they
trace. To investigate this dimension, we make use of a technique introduced
into dynamical systems theory by Broomhead and King (1986), based on
ideas from singular systems theory (Bertero and Pike 1982). The main ingredient of this method is reducing the dimensionality of the embedded

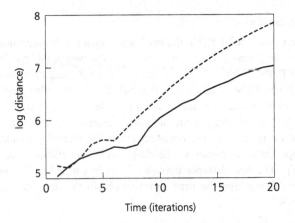

Figure 12.5 The divergence over time of two initially close delay vectors of the resource time series. The dashed curve shows a typical sample. The continuous curve gives the logarithm of the average divergence over time. Both graphs are for an embedding dimension $E = 4$ and a time delay $\tau = 5$. The delay vectors were extracted from a time series of length 10 000 for a simulation run on a 100×100 lattice. The continuous curve was calculated by taking the first vector in the time series, locating the nearest neighbor (in the Euclidean sense), and recording their distance apart at various later times. The next vector in the time series is then identified, its nearest neighbor located, and their distance apart at later times recorded. This process is repeated for all the vectors in the time series. Once the distances at later times between a large number of initially close vectors have been recorded, the logarithm of the average distances apart are calculated and plotted.

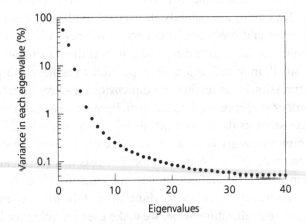

Figure 12.6 The normalized eigenspectrum of the resource time series. The lattice size is 150×150, the window size used in the singular value decomposition procedure (see Appendix 12.A) is 40, the number of points is 10 000, and the time delay, τ, is 1.

Box 12.3 Principal components analysis

Principal components analysis, or PCA, is the search for a new, lower-dimensional coordinate system to describe a set of variables (see Kendall 1980; Gauch 1982). The new coordinate axes have their origin at the center of gravity (mean) of the cloud of points, are orthogonal (independent), and are chosen so as to maximize the variance of the variables projected onto each of the axes (equivalent to minimizing the sum of the squared projection distances from the axis). The first axis has the greatest projected variance, the next axis has the next largest variance, and so on. The first axes therefore explain the greatest amount of the information in the variables and are thus termed the principal components. If projected variances onto the later axes are negligible, then the state of the system can be described by a lower-dimensional coordinate system than before. Thus PCA is a technique for projecting a multidimensional cloud of points onto a space of fewer dimensions that maximizes the variance accounted for and preserves as much of the original structure as possible.

The PCA technique starts with n vectors, with each vector representing m measurements. We have, therefore, n points in an m-dimensional space, which can be represented as follows: x_{ij}, $i = 1, \ldots, m$, and $j = 1, \ldots, n$. The points are first centered at the origin by subtracting the mean of each measurement, so that $\sum_{j=1}^{n} x_{ij} = 0$. The covariance matrix is then formed by multiplying the matrix whose elements are x_{ij} by its transpose, so that the elements of the new matrix are $y_{ij} = \frac{1}{n} \sum_{k=1}^{n} x_{ik} x_{jk}$. Eigenanalysis is then performed on this matrix. The eigenvectors are the new coordinate axes, and the associated eigenvalues are the variances accounted for by each axis. If some of the eigenvalues are negligible, then a smaller number of dimensions is needed to describe the data set.

PCA is available in many statistical packages. However, to use PCA in conjunction with the other analysis techniques introduced in this chapter, it is usually best to first center the delay vectors by subtracting their overall mean and then apply a singular value decomposition (see Appendix 12.A).

---

state space by performing a principal components analysis to the set of observed delay vectors. (The dynamical systems literature speaks here of singular value decomposition; the interrelations between the two concepts are explained in Box 12.3 and in Appendix 12.A.) The result is a decomposition of the embedded vectors into orthogonal components, such that the first component carries the largest possible variance, the second component carries the largest possible remaining variance, etc. These component variances are known separately as eigenvalues and collectively as an

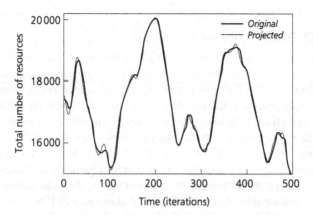

Figure 12.7 The resource time series projected onto the four eigenvectors with the highest eigenvalues. The parameters in the singular value decomposition procedure are as in Figure 12.6. The original time series is shown for comparison.

eigenspectrum due to the technique by which they are calculated. Moreover, the projection of the embedded vectors on their first k components, whatever the value of k, is optimal in the sense that it induces the minimal possible mean square error. If the attractor really were to trace a low-dimensional linear space and if our data were perfect, the dimension of this linear space would be equal to the number of nonzero eigenvalues.

In the presence of noise, however, all eigenvalues will be nonzero since noise tends to fill out the state space. By inspecting the eigenvalue spectrum, the eigenvalues associated with a deterministic component can be distinguished from those associated solely with noise. Those associated with noise tend to show up as a constant noise floor. Figure 12.6 is a typical eigenspectrum of a resource time series. We see that 94% of the variance of the time series is captured in the first four eigenvalues. This means that virtually all the dynamics in the original signal occur in a four-dimensional embedded space. Such a result is a strong indication of near-determinism. It is also surprising considering the high dimensionality of the original spatial system.

To reconstruct the reduced-noise version of the time series, the observed time series x_t can be projected onto the first four eigenvectors (see Figure 12.7 and Appendix 12.A for details). The fact that the time series can be reconstructed so accurately indicates how much of the original information is captured within the first four eigenvectors. Projecting the time series in this manner effectively compresses the information from the underlying complex spatio-temporal process into a time series of low-dimensional vectors. This information can then be used for analyzing ecosystems. It is important to note that these compressed data only encode the statistical structure of the patterns, not the actual patterns. It is not possible, therefore, to reconstruct the actual spatial patterns from the compressed data. However, in many applications only the statistical structure is of importance and the extra information is redundant. The technique is thus a method for removing this redundant information to focus only on the diagnostic statistical features.

12.5 Concluding Comments

A number of tools for analyzing spatial systems have been discussed. In the first part of this chapter a *fluctuation diagram* that identifies the spatial scale at which parts of the system begin to act independently was introduced. This diagram also indicates scales of aggregation. This methodology has been applied to two model systems for which it appears to work extremely well. Elsewhere, this technique has been applied to many more systems with equal success (Keeling *et al.* 1997b). The technique presupposes that the system under consideration has some special properties. These are discussed in heuristic terms in Section 12.3. They are probably commonly satisfied but difficult to prove for a particular system. Examples of such proofs, in each case based on a special feature of the system under consideration, can be found in Keeling *et al.* (1997b).

In the second part of the chapter we demonstrated how complex, stochastic spatial models can be characterized by low-dimensional vectors. This technique allows one to synthesize complex spatio-temporal patterns into a relatively small amount of essential information that can then be used for analysis. This procedure has been shown elsewhere to be very effective in detecting small amounts of change in an ecosystem (Rand and Wilson 1995). The next step is to develop additional tools based on such low-dimensional parameterization.

Appendix 12.A   Singular Value Decomposition

The methodology for this approach is outlined in Broomhead and King (1986) and has been extended or discussed by Broomhead *et al.* (1987, 1988), Mees *et al.* (1987), Mees and Rapp (1988), and Albano *et al.* (1988). It has also been briefly summarized by Grassberger *et al.* (1991) and Sauer *et al.* (1991) and has been applied to a number of problems (e.g., Fraedrich 1986; Vautard and Ghil 1989). The approach has also been used to study local attractor structure (e.g., Hediger *et al.* 1990; Broomhead *et al.* 1991).

If X_t is a centered (see Box 12.3) embedded time series (where τ and E are suitably chosen), we let Y be the matrix whose ith row is the vector X_i for $i = 0, 1, \ldots, N$ and let the trajectory matrix $Z = Y/\sqrt{N}$. If all the data points are used, then $N = N_T - (E-1)$, where N_T is the number of data points in our original time series. The singular value decomposition of this matrix is given by $Z = SHC^T$, where S and C^T contain the left and right singular vectors and the entries of the diagonal matrix H are the singular values $\eta > \eta > \cdots > \eta$. The singular vectors form an orthogonal basis set that spans the embedding space. Singular value decomposition is used because it is a robust methodology for solving linear least-squares problems and there are a number of readily available algorithms for decomposing a matrix and determining S and C.

Recall the matrix of embedded vectors and its decomposition, $Z = SHC^T$. We can decompose the matrix of singular vectors C into two subspaces:

$$C = C^D + C^\varepsilon, \quad \text{where } C^D = [C_1, \ldots, C_D, 0, \ldots, 0],$$
$$\text{and} \quad C^\varepsilon = [0, \ldots, 0, C_{D+1}, \ldots, C_E]. \tag{12.6}$$

Then

$$ZC(= SH) = ZC^D + ZC^\varepsilon$$
$$\Rightarrow \quad Z = ZC^D C^T + ZC^\varepsilon C^T. \tag{12.7}$$

Thus D is the number of nonzero eigenvalues associated with the deterministic component of the signal and $Z^D = ZC^D C^T$ is the projection onto the subspace associated with the deterministic dynamics. Each row, U_i, of Z^D is a D-dimensional vector that gives us a characterization of the state of the system at time $t = i$. In Figure 12.7 we show the reconstructed signal (the first column of Z^D) on top of the original and note their similarity.

If one wants to maintain the special role of the origin of the coordinate system, the same technique may be applied without first centering the embedded vectors. The result is a least-squares projection of the time series on a linear space that has its origin at the same place as the original data instead of at their center of gravity.

13

Lattice Models and
Pair Approximation in Ecology

Yoh Iwasa

13.1 Introduction

Ecological interactions such as predation, resource competition, parasitism, epidemic transmission, and reproduction often occur at spatial scales much smaller than that of the whole population. For organisms with limited mobility, it is extremely important to explicitly consider the spatial pattern of individuals when predicting population dynamics.

Lattice or cellular automaton models are useful for modeling spatially structured population dynamics. These models are most suitable for population dynamics of terrestrial plants (Chapters 2 to 4, 6), including models for wave regeneration of subalpine forest (fir waves: Iwasa *et al.* 1991; Satō and Iwasa 1993), forest gap dynamics (Smith and Urban 1988; Green 1989; Nakashizuka 1991; Kawano and Iwasa 1993), and population dynamics of perennials capable of vegetative propagation (Crawley and May 1987; Silvertown 1992; Harada and Iwasa 1994; see also Chapters 3 and 4). Lattice models have also been developed for other systems, such as marine invertebrate communities (Caswell and Etter 1992; Etter and Caswell 1994), predator–prey dynamics (Tainaka 1988; de Roos *et al.* 1991; Wilson *et al.* 1993), host–pathogen dynamics (Ohtsuki and Keyes 1986; Satō *et al.* 1994), and host–parasitoid dynamics (Hassell *et al.* 1991a, 1994), as well as for spatially explicit evolutionary games (e.g., Herz 1994; Boerlijst and Hogeweg 1995a; Grim 1995; see also Chapter 8).

Most analyses of lattice models have used computer simulations of spatial stochastic processes. The results of these computer simulations can be compared with those of *mean-field approximations*, the traditional population dynamic models which neglect spatial structure and assume perfect mixing of populations.

Matsuda *et al.* (1992) developed the *pair-approximation* method for lattice population dynamics, a system of ordinary differential equations that yields average densities and local densities, with the latter describing the correlation of states of nearest-neighbor cells. Pair approximations are capable of predicting the behavior of lattice models even when mean-field approximations fail (Harada and Iwasa 1994; Satō *et al.* 1994; Harada *et al.* 1995; see also Chapters 18 and 19).

In this chapter, the use of pair-approximation analysis is illustrated by means of several examples from ecology; these examples have the common property that the behavior of the lattice model is qualitatively different from that of the corresponding mean-field dynamics.

13.2   Plants Reproducing by Seed and Clonal Growth

The population dynamics of terrestrial plants have often been analyzed using matrix models, which classify individuals according to age or size and trace over time changes in the number of individuals in each class (Silvertown 1987; Kawano *et al.* 1987; Caswell 1989). These analyses neglect the spatial pattern of individuals and do not distinguish a daughter individual growing far from the mother from one growing next to it. This simplification has been fairly successful and is certainly a practical method for tracing dynamics of field populations, provided that plants propagate by producing seeds with a relatively long dispersal range.

However, many terrestrial plants reproduce vegetatively by producing runners, rhizomes, or root sprouts. Vegetative propagation is a characteristic of many perennials (Cook 1983), especially those in harsh environments such as dark forest floors, tundra, and alpine areas (Raven *et al.* 1981; Callaghan 1988). Because of the resulting short distance between offspring and parent, neglecting spatial structure is much more problematic when considering plants with vegetative propagation than when considering those that reproduce predominantly by seed.

Spatial clumping is a spontaneous result of spatially limited reproduction and local ecological interaction. Yet, it reduces the opportunity for vegetative propagation because the plants lack vacant space to reproduce into, creating an advantage for seeds with a long dispersal range. A high local density (or crowding) causes strong competition or cooperation between individuals. Hence we need to follow the dynamics of abundance and of spatial pattern simultaneously (Harada and Iwasa 1994, 1996).

Model description

I consider a lattice-structured habitat consisting of infinitely many sites. Each lattice site has two neighborhoods and is either occupied by an individual $(+)$ or vacant (0). Three kinds of transition in the state of lattice sites are allowed (Harada and Iwasa 1994):

- Death of individuals occurs at rate d.
- Offspring produced by vegetative propagation are placed into sites adjacent to the parent. The maximum birth rate, denoted by b, is realized when all the neighbor sites of the parental site are empty. The birth rate is lower if some of the neighbor sites are already occupied: the birth rate of an occupied parental site increases by b/z for each adjacent site.
- Seeds are dispersed over the entire lattice but can germinate and grow only if they land on a vacant site. Here we assume that an individual with n_+ neighboring plants has a seed production rate $m = s + \beta n_+/z$, where s is an intrinsic rate of seed production per plant, n_+ is the number of adjacent occupied sites, and β measures the strength of neighbor effects. To calculate the rate of transition from an empty site to an occupied site by sexual reproduction, we multiply this by ρ_+, the density of occupied sites in the entire lattice, and by ρ_0, the fraction of empty sites. A positive parameter β implies that an individual increases its seed production in the presence of neighbors; a negative β implies a decrease. For example, neighbors may enhance the seed production rate by protecting the plant from wind, attracting pollinators, or forcing the plant to redirect resources from vegetative reproduction to seed production. Alternatively, neighbors may decrease seed production by shading the plant, depleting the soil of nutrients, or attracting herbivores and pathogens.

I assume that each transition event occurs instantaneously at the rate given above. The model is a continuous-time Markov chain.

Figure 13.1a gives an example of spatial pattern at equilibrium generated by computer simulation of the model for a plant that propagates vegetatively. Occupied sites are indicated by black squares. Although initially the pattern of occupied sites is random, it converges to a clumped distribution due to limited dispersal of vegetative propagules. Occupied sites within a tight cluster may have no vacant sites in their neighborhood, which are necessary for vegetative propagation. In contrast, Figure 13.1b shows a population generated by the model for plants that propagate only by seed. In this case, the spatial pattern is random and there are vacant sites in the vicinity of most occupied sites. The difference in spatial pattern in turn affects the population dynamics.

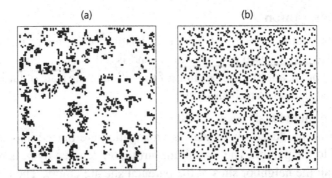

Figure 13.1 Spatial patterns generated by vegetative propagation and propagation by seed. Both types of propagation start from the same random initial condition and both reach equilibrium. (a) Plants reproducing only vegetatively via runners, rhizomes, roots, etc., that invade neighboring vacant sites. (b) Plants reproducing only by seeds dispersed over long distances. Only those seeds that happen to land on vacant sites can germinate and occupy the site. Parameters: (a) $d = 0.005$, $b = 0.0086$, $s = 0$, $\beta = 0$, $z = 4$; (b) same as in (a) except $b = 0$ and $s = 0.006$. These parameters were chosen so that the global density is the same in both figures ($\hat{\rho}_+ = 0.15$). *Source*: Harada and Iwasa (1994).

Dynamics of overall density and mean crowding

From these transition rules we can construct the dynamics of the overall population density. We call the probability that a randomly chosen site is occupied by an individual ρ_+ the *global density* of state $+$ (see Box 13.1). The probability that a randomly chosen site is empty is $\rho_0 = 1 - \rho_+$. The dynamics of the global density ρ_+ are written as follows:

$$\frac{d\rho_+}{dt} = -d\rho_+ + b(1 - q_{+/+})\rho_+ + (s + \beta q_{+/+})\rho_+(1 - \rho_+) . \qquad (13.1)$$

The first term on the right-hand side of Equation (13.1) represents the death of individuals. The second term corresponds to birth by vegetative propagation. Because vegetative reproduction is possible only in a neighboring vacant site, we must consider the states of the nearest neighbors of occupied sites. Let $q_{+/+}$ be the conditional probability that a randomly chosen nearest neighbor of an occupied site is occupied. The quantity $q_{+/+}$, called *local density*, thus represents the average degree of crowding in the neighborhood of individuals. Factor $1 - q_{+/+}$ in Equation (13.1) is the average fraction of vacant sites in the neighborhood of occupied sites; multiplying this factor by the maximum reproduction rate b gives the per capita rate of vegetative propagation. The third term on the right-hand side represents birth by seeds dispersed over the entire area. The seed production of an

---

Box 13.1 Pair-approximation variables

Global densities	Local densities
∎ Singlet ρ_i	
∎ Doublet ρ_{ij}	∎ Conditional doublet $q_{i/j}$
∎ Triplet ρ_{ijk}	∎ Conditional triplet $q_{i/jk}$

---

occupied site depends on the number of occupied sites among its nearest neighbors, and n_+/z in the seed production rate is replaced by $q_{+/+}$.

If $q_{+/+}$ is close to 1, the individuals form tight clusters: most occupied sites are likely to be surrounded by occupied sites. This does not mean that the average total density ρ_+ is high. Local density is about three times higher than global density in Figure 13.1a ($q_{+/+} = 0.44$, and $\rho_+ = 0.15$), but both densities are the same in Figure 13.1b ($q_{+/+} = \rho_+ = 0.15$).

The local density $q_{+/+}$ is closely related to statistics for spatial clumping based on quadrat sampling. In particular, Harada and Iwasa (1994) have demonstrated that the local density $q_{+/+}$ is equivalent to the mean crowding index m^* (Lloyd 1967; Iwao 1968) if each sampling quadrat includes only a nearest-neighbor pair. Mean crowding, defined as the average density experienced by individuals, has been developed as a useful method for statistical analysis of spatial distribution (Iwao and Kuno 1971; Taylor 1984). As the ratio of mean crowding to average density (denoted by m^*/m) is called the patchiness index (Iwao 1968), the ratio $q_{+/+}/\rho_+$ can be regarded as measuring the patchiness of a spatial pattern.

Equation (13.1) implies that, in order to determine the dynamics of the plant population, we need to know the local density – that is, we need to have some information about the spatial pattern. There are two ways to derive approximations for the above model: mean-field approximation and pair approximation.

Mean-field approximation. A common method for simplifying spatial population dynamics is to neglect spatial structure altogether by ignoring the correlation between neighboring sites on the lattice. This is done simply by assuming that local density is the same as global density: $q_{+/+} = \rho_+$. This mean-field approximation becomes exact if the spatial pattern is random (as in Figure 13.1b). However, if the plants reproduce vegetatively, spatial patterns obtained from computer simulations are clumped and local density can be much higher than global density (as in Figure 13.1a).

Pair approximation. Pair approximation is a method for constructing a system of ordinary differential equations for global and local densities and dealing with them as separate state variables that change over time. A detailed introduction to the pair-approximation technique is given in Chapter 18; for the purposes here, I briefly outline the most important general results. Let ρ_{++} be the probability that sites of a randomly chosen nearest-neighbor pair (a doublet) are both occupied. Because $q_{+/+} = \rho_{++}/\rho_+$, the dynamics of local density are given by

$$\frac{dq_{+/+}}{dt} = -\frac{\rho_{++}}{\rho_+^2}\frac{d\rho_+}{dt} + \frac{1}{\rho_+}\frac{d\rho_{++}}{dt}, \qquad (13.2)$$

and we can derive the dynamics of local density from those of doublet density ρ_{++}. The dynamics are given by the following equation:

$$\begin{aligned}
\frac{dq_{+/+}}{dt} = &- q_{+/+}\left[-d + b(1 - q_{+/+}) + (s + \beta q_{+/+})(1 - \rho_+)\right] \\
&- 2dq_{+/+} \\
&+ 2\{\left[z^{-1} + (1 - z^{-1})q_{+/0+}\right] + (s + \beta q_{+/+})\rho_+\}(1 - q_{+/+})
\end{aligned} \qquad (13.3)$$

(see Harada and Iwasa 1994; see also Chapter 18), where $q_{+/0+}$ is a conditional probability that a randomly chosen neighbor of the empty site of a 0+ pair other than the occupied site of the pair is an occupied site. This probability concerns the state of three connected sites +0+ (a triplet). To obtain $q_{+/0+}$, we must calculate the dynamics of triplet probabilities such as ρ_{+0+}, but, unfortunately, these dynamics will contain conditional probabilities of still higher orders. Because the two occupied sites in the +0+ triplet of ρ_{+0+} are not each other's nearest neighbors, their direct correlation might be weak. If so, we can construct a closed dynamical system of global and local densities by neglecting correlation beyond nearest neighbors; it is this assumption that gives rise to the pair approximation (Matsuda *et al.* 1992). As the name indicates, this is just an approximation and sometimes does not hold (see Iwasa *et al.* 1998), but it is more accurate than mean-field approximation. We thus neglect the occupied site attached to the empty site and replace $q_{+/0+}$ with $q_{+/0}$ (Harada and Iwasa 1994):

$$q_{+/0+} \approx q_{+/0} = \frac{(1 - q_{+/+})\rho_+}{1 - \rho_+}. \qquad (13.4)$$

Using this approximation, Equations (13.1) and (13.3) provide a closed dynamical system of two variables, ρ_+ and $q_{+/+}$. We can now apply standard methods for analyzing this pair of nonlinear differential equations.

The equilibrium of the dynamical system given by Equations (13.1) and (13.3) can be calculated by setting $d\rho_+/dt = 0$ and $dq_{+/+}/dt = 0$. A trivial solution given by $\hat{\rho}_+ = 0$ (extinction) always exists. In addition, there may be one or two positive equilibria, satisfying $0 < \hat{\rho}_+ < 1$ and $0 \le \hat{q}_{+/+} \le 1$. Analytical results for the equilibrium (a long and messy expression) and for its local stability can be obtained. From this result it can be shown that a positive feasible equilibrium is either a stable node or an unstable saddle, but never a focus.

There are three parameter regions for the results obtained by both mean-field and pair approximation: (1) global persistence, in which there is one positive stable equilibrium; (2) extinction, in which there are no positive stable equilibria; and (3) bistability, in which there are two positive equilibria, one stable and the other unstable. If the system is bistable, equilibrium values for global and local densities are determined by their initial conditions (for details, see Harada and Iwasa 1994).

Trade-offs between modes of reproduction

Parent plants must allocate limited reproductive resources to vegetative propagation and to seed production (constructing and maintaining flowers and fruits, rewarding pollinators and seed dispersers). We now consider a trade-off between vegetative reproduction and seed production, $b + \alpha s = K$, in which α is the cost of seed production relative to vegetative propagation and K is a constant proportional to the plant's total reproductive investment.

Figure 13.2 illustrates the equilibrium global density $\hat{\rho}_+$ for different fractions of resources invested in vegetative reproduction as part of total reproductive investment. The left end of the horizontal axis represents pure seed production (no vegetative propagation); the right end represents pure vegetative reproduction (no seed production). The dashed line is the equilibrium global density $\hat{\rho}_+$ predicted by mean-field approximation; the solid curve is $\hat{\rho}_+$ as predicted by pair approximation. Filled circles show results from computer simulations on a 100×100 square lattice with periodic boundaries. According to computer simulations and pair approximation, there exists an optimal fraction of vegetative reproduction that maximizes the equilibrium population density. The mean-field approximation fails to reveal such an intermediate optimum, however. An intermediate maximum occurs only for spatially structured populations and can be successfully explained by the dynamics based on pair approximation.

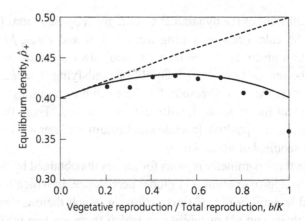

Figure 13.2 Trade-offs between vegetative reproduction and seed production. The horizontal axis indicates the fraction of total reproduction investment allocated to vegetative reproduction. The dashed line is the equilibrium density $\hat{\rho}_+$ predicted by mean-field approximation; the solid curve shows the global density $\hat{\rho}_+$ predicted by pair approximation. Filled circles are the results from direct computer simulation. Parameters: $d = 0.005$, $K = 0.01, \alpha = 1.4$. *Source*: Harada and Iwasa (1994).

Size distribution of clusters

The total number of individuals in Figures 13.1a and 13.1b is almost the same, but the spatial pattern is more clumped in the former. The population of Figure 13.1a includes fewer and larger clusters than that of Figure 13.1b. A cluster of adjacent occupied sites is regarded as a single large cluster of plants, and the number of individual sites included in a cluster is called the size of the cluster (see Figure 13.3 for an illustration of cluster size).

Let k be the cluster size and n_k be the number of clusters of size k within a large area consisting of L lattice sites. Because the total number of individual sites included in a cluster of size k is kn_k, the probability that a randomly chosen site is a member of a cluster of size k is kn_k/L.

The expected frequency distribution of cluster sizes can be calculated approximately by using nearest-neighbor correlation, that is, by using global and local densities as derived above.

First, consider the probability that a randomly chosen site is an isolated site – that is, a cluster of size 1. This case was studied for a lattice model of cell sorting in developmental biology (Mochizuki *et al.* 1996). If we consider only nearest-neighbor correlation, this is simply $\rho_+ (1 - q_{+/+})^4$, where the site is occupied with probability ρ_+ and all four of its neighbors are empty with the conditional probability $(1 - q_{+/+})^4$. This calculation is in the spirit of pair approximation because we neglect higher-order correlations. Similarly, the probability that a site is included in a cluster of size 2 is

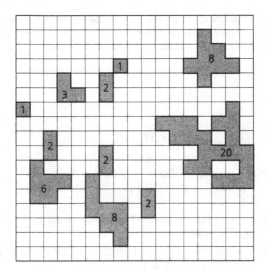

Figure 13.3 Definition of clusters of occupied sites. Eleven clusters of different sizes are shown, based on a regular two-dimensional square lattice and a von Neumann neighborhood. An aggregate of individual sites that are connected to each other is called a cluster; the cluster size is given by the number of sites included. Notice that, for example, the aggregates labeled 2 and 6 on the left do not belong to the same cluster.

$4\rho_+q_{+/+}(1-q_{+/+})^6$, where the factor 4 comes from the number of configurations and the power 6 indicates the number of neighboring empty sites surrounding a ++ pair.

Similar considerations apply to larger clusters (Kubo *et al.* 1996). For this, we need to take into account various configurations having the same number of occupied sites. Consider an individual site that belongs to a cluster of size k. Let c_k be the total number of possible configurations of clusters with size k, and let (i, k) be the ith configuration among c_k possibilities. Let $P(i, k)$ be the number of empty sites on the perimeter of the configuration (i, k). Then, based on pair approximation – that is, neglecting higher-order correlations beyond nearest-neighbor pairs – the probability that a single site is included in a cluster of size k is

$$\frac{kn_k}{L} = \rho_+q_{+/+}^{k-1}\sum_{i=1}^{c_k}\left(1-q_{+/+}\right)^{P(i,k)} . \tag{13.5}$$

In physics, the case in which the sites are independent ($\rho_+ = q_{+/+}$) is called a two-dimensional "site problem" in percolation models and has been studied extensively. Polynomials $f_k(x)$, defined as $f_k(x) = \frac{1}{k}\sum_{i=1}^{c_k} x^{P(i,k)}$, are called "perimeter polynomials" and are shown in Table 13.1 (Stauffer 1985; Sykes and Glen 1983).

Table 13.1 Perimeter polynomials for the site percolation process on a two-dimensional regular square lattice. *Source*: Sykes and Glen (1983).

k	$f_k(x)$
1	x^4
2	$2x^6$
3	$4x^7 + 2x^8$
4	$9x^8 + 8x^9 + 2x^{10}$
5	$x^8 + 20x^9 + 28x^{10} + 12x^{11} + 2x^{12}$
6	$4x^9 + 54x^{10} + 80x^{11} + 60x^{12} + 16x^{13} + 2x^{14}$
7	$22x^{10} + 136x^{11} + 252x^{12} + 228x^{13} + 100x^{14} + 20x^{15} + 2x^{16}$
8	$4x^{10} + 80x^{11} + 388x^{12} + 777x^{13} + 818x^{14} + 480x^{15} + 152x^{16}$ $+ 24x^{17} + 2x^{18}$
9	$28x^{11} + 291x^{12} + 1152x^{13} + 2444x^{14} + 2804x^{15} + 2089x^{16} + 856x^{17}$ $+ 216x^{18} + 28x^{19} + 2x^{20}$
10	$4x^{11} + 154x^{12} + 986x^{13} + 3676x^{14} + 7612x^{15} + 9750x^{16} + 8192x^{17}$ $+ 4330x^{18} + 1416x^{19} + 292x^{20} + 32x^{21} + 2x^{22}$

Figure 13.4 illustrates the cluster size distribution as derived from computer simulations (filled circles) and as predicted by Equation (13.5) with $\hat{\rho}_+$ and $\hat{q}_{+/+}$ calculated using pair approximation (solid line). It also shows the predictions by Equation (13.5) using only $\hat{\rho}_+ = \hat{q}_{+/+}$ calculated by a mean-field approximation (neglecting nearest-neighbor correlation, i.e., replacing $q_{+/+}$ with ρ_+; dashed line). If the population reproduces mostly by seed, the predictions of all three methods are the same and are quite accurate because the spatial pattern then is random ($\hat{\rho}_+ = \hat{q}_{+/+}$). In contrast, if the population reproduces mostly by vegetative propagules (as in Figure 13.4), the calculation based on mean-field approximation systematically overestimates the number of small clusters, but calculations based on pair approximation accurately predict the patterns. Pair approximation is quite accurate except when the plants reproduce only by vegetative propagules or when the population density is very low, in which case the correlation between sites over a fairly long distance becomes important (see Harada and Iwasa 1996; Kubo *et al.* 1996).

13.3 Forest Gaps

Over the past 20 years, many ecologists working on forest dynamics have focused their attention on "gaps," that is, openings encountered in forest canopies (Whitmore 1975; Yamamoto 1992). Gaps range in size from small openings created by the death of a single branch to large-scale blowdown caused by catastrophic disturbances such as storms; gaps may also

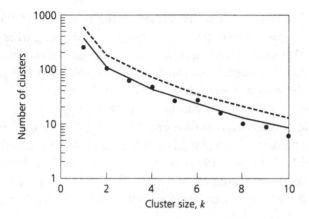

Figure 13.4 Cluster size distribution. Filled circles are from computer simulation of the model. The logarithmic vertical axis shows the number of clusters of a given size in a two-dimensional lattice of size $L = 100 \times 100$ with periodic boundaries; the horizontal axis indicates the cluster size k. The dashed line is the size distribution of clusters expected for a random spatial pattern with the same ρ_+ as the mean-field model. The solid line is the cluster size distribution predicted by Equation (13.5) using pair approximation. Parameters: $b = 1.78571 \times 10^{-3}$; $d = 5 \times 10^{-3}$; $s = 7.14286 \times 10^{-4}$. *Source*: Harada and Iwasa (1996).

be created by fires or aggregated insect outbreaks. The size, number, and spatial distribution of gaps, and the processes of gap formation and canopy recovery have been studied extensively (e.g., Runkle 1984; Nakashizuka 1987, 1991; Masaki *et al.* 1992; Yamamoto 1992, 1993; Kanzaki *et al.* 1994).

Studies of forest gap dynamics have shown that new gaps are more likely to occur adjacent to pre-existing gaps, for example, in subalpine spruce-fir forest (Lawton and Putz 1988), in deciduous old growth forest (Runkle 1984), and in tropical forest (Hubbell and Foster 1986; Kanzaki *et al.* 1994).

In this section, I use data from a tropical seasonal forest (Kubo *et al.* 1996) to model gap dynamics with gap expansion. Again, pair-approximation analysis is very accurate, whereas mean-field approximation is not.

A transition model for canopy gaps

Consider a habitat consisting of infinitely many sites arranged on a regular lattice. Each lattice site corresponds to an area of forest stand approximately 5 m × 5 m in size, which can be occupied by a large canopy tree (for an illustration see Figure 13.6).

Gap sites are created when canopy trees fall due to disturbance. It takes a number of years until young trees from seedlings or seeds grow to fill the gap or until the canopy becomes closed owing to growth of surrounding trees. During this recovery period, the height of the young trees is clearly lower than that of surrounding mature canopy trees and the site is recognized as a gap. In general, one can classify a site as being in a gap phase, a building phase, or a mature phase (Whitmore 1975; Kanzaki *et al.* 1994; see also Figure 4.5). Here, I consider only two categories: gaps and non-gaps. A site is considered to be a gap if the height of its trees is below a certain threshold, say 20 m (Runkle 1984; Hubbell and Foster 1986), 10 m (Runkle 1981; Yamamoto 1993), or 3 m (Lawton and Putz 1988), depending on the height of mature canopy trees. Transition from a newly formed gap to a non-gap may occur through canopy closure rather than growth by new recruiting individuals.

I denote a gap by the symbol 0 and a non-gap by the symbol + and assume that each site experiences transitions between the two states over time. Assume that the rate of transition from a gap to a non-gap site occurs at rate b per year. The inverse, $1/b$, is the average length of time that sites remain in the gap phase (i.e., the recovery time). The transition rate from a non-gap to a gap site increases with the number of neighbors that are gap sites. Let n_0 be the number of 0 sites among the nearest neighbors, and let us assume that an individual with n_0 gap neighbors has a death rate $d + \delta n_0 / z$. Here, δ measures the strength of neighbor effects in the gap formation process. The rate of transition from a non-gap site to a gap site is highest when all surrounding nearest neighbors are gap sites; in this case, the rate is equal to $d + \delta$.

Dynamics of gap areas

Consider the total gap area, or the fraction of gap sites in forests at equilibrium. The total average fraction of gap sites ρ_0, or global density of gaps, can be calculated by mean-field and pair approximation, as explained in Section 13.2. Here, however, the number of gap sites (0) are traced instead of the canopy sites occupied by tall trees (+).

By neglecting the correlation between neighbors – that is, by assuming that the local density is the same as the global density, or that $q_{0/0} = q_{0/+} = \rho_0$ – we obtain the mean-field approximation for the dynamics of the total fraction of gap sites:

$$\frac{d\rho_0}{dt} = (d + \delta\rho_0)(1 - \rho_0) - b\rho_0 . \tag{13.6}$$

The first term on the right-hand side of this equation is the gap formation rate; the second term is the gap recovery rate. To be exact, the per capita rate of gap formation is $d + \delta q_{0/+}$, which under mean-field conditions is approximated by $d + \delta \rho_0$. Equation (13.6) leads to the equilibrium fraction of gaps:

$$\hat{\rho}_0^M = \frac{-(b+d-\delta) + \sqrt{(b+d-\delta)^2 + 4d\delta}}{2\delta}, \qquad (13.7)$$

where the superscript M indicates mean-field approximation. In contrast, by distinguishing the local and global densities of gaps, the dynamics based on pair approximation can be derived (Kubo *et al.* 1996):

$$\frac{d\rho_0}{dt} = d - \left[b + d - \delta\left(1 - q_{0/0}\right)\right]\rho_0 , \qquad (13.8a)$$

$$\begin{aligned}
\frac{dq_{0/0}}{dt} = &- q_{0/0}\left\{\frac{d}{\rho_0} - \left[b + d - \delta(1 - q_{0/0})\right]\right\} - 2bq_{0/0} \\
&+ 2\left(1 - q_{0/0}\right)\left\{d + \delta\left[\frac{1}{z} + \frac{z-1}{z}\left(1 - q_{0/0}\right)\frac{\rho_0}{1 - \rho_0}\right]\right\} .
\end{aligned} \qquad (13.8b)$$

These equations constitute a closed dynamical system of two variables, ρ_0 and $q_{0/0}$. The equilibrium of Equations (13.8) is

$$\hat{q}_{0/0}^P = 1 - b\frac{\left(b+d+\delta\frac{z+1}{z}\right) - \sqrt{\left(b+d+\delta\frac{z+1}{z}\right)^2 - 4\delta\left(b+\frac{d}{z}+\frac{\delta}{z}\right)}}{2\delta\left(b+\frac{d}{z}+\frac{\delta}{z}\right)} , \qquad (13.9a)$$

$$\hat{\rho}_0^P = \frac{d}{b + d - \delta(1 - q_{0/0}^P)} , \qquad (13.9b)$$

where the superscript P indicates pair approximation.

In the limit when z tends to infinity, $\hat{q}_{0/0}^P$ and $\hat{\rho}_0^P$ in Equations (13.9) converge to $\hat{\rho}_0^M$ given by Equation (13.7). This implies that, when the spatial range of ecological interactions is large, the dynamics become similar to those under complete mixing. If $\delta > 0$, the local gap density is larger than the global gap density and the density predicted by mean-field approximation is between the two ($\hat{\rho}_0^P < \hat{\rho}_0^M < \hat{q}_{0/0}^P$; see Kubo *et al.* 1996).

Figure 13.5a illustrates the results of computer simulations for a two-dimensional lattice; the equilibrium global density ρ_0 (open circles) and local density $q_{0/0}$ (open squares) are shown for different neighbor-dependent mortalities δ. The local density of gaps is higher than the global density, indicating a clumped spatial distribution of gap sites.

The curves given by pair-approximation dynamics [Equations (13.8)] are quite accurate, fit well with the results of computer simulations, and correctly predict that $\hat{q}_{0/0}$ is larger than $\hat{\rho}_0$. Because clumping of gaps makes their spread less effective, by neglecting spatial structure the mean-field approximation $\hat{\rho}_0^M$ overestimates the equilibrium abundance of gaps and predicts neither the difference between local and global density nor the bistability that can occur for an intermediate interval of δ values; see below and Figure 13.5b.

Simulations in a one-dimensional system on a 100×100 lattice with periodic boundaries were qualitatively equivalent to those in the two-dimensional system discussed above, but the discrepancies between different methods of approximation were more pronounced (Kubo $et\ al.$ 1996).

Supply-dependent recruitment and bistability

So far, I have assumed that seeds or seedlings needed for recruitment are always available and that the rate of transition from newly formed gaps to non-gaps is independent of the number of parent trees that supply seeds. In reality, recruitment is likely to limit the recovery process. Specifically, suppose that the recovery rate of a gap site is proportional to the fraction of non-gap sites in the whole forest, so that b is replaced by $\alpha\rho_+$, where α is a positive constant. For some parameter values, the dynamics of pair approximation are then bistable, having two simultaneously stable equilibria: one positive equilibrium with a high density of trees and one trivial equilibrium with $\rho_0 = 1$, implying the absence of trees. Figure 13.5b illustrates the equilibrium global density ρ_0 and local density $q_{0/0}$ of a two-dimensional system for different neighbor-dependent mortalities δ. Filled circles are the results of a computer simulation, which are very close to the prediction by pair-approximation dynamics. In contrast, mean-field dynamics failed to predict the bistability (Kubo $et\ al.$ 1996).

Application to data from a tropical forest

Kubo $et\ al.$ (1996) applied the model of supply-dependent recruitment to the data from Hubbell and Foster (1986) for a neotropical forest on Barro Colorado Island, Panama. They chose $5\,m \times 5\,m$ for the size of each lattice site; Figure 13.6 gives the spatial pattern of these sites on the lattice. Sites are divided into three categories: (1) sites with trees below 20 m in height in the 1983 census (gray); (2) sites with trees at or above 20 m in height

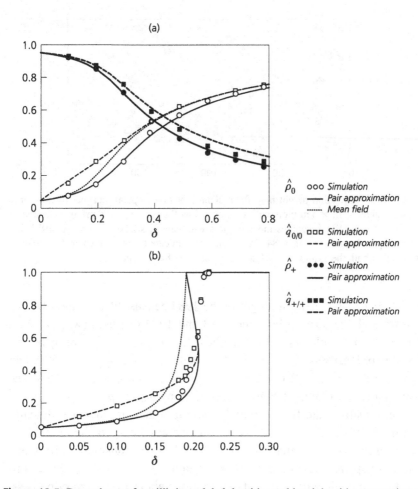

Figure 13.5 Dependence of equilibrium global densities and local densities on maximum gap expansion rate δ. (a) Gap recovery rate is constant. Circles and squares represent the results of computer simulations: open circles for the global density of gaps, $\hat{\rho}_0$; open squares for the local density of gaps, $\hat{q}_{0/0}$; filled circles for global density of non-gaps, $\hat{\rho}_+$; and filled squares for the local density of non-gaps, $\hat{q}_{+/+}$. These results are compared with those from pair approximation: thin continuous curves for $\hat{\rho}_0$, thin dashed curves for $\hat{q}_{0/0}$, thick continuous curves for $\hat{\rho}_+$, and thick dashed curves for $\hat{q}_{+/+}$. The mean-field approximation does not distinguish between global and local densities, and $\hat{\rho}_0^M$ is shown as a dotted curve. Parameters: $b = 0.20$; $d = 0.01$. (b) Gap recovery rate is proportional to global density of non-gap sites. Open circles and squares, and thin continuous and dashed curves show results for $\hat{\rho}_0$ and $\hat{q}_{0/0}$ as obtained, respectively, from computer simulations and pair approximation. Notice the hysteresis effect predicted by pair approximation: for a small interval of values for δ, two equilibria coexist. Results from mean-field approximation are shown as a dotted curve. Parameters: $\alpha = 0.20$; $d = 0.01$. For both figures, the system is simulated on a two-dimensional square grid of size 100×100 with $z = 4$. *Source*: Kubo *et al.* (1996).

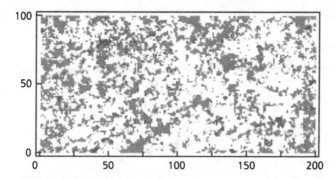

Figure 13.6 Vegetation height map for a 50 ha plot in a tropical seasonal forest. Gray squares indicate 5 m × 5 m plots with canopy below the threshold (20 m) as measured in the 1983 census. Black squares indicate plots with canopies 20 m or above in 1983, but with canopies below 20 m in 1984. White squares indicate canopies greater than 20 m as measured in both the 1983 and 1984 censuses. *Source*: Kubo *et al.* (1996).

in 1983 but below 20 m in height in the 1984 census (black); and (3) sites with trees above 20 m in both the 1983 and 1984 censuses (white). With 20 m as the threshold vegetation height separating gaps from non-gaps, the global gap density is $\rho_0 = 0.331$, local gap density is $q_{0/0} = 0.580$, and $q_{+/+} = 0.801$, $q_{0/+} = 0.199$, and $q_{+/0} = 0.420$ at the census in 1983. The spatial pattern is clumped, which is reflected by $q_{0/0} > \rho_0$.

The transition rate from a non-gap site to a gap site increases linearly with the number of gap sites that surrounded any non-gap site in 1983, as shown in Figure 13.7. Because the spatial data for transitions from a gap to a non-gap (recovery) site are not available, Kubo *et al.* (1996) examined three models with different assumptions about the recovery process: (1) constant recovery rate; (2) recovery rate proportional to global density of non-gap sites; and (3) recovery rate proportional to local density of non-gap sites. The authors concluded that the observed values of global and local densities of gaps can be explained by a combination of the first and third cases. For example, by choosing the recovery rate to be $b + \beta q_{+/0}$ with $b = 0.135$ and $\beta = 0.100$, pair-approximation dynamics predict an equilibrium pattern with $\rho_0 = 0.331$ and $q_{0/0} = 0.580$, as observed in the 1983 census.

Although the above analysis assumes that the observed pattern is at equilibrium, it is also likely that the environment may not be perfectly constant and the one-year transition period may not be long enough to average out the fluctuations of growth and disturbance rates, as many disturbance events in forests are episodic.

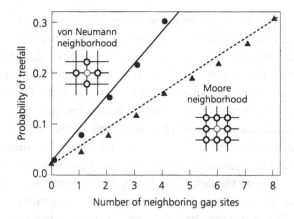

Figure 13.7 Probability of a canopy fall in a 5 m × 5 m plot for various numbers of surrounding 5 m × 5 m plots with low canopy height. Data are from Figure 13.6. Both the von Neumann neighborhood (where each site has four neighbors: $z = 4$) and the Moore neighborhood (where each site has eight neighbors: $z = 8$) are analyzed; both give a very good fit to a straight line, with regression $y = 0.069x + 0.024$ for the von Neumann neighborhood and $y = 0.035x + 0.018$ for the Moore neighborhood. According to the likelihood analysis, the von Neumann neighborhood gives a better fit. *Source*: Kubo *et al.* (1996).

13.4 Colicin-producing and Colicin-sensitive Bacteria

The outcome of competition between different species can also be greatly affected by their spatial structure, but experimental demonstrations using large animals and plants are often very difficult. A clear example of spatial affects on interspecific competition, however, is given by experiments using two strains of bacteria: such experiments can be done easily and are reproducible. Adams *et al.* (1979) studied the population dynamics of two strains of the *Escherichia coli* bacterium: one strain contains a plasmid that produces and is immune to a toxin called colicin; the other (without the plasmid) does not produce colicin and is sensitive to it. In the absence of colicin in the environment, the colicin-sensitive strain has a faster growth rate than the colicin-producing strain. In a well-mixed liquid culture system, if the colicin-producing strain is initially very rare, it will be defeated by the fast-growing colicin-sensitive strain. However, if the initial frequency of the colicin-producing strain is sufficiently high it can win the competition because the toxin it produces effectively suppresses the growth of the colicin-sensitive strain. Thus the outcome of competition depends on the initial frequency, giving rise to bistability. In contrast, bacteria competition experiments on soft agar did not show the frequency-dependent competitive advantage (Chao and Levin 1981), suggesting the importance of spatial structure for this interaction of strains. Durrett and Levin (1997) studied bacterial competition on soft agar as a lattice-structured population model and demonstrated through computer simulations that the parameter space is separated into just two regions: in one the colicin-sensitive strain wins, and in the other the colicin-producing strain wins. No bistability or coexistence is observed, in accordance with the experiments.

The two-strain lattice model of Durrett and Levin (1997) has dynamics almost identical to the evolution of social interaction studied by Matsuda (1987) and Matsuda *et al.* (1987), in which the mortality of each individual is modified by the behavior of neighbors (increasing with the number of "attacking" neighbors and decreasing with the number of "helping" neighbors). One difference between the two models is that in Matsuda's model players of both types are equally affected by these social interactions, whereas the colicin-producing type is immune to the colicin in Durrett and Levin's model. Another, less important, difference is that, to describe the cost of social interaction, Matsuda assumed a differential basic mortality, whereas Durrett and Levin instead assumed a differential basic reproductive rate. In this section I show how pair-approximation analysis can be applied to the colicin model (Iwasa *et al.* 1998).

A three-state model

Consider two strains of bacteria that live in a habitat composed of a number of sites, each of which is either occupied by one of the two strains or is vacant. These sites are arranged on a two-dimensional regular square lattice, imitating the experimental setting of competition on soft agar. Let ρ_1 be the fraction of sites occupied by the colicin-producing bacteria (type 1), ρ_2 be the fraction of sites occupied by the colicin-sensitive strain (type 2), and ρ_0 be the fraction of vacant sites. These global densities satisfy $\rho_0 + \rho_1 + \rho_2 = 1$. For the sake of brevity, I call a site occupied by the type-i strain an i-site ($i = 1, 2$) and a vacant site a 0-site.

Let β_i be the intrinsic birth rate of type i – in other words, the birth rate from an i-site into a neighboring 0-site is β_i/z. The maximum rate of reproduction β_i is realized when the i-site is surrounded by vacant sites. Let δ_i be the natural death rate (without the effect of colicin), and let γ be the additional mortality of the colicin-sensitive strain caused by the colicin-producing strain in its neighbor. Because colicin production is costly, the colicin-producing strain (type 1) should have a higher mortality or a lower reproductive rate than the colicin-sensitive strain (type 2): that is, $\beta_1 \leq \beta_2$ and/or $\delta_1 \geq \delta_2$.

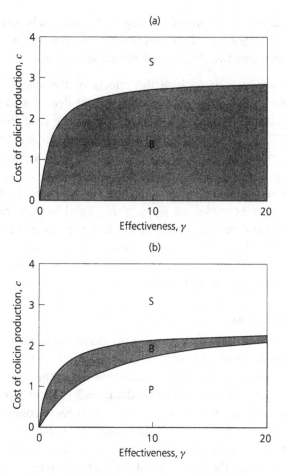

Figure 13.8 Phase diagram for the colicin bacteria model. (a) Well-mixed system. Horizontal axis shows the mortality enhancement γ of the colicin-sensitive strain due to colicin; vertical axis shows the cost of colicin production, $c = \beta_2 - \beta_1$. Other parameters: $\beta_2 = 4$, $\beta_1 = 4 - c$, $\delta_1 = \delta_2 = 1$. The colicin-sensitive strain always wins in region S; bistability is observed in region B, where the outcome of competition depends on the initial relative frequency. (b) Pair-approximation dynamics. The colicin-sensitive strain always wins in region S; the colicin-producing strain always wins in region P; and bistability is observed in the narrow region B.

Mean-field approximation. The mean-field approximation of the lattice model can be used in cases with an additional external process of mixing, as in the experiment in well-mixed liquid. In the mean-field dynamics, the colicin-producing strain cannot invade the equilibrium population dominated by the colicin-sensitive strain. Figure 13.8a illustrates the effects of varying β_1 and γ holding other parameters fixed ($\beta_2 = 4$, $\delta_1 = \delta_2 = 1$). The parameter space is separated into two regions with qualitatively different dynamics. In region S, the colicin-sensitive strain always wins. In region B, the dynamics have an unstable equilibrium (a saddle) that separates two domains of attraction for fixation of either type.

Pair approximation.   To analyze the effects of the spatial localization of interactions, I introduce the conditional local densities to supplement consideration of the global densities ρ_i. There are nine local densities $(q_{1/1}, q_{2/1}, q_{0/1}, q_{1/2}, \dots)$, but not all of them are independent because there are three equations such as $q_{1/1} + q_{2/1} + q_{0/1} = 1$ and another three such as $q_{1/2}\, \rho_2 = q_{2/1}\, \rho_1$ that always hold. The dimensionality may be reduced through these relationships, leaving five independent variables without loss of generality. For example, ρ_1, ρ_2, $q_{1/1}$, $q_{2/1}$, and $q_{2/2}$ can be chosen as the five independent variables and all the others expressed in terms of these five (Iwasa *et al.* 1998). Based on pair approximation, a system of ordinary differential equations of these five variables has been derived as a closed dynamical system (Iwasa *et al.* 1998).

Invasion conditions

Pair-approximation dynamics allow us to derive approximate analytical conditions for successful invasion of one strain into a population dominated by the other. Here, I consider the case in which the lattice population is near the equilibrium at which only the type-2 (colicin-sensitive) strain exists and ask whether the density of the rare type-1 (colicin-producing) strain can increase in the system. At this equilibrium, lattice sites are either type 2 or type 0 (vacant), with stabilized abundances and nearest-neighbor correlations. Because the type-1 invader is rare, the dynamics can be simplified considerably via linearization. The abundance ρ_2 and the nearest-neighbor correlation of the resident type $q_{2/2}$ are unlikely to be affected by the rare invader. Thus, in effect, the invader (type 1) increases or decreases in an environment defined by the resident (type 2). However, it is important to note that the invader typically develops a clumped spatial distribution and hence the density of invaders in the neighborhood of an average invader (i.e., the local density $q_{1/1}$) is not low.

Whether or not the rare invader can increase in global density is determined by the sign of its per capita growth rate. The sign of the Malthusian parameter of the rare invader is dependent solely on the local densities $q_{1/1}$ and $q_{2/1}$. Dynamics derived by pair approximation can be used to compute these two quantities near the equilibrium. These equations, however, include three other variables, which can be estimated by their values at the equilibrium.

Thus the procedure is as follows. First, the equilibrium global and local densities of the residents (ρ_2, $q_{2/2}$) are calculated by neglecting the rare invader ($\rho_1 = 0$). Using these solutions and $\rho_1 = 0$, we have an autonomous system of differential equations of two variables ($q_{1/1}$, $q_{2/1}$), indicating the local environments for an average rare invader. Using the equilibrium of this system of differential equations, the invasibility condition is calculated for the rare type (ρ_1) from the sign of its per capita growth rate. We can then derive the condition for successful invasion of the colicin-producing strain (Iwasa *et al.* 1998). This specifies the boundary between regions P and B in Figure 13.8b. To compute the invasibility of the alternative equilibrium in which the colicin-producing strain is resident and the colicin-sensitive strain is rare, a different choice of five independent variables (ρ_1, $q_{1/1}$, $q_{1/2}$, $q_{2/2}$, ρ_2) needs to be made to calculate the boundary between regions S and B in Figure 13.8b.

There is a wide parameter region labeled P in which the colicin-producing strain can invade, even if it is very rare, and where it is expected to spread and eliminate the opponent. This is in sharp contrast to the model of a well-mixed system, in which invasion by the rare colicin-producing strain is never possible in a population dominated by the colicin-sensitive strain. The parameter region S, in which the colicin-sensitive strain always wins, is broader in Figure 13.8b than in Figure 13.8a. Note also that the parameter region B of bistability is smaller in Figure 13.8b than in Figure 13.8a.

Computer simulations of the model in a large lattice conducted by Durrett and Levin (1997) suggest that there is no parameter region for bistability, which is in conflict with the pair-approximation prediction. Computer simulations of the spatial stochastic model, carried out on small lattices (30×30 and 50×50) with periodic boundaries, however, showed some tendency toward bistability (see Iwasa *et al.* 1998).

13.5 Limitations, Extensions, and Further Applications

This chapter describes three cases in which the consideration of spatial clumping improves model predictions. In many cases the dynamics based on pair approximation successfully explain the results: for example, the high equilibrium density given by an intermediate optimal mixture of vegetative propagation and reproduction by seed; the gap size distribution in the population dynamics of perennials; the clumping tendency of gap sites and the total gap area at equilibrium in the forest gap dynamics with a gap expansion process; the bistability behavior of the forest gap dynamics with supply-dependent recruitment; and the invasion by a rare colicin-producing bacteria strain. None of these results are explained by mean-field approximations in which spatial structure is neglected.

Spatial segregation and pair-edge approximation

There are mathematical arguments suggesting that bistability is impossible on an infinitely large lattice, given that interaction between states occurs over a limited spatial range (Liggett 1978; Durrett 1982; Gray 1982; Durrett and Levin 1994b; Durrett and Neuhauser 1994). These results are independent of the dimensionality of the lattice. For very large lattices and generic initial conditions, each type of state will initially have some local areas of dominance if the mean-field model exhibits bistability. In this case, the dynamics become determined simply by the movement of the boundary between two areas, in each of which one of the two types dominates. This front cannot stabilize and eventually the type with the advancing boundary always wins. This argument does not apply to the two models discussed in this paper (perennials with vegetative propagation and gap dynamics with supply-dependent recovery) because in both models the seeds have an infinitely long spatial range, and in fact these models produce bistability (see Figure 13.9). However, the argument does apply to the bacteria competition model and thus is in conflict with some of the predictions made by the pair-approximation dynamics.

Computer simulation of the bacteria competition model shows that, starting from a random spatial pattern, the lattice is segregated into two kinds of areas, some dominated by types 1 and 0, and others dominated by types 2 and 0. The fate of the system can then be predicted by examining the movement of the boundary between these two areas; this tells us whether type 1 or type 2 can win in the lattice population. Pair approximation assumes that the correlation between sites can be approximated by considering correlations between neighbors. It is plausible that pair approximation fails to predict bistability because the whole system is segregated into large subareas in which the proportions of the colicin-producing strain differ greatly. For small lattices, the model shows dependency on the initial frequency (the dynamics show bistability), though the pattern is blurred by additional stochasticity caused by the finiteness of the model (Iwasa *et al.* 1998).

By assuming that the equilibrium of the pair-approximation dynamics can characterize the density within each area, and by calculating transitions of various configurations of the boundary, Ellner *et al.* (1998) succeeded

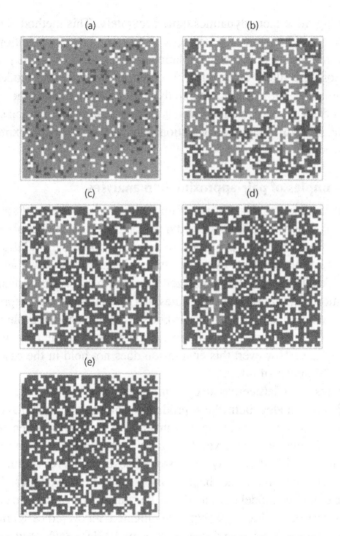

Figure 13.9 The spatial pattern of a single simulation run for the colicin-bacteria model. Black, gray, and white squares represent 1-, 2-, and 0-sites, respectively. Parameters: $\gamma = 2$; $c = 0.5$; the initial frequencies of the 1- and 2-sites are 0.1 and 0.8, respectively. The spatial pattern is shown at times (a) $t = 0$, (b) $t = 50$, (c) $t = 100$, (d) $t = 150$, and (e) $t = 200$. Starting from a random spatial pattern (a), in the first 50 steps the whole lattice becomes separated into areas in which 1- and 0-sites are dominant and areas in which 2- and 0-sites are dominant (b). Subsequently, the areas occupied by types 1 and 0 increase and those occupied by types 2 and 0 decrease (b, c, and d). Finally, the colicin-sensitive strain is eliminated (e).

in predicting the system dynamics quite accurately. This method is called pair-edge approximation. Therefore, if there is no clear segregation into two or more areas in computer simulations of the model, then the pair approximations are likely to be quite accurate, as shown in the models for vegetative propagation and forest gap dynamics. In contrast, if the spatial pattern tends to be segregated into domains, then pair approximation can be inaccurate and a different approximation, such as pair-edge approximation may be needed.

Other examples of pair-approximation analyses

Pair-approximation analysis is useful for many other examples of spatially explicit population and evolutionary dynamics. Based on analytical expressions for invasibility, it is possible, for example, to derive evolutionarily stable life histories of plants living in lattice habitats based on pair approximation. Takenaka *et al.* (1997) analyzed a case with two types of plant that differ in mortality and reproductive rates, but that both can only reproduce into neighboring sites. Here, the invasibility condition and the evolutionarily stable life history under a trade-off are the same as in the case without spatial structure. However, this conclusion does not hold in the case with more general modes of interaction.

Dynamics of a lattice-based epidemic were analyzed by Satō *et al.* (1994); in this model, both the reproduction of healthy host plants and the transmission of pathogens were assumed to occur only between nearest neighbors. The authors observed that if the pathogen transmission rate is high, extinction of both healthy and diseased hosts occurs for a wide range of parameters, which does not happen under mean-field dynamics. Hence the lattice epidemic model does not have the "threshold" susceptive density that characterizes traditional susceptible–infected–recovered (SIR) models without spatial structure (Anderson and May 1991). Satō *et al.* (1994) show that pair approximation calculated in a modified form can explain this behavior quite accurately (see Chapter 18).

Lattice models and pair-approximation analysis are also used for modeling evolution. For example, Nakamaru *et al.* (1997) analyzed the effect of spatial structure in the evolution of cooperation. The model was based on the iterated Prisoner's Dilemma game played on the vertices of a lattice; accumulated payoffs affected mortality, and the vacant sites created by deaths were filled by a random copy of a neighbor. Nakamaru *et al.* (1997) analyzed the model using pair-approximation dynamics and concluded that the spatial structure allowed for an initial increase in the abundance of the rare cooperative strategy (represented by *Tit For Tat*) in a population dominated by a defection strategy (*Always Defect*), which is not possible in the completely mixed model (see Chapter 8). The effect of spatial structure was ambivalent, since there are parameter regions in which the cooperative strategy (*Tit For Tat*) can become fixed in the completely mixed population but not in the lattice populations. The ambivalent effect of spatial structure in facilitating and inhibiting the evolution of cooperation has been pointed out previously (Matsuda 1987; Matsuda *et al.* 1987; Wilson *et al.* 1992; Taylor 1992; Ferrière and Michod 1996; see also Chapter 17).

Pair approximation is a useful approximate analytical tool for lattice models that otherwise can be explored only by direct computer simulation. Analytical solutions are much more powerful than direct computer simulation for examining the model's dependency on parameters and initial conditions. Thus, both the analysis based on pair approximation and computer simulation are useful tools for improving our understanding of lattice models.

Acknowledgments This chapter was written during the academic year of 1996/1997 while I was a Fellow at the Institute for Advanced Study, Wissenschaftskolleg zu Berlin. I thank Professor Hirotsugu Matsuda, who developed the pair-approximation analysis in ecology and evolution. I also thank my collaborators in the works summarized in this article: H. Ezoe, N. Furumoto, Y. Harada, T. Kubo, H. Matsuda, A. Mochizuki, S.A. Levin, M. Nakamaru, K. Satō, Y. Takeda, and Y. Takenaka.

14

Moment Approximations of Individual-based Models

Richard Law and Ulf Dieckmann

14.1 Introduction

This chapter illustrates insights into individual-based spatial models of ecological communities that can be gained from deterministic approximations.
To do this we revisit some of the issues raised in Chapter 1 and show how
approximations can help to

- separate the signal of an ecological stochastic process from intrinsic
  random variation;
- clarify qualitative dependencies that underlie the ecological stochastic
  process;
- determine how ecological stochastic processes depend on their
  parameters.

The deterministic approximations we use are the dynamics of spatial
moments. These are closely related to pair-approximation methods (Matsuda *et al.* 1992; Harada and Iwasa 1994; see also Chapters 13, 18, and 19),
but are constructed in a continuous rather than a discrete space. They also
differ from diffusion approximations (see Chapters 16, 17, 22, and 23) in
that they deal specifically with both structure at small spatial scales and the
discrete nature of individual plants and animals. Moment methods represent a new departure in ecology for understanding the effects of interactions
and movements of individuals in small neighborhoods (Bolker and Pacala
1997; Dieckmann *et al.* 1997; Law and Dieckmann, in press), and we argue
that they hold promise for gaining understanding of ecological processes
where the mean-field assumption breaks down. Moment methods are particularly helpful for providing insight into dynamics of plant communities,
because interactions mostly occur with immediate neighbors (see Chapter 2). The methods also have potential for describing ecological dynamics
in certain kinds of structured landscapes (Wiens *et al.* 1993; Dunning *et al.*
1995).

Formal treatments of moment methods are given in Chapters 20 and 21. The equations used here are derived in Chapter 21 and differ in certain respects from those in Chapter 20. The motivation, however, is the same: to gain insight into the complex behavior of individual-based stochastic processes of ecological communities. We develop ideas in the context of plant communities comprising one or two competing species (see also Chapter 20), but the formal structures can be applied more widely to ecological systems with spatial structure.

14.2 Spatial Patterns and Spatial Moments

Consider a community living in a large, two-dimensional space, with individuals located at points $x = (x_1, x_2)$ in this plane. The abiotic environment is homogeneous in space, and any spatial structure that develops is generated internally by the community. Locations of individuals of species i at some point in time t are given by a function $p_i(x)$, and these are collected into a vector of density functions $p(x) = (p_1(x), p_2(x), \dots)$ to give what we call the *spatial pattern* of the community. If individuals are not located at random in the plane (i.e., if their pattern is not described by a homogeneous Poisson process: see Chapter 5), we refer to the community as having *spatial structure*. As explained in Boxes 21.2 and 21.3, an individual is represented as a Dirac delta function, and the function $p_i(x)$ is the sum of all these individual contributions.

The community changes through three primary, stochastic events acting on individuals: birth, death, and movement. Whenever an event occurs, a new spatial pattern $p(x)$ is generated. Depending on how the events take place, a myriad of different spatial patterns can develop over the course of time; Figure 14.1 shows results at two points in time from just one realization of two competing species. The upper spatial pattern corresponds to a random layout of individuals of two species at time 0. During the realization, the spatial pattern is repeatedly updated, and the lower spatial pattern in Figure 14.1 shows the pattern that has emerged by time 15, after approximately 2000 events have taken place. By this time, species 1 has developed aggregations of individuals and there is some spatial segregation such that where species 1 occurs, species 2 tends to be absent.

The problem with such realizations is that they are time consuming to generate and difficult to understand. Approximations based on moments try to avoid such drawbacks by replacing the spatial pattern with statistics summarizing its main features and then describing dynamics in terms of these statistics. In a sense this approach has been used in ecology for

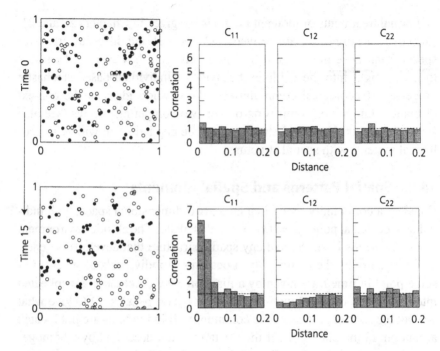

Figure 14.1 Spatial patterns at time 0 (top row) and time 15 (bottom row) from a realization of a stochastic model of two competing species. Locations of species 1 are given by filled circles; those of species 2, by open circles. On the right-hand side are the three radial pair correlation densities that summarize important aspects of spatial information.

many years, because the spatial mean density, widely used by ecologists, is the lowest-order spatial moment. The new departure is putting into place dynamics of a second-order spatial moment that carries information about spatial structure over time.

The two spatial moments we use are defined as follows. The first moment,

$$N_i(p) = \frac{1}{A} \int p_i(x)\, dx \,, \tag{14.1}$$

is the familiar mean density of species i over a spatial region A. The second moment, holding information on spatial structure, we call the *pair correlation density* function. This is a product of pairs of densities of individuals of species i and j, averaged over the spatial region A:

$$C_{ij}(\xi, p) = \frac{1}{A} \int p_i(x) \left[p_j(x + \xi) - \delta_{ij}\, \delta(\xi) \right] dx \,, \tag{14.2}$$

where j is located at ξ relative to i. [The term $\delta_{ij}\, \delta(\xi)$ is needed to remove self-pairs, and comprises the Kronecker symbol δ_{ij}, which takes the value 1 when $i = j$ and 0 otherwise, and the Dirac delta function $\delta(\xi)$; see Boxes 21.2 and 21.3 and Section 21.3, Corrected correlation densities.] We scale the pair correlation densities here and below by dividing them by the product $N_i N_j$; in the absence of spatial structure they then take the value 1; aggregated (respectively regular) patterns are given by values greater (respectively less) than 1. To make the functions still easier to depict, we display them in a radial form; this entails integrating over the angle around the circle, which results in no loss of information if the processes, like many of those in real ecological systems, are isotropic. The pair correlation density is not a central moment (see Chapter 20); we use it because the dynamic equations are then simpler (Chapter 21). It is important to understand that the second moment is a function rather than a scalar quantity, giving information on spatial structure over a range of spatial scales, in keeping with measures of spatial structure used in the past in plant ecology (Pielou 1968; Greig-Smith 1983).

To illustrate how pair correlation densities capture information about spatial structure, we have computed the second moments for the spatial patterns shown in Figure 14.1. In this two-species system there are three correlation functions associated with any spatial pattern: two auto-correlation functions $C_{11}(\xi, p)$, $C_{22}(\xi, p)$, and a cross-correlation function $C_{12}(\xi, p)$ (Figure 14.1). At time 0, there is no spatial structure and the functions are all flat. At time 15, the tendency for species 1 to form aggregations causes an excess of pairs at small distances, giving values of $C_{11}(\xi, p)$ much greater than 1 at short distances. Conversely, the segregation between species 1 and 2 causes a shortage of pairs at small distances, giving values of $C_{12}(\xi, p)$ much less than 1 at small distances.

During a realization of a stochastic process, the spatial moments (14.1) and (14.2) take new values each time the pattern changes. But, a realization is, of course, just one of an infinite ensemble that can be generated from the stochastic process, and it is the generic properties of the ensemble, not the properties of an individual realization, that we need to base our understanding on. We therefore replace the moments of the pattern at time t with their averages over all realizations at this time:

$$N_i = \int N_i(p)\, P(p)\, dp \ , \qquad (14.3)$$

$$C_{ij}(\xi) = \int P(p)\, C_{ij}(\xi, p)\, dp \ , \qquad (14.4)$$

where $P(p)$ is the probability density for patterns p at time t, and the integration dp is over the space of functions p.

The quantities in Equations (14.3) and (14.4) comprise the state variables of the moment dynamics we use below. We derive a system of equations for these dynamics in Chapter 21, and these equations couple the changes in the first and second moments. In this way there is feedback from the spatial structure to the dynamics of the mean density (and vice versa) as the spatial system unfolds over time.

14.3 Extracting the Ecological Signal from Stochastic Realizations

Stochastic processes are a good framework in which to formalize ideas about ecological events acting on individuals (e.g., Pacala *et al.* 1996). In particular, they have the advantage of not glossing over the effects fluctuations have on local and global states of ecological systems. They have the drawback that it may not be obvious from individual realizations what the generic behavior of the process is. To illustrate this problem, and to show how moment dynamics can help to overcome it, we give an example of two competing plant species from Law and Dieckmann (in press), essentially a spatial version of the familiar Lotka–Volterra model of competition:

$$\frac{d}{dt}N_i = (b_i - d_i)\,N_i - \sum_j d'_{ij}\,N_i\,N_j \quad \text{for } i, j = 1, 2 \ . \qquad (14.5)$$

The equations are parameterized here with b_i (respectively d_i) as a density-independent birth (respectively death) rate. The term d'_{ij} is a component of the death rate that depends on the density of the competing species j, competition being intraspecific when $i = j$, and interspecific when $i \neq j$.

The spatial extension arises in that individuals are indexed by location in space, and the spatial model keeps track of (1) competition among individuals located close enough together and (2) movements of individuals. As we are dealing with plant species, we assume that movements are always associated with seed dispersal (birth of a new plant). Parameters of the community are set such that, on the one hand, species 1 is the stronger competitor and would replace species 2 in the absence of spatial structure. On the other hand, seeds of species 2 disperse over longer distances. Once the dynamics allow for spatial structure, whether species 2 goes to extinction is an open question.

Stochastic process. From the general birth–death–movement process given in Chapter 21 [Equations (21.1) and (21.2)], we can write down a stochastic process explicitly for two competing species. The probability per unit time of the transition from a pattern $p(x)$ to another pattern $p'(x)$ is fully defined in terms of two independent events.

■ *Birth.* The probability per unit time $B_i(x, x', p)$ that an individual of species i, located at point x in a pattern $p(x)$, gives rise to a daughter at location x' is given by

$$B_i(x, x', p) = b_i \, m_i^{(b)}(x' - x) , \tag{14.6}$$

where b_i, the birth rate, is multiplied by a dispersal term $m_i^{(b)}(x' - x)$ independent of the birth rate, placing the daughter at location x' with probability density $m_i^{(b)}(x' - x)$.

■ *Death.* The probability per unit time $D_i(x, p)$ that an individual of species i, located at point x in a pattern $p(x)$ dies is given by

$$D_i(x, p) = d_i + \sum_j d'_{ij} \int w_{ij}^{(d)}(x' - x)\big[p_j(x') - \delta_{ij}\delta_x(x') \big] dx' . \tag{14.7}$$

Here, d_i is the neighbor-independent component of the death rate common to all individuals, and the remaining terms on the right-hand side deal with the effects of competition with neighbors. Specifically, $w_{ij}^{(d)}(x' - x)$ is a competition kernel (see Box 20.1) weighting the effect of a neighbor of species j at location x' according to its distance from x; this is multiplied by the density $p_j(x')$ of plants of species j at x', integrated over all locations x', and scaled by the parameter d'_{ij}. The expression $\delta_{ij}\delta_x(x')$ removes the target plant of species i at location x from the competition kernel; it is needed because the target plant does not compete with itself.

We use Gaussian functions to describe dispersal distances and the effect of distance on competition, and characterize the functions with two parameters. The first parameter $s_{ij}^{(b)}$ (respectively $s_{ij}^{(d)}$) is the standard deviation for dispersal (respectively competition). The second parameter truncates the functions at maximum radius $r_{ij}^{(b)}$ (respectively $r_{ij}^{(d)}$) for dispersal (respectively competition), and can be thought of as setting an upper limit on the distance over which dispersal (respectively competition) takes place. The functions are normalized so that their integrals are equal to 1. Table 14.1 gives the parameter values; notice the competition parameters are set such that species 1 is a stronger competitor than species 2, but disperses over shorter distances than species 2.

Four realizations of the stochastic process, run from time 0 to 15, are shown in Figure 14.2. (Here, and in all realizations below, the stochastic process is implemented in a space of unit area with periodic boundaries.) The graphs give the paths in the plane of population densities of the two species, and it is important not to confuse this with the physical space in

Table 14.1 Parameter values for a community of two competing species. *Source*: Law and Dieckmann (in press).

	Parameter	Value for species i		Explanation
		$i = 1$	$i = 2$	
Death	d_i	0.2	0.2	
	d'_{i1}	0.001	0.002	Species 1 is a stronger
	d'_{i2}	0.0005	0.001	competitor than species 2
	$s^{(d)}_{i1}$	0.03	0.03	
	$s^{(d)}_{i2}$	0.03	0.03	
	$r^{(d)}_{i1}$	0.12	0.12	
	$r^{(d)}_{i2}$	0.12	0.12	
Birth	b_i	0.4	0.4	
	$s^{(b)}_i$	0.03	0.2	Species 2 disperses
	$r^{(b)}_i$	0.12	0.5	farther than species 1

Note: Simulations were done on the unit square with periodic boundaries.

which the two species live and interact. The generic behavior, or signal, of the stochastic process is not clear from inspection of the paths. Demographic stochasticity masks any obvious trend: the paths differ from one another and have the appearance of tangled webs. As one would expect, matters are improved by taking the mean path of some realizations, but the number of realizations needed may be quite large. This can be seen in Figure 14.2e, which shows the mean of 20 realizations; a distinct curl to the path is now evident.

Moment dynamics. From the general equations for the dynamics of the first and second spatial moments (21.9) and (21.10), we can write down the equations as they apply to two competing species. The dynamics of the first moments (mean densities) are given by

$$\frac{d}{dt} N_i = (b_i - d_i) N_i - \sum_j d'_{ij} \int w^{(d)}_{ij}(\xi') C_{ij}(\xi') \, d\xi' , \qquad (14.8)$$

for $i, j = 1, 2$. The first term on the right-hand side is the neighborhood-independent component of births and deaths, and the second term is the neighborhood-dependent component of deaths. Evidently, in turning from the nonspatial version of the Lotka–Volterra equations to the spatial version, the products $N_i N_j$ of the Lotka–Volterra equations are replaced by integrals involving the pair correlation densities, compare Equations (14.5) and (14.8). The effect of competition depends fundamentally on the relative

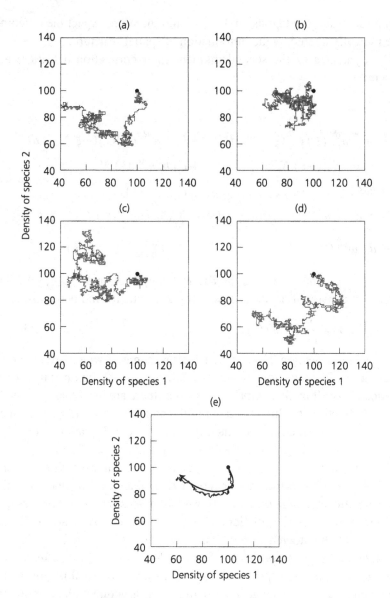

Figure 14.2 Plane of population density of two competing species. Graphs (a), (b), (c), and (d) show the path mapped out by individual stochastic realizations over 15 time units. (e) Mean path of 20 realizations, with the deterministic approximation (found by solving the equations for the dynamics of the first and second spatial moments) superimposed as a smooth curve. The filled circle marks the starting point of each path.

magnitude of these integrals, and we should therefore expect the outcome to be radically altered by the introduction of spatial structure.

The dynamics of the second moment (pair correlation densities) are given by

$$\frac{d}{dt} C_{ij}(\xi) =$$

$$
\begin{aligned}
&+ b_i \int m_i^{(b)}(\xi') \, C_{ij}(\xi + \xi') \, d\xi' && + b_j \int m_j^{(b)}(\xi') C_{ji}(-\xi + \xi') \, d\xi' \\
&+ \delta_{ij} \, b_i \, m_i^{(b)}(-\xi) \, N_i && + \delta_{ji} \, b_j \, m_j^{(b)}(\xi) \, N_j \\
&- d_i \, C_{ij}(\xi) && - d_j \, C_{ji}(-\xi) \\
&- \sum_k d'_{ik} \int w_{ik}^{(d)}(\xi') \, T_{ijk}(\xi, \xi') \, d\xi' && - \sum_k d'_{jk} \int w_{jk}^{(d)}(\xi') \, T_{jik}(-\xi, \xi') \, d\xi' \\
&- d'_{ij} \, w_{ij}^{(d)}(\xi) \, C_{ij}(\xi) && - d'_{ji} \, w_{ji}^{(d)}(-\xi) \, C_{ji}(-\xi) \, ,
\end{aligned}
\tag{14.9}
$$

for $i, j = 1, 2$. The term δ_{ij} is the Kronecker symbol, and $T_{ijk}(\xi, \xi')$ is the correlation density of triplets, for which we assume the moment closure

$$T_{ijk}(\xi, \xi') = \frac{1}{N_i} C_{ij}(\xi) \, C_{ik}(\xi') \, , \tag{14.10}$$

as explained in Section 21.4. With 10 terms, some of which are integrals, Equation (14.9) looks somewhat forbidding. But the complexity of the equation is not altogether surprising, because there are five types of event in species i affecting the flux in and out of the pair density $C_{ij}(\xi)$, as given by the terms in the first column on the right-hand side of Equation (14.9). For each type of event affecting i, there is an equivalent one affecting j, given by the terms in the second column. It can be seen from inspection of Equations (14.8) and (14.9) that they are coupled, so changes in the mean density affect the rate of change of the pair correlation densities, and vice versa.

Numerical integration of Equations (14.8) and (14.9) clearly indicates the signal of the stochastic process, shown as the smooth curve in Figure 14.2e. At the start, species 1 is favored over species 2, but, after a short period of time, this is reversed. The reason for this reversal is quite interesting and results from the generation of spatial structure in the community (Law and Dieckmann, in press). To begin with, individuals are situated at random locations in the plane. Species 1, the stronger competitor, is at an advantage in these circumstances and starts to increase in density, whereas species 2 decreases; thus the deterministic path starts by pointing down and to the right. But species 1 is also a poorer disperser and begins to develop a clumped spatial pattern which inflates the strength of intra-specific competition. Eventually, clumping becomes strong enough to place species 1 at a disadvantage in the community, causing the deterministic path to swing round in the direction of increasing density of species 2 and decreasing density of species 1.

The deterministic path in Figure 14.2e provides information on a small part of the plane of population densities. To illustrate the dynamical behavior more broadly, we have taken a larger sample of paths and run them for a longer period of time, as shown in Figure 14.3 (Law and Dieckmann, in press). Figure 14.3a gives the mean path, averaged over 20 realizations, from a grid of starting points of the stochastic process. This shows that species 1, despite its competitive advantage, is eventually always driven to extinction due to the excess intraspecific competition it generates by spatial aggregation. Figure 14.3b gives the corresponding deterministic paths obtained by solving Equations (14.8) and (14.9). There is a close match between the paths in Figure 14.3a and 14.3b: evidently the moment dynamics give a good approximation to the behavior of the stochastic process. The same cannot be said of the nonspatial Lotka–Volterra competition equations (14.5), shown in Figure 14.3c. These equations neglect the effects of dispersal and the small neighborhoods within which competition occurs, and lead one to expect that species 2 should be driven to extinction. Space is clearly crucial here – to ignore it is to be qualitatively in error.

The message from this analysis of two competing species is threefold. (1) The signal of the stochastic process often can barely be seen from looking at individual realizations. (2) The method of moments provides a close approximation to the average behavior of the stochastic process and can be used to gain insight into its generic features. (3) The spatial extension is fundamental to understanding the dynamics of plants competing and dispersing in small neighborhoods.

14.4 Qualitative Dependencies in a Spatial Logistic Equation

Here we show some qualitative properties of stochastic realizations that can be better understood by means of deterministic approximations. We do this in the context of a spatial extension of the familiar logistic model of single-species population growth

$$\frac{d}{dt}N = (b - d)N - d'N^2 , \tag{14.11}$$

parameterized in the same way as Equations (14.5). The spatial extension turns out to have some qualitative features that are surprising, at least at first

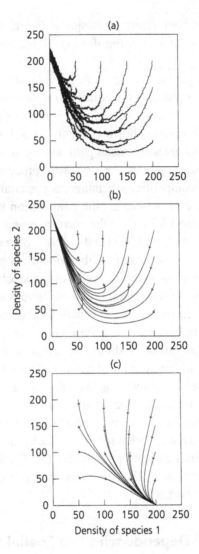

Figure 14.3 Plane of population density of two competing species. Paths are given from a grid of 16 combinations of starting densities. Individuals are placed at random locations in space at the start, and paths are allowed to run for 100 time units. (a) Mean paths of 20 realizations of the stochastic process. (b) Deterministic paths given by the dynamics of the first and second moments. The paths are projections from higher-dimensional dynamics, Equations (14.8) and (14.9), and may therefore intersect one another. (c) Paths given by the nonspatial, Lotka–Volterra competition equations. *Source*: Law and Dieckmann (in press).

sight. Yet these features follow in a simple, natural way from the moment dynamics, as we show below.

Stochastic process. The stochastic process has much in common with the spatial Lotka–Volterra competition equation described in Section 14.3, allowing (1) an increased risk of death of individuals located close enough together and (2) dispersal of seeds. Notationally, the stochastic process is easier to handle because, with only one species, terms do not need to be indexed by species. As before, there are two stochastic events.

■ *Birth.* The probability per unit time $B(x, x', p)$ that a plant, located at point x in a pattern $p(x)$, gives rise to a daughter plant at location x' is given by

$$B(x, x', p) = b\, m^{(b)}(x' - x) \,. \tag{14.12}$$

■ *Death.* The probability per unit time $D(x, p)$ that a plant, located at x in a pattern $p(x)$, dies is given by

$$D(x, p) = d + d' \int w^{(d)}(x' - x) \left[p(x') - \delta_x(x') \right] dx' \,. \tag{14.13}$$

Terms are as defined in Equations (14.6) and (14.7), and parameter values that remain fixed in the simulations below are $b = 0.4$, $d = 0.2$, and $d' = 0.001$.

Three sets of realizations of this stochastic process are given in Figure 14.4. The sets differ in the size of the competition neighborhoods and in the distances over which seeds disperse. The results are surprising. Depending on the choice of parameter values for local competition and seed dispersal, the populations may grow to densities substantially larger or smaller than one would expect from the equilibrium density of the logistic equation (14.11), which equals 200 with the parameter values above. This is a good example of the new phenomena that emerge when the spatial extension to population dynamics is introduced, as has been stressed at many places in Part B.

Moment dynamics. Why does the asymptotic density depend on the competition neighborhood and distances over which seeds disperse? The answer becomes apparent once the equation for the dynamics of the first spatial moment of the stochastic process is put into place; this is a simplified version of Equations (14.8),

$$\frac{d}{dt} N = (b - d)N - d' \int w^{(d)}(\xi') C(\xi') \, d\xi' \,. \tag{14.14}$$

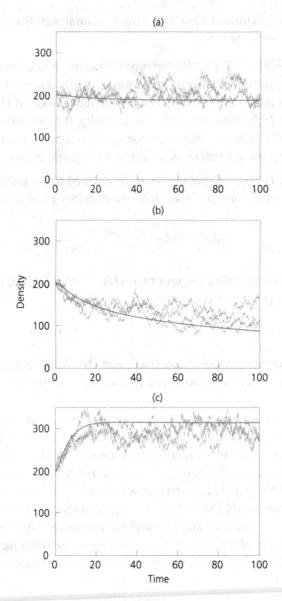

Figure 14.4 Stochastic realizations of a spatial, logistic model of population growth with contrasting dispersal and competition neighborhoods. In each case three realizations are run, starting with 200 individuals located at random (200 is the equilibrium value of the corresponding nonspatial logistic model). (a) Competition and dispersal take place over large distances: $s^{(d)} = 0.15$, $s^{(b)} = 0.15$. (b) Competition and dispersal take place over intermediate distances: $s^{(d)} = 0.05$, $s^{(b)} = 0.05$. (c) Competition takes place at a smaller scale than dispersal: $s^{(d)} = 0.015$, $s^{(b)} = 0.15$. The maximum radii are three times the standard deviations. Superimposed on the realizations is the deterministic approximation obtained from solving the equations for the dynamics of the first and second moments.

The term N^2 of the nonspatial model is replaced by an integral involving the pair correlation densities, compare Equations (14.11) and (14.14). Evidently, the outcome depends on the spatial structure that develops over time, as given by $C(\xi')$, and on the extent to which plants experience this structure, as given by $w^{(d)}(\xi')$. If the integral is greater than N^2, the population comes to equilibrium at a density less than that of the nonspatial version. Conversely, if the integral is less than N^2, the equilibrium density is greater than that of the nonspatial version. Clearly, details of the spatial structure really do matter.

To understand how the spatial structure develops, we need the dynamics of the second moment:

$$
\begin{aligned}
\frac{1}{2}\frac{d}{dt} C(\xi) = \ & + b \int m^{(b)}(\xi'') \, C(\xi + \xi'') \, d\xi'' \\
& + b \, m^{(b)}(\xi) \, N \\
& - d \, C(\xi) \\
& - d' \, C(\xi) \int w^{(d)}(\xi') \, C(\xi')/N \, d\xi' \\
& - d' \, w^{(d)}(\xi) \, C(\xi) \, .
\end{aligned}
\tag{14.15}
$$

This equation is simplified from Equation (14.9). The two columns on the right-hand side of Equation (14.9) can be added together, because there is only one species and the process is isotropic; we then divide both sides by 2, getting the factor $1/2$ on the left-hand side of Equation (14.15).

It turns out that the spatial structure that develops over time depends on the size of the competition neighborhood and dispersal distances. Figure 14.5 illustrates this with stochastic realizations using the three combinations of parameter values given in Figure 14.4. All three start with the same random spatial pattern and show the spatial pattern of a realization of the stochastic process after a long period of time has elapsed. In Figure 14.5a, seeds disperse over relatively large distances and the correlation function shows rather little spatial structure. In fact, it would not make much difference if there were more small-scale spatial structure, because the neighborhoods of competition integrate over relatively large areas and would be blind to such structure. In Figure 14.5b, where both dispersal distances and competition neighborhoods are smaller, matters are different. Shorter dispersal leads to more aggregation, and the smaller neighborhoods of competition make the plants sensitive to this structure. Thus the plants experience a density much greater than the spatial average in the interaction neighborhoods, and population growth is stopped at lower mean density. In

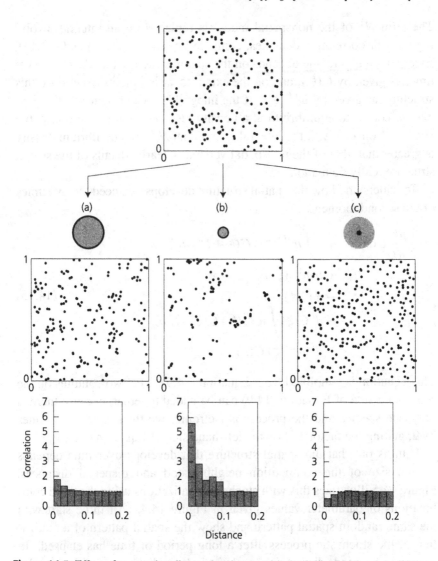

Figure 14.5 Effect of contrasting dispersal and competition neighborhoods on spatial pattern in a logistic model of population growth. The spatial pattern at the top shows the initial state, comprising 200 individuals placed at random locations. Columns (a), (b), and (c) have parameter values matching Figures 14.4a, 14.4b, and 14.4c, respectively; the radius of the heavy circle is the standard deviation $s^{(b)}$ of the competition neighborhood, and the radius of the shaded area is the standard deviation $s^{(d)}$ of the dispersal neighborhood. The columns show the state of sample realizations at time 100 in terms of the spatial pattern and the corresponding auto-correlation function.

Figure 14.5c, the competition neighborhoods are made still smaller (while the dispersal neighborhood is reset to a larger size), and plants that lie too close together have very high mortality rates. This high mortality of plants with close neighbors leads to a regular spatial pattern at short distances. The correlation function is now much less than 1 in the immediate neighborhood of plants, and the plants experience a local density much lower than the spatial average. The population therefore continues to grow to a greater mean density.

How good an approximation to the stochastic process do the moment dynamics provide? A partial answer is given by placing the time course of mean density on the stochastic realizations in Figure 14.4. This shows that the moment dynamics do correctly capture the qualitative outcome and give a reasonable approximation to the quantitative properties of the dynamics.

The message from this example is that qualitative features of the stochastic realizations, at first sight surprising and non-intuitive, become much easier to understand once dynamics have been reduced to the first and second spatial moments. As an ecological footnote, the logistic equation, especially its equilibrium density, has played a central role in developing ideas in ecology. The results here suggest that we may be seriously misled in applying the nonspatial logistic equation to systems with strong spatial structure and small neighborhoods of interaction (see Roughgarden 1997). We should stress though that moment models have still to address some important properties of population growth, such as the dependence of competition neighborhood on size of individuals (Chapter 2).

14.5   Exploration of Parameter Space

The results in Section 14.4 suggest that the spatial extension of the logistic equation has dynamics that depend sensitively on the size of the neighborhoods over which individuals compete and the distances over which dispersal occurs. But so far we have only investigated three fixed points in parameter space and have little idea about the overall form of the dependence. Below, we use the deterministic approximation based on the dynamics of the first and second spatial moments, Equations (14.14) and (14.15), to show how more knowledge about the equilibrium density can be gained.

A convenient method for obtaining equilibrium densities of the moment dynamics is

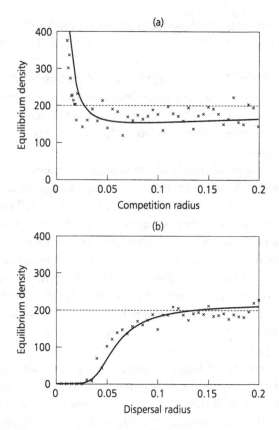

Figure 14.6 Equilibrium densities of a spatial logistic equation, expressed as a function of competition radius and dispersal distance. (a) Competition radius $s^{(d)}$ is varied between 0 and 0.2 for $s^{(b)} = 0.08$; (b) shows the effect of varying $s^{(b)}$ over the same interval for $s^{(d)} = 0.05$. While outcomes of stochastic realizations (crosses) show considerable scatter, and mean-field results (dashed lines) are sometimes qualitatively in error, the deterministic approximation based on spatial moments (continuous curves) provides satisfactory descriptions of how equilibrium densities depend on the radii.

■ to determine, with $\hat{c}(\xi) = \hat{C}(\xi)/\hat{N}^2$, the equilibrium of the first spatial moment for a given second moment from Equation (14.14),

$$\hat{N} = \frac{b - d}{d' \int w^{(d)}(\xi') \hat{c}(\xi') \, d\xi'} , \tag{14.16}$$

■ and then, using that solution in Equation (14.15), to solve numerically for the equilibrium of the (normalized) second moment, $\frac{d}{dt}\hat{c}(\xi) = 0$.

Figure 14.6 shows how sensitive equilibrium densities are to changes in local competition and dispersal. Evidently, the equilibrium density given by the mean-field results of the nonspatial logistic equation applies only in the limits as the interaction neighborhood becomes large and dispersal distances become large. If the competition neighborhood is small relative to dispersal, densities much in excess of the mean-field value occur due to the tendency of other individuals to be absent in this small neighborhood (Figures 14.5c and 14.6a). At the other end of the scale, if dispersing offspring tend to fall within the competition neighborhood, and the competition neighborhood is itself small, the population can go extinct (left range in Figure 14.6b). Extinction comes about as a result of dense aggregations of individuals within the competition neighborhood. The denominator in Equation (14.16) is then large; for finite populations, this leads to population sizes so small that accidental extinction by demographic stochasticity becomes very likely.

Clearly, the spatial version of the logistic equation has some new and interesting features. But the new features that emerge are not easily accessible from numerical studies of the stochastic process itself. It is through approximation schemes, here the moment dynamics, that we gain understanding of how the dynamics depend on model parameters. Once a dynamical system that gives an acceptable approximation to the stochastic process is in place, a battery of analytical techniques (such as methods from bifurcation theory) is available to gain understanding of the consequences of spatial heterogeneity.

14.6 Concluding Comments

If a strong foundation for studying the dynamics of spatially structured communities is to be developed, we need methods that give good approximations to the underlying, intricate, individual-based processes. We think methods based on the dynamics of spatial moments hold promise in this regard. The approximations evidently work well. Moment models enable the signals from ecological stochastic processes to be extracted reliably and give new insights into the rich dynamics of spatial ecological processes. The models are, of course, no more than approximations and work best where spatial structure applies at a small spatial scale and is not too extreme in intensity; the range of spatial structures over which the method works has yet to be determined in detail. Structure at large spatial scales is better dealt with by diffusion approximations (Chapters 16, 17, 22, and 23); to account for spatial structure at both small and large scales simultaneously, it may be possible to amalgamate moment and diffusion approximations.

A recurring feature of the moment dynamics above is the use of a spatial integral of the pair correlation densities weighted by some function of distance, Equations (14.8) and (14.9), in place of the product of the average densities $N_i N_j$. We are replacing the spatial average of the densities with what might be termed an average neighborhood that carries information about local spatial structure. The integral, in effect, formalizes a notion of the "plant's-eye view" of the community that has been in the plant-ecological literature for many years (Turkington and Harper 1979; Mahdi and Law 1987). In switching the focus to the average neighborhoods of plants, some basic changes in plant community dynamics are to be expected. For instance, it is well established in natural plant communities that conspecifics tend to be aggregated (e.g., Greig-Smith 1983; Mahdi and Law 1987). As shown in Chapter 20, spatial structure may make it much easier to achieve coexistence of plant species than has previously been thought.

Evidently, theoretical ecologists have some fundamental thinking to do about the effects of spatial structure. Many basic ideas in ecology come from theory based on the mean-field assumption: the maximum sustainable yield, the competitive exclusion principle, community stability, and so on. But terrestrial communities are spatially structured, and mean-field dynamics are often inappropriate. It is clear from the examples given here and elsewhere in the book that spatial structure needs to be properly incorporated into ecological dynamical systems if we are to avoid coming to seriously mistaken conclusions about ecological processes. Moment approximations provide a key for opening the door into spatial structure.

15

Evolutionary Dynamics in Spatial Host–Parasite Systems

Matthew J. Keeling

15.1 Introduction

Simple mathematical models that describe the progression of a disease through a well-mixed population have been the subject of vast amounts of study in recent years (Anderson and May 1991). However, despite our fairly clear understanding of the dynamics produced by these simple epidemiological systems, many problems arise when the underlying assumption of spatial homogeneity no longer holds, as unfortunately is usually the case. Although many studies have realized the importance of space to disease spread (Cliff *et al.* 1981; Durrett 1988a; Dwyer 1992b; Bolker 1995; Bolker and Grenfell 1995; Cliff 1995; Durrett 1995a; Grenfell *et al.* 1995a; Metz and van den Bosch 1995; and Mollison and Levin 1995), generic phenomena are difficult to extract due to the vast amount of data produced by spatial simulations and the computationally expensive nature of the problem.

In this chapter I use a simple probabilistic cellular automaton (PCA) to investigate the effects of including discrete, spatially distributed populations, limited local interactions, and stochastic processes in the modeling framework. PCAs are the simplest form of stochastic, spatial models, yet they illustrate how many of our common mathematical techniques break down for this type of system. The PCA introduced here is a caricature model. I attempt to define the simplest form of a spatial disease model and neglect any detail or nongeneric behavior that would accompany the precise simulation of a disease. PCA models (or interacting particle systems) have been used with great success in a wide variety of ecological and epidemiological applications (Mollison 1977; Ermentrout and Edelstein-Keshet 1992; Durrett and Levin 1994a, 1994b; Rhodes and Anderson 1996).

Attempts are also made to capture some of the basic features of the cellular automaton using a nonspatial model (the PATCH model) that incorporates intermediate-scale spatial structures but neglects local correlations and large-scale stochastic fluctuations.

15.2   Dynamics of the Spatial Host–Parasite Model

In this section, the simplest version of the PCA model is described to help clarify the type of system we are dealing with (refinements will be explained later as they are needed).

A probabilistic cellular automaton model

The system consists of a two-dimensional square lattice of sites with periodic boundary conditions, so that all activity takes place on a torus – the sides are wrapped around to meet each other. This boundary condition is chosen merely for simplicity and has little effect on the dynamics of the system: we will be working with lattices large enough that most sites experience no influence from the boundary. Each site is in one of three basic states: *empty* (E), occupied by a healthy *host* (H), or occupied by a *parasitized* host (P). The behavior of each site is stochastic; the probability of transition to another state is determined by the four nearest neighbors (north, south, east, and west, the so-called von Neumann neighborhood). The exact form of the neighborhood has little qualitative effect, so the simplest and computationally quickest neighborhood type was chosen. However, as the neighborhood becomes sufficiently large, so that each site experiences a global average, the dynamics approach the mean-field behavior (Keeling and Rand, in press).

A healthy host grows into an empty site in its neighborhood with probability g (infected hosts cannot grow) and is infected by a parasitized host in its local neighborhood with probability of transmission T. The parasite is assumed to be fatal, that is, sites occupied by parasitized hosts become empty at the next iteration. Table 15.1 summarizes the transition probabilities, with E, H, and P being the number of empty, host, and parasite sites in the surrounding neighborhood, respectively. The entire lattice is updated synchronously, so if η_t is the configuration (or state of the entire lattice) at time t, then the probability that the configuration at the next iteration is η_{t+1} is the product of the transition probabilities at each site. Although it can be argued (Brown 1993) that the computationally slower asynchronous updating is more natural, no qualitative difference between the two methods has been observed for this model.

Table 15.1 Transition probabilities for the host–parasite probabilistic cellular automaton.

From	Probability of transition to		
	Empty	Host	Parasitized
Empty	$(1-g)^H$	$1-(1-g)^H$	0
Host	0	$(1-T)^P$	$1-(1-T)^P$
Parasitized	V	0	$1-V$

This model was first introduced by Rand *et al.* (1995). It has since been used as a basis for studying the effect of space and stochasticity in multiple-host multiple-parasite systems and more complex evolutionary scenarios, and as a test for pair-wise models and other nonspatial approximations (Keeling and Rand 1995, in press; Keeling 1995).

Throughout the simulations the growth rate of the host into empty cells g was fixed at 0.05, which was slow enough to be realistically below the transmission rate T but fast enough so that multiple simulations were feasible. The virulence V was also fixed (at 1, implying that infections are lethal) so that attention could be focused on changes in the behavior with respect to transmissibility. The behavior with respect to changes in the virulence of the pathogen is discussed by Rand *et al.* (1995), but the same qualitative phenomena hold.

Spatial statistics and extinction

When running the PCA model, it quickly becomes clear that the exact spatial state of the system is far too sensitive to stochastic events for any precise prediction (Figures 15.1 and 15.2). However, once the initial transient behavior has disappeared (after 100 or so iterations), the proportions of hosts, parasites, and empty sites, as well as the spatial statistics, do not appear to vary significantly. (A more mathematical and precise definition for infinitely large systems would be that, asymptotically in time, the distribution of configurations η_t converges to an invariant ergodic measure.) On a finite lattice of size $\ell \times \ell$, we would expect order ℓ^{-1} fluctuations in all the statistical quantities associated with population densities.

These observations were confirmed by examining the numbers of each type of site (empty, host, or parasite) and the proportion of neighborhood types, which remain fairly constant, affected only by stochastic fluctuations (Rand *et al.* 1995). Therefore, not only are the proportions of each species constant but so are the local correlations, and hence the interaction probabilities, because the possible interactions are determined by the local neighborhoods.

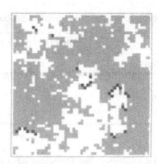

Figure 15.1 Example of the behavior of the host–parasite PCA for $g = 0.05$ and $T = 0.55$. Empty sites are white; hosts, gray; and parasites, black. The spatial distribution is patchy with small broken waves of parasites.

Figure 15.3 shows the average percentage of sites occupied by parasites versus the transmissibility T. The dashed line is the average given that the parasite population survives, the solid line is the average including extinctions. It should be noted that the number of parasites in the cellular automaton model is lower than predicted using the mean-field equations (Box 15.1) by a factor of about three. Also, once the transmissibility has risen above 0.55, the number of parasites in the extant systems remains fairly constant. It therefore appears that an increasing proportion of the systems suffer extinction of the parasites, but in the few simulations that remain, the level of parasitism is maintained.

As can be seen from Figure 15.4, the number of hosts per parasite is a monotonically decreasing function of the transmissibility, even though the proportion of parasites remains constant (Figure 15.3), so there must be an increased number of empty sites. It is this increase in the proportion of empty sites, which act as barriers to parasite transmission, that eventually leads to parasite extinction.

Identifying the resulting spatial scale

With many spatial models the scale at which the dynamics are observed (or the size of the model itself) can have a great impact on the perceived behavior. The same is true for this PCA model. It is vital that we consider the dynamics at the correct spatial scale; if the lattice is too small, stochastic fluctuations dominate; if the lattice is too large, interesting behavior may be averaged out. Using the method developed by Keeling *et al.* (1997b; see also Chapter 12), the length scale of this system was found to be between 80 and 90 cells. This scale agrees well with the observation that the parasite quickly dies out if a lattice size of 70×70 or less is used, while persistence is only weakly dependent on the lattice size once this size exceeds 100×100 cells.

Figure 15.2 Example of the behavior of the host–parasite PCA for $g = 0.05$ and $T = 0.6$. The waves of parasites are much less broken, leaving isolated patches of hosts and a higher proportion of empty sites.

Existence of a critical transmissibility

This section contrasts the behavior of the homogeneous mean-field system (given in Box 15.1) with that of the cellular automaton, demonstrating the existence of a critical transmissibility T_c above which the spatially distributed parasite cannot survive. In the homogeneous model given by Equation (a) in Box 15.1, each cell is affected equally by every other cell, irrespective of its location. The homogeneous model also takes the space as infinite, so the parasites can survive down to unrealistically low levels (atto-parasites) and still recover to a reasonable population size. With the cellular automaton, neither of these simplistic assumptions is true and a new phenomenon, the critical transmissibility, emerges due to spatial structuring.

In the PCA model each lattice site is influenced only by its immediate neighborhood, so a parasite with a high transmissibility will infect all surrounding hosts. This happens so quickly that the host population has no time to recover, thus in subsequent iterations too few healthy hosts are located next to parasites to sustain the parasite population. The parasites "burn themselves out." This can be compared with the so-called forest-fire models (Bak *et al.* 1990; Paczuski and Bak 1993). This effect is enhanced on a finite lattice because population densities near $1/$(number of sites) usually die out.

When the parasite is fatal ($V = 1$) and the growth rate is small ($g = 0.05$), the most obvious spatial effect at the local scale has to do with the number of healthy hosts that can surround a parasitized host. As the currently parasitized host must have been infected by one of its neighbors (which was subsequently killed by the parasite), it can have at most three healthy hosts to which it can spread the infection. Therefore, we should expect the lower threshold, T_t, to have increased from $\frac{1}{4}$ (see Box 15.1)

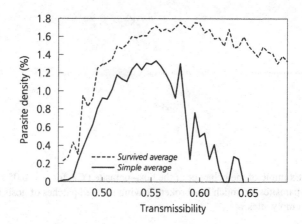

Figure 15.3 The percentage of sites occupied by parasites versus transmissibility in the PCA model. The solid line shows sites occupied for all simulations together. The dashed line shows sites occupied when the parasite is known to have survived. Notice the decline in the simple average for $T > 0.6$.

to $\frac{1}{3}$. In fact, due to other spatial considerations the actual threshold value of T_t is approximately 0.46. This result compares well with the threshold known from percolation theory (Stauffer 1994) for the occurrence of long-range connections across a grid given that neighboring cells have a fixed probability of being connected.

When running the model for transmissibilities larger than 0.46, characteristic behavior and patterns soon emerge, although the precise configurations or dynamics are unpredictable. Fronts of parasites move through the hosts, leaving in their wake empty sites that are slowly colonized by the remaining hosts (see Figure 15.1). For larger transmissibilities the fronts are more complete, so the number of empty cells, and therefore the time for recolonization by a host, is increased (Figure 15.2) until at large transmissibilities the parasites die out due to insufficient availability of hosts.

When the transmissibility is increased to values much greater than 0.6, the parasites die out with an ever-increasing probability. This brings us to the hypothesis that there exists a critical transmissibility T_c above which, for a finite lattice, the probability of survival is zero (Durrett 1995a). If the transmissibility T is close to 1, the parasites spread as a continuous wave front, leaving only empty sites behind; thus on any finite lattice the parasites (and the hosts) die out fairly rapidly. As the transmissibility decreases there are more breaks in the wave front and some hosts remain to recolonize the space behind. For a small enough T, recolonization occurs rapidly enough to prevent the parasites from fully exploiting their local environment.

Box 15.1 The mean-field approximation for the host–parasite probabilistic cellular automaton

By converting the basic set of rules for the probabilistic cellular automata into a set of coupled difference equations, we obtain the following:

$$
\begin{aligned}
\text{Hosts} \quad H' &= H(1 + 4g(1 - H - P) - 4TP) \\
\text{Parasites} \quad P' &= P(1 + 4TH - V) \, .
\end{aligned}
\tag{a}
$$

When T is greater than the threshold $T_t = V/4$, the nontrivial stationary point is found to be

$$
\begin{aligned}
H^* &= V/4T \\
P^* &= g(4T - V)/4T(g + T) \, .
\end{aligned}
\tag{b}
$$

The stationary point is stable for all values of growth rate g that are small compared with the transmissibility T. This fixed point exists for all transmissibilities above $V/4$; for transmissibilities below this, fewer than one secondary case is produced on average per parasite ($R_0 < 1$), so the parasites become extinct. There is no upper bound on the transmissibility T for which the fixed point exists.

It should be noted that any real-life situation will always have a finite population (which corresponds to a finite lattice size), so the parasite cannot be expected to persist forever. If world populations and evolutionary time spans are to be studied, however, then the value of T_c is a close approximation to the largest transmissibility we should expect.

Estimating the critical transmissibility

Unfortunately, because of the stochastic nature of the model the critical transmissibility T_c cannot be easily estimated. In any finite system, for any finite amount of time, there is always a nonzero probability of survival and of extinction. We need to examine the probability of survival on an infinite lattice for an infinitely long period of time. To accomplish this, we should estimate the chance of surviving for t iterations starting with a small finite patch of hosts and parasites (in an infinite lattice), while allowing t to tend to infinity. The parasites always spread faster than the hosts, so at some time $t = t_0$ there exists a region of hosts and parasites that is homogeneous at large scales, and this will expand at a near constant velocity v (compare with Cox and Durrett 1988; Zhang 1993). Let $P(\ell, T)$ be the probability that the parasites, with transmissibility T, survive on an $\ell \times \ell$ lattice for ℓ/v iterations. We then estimate the critical transmissibility T_c as the minimum transmissibility such that $P(\ell, T) \to 0$ as $\ell \to \infty$.

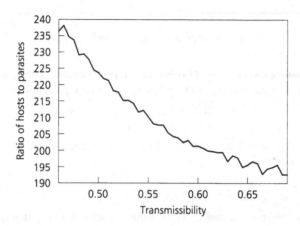

Figure 15.4 A comparison of the average number of hosts per parasite for a range of transmissibilities. Values for $T > 0.65$ were difficult to obtain due to the frequent extinction of the parasite.

We now use the method outlined above to concretely estimate the critical transmissibility T_c for our model. For all values of transmissibility up to T_c, the velocity v of the spreading disk of hosts and parasites was found to be close to 0.1, with a slight decrease with transmissibility over the whole range. A transmissibility T was considered to lie above T_c if the parasite failed to survive for the appropriate number of iterations ($10 \times$ lattice width) in all 200 simulations. Starting from an initial random spread of hosts and parasites, approximations to T_c were found for lattice sizes from 80×80 to 1000×1000. The results clearly predict a critical transmissibility of about 0.67. This value corresponds well with qualitative observations of parasite persistence (see Figure 15.3).

The monotonicity of the system (as illustrated in Figure 15.4) was used to find T_c by means of a bisection algorithm. With this method, if T_c lies within an interval of transmissibilities, then simulations at the middle transmissibility tell us whether the critical value is within the upper or lower half of the interval. In this way the interval is halved at each step. Although it would only take a few extra steps of the bisection algorithm to obtain a more accurate approximation (to three significant figures, for example), it was felt that with only 200 simulations and a maximum lattice of 1000×1000, greater accuracy could not be justified. Also, as indicated above, the exact value of the critical transmissibility is fairly irrelevant when dealing with any real population: it is the existence of such a threshold that is important.

15.3   A Difference Equation for the Dynamics of Local Configurations

One of my main aims is to use these PCA systems to find new methods and to obtain insights into the modeling of more realistic ecosystems. In cellular automaton models, hosts generally grow in patches separated by empty cells (see Figures 15.1 and 15.2). These host patches also have their own dynamics – expanding, breaking up, and amalgamating with others. The entire behavior at each step can be broken into four basic processes, which were used to create the PATCH (Partitioning, Allocation, Transmission and Coalescing of Habitats) model. In contrast to the standard patch or meta-population model, the number and size of the patches and the interaction between patches and within patch populations are allowed to be dynamic variables.

Description of the PATCH model

We can define nine types of patches according to whether empty sites, hosts, and parasites are present. It is necessary to use some set notation to simplify the subsequent equations. Let $Q = \{E, H, P\}$, the set of species, and let G be the set of all sets contained within Q. Ignoring the empty set, there are nine elements in G, each corresponding to a type of patch. This means that for $i \in G$, if $H \in i$ then a patch of type i contains hosts. Moreover, if $A \in Q$ and $A \in i$, then we denote by A_i the average proportion of species A in patch type i; if $A \notin i$, then $A_i = 0$. Also associated with each patch type are a size $S_i^2 \geq 1$ (the number of cells within a patch) and a density D_i of such patches. Hence the total density of species A can be calculated as

$$A = \sum_{i \in G} A_i D_i S_i^2 \,. \tag{15.1}$$

Within any patch the distribution of species is assumed to be random. Thus we ignore all local spatial effects, such as those obtained using pairwise models (see Chapters 18 and 19). Also, for simplicity, all patches of type i are assumed to be identical squares containing proportions E_i, H_i, and P_i of each species. In the derivation of the difference equations for the PATCH model, we wish to eliminate the problems of dealing with atto-individuals (i.e., unrealistically small densities). It is insisted that each patch contain at least one of each species present, therefore, if $A \in i$ then $A_i \geq 1/S_i^2$; values of A_i less than $1/S_i^2$ are interpreted as patches losing species A.

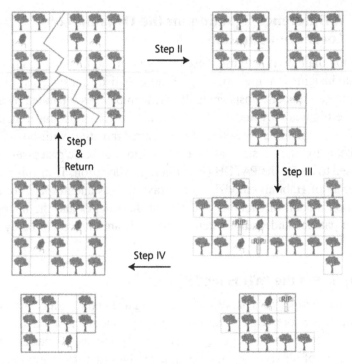

Figure 15.5 Pictorial representation of the PATCH model.

As shown in Figure 15.5, the dynamics of the PATCH model can be described by four distinct phases. In each step, all quantities are replaced by their average values; no consideration is given to the variance between patches of the same type. The first phase – partitioning the habitat – breaks the patch squares along lines of empty cells, forming two or more smaller square patches. In the next phase, allocation, species are redistributed randomly between these new smaller patches. In this procedure not all patches obtain the full complement of species, and therefore different types of patches may be formed. In Figure 15.5, a large patch of type $\{E, H, P\}$ forms three different types of patches: an $\{H, P\}$ patch, an $\{E, H\}$ patch, and an $\{E, H, P\}$ patch. The third phase, transmission, deals with the dynamics at the level of individuals (as opposed to those at the level of patches). Hosts grow, colonizing new territory; infection is transmitted between hosts; and infected hosts die. This step can be compared with the standard mean-field model. The final step is to coalesce patches. As hosts grow into the surrounding cells, many patches may become fused, forming a new, larger patch, possibly of a different type. The formal details of the transitions occurring during these phases are given in Appendix 15.A.

This PATCH model is an extension of the model given by Rand *et al.* (1995) in that it can be made to accommodate more parasite or host species and accounts for all nine possible patch types. It should be noted, however, that we are only interested in three main patch types, $\{H\}$, $\{E, H\}$, and $\{E, H, P\}$, as the others are usually rare and of short duration. For example, patches containing only parasites will disappear at the next iteration, and patches composed entirely of empty sites obviously cannot exist.

The PATCH model does not give exactly the same results as seen in the PCA model, and we should not expect it to. The discrepancies can be mainly attributed to the assumption of homogeneity within patches and the homogeneity of patches over space (i.e., only intermediate-scale patterns are taken into account). Many modifications could be made to improve the accuracy of the PATCH model, but we are searching for a simple cause for the qualitative difference between the PCA and the standard mean-field equations of Box 15.1. In particular, we wish to understand the phenomenon of a critical transmissibility and the behavior of the system as we approach this limit.

It is in this role that the PATCH model gives the greatest benefits. The PATCH model is a set of nonspatial deterministic equations; therefore, the results from simulations are achieved far faster, and the many difficulties due to stochasticity are eliminated. This model is only concerned with large-scale patterns (at the scale of patches), so features of the cellular automaton not captured by this model may be attributed to local correlations. Many of the features of the cellular automaton model can be reproduced by pair-wise correlation models (Keeling and Rand, in press; see also Chapters 18 and 19), but this is believed to be because the local correlations of these models naturally produce some large-scale spatial structures.

Dynamics of the PATCH model

The PCA and PATCH models have qualitatively similar behavior for changes in transmissibility T (see Figure 15.6): both have a minimum T_t and maximum T_c transmissibility for which parasites persist, and both show an increasing proportion of parasites with increased transmissibility. However, the significantly lower levels of parasitism observed for the PCA model, compared with the mean-field model, are not reproduced by the PATCH model. The levels of parasitism in the PATCH model are slightly below those in the mean-field models; this difference can be attributed to the isolation of hosts in patches without parasites. We can therefore assume that it is the fracturing of the habitat into isolated patches, preventing spread

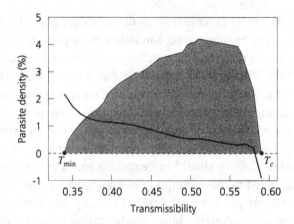

Figure 15.6 Percentage of parasites versus transmissibility (for the PATCH model). The thick solid line indicates the estimated level of the selective pressure toward higher transmissibility (see Section 15.4).

of the parasite, that is responsible for parasite extinction above the critical transmissibility. However, the lower levels of parasitism are primarily due to negative spatial correlations between hosts and parasites at a local scale.

We can use the PATCH model to test our intuitive understanding of the large-scale behavior. Figure 15.7 shows the size of the patches containing empty sites, hosts, and parasites ($S_{\{E,H,P\}}$). For transmissibility $T < 0.5$, the patches are very large (more than $10\,000$ cells) and thus the persistence and invasion behavior obey mean-field assumptions. However, for larger transmissibilities there is a decline in patch size, and a parasite in a small patch can quickly infect all available hosts, destroying the patch. This confirms our understanding from the cellular automaton model that it is the rapid exploitation of hosts together with the small isolated patches containing parasites that lead to the critical transmissibility.

15.4 Evolution to Critical Transmissibility

We have observed how the dynamics of the PCA model differ from those of its deterministic, homogeneous counterpart at the ecological time scale. Here, we examine how the parasite evolves over longer time scales. To do so, we include a mutant strain of parasite. As we are mainly concerned with evolution of transmissibility, the mutant strain will be assumed to differ only in the transmission parameter.

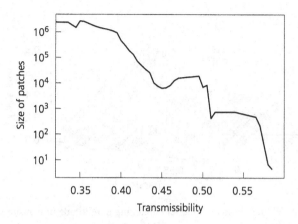

Figure 15.7 The size of patches containing empty sites, hosts, and parasites $(S_{\{E,H,P\}})$ for a range of transmissibilities T. The patch size decreases dramatically toward zero as the critical transmissibility is approached.

Results from the full population dynamics

One extension of the PCA that is easily accomplished and has simple, informative behavior is to allow each parasite to have its own transmissibility. The full evolutionary behavior of the system can then be modeled by allowing the transmissibility of each parasite to "mutate" by an amount $\pm\mu$ every time it infects a new host. At any one time there will be a range of transmissibilities competing with each other. The results from a single simulation of 10^5 iterations with $\mu = 10^{-4}$ is shown in Figure 15.8. From this one simulation it appears that the parasites evolve to a transmissibility close to the critical transmissibility T_c. This assumption is supported by results from many other simulations.

I propose that this phenomenon can be ascribed to two conflicting forces. In any situation a parasite will always reap an immediate benefit from having a higher transmissibility, as it will spread faster. However, for transmissibilities close to or above the critical transmissibility T_c, the parasite will have decreased the number of hosts in its local environment and therefore will drive itself to extinction. The two competing forces of immediate benefit and long-term persistence are equal at the critical transmissibility T_c.

If the mutation rate μ is increased, then the long-term average transmissibility increases because the spatial effects that disadvantage high transmissibilities require a long time to take effect; they are weak compared with the mutation rate. (For a mutation rate $\mu = 10^{-3}$, the long-term average transmissibility increases to approximately 0.7.)

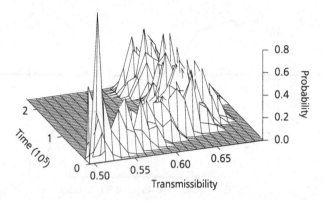

Figure 15.8 A cellular automaton model containing a multitude of parasite strains shows rapid convergence to T_c.

An approximating random walk in the transmissibility

To consider evolution in a more quantitative framework, it is necessary to examine the behavior of the selective pressure, a measure of the speed of evolution in a particular direction. For transmissibility, the selective pressure s is given by

$$s = \lim_{\varepsilon \to 0} \frac{1}{\varepsilon} \ln(\text{Initial growth ratio of mutant}) ,$$

$$\varepsilon = \text{Transmissibility of mutant} - \text{Transmissibility of resident parasite}$$

(15.2)

at the stationary point when the mutant parasite population is small. This formulation for the evolutionary speed assumes that the mutation rate is so low that only one mutant strain is present at any time. Box 15.2 calculates the selective pressure for the simple mean-field model. This selective pressure predicts that we should observe a succession of invasions by mutants with higher transmissibilities tending to the maximum value of 1.

The complication of a mutant species of parasite with a different rate of transmission (T_M as opposed to T_P for the resident parasite) must now be added to our PCA model. For simplicity, we assume that each host can be infected by at most one parasite. If two different strains of parasite compete for the same host, the victor is determined probabilistically according to the transmission rates given in Table 15.2. To determine the evolutionary behavior, the standard PCA model (containing just empty sites, hosts, and normal parasites) is run for a sufficient number of iterations to remove transients. Mutants are then introduced by changing a small proportion of the parasites to a mutant strain with a different transmissibility.

In a manner similar to Equation (15.2), I now wish to define the selective pressure for a stochastic spatial system. But we do not have a lattice

---

Box 15.2 Selective pressure in the mean-field equations

To consider evolution, a mutant strain must be added to the homogeneous mean-field equations [see Equation (a) in Box 15.1],

$$H' = H + 4g(1 - H - P - M)H - 4T_P HP - 4T_M HM$$
$$P' = 4T_P HP \tag{a}$$
$$M' = 4T_M HM ,$$

where M is the mutant population, and T_P and T_M are the transmissibilities of the resident parasite and mutant, respectively. Equation (a) has a stationary point $H^* = \frac{1}{4T_P}$, $P^* = \frac{g(4T_P-1)}{4T_P(T_p+g)}$, $M^* = 0$, from which the selective pressure $s(T_P)$ can be found:

$$
\begin{aligned}
s(T_P) &= \lim_{T_M \to T_P} \frac{\ln(4T_M H^*)}{T_M - T_P} \\[2mm]
&= \lim_{T_M \to T_P} \frac{\ln(\frac{T_M}{T_P})}{T_M - T_P} \\[2mm]
&= \lim_{T_M \to T_P} \frac{\ln(1 + \frac{T_M - T_P}{T_P})}{T_M - T_P} \\[2mm]
&= \frac{1}{T_P} .
\end{aligned}
\tag{b}
$$

---

counterpart for the initial growth ratio of a mutant population, or it should be the speed of the wave in which the mutant population spreads over space. However, by delving a little deeper we see that the speed of evolution is determined by the probability that advantageous mutants pass through the boundary layer, where numbers are still so low that the stochastic aspects of individual behavior dominate. The initial growth ratio is important only in so far as it is proportional to this probability (Dieckmann and Law 1996). Therefore, instead of referring to the initial growth ratio, we may directly refer to the probability of invasion and subsequent takeover. For the infinite lattice counterpart of our model, we expect that evolution will only be successful in at most one direction. A substitution then consists of three distinct steps:

- The mutant overcomes the initial, relatively large, stochastic fluctuations when its numbers are still small.
- A wavelike spread of the newly introduced mutant over the spatial domain occurs.
- The former resident is eventually ousted.

Table 15.2 Transition probabilities for the extended host–parasite probabilistic cellular automaton.

Probability of transition from Host to:

Host	$(1 - T_P)^P (1 - T_M)^M = q$
Parasite	$\dfrac{[1 - q](1 - (1 - T_P)^P)}{1 - (1 - T_P)^P + 1 - (1 - T_M)^M}$
Mutant	$\dfrac{[1 - q](1 - (1 - T_M)^M)}{1 - (1 - T_P)^P + 1 - (1 - T_M)^M}$

The last two processes follow automatically as soon as the first has taken place. However, the simulations deal with a finite lattice where there is a positive probability of a substitution succeeding both for higher and lower transmissibilities, thus we need to consider evolution in both directions. The best possible extension to this case is to require the selective pressure to at least capture the mean rate of evolutionary change through the subsequent gene substitutions.

This leads to the following definition, with $\varepsilon = |T_M - T_P|$ (the difference between the mutant and resident transmission rates):

$$s^+ = \text{Prob}\{T_P + \varepsilon \text{ invades}\}, \tag{15.3}$$

$$s^+ = \text{Prob}\{T_P - \varepsilon \text{ invades}\}, \tag{15.4}$$

$$s = \tfrac{1}{2} \lim_{\varepsilon \to 0} \tfrac{1}{\varepsilon}(s^+ - s^-). \tag{15.5}$$

So s represents the difference between the probabilities of a successful invasion for higher and lower transmissibilities.

As we are dealing with small probabilities, the numerics are expensive because many simulations are necessary. However, the results indicate that the selective pressure is positive up to T_c, so there is always evolution toward the critical transmissibility. Unfortunately, complications may arise close to T_c, because the system often collapses so that both mutants and normal parasites die out.

Using the values of the two selective pressures (s^+ and s^-), a Markov process can be formed to model the behavior of a slowly mutating population of pathogens. The probability that the population moves to a higher or lower transmissibility is proportional to s^+ and s^-, respectively, As this does not rely on the past history of the population, it is a Markov process. The population is considered to be slowly mutating so that there is only one mutant strain present at any time. The probability $\mathscr{P}(T)$ of the resident parasite having transmissibility T evolves according to the following:

$$\mathscr{P}(T)' = (1 - s^+ - s^-)\mathscr{P}(T) + s^+ \mathscr{P}(T - \varepsilon) + s^- \mathscr{P}(T + \varepsilon). \tag{15.6}$$

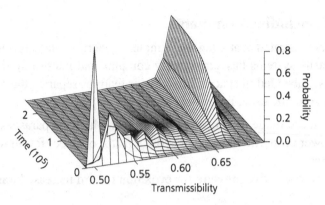

Figure 15.9 Convergence of distribution of parasite transmissibilities to an asymptotic limit at $T = 0.67$. This outcome is the result of more transmissible parasites having a local advantage, with those having transmissibilities greater than T_c being doomed to extinction.

That is, parasites with transmissibility T evolve to a transmissibility $T + \varepsilon$ at a rate proportional to s^+, and evolve to a transmissibility $T - \varepsilon$ at a rate proportional to s^-. Starting with a delta function at $T = 0.51$ and iterating the map (15.6), an asymptotic distribution of probabilities is reached with the maximal value occurring at $T = 0.67$, which is the critical transmissibility (Figure 15.9). (Note that Figure 15.9 shows the probability distributions for the different realizations of the stochastic monomorphic model, whereas Figure 15.8 gives the relative frequencies of the various types within a given simulation.)

One feature that is consistently found with this and other spatial models is the far lower values for selective pressure compared with results from homogeneous mean-field equations. This difference can be attributed to the smaller number of possible contacts due to the local interactions and nontrivial spatial structure. The buildup of strong spatial correlations means that nonlinear effects (the presence of other mutants) are felt far earlier, which in turn leads to lower selective pressure. Another factor is the high risk of stochastic extinction in the initial stages of invasion due to the relatively large fluctuations experienced in an individual's environment.

The level of the selective pressure for greater transmissibility for the PATCH model is also shown in Figure 15.6, which again predicts evolution toward criticality. The reduced selective pressure, which is a feature of many spatial systems, is not observed in this PATCH model. Again, this phenomenon is attributable to local interactions: in the cellular automaton resident and mutant parasites can only compete directly across a narrow boundary, whereas in the PATCH model there is direct competition throughout any patches that contain both parasite strains.

15.5 Concluding Comments

I have demonstrated that even the simplest generic caricature model for host–parasite systems has surprisingly complex and interesting dynamics when it is embedded in space. Three major features separate the results of this PCA from the usual mean-field theory:

- The reduced range of transmissibilities over which the parasites persist
- The lower proportion of parasites (by a factor of three) for all transmissibilities
- The lower selective pressure for evolution toward increased transmissibility

Despite the gross simplifications made in each step of the PATCH model, the overall results are in good qualitative agreement with those of the PCA. The results from the PATCH model lend supporting evidence to the idea that it is the intermediate-scale spatial structure that causes many of the departures from the mean-field theory. The fracturing of the habitat into small isolated patches of hosts and parasites is what leads to the extinction of the parasite at high transmissibilities and, therefore, to the existence of a critical transmissibility.

Many other scenarios have also been considered (Keeling and Rand, in press), but because of a profusion of parameters, generic statements about behavior or a comprehensive sweep of parameter space are infeasible.

The three features mentioned above can have a profound effect on modeling the spread of disease through spatially distributed populations. The existence of a critical transmissibility T_c implies that we need not rely on relationships between transmissibility and virulence to explain why not all pathogens have evolved to $T = 1$, $V = 0$ (the evolutionary attractor for simple mean-field models). The decreased levels of selective pressure mean that evolution in the natural world may not be as rapid as would be expected from mean-field models. Finally, the persistence of the parasite at far lower levels of prevalence than predicted by homogeneous models may well be a ubiquitous feature of many systems where ephemeral spatial refuges and strong local correlations exist. In that case, the parameterization of epidemiological models from simple global averages will produce misleading results, as interactions will appear to be far weaker than when space is taken into account.

Appendix 15.A Mathematical Specification of the PATCH Model

The specification of the PATCH model can be divided into the following four operations (see Figure 15.5).

Partitioning the habitat. Before partitioning a patch, we first allow hosts to grow into empty sites, which they do with probability g. The probability that an empty site remains uncolonized by a given neighbor is $1 - gH_i$, therefore, with four neighbors

$$E_i' = E_i(1 - gH_i)^4 ,$$
$$H_i' = H_i + E_i\left[1 - (1 - gH_i)^4\right] .$$
$$(15.7)$$

If the proportion of empty sites E_i is greater than one-half, the patch is assumed to be fragmented into individual sites. Otherwise, we need to approximate the number of breaks along lines of empty sites. Let p_j be the probability that a site in row j is empty and connected to row 1 by a path of empty sites. This will occur if the site is empty and either the site directly below in row $j - 1$ is empty and connected to the base or a site on either side is empty and connected to the base. Therefore,

$$p_{j+1} = E_i\left[p_j + (1 - p_j)(2p_{j+1} - p_{j+1}^2)\right]$$
$$\Rightarrow p_{j+1} = \frac{2E_i(1 - p_j) - 1 + \sqrt{1 - 4E_i(1 - p_j)(1 - E_i)}}{2E_i(1 - p_j)} ,$$
$$(15.8)$$

where $p_1 = E_i$. Each patch can be fragmented into smaller patches through unbroken lines of empty sites linking row 1 to either an edge or to the top row. The probability that an edge on row j is linked to row 1 by empty sites is

$$e_j = (1 - e_{j-1})p_j .$$
$$(15.9)$$

Therefore, the expected number of fragmenting paths is

$$F_i \approx S_i p_{S_i} + 2 \sum_{j=2}^{S_i-1} e_j .$$
$$(15.10)$$

This operation changes the variables for patch type i as follows: the density of sites D_i is increased by a factor $F_i + 1$; the empty sites that are part of the fragmenting paths are removed (there are assumed to be $\frac{S_i F_i}{2}$ of these sites) and the proportion of each species is recalibrated to sum to 1; finally, the average patch size is altered to account for the fragmenting and loss of empty sites,

$$S_i' = \sqrt{\frac{S_i^2}{F_i + 1} - \frac{S_i F_i}{2(F_i + 1)}} .$$
$$(15.11)$$

It should be noted that this is a very crude approximation, as all breaks must be between distinct edges. However, all that is important for working the model is that as the proportion of empty sites increases there is a rapid increase in the fragmentation of the patch. Dropping the primes, we consider the next phase.

Allocating the species.   As the patches are now reduced in size, it is likely that they no longer contain all the species of the original habitat type. For example, if a patch of type $\{E, H, P\}$ has been subdivided, there is a good chance that not all the new smaller patches contain parasites; thus these sites revert to type $\{E, H\}$. By assuming a random distribution of species in the new patches, they can be reallocated a habitat type.

The proportion of patches of type i that after allocation revert to type j ($j \subseteq i$) is $P_{i,j}$,

$$P_{i,j} = \left(1 - \sum_{A \notin j} A_i \right)^{S_i^2} - \sum_{k \subset j} P_{i,k} \, . \tag{15.12}$$

In other words, it is the probability that all the S_i^2 cells contain species found in j minus the probability that the patch produced is of type $k \subset j$; that is, it contains only those species found in j and all the species found in j. We can calculate the density of all the new patches of type j that have been formed,

$$D_j' = \sum_{i \supseteq j} P_{i,j} D_i \, . \tag{15.13}$$

Therefore, the average size is

$$S_j' = \sqrt{\frac{1}{D_j'} \sum_{i \supseteq j} P_{i,j} D_i S_i^2} \, . \tag{15.14}$$

The average proportion of species within each patch must now have also changed, as all individuals have been redistributed. We calculate the proportion of species in each patch, insisting that if a patch contains a given species then it must contain at least one individual of that species (i.e., no fractional individuals are allowed). The proportion of species A found in a patch of type j formed from a patch of type i is found as follows. Let ψ be the proportion of species A remaining after all the patches (of type k) that are subsets of j have been removed; then

$$\psi_{i,j}(A) = \frac{A_i - \sum_{k \subset j} P_{i,k} \psi_{i,k}(A)}{1 - \sum_{k \subset j} P_{i,k}} \qquad \text{if } A \in j \, , \tag{15.15}$$

or is zero otherwise. The values of $\psi_{i,j}(A)$ can then be calculated using an iterative procedure. Therefore, each new patch of type j, which was formed from a patch of type i, on average contains $\psi_{i,j}(A) S_j^2$ individuals of type A. It is simple to confirm that $\psi_{i,j}(A) S_j^2 \geq 1$ for all $A \in j$, that is, all patches contain at least one individual of each species present. Therefore, after the redistribution of individuals, the average proportion of species A found in all the new patches of type j is

$$A_j' = \frac{1}{S_j'^2} \sum_{i \supseteq j} P_{ij} D_i S_i^2 \psi_{i,j}(A) \, . \tag{15.16}$$

Transmission of parasites and growth of patches. This operation occurs according to the standard homogeneous mean-field approximation to the dynamics, with hosts growing into unoccupied sites around the edge of each patch and the parasite being transmitted to healthy hosts. The only concession we make to local-scale correlations is that each parasitized host is assumed to have a maximum of three uninfected host neighbors:

$$P_i' = \left[1 - (1 - T_P P_i - T_M M_i)^3\right] \frac{\left[1-(1-T_P P_i)^3\right]}{\left[1-(1-T_P P_i)^3\right]+\left[1-(1-T_M M_i)^3\right]}$$
$$M_i' = \left[1 - (1 - T_P P_i - T_M M_i)^3\right] \frac{\left[1-(1-T_M M_i)^3\right]}{\left[1-(1-T_P P_i)^3\right]+\left[1-(1-T_M M_i)^3\right]} .$$

(15.17)

Because it is assumed that all the patches are square, as long as $S_i \geq 2$, there are $4S_i$ empty sites around each patch into which hosts can grow. The size of each patch is increased by host growth (S_i^+) but decreased by deaths due to parasitism (S_i^-) around the edge of the square,

$$S_i' = \sqrt{S_i^2 + S_i^+ + S_I^-} ,$$

(15.18)

where $S_i^+ = 4g H_i S_i$ and $S_i^- = 4(P_i + M_i)S_i$. The density of each species within the patch must now be modified to take this growth into account, such that all the densities sum to 1.

Coalescing the habitats. The final operation accounts for combining of patches due to the bridging of gaps by the growth of hosts into the empty spaces between the patches. The density of the new patch formed when patches i and j coalesce is

$$C_{i,j} = D_i D_j \frac{\text{Growth into empty sites}}{\text{Number of empty sites}}$$
$$= D_i D_j \frac{S_i^+ + S_j^+}{1 - \sum_{j \in G} D_k (S_k + 1)^2} .$$

(15.19)

The new patch is of type $i \cup j$ and of size $\sqrt{S_i^2 + S_j^2}$.

16

Foci, Small and Large: A Specific Class of Biological Invasion

Jan-Carel Zadoks

16.1 Introduction

A particularly clear and often observed spatial pattern has multiple, roughly circular patches of a single species standing out against a different background. More often than not, this pattern is the result of the outgrowth of local colonies from sparse, randomly distributed inocula. This chapter analyzes this phenomenon from the perspective of plant pathology.

In the context of plant pathology, a focus is defined as a patch of crop with disease limited in space and time (Anonymous 1953). Here, I deal with foci due to infectious disease of plants caused by fungi. I primarily discuss polycyclic fungal diseases (i.e., diseases with many reproduction cycles per season; Zadoks and Schein 1979) of foliage in annual crops. The literature contains various publications with illustrations of foci, sometimes from aerial photography (Colwell 1956; Brenchley and Dadd 1962; van der Werf 1988). The reality of foci cannot be doubted. Their shape is usually roundish, unless the wind has exerted a great influence on spore dispersal (Brenchley and Dadd 1962; Zawolek 1993). Foci are not limited to fungal disease of foliage; they are also known for bacterial and viral diseases, plant pathogenic nematodes, and insects.

The earliest mention of "foci" caused by a soil-borne fungus dates from 1728 (Duhamel de Monceau, in Zadoks 1981). Foci reappear in the literature late in the eighteenth century for black stem rust (*Puccinia graminis*) of wheat or rye, and disappear again until 1896 (Eriksson and Henning 1896). Systematic studies on black stem rust of wheat and oats and on crown rust (*Puccinia coronata*) of oats were initiated in the United States during the 1920s and 1930s. My own work on yellow stripe rust (*Puccinia striiformis*) of wheat (Zadoks 1961) induced a career-long interest in the phenomenon of foci, small and large. Initially, mathematics could only handle disease progress in time (Vanderplank 1963). When dynamic simulation became available, disease progress in space could be modeled. Analytical approaches now also produce a good match between predicted and observed focal expansion rates (van den Bosch *et al.* 1988c, 1990b; Zadoks and van den Bosch 1994; see also Jeger 1983, 1989, and the references in Metz and van den Bosch 1995).

Spreading foci are just one means by which plant pathogens can conquer space. Other means are available, such as the formation of spore clouds either spreading as Gaussian plumes in contact with the Earth's surface or traveling in the upper air detached from the soil surface. In the latter case, rain deposits the spores on the crop. This chapter emphasizes foci spreading at a constant rate ("running wave" models); theory also allows for "dispersive wave" models in which the speed continues to increase (see Appendix 16.A, Discrete-time kernel-based models; and Section 23.3, Spatial scales). The latter alternatives are not considered here, but they merit more attention.

This chapter "focuses" on the development of foci, small and large, as down-to-earth and dirty empirical counterparts to neat and high-brow theoretical concepts. Section 16.2 concentrates on the different degrees of complexity of the process of epidemic spread, or "epidemic orders." Section 16.3 develops the theory of foci from a mixed descriptive and theoretical perspective; the more mathematical Chapter 23 provides a technical counterpart. Finally, Section 16.4 discusses generalizations as well as extensions to more complex forms of spatial spread.

16.2 Epidemic Orders

As foci can be small or large, we organize our thoughts by distinguishing three orders of epidemics (Heesterbeek and Zadoks 1987), called zero-, first-, and second-order epidemics. The criteria on which the distinction is made are size, complexity, and time scale. In principle, all three orders of focal epidemics obey the same general rule, that of constant radial expansion.

Zero-order epidemics

These are small-scale foci of about 1 m in diameter (Figures 16.1 to 16.6), as often seen in potato late blight (*Phytophthora infestans*), in yellow rust (*P. striiformis*) of wheat (Zadoks 1961), in rust (*Puccinia arachidis*; Savary 1986) and late leaf spot (*Cercosporidium personatum*) of peanut (Savary

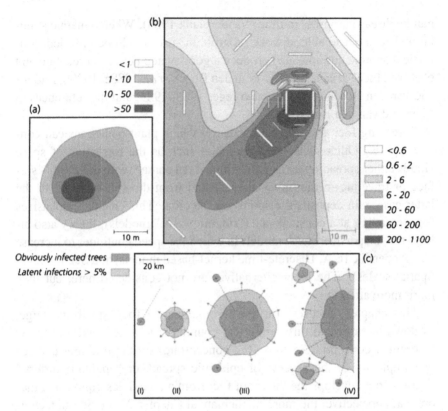

Figure 16.1 Zero-order epidemics, static isopath representations of foci. (a) Potato late blight (*Phytophthora infestans*) in the Netherlands, 1954. Wind caused a strong anisotropy. Legend: diseased leaf area (%). *Source*: van der Zaag (1956). (b) Loose smut (*Ustilago tritici*), a monocyclic disease of wheat. In ordinary foci, some spread goes against the predominant wind direction. Legend: infected wheat heads per $100\,\mathrm{m}^2$. *Source*: Oort (1940). (c) Sketch of focal spread of swollen shoot disease (a viral disease of tree crops) in cocoa, with transition from zero- to first-order epidemics. (I) Infection from outside source, (II) local spread, (III) appearance of satellite outbreaks, and (IV) area of extensive infection. Short arrows represent spread of virus by mealybugs moving through the canopy. Dashed arrows represent spread of virus by wind-borne mealybugs. *Source*: Vanderplank (1963; after Thresh 1958).

and van Santen 1991), and in many other pathosystems (Robinson 1976). These are the foci of the definition in Section 16.1. They typically begin with a single lesion (Lambert 1929; Zadoks 1961) and attain an easily detectable size after four to six generations of multiplication, and they are typically about 0.5–2 m (and seldom over 20 m) in diameter. They are mentioned extensively in the literature (Table 16.1). Such zero-order epidemics are not restricted to fungal diseases but also occur in plant diseases caused

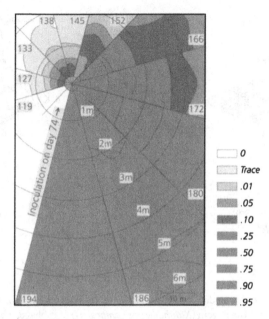

Figure 16.2 Zero-order epidemic, dynamic spider web representation of focus progress in time and space. Yellow stripe rust (*Puccinia striiformis*) on wheat. Days are counted from January 1. Legend: diseased leaf area (%). *Source*: Zadoks and Schein (1979).

by bacteria (Kocks and Zadoks 1996; Yang *et al.* 1991) and viruses (Thresh 1958; van der Werf 1988), nematodes (Been and Schomaker 1998), and insects (e.g., brown planthopper, *Nilaparvata lugens*, on rice; J.-C. Zadoks, personal observations).

First-order epidemics

When weather conditions are adverse to spread of the pathogen, the foci become stationary or even disappear with further growth of the canopy. But when the weather is favorable, the disease spreads. The primary focus enlarges, secondary or daughter foci appear in the same field, and fields in the neighborhood of the first infected field show evidence of the disease, maybe even in the form of small secondary and tertiary foci (Brenchley and Dadd 1962; Thresh 1958; Zadoks 1961; Zawolek 1993; Sache and Zadoks 1995). This process of focal spread of a disease over a larger area within a single growing season is called a first-order epidemic. The size of a first-order epidemic may be restricted to one field only, but cases exist where it attains a diameter in the order of 1000 km (Figures 16.7 to 16.9; Table 16.2).

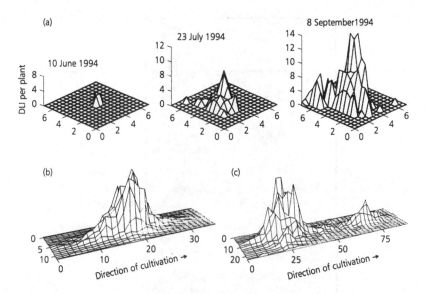

Figure 16.3 Zero-order epidemic (focus), static hat representations. For horizontal axes, 1 unit equals 1 m. (a) Black vein (*Xanthomonas campestris*) of cabbage, a bacterial disease. Note the incipient daughter foci around the mother focus. Vertical axes show diseased leaf incidence (DLI) per plant. *Source*: Kocks and Zadoks (1996). (b, c) Potato cyst nematode (*Globodera rostochiensis*), a monocyclic pest. The anisotropy was caused by soil tillage. Note the daughter focus in (c). Vertical axes show relative measure for number of cysts per kilogram of soil. *Source*: Been and Schomaker (1998).

Special cases of zero- and first-order epidemics are those beginning with a localized mass inoculum instead of a single spore or lesion. Such is the case of the black stem rust of wheat (*P. graminis*), starting from barberry (*Berberis spp.*) bushes, so well documented in the American literature of the 1920s, but also documented around 1810 in Denmark for black stem rust of rye (reviewed in Hermansen 1968). If the focus begins with a mass inoculum – for example, 1 m² of lesions instead of a single lesion – the buildup time before the outward movement of gradients begins is reduced, and the focus "gains time" (compare Figures 16.6, 16.8, and 16.12). Zero- and first-order foci may start from refuse piles containing infectious material, as in potato late blight (*P. infestans*; Bonde and Schultz 1942/1943; Zwankhuizen *et al.* 1998) and in black rot of cabbage (*Xanthomonas campestris*; Kocks and Zadoks 1996). These special cases are of great economic importance, and it pays to eradicate the pathogen from refuse piles.

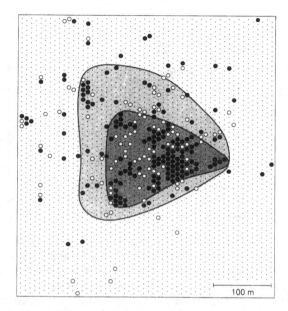

Figure 16.4 Zero-order epidemic, dynamic dot representation of focal expansion. Spear rot (cause unknown) of oil palm in a plantation in Surinam. Grid points represent individual trees, 9 m apart. An anisotropic focus develops in a field with scattered background infections. Filled circles and dark shading indicate infection in January 1986; open circles and light shading indicate infection in January 1987. *Source*: van de Lande and Zadoks (in press).

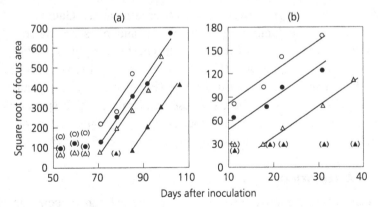

Figure 16.5 Zero-order epidemics, dynamic "area within the isopath" representation. The square root of the focus area increases linearly with time, irrespective of the disease severity level along the isopath of measurement. Open circles represent 1% severity; filled circles, 5%; open triangles, 10%; and filled triangles, 50%. The straight line fits (slopes forced to be equal) to the unbracketed data points provide an estimate of the spatial expansion rate. (a) Yellow stripe rust (*Puccinia striiformis*) of wheat, average data of five replicated artificial foci in a commercial field. (b) Downy mildew (*Peronospora farinosa*) on spinach in the field. *Source*: van den Bosch *et al.* (1988c).

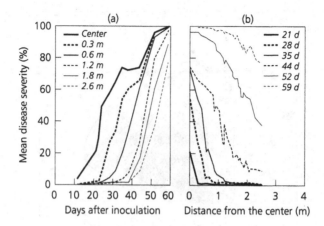

Figure 16.6 Zero-order epidemic, dynamic running wave representation of single focus. Rust fungus (*Uromyces viciae-fabae*) on faba bean. The front of the epidemic takes a certain shape and moves outward with approximately constant speed (running wave). (a) The disease growth curves at different distances from the center of the focus are more or less parallel sigmoid curves, showing the delay of disease development with increasing distance from the center. (b) Gradual buildup of the disease profile from approximately exponential to sigmoidal, followed by an outward movement maintaining the sigmoidal shape. *Source:* Sache and Zadoks (1995).

Second-order epidemics

Epidemics often progress over time to cover large areas. Gäumann (1946) then speaks of pandemics. Usually such pandemics are the result of new introductions, as was the case with chestnut blight (*Endothia parasitica*) in the United States (Heald and Studhalter 1914), tobacco blue mold (*Peronospora tabacina*) in Europe (Populer 1964), and peanut rust (*P. arachidis*) in Africa. We call these pandemics, which spread over a large area (millions of square kilometers) over a certain number of years, second-order epidemics (Figures 16.10 to 16.12 and Figure 16.15).

16.3   A Theory of Foci

The mathematical theory for development of foci can be constructed in steps, from description to deterministic simulation to mathematical analysis. The analytical approach allows quick quantitative comparisons between models, and between models and field data, at the cost of a stylized representation of the messy biological reality (although less stylized than in the early phases of the simulation approach). As a fourth step, the direct numerical integration of physical equations for spore dispersal

Table 16.1 Some records of zero-order epidemics.

Organism	Host	$c_0{}^a$	References
Fungi			
Botrytis elliptica	Lily	–	J. Köhl, personal communication
Cercospora beticola	Sugar beet	–	J.-C. Zadoks, personal observations
Peronospora farinosa	Spinach	2.3	van den Bosch *et al.* 1988c
Phytophthora infestans	Potatoes	–	Brenchley and Dadd 1962;[b] Hirst and Stedman 1960;[b] Zwankhuizen *et al.* 1998[b]
Puccinia arachidis	Peanut	–	Savary 1986
Puccinia coronata	Oats	≈50	Berger and Luke 1979[b]
Puccinia graminis	Rye/wheat	8	Colwell 1956;[b] Hermansen 1968
	Wheat	≈20	Eriksson and Henning 1896; Joshi and Palmer 1973;[b] Roelfs 1972
Puccinia recondita	Wheat	9	J.-C. Zadoks, unpublished data
Puccinia striiformis	Wheat	–	Joshi and Palmer 1973; Zadoks 1961; Zadoks and Schein 1979
	Wheat	8	van den Bosch *et al.* 1988c, 1990b
Uromyces appendiculatus	French bean	14	Habtu *et al.* 1995[b]
Uromyces viciae-fabae	Broad bean	–	Sache and Zadoks 1995[b]
Virus			
Sugar beet yellows	Sugar beet	–	van der Werf 1988[b]
Swollen shoot	Cocoa	–	Thresh 1958;[b] J.-C. Zadoks, personal observations
Unknown etiology			
Spear rot	Oil palm	–	van de Lande and Zadoks, in press[b]

[a] c_0 = radial expansion rate in cm per day.
[b] Picture given in reference.

is considered together with a relatively realistic model for pathogen reproduction, stochastic initiation of new infections, and inhomogeneity of the environment.

Description

Description entails mapping the epidemic in such a way that a single, more or less circular curve separates the diseased area from the non-diseased area. This curve, called an "isopath" (Berger and Luke 1979), connects all points with the same, usually low, disease level (see Figures 16.1a and 16.1b). When we make such maps at regular time intervals, we may see that the focus expands and the isopath is displaced in an outward direction (see Figure 16.6).

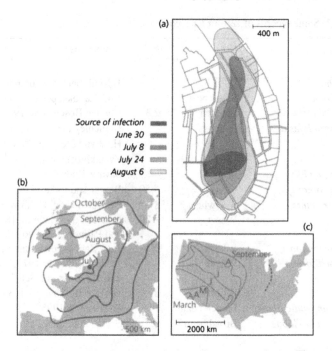

Figure 16.7 First-order epidemics, dynamic isopath representations. The spatial scale may vary considerably according to crop, pathogen, and weather. (a) Downy mildew fungus (*Peronospora destructor*) of onions in the Netherlands. *Source*: van Doorn (1959). (b) Potato late blight (*Phytophthora infestans*) in Europe, 1845. The pandemic began in Flanders. The weather was extremely favorable, the crop was excessively susceptible, and the radial expansion rate was practically constant for over three months and nearly isotropic. *Source*: Heesterbeek and Zadoks (1987, based on Bourke 1964). (c) Sugar beet mildew (*Erysiphe betae*) in the western United States. Note the delay caused by the Rocky Mountain chain. Dashed contour indicates region of farthest spread of disease. *Source*: Heesterbeek and Zadoks (1987).

More interestingly, displacement of the isopath and of the focal "front" (Vanderplank 1975) – or, in other words, the radial expansion of the focus – often occurs in a regular way, proceeding with a more or less constant speed denoted by c_0. The period of constant radial expansion rate is limited, because in the beginning of the epidemic the radial expansion rate must build up from zero and at the end of the epidemic the rate will decline as the weather conditions change or the crop fails. This dynamic isopath mapping has been done for all three orders of epidemic (see Figures 16.4, 16.7, 16.10, and 16.11).

For zero-order epidemics, studies were made by van den Bosch *et al.* (1988c) on *P. striiformis* of wheat and *Peronospora farinosa* of spinach.

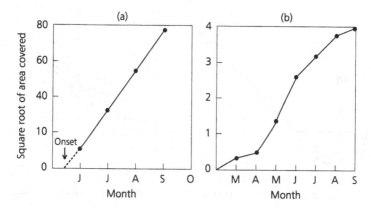

Figure 16.8 First-order epidemics, "area within the isopath" representation. *Source*: Heesterbeek and Zadoks (1987). (a) Potato late blight from Figure 16.7b. (b) Sugar beet mildew from Figure 16.7c.

These epidemics demonstrate that a constant radial expansion rate of a focus can indeed be measured (see Figures 16.5 and 16.6, and Table 16.1). Work on linear "foci" of *P. infestans* of potatoes has pointed to the same effect (Minogue and Fry 1983a, 1983b; van Voorst *et al.* 1987).

First-order epidemics have been studied by Zadoks and Kampmeijer (1977), Heesterbeek and Zadoks (1987), and Zadoks (1988), using *P. infestans* of potatoes in Europe and *Erysiphe betae* of sugar beet in the United States as examples. A small-scale first-order experiment was performed by Sache and Zadoks (1995) using faba bean rust (see Table 16.2).

The same ideas can be applied to second-order epidemics (Figure 16.15), such as the pandemic of *Peronospora tabacina* in Europe (Populer 1964) and the epidemic of *E. parasitica* on American chestnut (*Castanea dentata*) in the United States (Heesterbeek and Zadoks 1987). Apparently, the phenomenon of the constant expansion rate is not limited to annual crops but can also be seen in tree "crops."

However, the reader should be cautious. Not all large-scale pandemics and small-scale epidemics behave in the way described. We recognize a pattern, but not necessarily the only possible pattern. Nonetheless, the broad pattern described here can be generalized. It applies to several insect pests, to some viral diseases, and to at least one human disease, the famous plague of 1348 due to *Yersinia (Pasteurella) pestis* (Noble 1974; see Table 16.3).

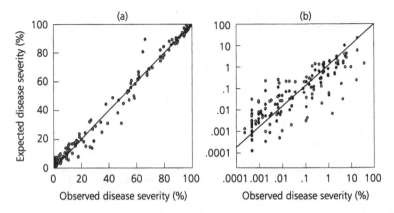

Figure 16.9 Zero- and first-order epidemics of the rust fungus (*Uromyces viciae-fabae*) on faba bean. Observed disease severity (horizontal) plotted against expected disease severity (vertical) according to descriptive model (iv) in Jeger (1983). This model represents a focus by a decreasing logistic function in the radius, with a time-dependent location parameter; in model (iv) the time dependence corresponds to an increasing expansion speed, as tends to be observed in the initial phase of focus buildup. Circles represent observations in space and time. The slopes of zero- and first-order epidemics are quite different. (a) Zero-order epidemic. (b) First-order epidemic on 30 trap plots spread over 45 m in a 45° sector downwind of the same focus. Note that observations excluding the two secondary foci (filled circles) fit the model better than observations including them. The disease severity in the trap plots was determined per square meter of crop, allowing a variation between 0.0001% and 20% to be observed. *Source*: Sache and Zadoks 1995.

Simulation approach

Dynamic simulation was introduced into plant pathology in 1968 (Waggoner 1968; Zadoks 1968). Spatio-temporal simulation began in 1974 (Kiyosawa and Shiyomi 1974). Kampmeijer and Zadoks (1974) published the first complete spatio-temporal simulation model, EPIMUL, written in FORTRAN, using a Gaussian dispersal kernel. Though the spatial model was relatively simple and the temporal model applied a simple form of Vanderplank's (1963) equation, the suggestion of focus expansion at a constant radial speed was made. EPIMUL has also been used to study the effect of mixtures of crop cultivars, mixtures of pathogen genotypes, and varietal patterns (Mundt and Leonard 1986; Mundt et al. 1986; Lannou and Mundt 1996; Lannou et al. 1994).

Vanderplank's equation reads as follows:

$$dx(t)/dt = R_c[x(t-p) - x(t-i-p)][1 - x(t)] , \qquad (16.1)$$

where R_c is the "corrected basic infection rate," p is the latent period, i is the infectious period, $x(t)$ is the fraction of diseased foliage, the term

Table 16.2 Some records of first-order epidemics.

Organism	Host	$c_0{}^a$		Reference
Fungi				
Erysiphe betae	Sugar beet	1.8	N	Zadoks 1988[b]
Helminthosporium maydis	Maize	11	N	Zadoks and Kampmeijer 1977[b]
Peronospora tabacina	Tobacco	16[c]	N	Zadoks and Kampmeijer 1977
Phytophthora infestans	Potato	6	E	Zadoks and Kampmeijer 1977[b]
	Potato	15	E	Jeger 1983[b]
	Potato	–	N	Zadoks and Kampmeijer 1977
	Potato	–	E	Zwankhuizen *et al.* 1998[b]
Puccinia graminis	Wheat	–	N	Schmitt *et al.* 1959; Underwood *et al.* 1959
Puccinia graminis	Wheat	50	I[d]	Nagarajan and Singh 1975; Nagarajan *et al.* 1976
Puccinia horiana	Chrysanthemum	–	E	Zandvoort 1968[b]
Puccinia striiformis	Wheat	9	E	Zadoks 1961[b]
Uromyces viciae-fabae	Faba bean	–	E	Sache and Zadoks 1995

Note: E = Europe, I = India, N = North America.
[a] c_0 = Radial expansion rate in km per day.
[b] Picture given in reference.
[c] 10–32 km per day.
[d] Some characteristics of Indian wheat rust epidemiology are typical for focal epidemics.

$[x(t - p) - x(t - i - p)]$ is the fraction of the diseased foliage infected between $t - p$ and $t - i - p$ time units ago, and the term $[1 - x(t)]$ gives the fraction of the foliage still available for infection. The equation produces a wavelike pattern of increase in the disease level x, with $0 \le x \le 1$. The higher the generation number, the more spread out the waves. Consequently, later waves start to overlap to a point where they are no longer visible as separate entities. The overall shape of the disease growth curve is sigmoidal in time. The equation has stimulated much research, even though it is essentially false because it lacks a spatial element. Although the equation cannot be solved analytically, it is well suited to dynamic simulation.

EPIMUL operated by numerically integrating the spatial extension of Vanderplank's equation on a rectangular grid. Although a simple, deterministic model, it showed general principles that were in accordance with field experience. Building on the experience with EPIMUL, more powerful models were developed, such as PODESS (Zawolek and Zadoks 1989) described below. Of course, a simulation model cannot give proof of a theorem such as the constant radial expansion rate of foci, but it can illustrate the principle.

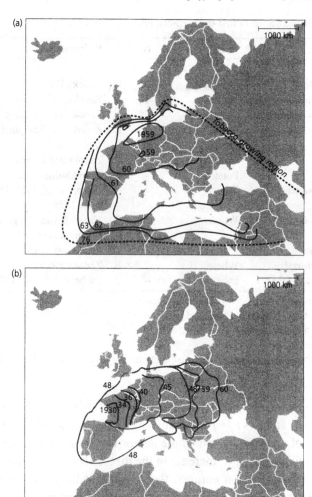

Figure 16.10 Second-order epidemics, dynamic isopath representation. (a) Tobacco blue mold (*Peronospora tabacina*) in Europe, 1958–1976. During the first year, the epidemic picked up speed; for two to three years it moved at a nearly constant radial expansion rate; and toward the end it slowed down. *Source*: Heesterbeek and Zadoks (1987). (b) Colorado beetle (*Leptinotarsa decemlineata*) on potatoes in Europe, 1930–1960. *Source*: Zadoks and Kampmeijer (1977).

Analytical approach

Wavelike population expansion has long been studied in the mathematical literature. The two basic approaches are reaction–diffusion equations (see Chapter 22) and integral equations (see Chapter 23). In the study of the spatio-temporal development of epidemics, integral equations quickly

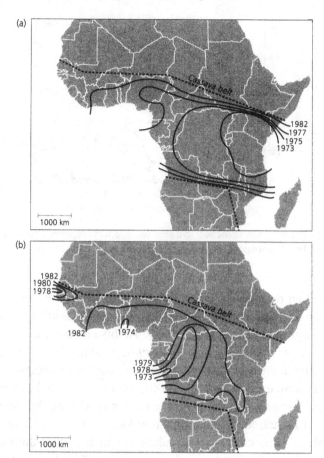

Figure 16.11 Second-order epidemics, dynamic isopath representation. Arthropods on cassava in Africa. The damage done by these arthropods was horrendous. *Source*: Osae-Danso (unpublished). (a) The cassava green mite (*Mononychellus tanajoa*) was inadvertently introduced into Africa, supposedly in Uganda, in 1971. Spread occurred within the cassava belt, primarily from east to west. (b) The cassava mealybug (*Phenacoccus manihoti*) was introduced into Africa in the early 1970s and spread from west to east. As is clear from the map, multiple introductions occurred.

gained the upper hand (Thieme 1977a, 1979a), in part inspired by EPIMUL (Diekmann 1977, 1978, 1979), with the integro-differential equation models of Kendall (1957, 1965) and Mollison (1972a, 1972b) as notable forerunners. The results confirmed earlier impressions: foci do indeed expand at constant radial velocity, after an initial buildup period.

Unfortunately, although capable of accommodating a good amount of biological detail, the integral equation models were of a general nonparametric form, not conducive to quantitative application in plant

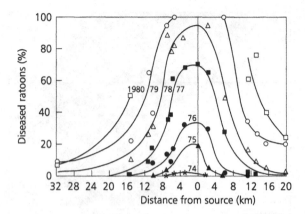

Figure 16.12 Second-order epidemic, dynamic running wave representation. Fiji disease, a virus spread by planthoppers, in sugarcane ratoons (new shoots) in Australia. *Source*: Ryan (1988).

pathology. Van den Bosch set himself the task of transforming the equations in such a manner that parameters appeared that could be measured by plant pathologists (van den Bosch 1990). Archive material of experiments with *P. striiformis* of wheat and *P. farinosa* of spinach was reexamined. Some experiments were used to estimate the necessary parameters, other experiments were used to measure the constant rate of radial expansion.

The input variables of the integral equation model are the time kernel, the dispersal kernel, and a multiplication factor. The time kernel shows the normalized time course of sporulation, which is zero during the latent period, rises quickly to a peak, and tapers off with time to become zero again (Figure 16.13); it can be graduated by a delayed gamma density (see Box 23.1). When the spore and lesion distributions are isodiametric, the dispersal kernel equals the normalized spatial distribution of second-generation daughter lesions around a mother lesion (the first generation, with the incoming spore initiating the mother lesion being the zeroth generation) along one radial axis (Figure 16.14). This one-generation gradient can be graduated by a Bessel distribution (see Box 23.1). [Note that Gregory's (1968) "primary gradient" excludes secondary spread but not necessarily the gradient effects of fourth- and higher-generation spore production within the source; in practice, however, many published gradients are actually third-generation gradients.] The multiplication factor R_0 equals the number of daughter lesions per mother lesion in the initial phase of the epidemic; its estimation is discussed in Box 16.1. The output variable of the

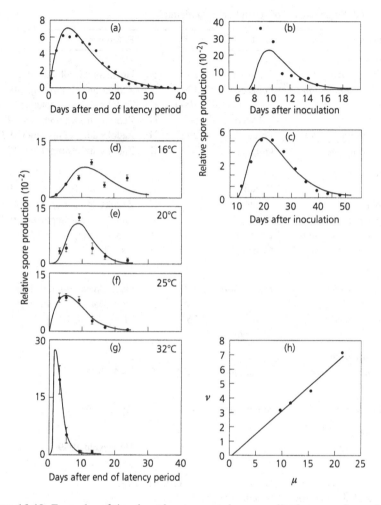

Figure 16.13 Examples of time kernels, represented as normalized curves of sporulation (fungus reproduction) versus time. The fitted curves are gamma densities starting at the estimated end of the latent period (see Box 23.1). (a) Brown leaf rust (*Puccinia recondita*) on wheat at a single temperature. *Source*: F. van den Bosch, personal communication, data from Mehta and Zadoks (1970). (b) Downy mildew (*Peronospora farinosa*) on spinach established in a growth chamber. *Source*: van den Bosch *et al.* (1988c). (c) Yellow stripe rust (*Puccinia striiformis*) on wheat, determined in the greenhouse at about 10°C. *Source*: van den Bosch *et al.* (1988c). (d to g) Blast (*Pyricularia oryzae*) on rice at four temperatures. *Source*: van den Bosch *et al.* (1988b). (h) Same data as in (d to g). The mean μ and standard deviation v of the normalized kernel (including the latent period) plotted against each other. The straight line relationship through the origin indicates that any temperature changes can be accommodated by a simple change of the overall speed of the spore production process. *Source*: F. van den Bosch, personal communication, data from Kato and Kozaka (1974).

Table 16.3 Some records of second-order epidemics.

Organism	Host	$c_0{}^a$		Reference
Fungi				
Endothia parasitica	Chestnut	26	N	Heald and Studhalter 1914
Peronospora tabacina	Tobacco	130	E	Zadoks and Kampmeijer 1977[b]
		1000	E	Zadoks and Kampmeijer 1977
	Tobacco	50	N	Zadoks and Kampmeijer 1977[b]
Phytophthora infestans	Potato	295	N	Zadoks and Kampmeijer 1977
Uncinula necator	Grape	375	E	Zadoks and Kampmeijer 1977[b]
Insects				
Leptinotarsa	Potato	29	E	Zadoks and Kampmeijer 1977[b]
decemlineata	Potato	50	N	Zadoks and Kampmeijer 1977[b]
Lymantria dispar	Trees	10	N	Zadoks and Kampmeijer 1977[b]
Phenacoccus manihoti[c]	Cassava	250[d]	A	Osae-Danso, unpublished
Mononychellus tanajoa[e]	Cassava	400[d]	A	Osae-Danso, unpublished
Parasitoid insects				
Epidinocarsis lopezi	*P. manihoti*	150[d]	A	Osae-Danso, unpublished
		75[d]	A	Osae-Danso, unpublished
Torynomus sinensis	*Dryocosmus kuriphilus*	25[f]	J	Moriya *et al.* 1989
Predatory mite				
Typhlodromalus aripo	*M. tanajoa*[e]	200[g]	A	IITA 1996
Virus				
Fiji disease virus	Sugarcane	≈4.5		Ryan 1988[b]
Vertebrates[h]				
Collared dove	–	44	E	Hengeveld 1989
Muskrat	–	11	E	van den Bosch *et al.* 1992[b]

Note: A = Africa, E = Europe, J = Japan, N = North America.
[a] c_0 = Radial expansion rate in km per year.
[b] Picture given in reference.
[c] *P.m.* = Cassava mealybug.
[d] First approximation, multiple introduction possible.
[e] *M.t.* = Cassava green mite.
[f] Velocity estimated, might reflect dispersive wave.
[g] Multiple introductions may lead to exaggerated estimate.
[h] See also Skellam (1951).

model is the focal expansion rate c_0, expressed as length over time. The procedure for calculating c_0 is described in Section 23.3.

Observed and independently calculated values of the focal expansion rate c_0 were in agreement (Table 16.4; van den Bosch *et al.* 1988c); the theory describes focus formation in plant pathology well, within the limits

---

Box 16.1 Estimating R_0

If latent periods are long relative to the effective infectious period and the initial infection is small and concentrated in time, R_0 can be estimated as the relative increase in disease severity from the zeroth to the first generation of infection.

If the initial generations cannot be clearly distinguished in the field, R_0 can be estimated from the initial exponential growth rate r_0 of an epidemic started from a very small homogeneously distributed inoculum, after the generation waves have subsided but before the exhaustion of uninfected leaf area has become noticeable, in combination with the time kernel a_1 using

$$R_0 = \left[\int_0^\infty \exp(-r_0\tau)\, a_1(\tau)d\tau \right]^{-1} .$$

R_c from Vanderplank's equation (16.1) is related to R_0 through $R_0 = i R_c$.

---

stated in the introduction. A new experiment was designed to corroborate this result, based on the prediction that in an ideal mixture of susceptible and resistant plants the constant rate of focus expansion should be approximately proportional to the logarithm of the proportion of susceptible plants in the mixture (see Box 23.4). A mixed wheat crop is a close approximation to such an ideal mixture. In the Netherlands, an experiment with four mixing proportions in three replicates did indeed confirm the theory (van den Bosch *et al.* 1990b). Also, in Ethiopia, experimental confirmation was found with common beans and bean rust (Habtu *et al.* 1995). The theory evidently holds for zero-order epidemics.

Incorporating realistic complications

The simulation approach can be extended by more realistically incorporating the physical processes involved in the development of foci. A phenomenological equation for turbulent diffusion combined with an extension of Vanderplank's equation provided the basis of a new and versatile simulation model, PODESS (Zawolek and Zadoks 1989).

The new simulation model is highly complex and very flexible. When set to a zero-order epidemic, it produces results very similar to those from EPIMUL in time and two-dimensional space. Two-dimensional space is compartmentalized and each compartment can be given a susceptibility level. Complicated geographical situations thus can be simulated, as can cultivar mixtures and mosaics (Kampmeijer and Zadoks 1974; Mundt and

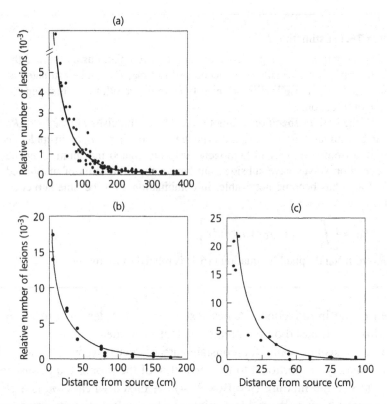

Figure 16.14 Examples of dispersal kernels (or contact distributions), measured in the field, represented as normalized curves of spore deposition density or lesion density with distance from the source. The dispersal kernel records the spores deposited or the lesions resulting therefrom during the first infectious period of the focal source. It is often nearly synonymous with the primary infection gradient of Gregory (1968). The fitted curves are Bessel densities (see Box 23.1). (a) Rice blast (*Pyricularia oryzae*) lesions, covering 4 m. *Source*: van den Bosch *et al.* (1988b). (b) Yellow stripe rust of wheat (*Puccinia striiformis*), field observations over 1.75 m. *Source*: van den Bosch *et al.* (1988c). (c) Downy mildew (*Peronospora farinosa*) of spinach, combined data from field and greenhouse experiments. *Source*: van den Bosch *et al.* (1988c).

Leonard 1986; Zawolek and Zadoks 1989). The model can be set to a first-order epidemic covering a large number of fields with different levels of susceptibility.

PODESS can also handle inhomogeneous three-dimensional space – for example, to simulate the development of a cereal rust focus on three leaf layers with different properties. By replacing low values of the spore density with a sample from a Poisson process with the mean being the deterministically calculated value, the model can generate daughter foci (Zawolek 1993). Daughter foci are often observed in nature (Thresh 1958;

Table 16.4 The constant rate of focus expansion c_0 (in cm per day) for stripe rust (*Puccinia striiformis*) on wheat and downy mildew (*Peronospora farinosa*) on spinach: Comparison of observed and predicted values and error analysis. *Source*: van den Bosch *et al.* (1988c).

	Stripe rust, wheat	Downy mildew, spinach
$c_{0,\text{predicted}}$	8.0 ± 1.5	3.0 ± 2.4
$c_{0,\text{observed}}$	9.4 ± 0.8	2.3 ± 0.2
Estimate of	Contribution to the variance of $c_{0,\text{predicted}}$	
Time kernel	0.05	1.15
Dispersal kernel	2.22	2.05
Growth rate r_0	0.11	
Gross reproduction R_0		2.53

Zadoks 1961; Brenchley and Dadd 1962; Sache and Zadoks 1995). Shaw (1995) obtained similar results through an efficient individual-based simulation procedure geared at handling so-called fat-tailed kernels (see Section 23.3, Spatial scales). Earlier models for the spatial spread of epidemics could not do so. A second finding was that simulated first-order epidemics with stochastic spore dispersal develop more slowly than those with deterministic spore dispersal.

The most fascinating results are obtained when the model is set to two-dimensional space with two different dispersal mechanisms, a "long" and a "short" one. The long dispersal mechanism is characterized by a relatively low frequency of dispersal events combined with a relatively large distance parameter. The short mechanism is characterized by a relatively high frequency combined with a relatively small distance parameter. Ideally, both should have stochastic spore dispersal, but it usually suffices to make only the long distance dispersal stochastic. The result is that the conquest of space by an epidemic starting from a single lesion proceeds with peak efficiency and maximum speed when the proportion of spores dispersing through the short mechanism has an optimum value of 0.8 (Zawolek and Zadoks 1989). This result has been corroborated by simplified analytical models, with the qualifier that the optimal dispersal fraction may depend on the other model parameters (Metz and van den Bosch 1995, Figure 8; van den Bosch *et al.*, 1999).

This finding corroborates an intuitive statement by Vanderplank (1975). Our short mechanism is equivalent to his short dispersal horizon, and our long mechanism is equivalent to his long dispersal horizon. He argued that any pathogen needs at least two dispersal mechanisms to survive: one to multiply up to the point of self-eradication, and one to find new hosts and fresh food. Arguably, a dual dispersal mechanism is a survival strategy, and the simulation model, in providing numerical support for Vanderplank's ideas, suggests an optimal deployment of spores by a dual dispersal mechanism.

Empirical evidence suggests that it may even be necessary to consider three dispersal scales (Leonard 1971). This suggestion was implemented in a recent simplified analytical model (van den Bosch *et al.* 1994, 1999). One result is that, as far as the dispersal kernel is concerned, only the tail end matters in determining the expansion velocity of a second-order epidemic.

16.4 Generalizations

The Kermack–McKendrick model (1927) from medical epidemiology was generalized and found to be applicable to botanical epidemiology (Zadoks and van den Bosch 1994). Diffusion models, used to describe epidemics of human disease (such as the great plague of 1348), are also applicable to botanical epidemiology. Several of the concepts developed for focal disease of foliage due to polycyclic fungi also seem to be applicable to bacterial and viral disease of foliage and even to insect pests (Zadoks and Kampmeijer 1977). Obviously, we touch upon some very general principles.

Ecologists preceded plant pathologists in their consideration of spatial population waves (Skellam 1951), and many examples exist of more or less circular spread at a constant expansion rate, including the muskrat (*Ondatra zibethica*; Williamson and Brown 1986) and the collared dove (*Streptopelia decaocto*; Hengeveld 1989). The formal aspects of these examples are very similar to those of second-order epidemics (see Table 16.3).

The principle of a constant expansion rate apparently is valid at the zero-order scale of about 1–1000 m and at the first-order scale of about 1–1000 km (both orders occurring within a single season), and at the second-order scale of about 1000 km (developing over several seasons). Current theory is based explicitly on a mechanistic approach of polycyclic fungal diseases of annual crop foliage. The theory can be "stretched" to plant diseases and even pests with slightly different underlying mechanics. But how much further can it be stretched? In other words, what is the domain of validity of the present theory?

Theoretical considerations

The foci of any order covered by the theory have circular symmetry. In real life, deviations from isotropy due to various causes are the rule. Crops are usually planted in rows and zero-order epidemics may spread faster along the rows than across them (Zadoks 1961). In first-order epidemics a dominating wind direction may cause anisotropy (Zawolek 1993). In second-order epidemics geography plays a role, as mountains, seas, and deserts without susceptible crops form physical and climatic barriers to epidemic spread (see Figures 16.10 and 16.11). Variations in sowing and harvesting time at a continental scale are additional reasons for lack of circular symmetry. Numerical models can handle many causes of anisotropy, but in so doing they become less general.

On the other hand, a dormancy period (or off-season) of the annual crop and/or its pathogen does not fundamentally change the conclusion of constant radial expansion rate, as theory (van den Bosch *et al.*, 1999) and empirical results (see Figure 16.10) show.

Practical examples

With bacteria (Kocks and Zadoks 1996; Yang *et al.* 1991), viruses (Thresh 1958; Ryan 1988; van der Werf 1988), and nematodes (Been and Schomaker 1998), we often see zero-order epidemics. The mechanics of such epidemics may differ from those of fungi, yet the phenomena are similar, though not necessarily identical. Unfortunately, the language used to describe the phenomena in these different subject areas is often quite dissimilar. Soil-borne pathogenic fungi, which often cause circular foci, as first described in detail in 1728 (reviewed in Zadoks 1981), are another interesting special case of zero-order epidemics. Foci of insects at the second trophic level – that is, insects eating weeds – have been photographed (Room and Thomas 1985).

The principle of focal spread applies to some diseases of perennials, but data are scant. I have observed such foci in pear orchards and hawthorn hedges affected by fire blight (a bacterial disease caused by *Erwinia amylovora*; Schouten 1991) and in oil palm plantations due to spear rot (a disease of unknown etiology; Mariau *et al.* 1992; van de Lande and Zadoks, in press) in Latin America. It should be noted that in tree crops the vegetation period covers many years instead of one season.

A most fascinating "pandemic" was caused by the transposable P-element in *Drosophila melanogaster*, which spread over the world in some 40 years after an arguable horizontal gene transfer event from *Drosophila willistoni* in North America (Capy *et al.* 1998). Gradual spread over the Eurasian continent from both west and east is indicated, but insufficient detail is available to calculate rates of spread (Figure 16.16).

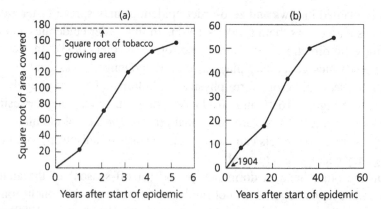

Figure 16.15 Second-order epidemics, dynamic "area within the isopath" representation. Ordinates in units of approximately 20 km. *Source*: Heesterbeek and Zadoks (1987). (a) Tobacco blue mold (*Peronospora tabacina*) in Europe, 1958–1976 (from Figure 16.10a). (b) Chestnut blight (*Endothia parasitica*) on American chestnut in the eastern United States, 1904–1950. The pandemic spread at an approximately constant radial expansion rate over more than four decades. This is one of the few recorded pandemics on trees in a natural, though logged, forest. No doubt birds were instrumental in the long-distance spread, with foci ahead of the epidemic front.

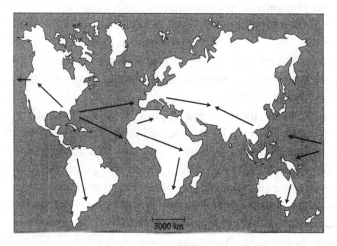

Figure 16.16 Second-order epidemic, arrow representation. Spread of the P transposon in *Drosophila melanogaster* beginning in the southeastern United States (before 1950). Invasion of Eurasia came from both west and east and proceeded gradually – from before 1960 to 1980 and between 1980 and 1990, respectively – but too few data points are available for more detailed mapping; tropical Africa was invaded between 1970 and 1990, and tropical Oceania and Indonesia, between 1980 and 1990. *Source*: Capy *et al.* (1998).

16.5   Concluding Comments

Botanical epidemics of all three orders represent instances of invasions of a biological entity into an environment previously free from that entity – for example, yellow stripe rust in a field of wheat. In principle, they spread at a constant radial expansion rate. For polycyclic fungal diseases of foliage there is a solid mechanistic foundation for the analytical and numerical models. In the case of zero-order epidemics, experimental verification of the analytical model has been achieved. In the case of first- and second-order epidemics, experimental verification is hardly possible and highly unethical. However, empirical confirmation is available and deviations can often be explained in terms of the analytical model. Continued development of the theory is needed for application in crop protection and in human and veterinary medicine. Increases in international traffic and the political emphasis on free trade will eventually lead to many more biological invasions. Good theory will be invaluable in predicting coming events and hopefully in successfully intervening to check the spread of invaders.

The phytopathological theory of "foci, small and large" is but a specific example of a more universal ecological phenomenon of biological invasions. For pathogens beyond those considered above – including soil-borne fungi, bacteria, viruses, nematodes, and insects, and possibly even insect transposable elements – the epidemic phenomena are very similar to our zero-, first-, and second-order epidemics, but the mechanistic foundation of the theory may need adjustment. Further experimental explorations of the overall domain of the theory's validity are called for. Section 23.2 explores this domain from an *a priori* perspective. Appendix 16.A surveys the quantitative application of the analytical approach from a wider biological perspective.

Appendix 16.A   Quantitative Applications of Models for Spatial Population Expansion
(by Johan A.J. Metz)

This appendix contains a necessarily incomplete survey of the quantitative uses of the mathematical theory of wavelike population spread as reviewed in this chapter (on empirical data from crop protection) and in Chapters 22 (on reaction–diffusion models) and 23 (on kernel-based models).

Single-species diffusion models

Skellam (1951), who introduced the reaction–diffusion theory into population dynamics proper, considered two examples, the spread of the muskrat in Europe and the return of oaks after the last ice age. In the latter case, he used an estimate of c_0 plus a rough *a priori* estimate of the local population growth rate in the virgin environment r_0 to estimate the diffusion coefficient D. He interpreted the large discrepancy between the estimated D and the observed spread pattern of acorns as a pointer to an alternative long-distance dispersal mechanism, probably the movement of acorns by jays. In the same spirit, Ammermann and Cavalli-Sforza (1973, 1984) found the speed of the neolithic transition in Europe to be in accordance with a rough *a priori* estimate of the parameters.

In the 1980s some detail was introduced. Lubina and Levin (1988) found different speeds for the return of Pacific coast sea otters in different parts of their range. They estimated the parameters from the overall population statistics in the initial phase of the population expansion, D from the increase in the variance of the population distribution over space, and r_0 from the overall population growth rate. These quantities proved good predictors of the speed of the full-fledged wave. Andow *et al.* (1990, 1993), noting that speeds are often different in different directions and areas, developed a technique for analyzing the progress of the wave front sector by sector. In addition, they predicted wave speeds for the spread of (1) the muskrat over Europe and (2) the small cabbage white and (3) the cereal leaf beetle over North America. Independent estimates of r_0 were obtained from either life-history data or the slope of the regression of ln(population size) against time; those for D were from mark–recapture experiments. The predictions for cases (1) and (2) compared well with the observations. In case (3) there was considerable discrepancy, probably due to an alternative dispersal mechanism such as hitchhiking on human transportation.

Multispecies diffusion models

Application of multispecies models began with the work of Murray and coworkers on the spread of rabies, published in various papers beginning with Kallén *et al.* (1985) and surveyed in Murray (1989). These papers are of considerable general interest. However, the assumption that only the infected foxes move, and moreover at a high speed, is not borne out by the radiotracking data of Andral *et al.* (1982) and is also in discord with data showing that the speed changes with the breeding cycle of the foxes (Moegle *et al.* 1974). A survey of alternative models can be found in Bacon (1985). Box 23.5 shows how the kernel approach of Chapter 23 helps to take a closer look at the dispersal mechanism, with surprising consequences. Okubo *et al.* (1989) modeled the spread in England of the gray squirrel at the expense of the red squirrel. Unfortunately, they could not obtain any good data on D, but working backward from the observed speed they gave a D within a reasonable range.

With Dwyer's work (1992a, 1992b, 1994; Dwyer and Elkinton 1995) on the spatial spread of insect viruses, real detail was incorporated. In every example, independent parameter estimates were used. Three of the cases gave a good match between predicted and observed speeds; in the fourth case a discrepancy was found, probably due to some mechanism of long-range dispersal not accounted for in the model.

Continuous-time kernel-based models

Van den Bosch *et al.* (1988a, 1988b, 1988c, 1990b) made the first quantitative application of integral-kernel-based models to focus expansion in fungal plant diseases. As a birth distribution, they used a delayed gamma age-at-birth distribution times a Bessel distribution, with parameters estimated from controlled field and greenhouse experiments. R_0 was estimated from population-level experiments. The resulting predicted expansion rates matched with the observed ones. An additional experiment in which R_0 was varied by mixing susceptible and resistant hosts also matched with theory (van den Bosch *et al.* 1990b; Habtu *et al.* 1995). Van den Bosch *et al.* (1990a) once again treated the spread of the muskrat (using the results given in Box 23.2) and of rabies (using the results given in Box 23.3). In both cases, independent parameter estimates led to adequate quantitative predictions of the observed spread rates. Van den Bosch *et al.* (1992) and Lensink (1997) considered the spread of various bird species using the formula given in Box 23.2. In all cases good predictions were obtained.

Discrete-time kernel-based models

In integro-difference models, generations are treated as non-overlapping and spatial dispersal is represented by a continuous kernel. In recent years integro-difference models have been the subject of intensive research, with much attention devoted to complications such as fat-tailed kernels (see Section 23.3, Spatial scales) and Allee effects (see Section 23.4, Interactions between individuals). So far, two attempts have been made at quantitative application of this body of theory. Veit and Lewis (1996) constructed a detailed model for the spread of house finches in North America, which combines an Allee effect with a fat-tailed displacement distribution. Despite these complications (which should make predictions from the model less robust) they found a fair overall agreement between predicted and observed speeds using independent estimates. Clark (1998) and Clark *et al.* (1998) considered the return of the oaks, and of various other trees, after the end of the last ice age. Clark's main objective was to explain the high speed of return in the light of what is known about the local features of the dispersal mechanism. He expended considerable effort estimating the fat-tailedness of displacement kernels. This is a difficult topic and the results are somewhat inconclusive. However, fat-tailed kernels lead to increasing expansion velocities. Therefore, a different way of checking the hypothesis of fat-tailedness is to see whether this acceleration has left traces in the subfossil records. R. Hengeveld and F. van den Bosch (personal communication) have scrutinized the published data for such traces but as yet have found no sign of acceleration.

17

Wave Patterns in Spatial Games and the Evolution of Cooperation

Régis Ferrière and Richard E. Michod

17.1 Introduction

Our understanding of the evolution of animal behavior has been greatly enhanced by the use of game theory (Maynard Smith 1982). Classical games assume that a given individual is equally likely to interact with any other member of the population and that the success of any individual depends on the frequency of all other strategies represented in the population. Yet natural environments possess a spatial dimension: individuals have limited mobility and interact locally with their neighbors. Only recently have attempts been made to incorporate this important property into the study of evolutionary games. Different approaches have been followed: numerical simulations of "games on grids" (Nowak and May 1992; Lindgren and Nordahl 1994; see Chapter 8); analytical study of correlation equations for games on lattices (Nakamaru *et al.* 1997; see Chapter 13); and analytical study of "replicator–diffusion" equations (e.g., Vickers 1989; Vickers *et al.* 1993; Ferrière and Michod 1995, 1996; see Chapter 22). In this chapter we restrict ourselves to the last of these methodologies and provide an introduction to its mathematical underpinnings and biological applications. Elements of a general theory of replicator–diffusion equations are expounded in detail in articles by Vickers (1989), Hutson and Vickers (1992), Vickers *et al.* (1993), and Cressman and Vickers (1997). We present an overview of these important results in Section 17.2. Sections 17.3 and 17.4 show how replicator–diffusion models can be used to study spatial versions of the iterated Prisoner's Dilemma game, a well-known metaphor for evolution toward cooperation between genetically unrelated individuals (Trivers 1971; Axelrod and Hamilton 1981; Maynard Smith 1982; Hofbauer and Sigmund 1998).

17.2 Invasion in Time- and Space-continuous Games

When considering the adaptive dynamics of long-term evolution, the crucial question is whether a mutant phenotype can invade resident phenotypes (Metz *et al.* 1992; Diekmann *et al.* 1996; Geritz *et al.* 1997; Dieckmann 1997). Obviously, the very nature of an invasion event requires individual mobility, while the game theoretical context requires interactions between individuals that share the same neighborhood. This creates the need for deriving invasibility criteria that explicitly account for spatial effects in the dynamics of mutant invasion. Much of the theory of invasion in time- and space-continuous games is relatively new. This section presents the basic ideas needed for subsequent applications and points to some unsolved questions.

Replicator–diffusion equations

Hereafter, the payoff matrix of a game is denoted by A. We set $A = [a_{ij}]$, where a_{ij} is the payoff to an individual who plays strategy i against an opponent who plays j. Let Q be a probability vector composed of q_i denoting the proportion of individuals who play strategy i. Taylor and Jonker (1978) have suggested incorporating continuous time into the dynamics of games using so-called replicator equations

$$\frac{dq_i}{dt} = q_i \left[(AQ)_i - Q^T AQ \right] \qquad (1 \le i \le k) \,, \tag{17.1}$$

where k is the number of strategies, $(AQ)_i$ denotes the ith coordinate of vector AQ, and the exponent T indicates vector transpose. This model stems from the idea that the growth rate of a strategy is equal to its absolute payoff $(AQ)_i$. Hence the abundance of strategy i is described by unlimited exponential growth, according to the equation $dn_i/dt = [(AQ)_i]n_i$. One can then write q_i as n_i divided by the total population density (i.e., the sum of the densities of all strategies) and use the latter equation to recover Equations (17.1). The replicator equation expresses the fact that the change in any strategy's frequency is determined by the population growth rate of that strategy compared with the average population growth rate, that is, the average payoff $Q^T AQ$. Equations (17.1) are important because there is a relationship between the stable equilibrium points of Equations (17.1) and the game's evolutionarily stable strategies, or ESSs (Zeeman 1980; Hofbauer and Sigmund 1998). [We recall that an ESS is a strategy that, when common in the population, cannot be invaded by any small group of individuals playing a different strategy; see, e.g., Maynard Smith (1982).]

The inclusion of continuous space is not so straightforward. If all individuals move at the same rate, we are on a well-worn trail (see Hadeler 1981). But it is quite reasonable to expect that the dispersal rate is affected by, or indeed part of, the strategy chosen. Using the standard diffusion approximation of a random walk (see Chapter 22), Vickers (1989) introduced the following "replicator–diffusion equations":

$$\frac{\partial n_i}{\partial t} = n_i \left(\frac{(AN)_i}{m} - \frac{N^T AN}{m^2} \right) + \mu_i \frac{\partial^2 n_i}{\partial x^2} \quad (1 \leq i \leq s) , \quad (17.2)$$

where each strategy is characterized by its own diffusion (or mobility) rate μ_i. Here, $n_i(x, t)$ is the density of the i-strategists at location x and time t and $m(x, t)$ is the total density at x and t $(m = n_1 + n_2 + \cdots + n_s)$. For simplicity, we drop the x and t variables in the equations when there is no ambiguity.

Like the replicator equations, this model assumes that an individual playing strategy i and located at x at time t receives the a_{ij} payoff if it interacts with a neighbor playing strategy j, which occurs with a probability approximated by the frequency n_i/m of j-players at x and t. A bookkeeping of the payoff contributions to strategy i from all different j strategies yields the first part of the growth rate of strategy i [the bracketed term in Equations (17.2)]. In addition there is a regulatory, negative term that accounts for the fact that local densities stay bounded. This regulatory term takes the form of a discount precisely equal to the average payoff earned at location x at time t (which is calculated by summing over i the average payoff to strategy i weighted by its frequency n_i/m at location x and time t). It must be stressed that this discounting term is a very special one, chosen only for mathematical convenience. No particular physiological or behavioral mechanism is known that lets individuals adapt their per capita birth and death rates to local circumstances so as to keep the local population averages of these quantities exactly zero at all times. Cressman and his coworkers (see Cressman and Dash 1987; Cressman and Vickers 1997) have elaborated on this issue by assuming that an individual's fitness is composed of its payoff in the contest with other strategies together with a background fitness that is common to all strategies. Their approach has the merit of relating the spatial dynamics of the game to individual life-history traits, but it is hampered by very demanding mathematics (Cressman and Vickers 1997). In Section 17.3, we take advantage of the great mathematical tractability of replicator–diffusion equations (17.2) to explore invasion issues in the context of spatial games between cooperative and selfish players.

Hutson and Vickers (1995; see also Chapter 22) have used the background fitness model of local population regulations to address the same problem. In Section 17.4 we see that their results are consistent with those obtained through the analysis of the simpler replicator–diffusion equations.

Replicator–diffusion equations form a distinct class of reaction–diffusion models because of their specific reaction term. Equations (17.2) assume that space is one-dimensional and reduces to an "x-axis." The formalism, however, straightforwardly carries over to higher dimensions (see Chapter 22 for a detailed account of the rationale of reaction–diffusion models). If the spatial domain is bounded, impermeable boundary conditions are imposed.

Invasibility and evolutionary stability

A few mathematical results are available to investigate the invasibility or evolutionary stability of a strategy in a spatial game described by replicator–diffusion equations. They all relate the dynamics of the spatial model given by Equations (17.2) to its nonspatial counterpart, Equations (17.1). Here, we state the mathematical theorems in a self-contained manner to make them unambiguously applicable to any particular model that falls under their scope. The spatial iterated Prisoner's Dilemma (IPD) offers an opportunity to operate this machinery, as we will see in Section 17.3.

Vickers (1989) provided the first stability analysis of the replicator–diffusion equations. He found that an *interior* ESS is so stable that it precludes any spatial dependence:

Proposition 17.1 If matrix A has an interior ESS, that is, an ESS for the replicator equations [Equations (17.1)] given by a frequency vector Q with all nonzero coordinates, then this ESS is also stable in the spatial game governed by Equations (17.2) for all choices of the diffusion coefficients μ_i.

The situation becomes much more complicated if there is no interior ESS in the homogeneous game. Hutson and Vickers (1992) addressed the case where there are only two strategies and each pure strategy is an ESS. In the absence of spatial effects each ESS is, by definition, stable. The inclusion of diffusion creates the possibility of a traveling wave that in effect replaces one ESS with the other. To state Hutson and Vickers' main theorem, we first recast the payoff matrix A as

$$A = \begin{bmatrix} \alpha & 0 \\ 0 & \beta \end{bmatrix} \tag{17.3}$$

by setting $\alpha = a_{11} - a_{21}$ and $\beta = a_{22} - a_{12}$, an operation that does not affect the dynamics. To ensure that each pure strategy is an ESS, $\alpha > 0$ and $\beta > 0$. Hutson and Vickers (1992) investigated the invasion of a region dominated by one strategy – for example, for $x > 0$ strategy 1 is played by almost all individuals and for $x < 0$ strategy 2 is prevalent.

Proposition 17.2 Assume that A has the form (17.3) with $\alpha > 0$ and $\beta > 0$. There then exists a function F such that, if

$$\frac{\beta}{\alpha} > F\left(\frac{\mu_1}{\mu_2}\right), \tag{17.4}$$

a traveling wave front with positive speed (i.e., moving from $x = -\infty$ to $x = \infty$ along the spatial axis) will connect the two homogeneous pure population equilibria. The function F is well approximated over the range 0.05–20 by $F(u) \approx u^{-0.61}$.

The following statements explain the importance of Proposition 17.2 for analyzing applications.

■ Proposition 17.2 states that if inequality (17.4) is satisfied, then a traveling wave replaces a pure strategy-1 population with a pure strategy-2 population. If the inequality is reversed, the sign of the wave speed becomes negative and strategy 2 replaces strategy 1 in a traveling wave. There is virtually no room for coexistence, except perhaps in the atypical boundary case $\beta/\alpha = F(\mu_1/\mu_2)$. Thus, in a generic two-strategy game where both strategies are ESSs, a traveling wave necessarily exists and replaces one strategy with the other.

■ If the payoffs α and β are not influenced by the mobility rates μ_1 and μ_2, condition (17.4) asserts that the dominating strategy must have large payoffs and small diffusion rates. (In the IPD, however, the payoffs do depend on the mobility rates.)

■ There is strong numerical evidence to support the claim that the propagation of a traveling wave replacing strategy 1 with strategy 2 is strictly equivalent to the growth of a localized clump of individuals playing strategy 2 amid a "sea" of players using strategy 1. Accepting this conjecture, inequality (17.4) reads as a criterion of invasibility.

While Proposition 17.1 assumes that the replicator equation admits a stable solution corresponding to an interior (mixed) ESS, Proposition 17.2 addresses the case where the (ESS) equilibria associated with each pure strategy are the only ones. Vickers *et al.* (1993) have shed some light on the case where the replicator equation admits an internal stable equilibrium which is not an ESS. This is an interesting case because, in the spatial context, it may lead to the formation of patterns.

Let $\hat{Q}$ be an internal, stable solution to the replicator equation. Assuming that $\hat{Q}$ is not an ESS means that it is invadible, and the simplest situation here arises when $\hat{Q}$ can be invaded by a pure strategy (say, strategy 1). Vickers *et al.* (1993) proved the following theorem.

Proposition 17.3 If strategy 1 can invade $\hat{Q}$, that is, $a_{11} > (\hat{Q}^T A)_1$, then there exists a combination of mobility rates μ_i $(1 \leq i \leq k)$ such that $\hat{Q}$ is not spatially stable.

This statement is important in light of Turing's (1952) well-known idea that spatial patterns are often associated with equilibria which are stable in the nonspatial system (i.e., without diffusion) and unstable with respect to spatially heterogeneous perturbations. This phenomenon is the so-called Turing instability (see Chapter 22). In the framework of replicator–diffusion equations, Vickers *et al.* (1993) have raised three important points.

- A bifurcation analysis shows that this pattern-formation mechanism is operative in spatial games under the conditions of Proposition 17.3.
- There must be at least three pure strategies in the game for Proposition 17.3 to apply.
- A converse of Proposition 17.3 holds when there are exactly three strategies. If $\hat{Q}$ resists invasion by any pure strategy, then it is spatially stable and no spatial pattern can be produced.

Patterns arising from the Turing instability vary in space but are constant in time. Yet variations in space and time are essential features of the dynamics of ecological systems. Vickers *et al.* (1993) have provided a numerical example of a three-strategy game that exhibits another kind of instability (namely, a Hopf bifurcation) that results in spatial patterns which are periodic in time. It should be noted that a general theory of the bifurcations of the three-strategy game is still pending.

17.3 Invasion of *Tit For Tat* in Games with Time-limited Memory

In Chapter 8 of this volume, Nowak and Sigmund expound on the basics of the IPD game. Here, we refer to concepts and notations introduced by these authors. Investigating the relative invasibility of well-known strategies like the cooperative *Tit For Tat* (*TFT*) strategy and the selfish *Always Defect* (*AD*) strategy serves to demonstrate some of the mathematical techniques introduced in the previous section.

The spatial struggle of *Tit For Tat* and *Always Defect*

The IPD has proved tremendously fruitful as a paradigm for studying the evolution of cooperation. Game theorists originally identified the *TFT* strategy as the most robust and stable strategy in the IPD (Axelrod and Hamilton 1981). Subsequent theoretical developments (Nowak and Sigmund 1992, 1993) emphasized that the *TFT* strategy could be the first step toward cooperation in a world of unconditional defectors playing *AD*. To explain the emergence of cooperation, it is therefore crucial to understand how *TFT* can gain a foothold in a population dominated by *AD*.

A major problem concerning the nonspatial IPD is that it fails to convincingly settle this issue. Depending on the probability w of continuing the game, either *AD* is the only ESS, hence *TFT* has no chance to invade, or both *AD* and *TFT* are ESSs (which happens when w is sufficiently large), implying that *TFT* can invade an established *AD* population only if the *TFT* frequency exceeds a certain threshold. Because the nonspatial IPD assumes an infinite population, this result means that an initially finite group of *TFT* newcomers will never spread. It has long been claimed that small clusters of finite size should still have a chance of spreading, because cooperators within a cluster experience a high probability of interacting with each other. To weigh this claim, one might compare the average payoff earned by a *TFT* within the cluster with the *AD* payoff averaged over the whole population. In doing so, however, one would overlook *TFT–AD* interactions which locally influence the payoff to *AD* players in the vicinity of the cooperative focus – a local payoff likely to be of critical importance to determining the eventual fate of the *TFT* population. Numerical examples (Nowak and May 1992, 1993) based on cellular automata demonstrate that local interactions have a significant effect on the outcome of the game between cooperators and defectors. For this reason, there has been much interest in setting up versions of the IPD that specifically account for spatial dynamics and local contests.

A replicator–diffusion model

We now assume that *TFT* and *AD* players are free to move. We wish to describe the game using a replicator–diffusion equation. This amounts to writing down the payoff matrix A taking into account the organisms' mobility and other individual traits (mortality and interaction time). In the nonspatial game, the parameters are the payoffs S, P, R, T (see Chapter 8), and the probability w that two particular interacting individuals continue their interaction in the next round of the game. In the spatial version of the game, that probability w is influenced by the individuals' traits (including the mobility rates), and some work is required to make this relationship explicit.

Microscopic description of interactions. We first describe an individual-based model of the population. We assume that each individual in the population occupies a position in space that is a function of time. The population is distributed along a one-dimensional axis: it can be thought of as spread along a coastline or a river bank; or if the environment is really two-dimensional, variations in the strategy mix may occur in one direction only. For the purpose of defining local interactions, we regard space as being divided into discrete contiguous cells of length Δl so that each cell contains two individuals at any time. Interactions are initiated between two individuals located in the same cell. Thus Δl defines the "interaction length," which we assume to be constant across space. Each interaction lasts Δt units of time, which we define as the "interaction time" of the game. Interactions occur consecutively, without any "rest time" in between.

The payoffs S, P, R, and T determine the per capita reproductive rate. Thus if the payoff is S, for example, to each individual of a group of size n, then their numbers increase at a rate Sn in the absence of all other effects. We assume that interactions have no direct effect on individual mortality. Let us consider a *TFT* player within a given cell, and let p_T and p_D be the probabilities that the partner is a *TFT* or an *AD* player, respectively. Then the reproductive success of the nominal *TFT* player during the small time interval Δt is

- $R\Delta t$ if the co-player is a *TFT*, which occurs with probability p;
- $P\Delta t$ if the co-player is a defector already encountered on the previous interaction, which occurs with probability $p_D w$;
- $T\Delta t$ if the co-player is a defector not encountered on the previous interaction, which occurs with probability $p_D(1 - w)$.

Here w denotes the probability that the same two individuals located in a given cell at time t were also sharing a cell at time $t - \Delta t$.

From microscopic interactions to macroscopic dynamics. Mobility is modeled by a random walk, and we make the classical diffusion approximation. Thus we define mobility (or diffusion) rates for *TFT* and *AD* players, denoted by μ_T and μ_D, respectively. Then the derivation of w is straightforward (see Ferrière and Michod 1996, for details) and yields

$$w = \left[4\sqrt{\pi}\sqrt{(\mu_T + \mu_D)\Delta t}\right]^{-1}$$

$$\times \iint_{u,v\in[-\Delta l/2, \Delta l/2]} \exp\left[-\frac{(u-v)^2}{4(\mu_T + \mu_D)\Delta t}\right] du\, dv \;. \tag{17.5}$$

For small cell length Δl, the following approximation holds [Equation (7) in Ferrière and Michod 1996]:

$$w \approx \frac{2}{\sqrt{\pi}} \frac{\Delta l}{\sqrt{\Delta t}} \frac{\mu_T \mu_D}{(\mu_T + \mu_D)^{5/2}} \;. \tag{17.6}$$

We derive a replicator–diffusion model of the population dynamics by letting Δl go to zero and rescaling time appropriately, such that $\Delta l/\sqrt{\Delta t}$ approaches a positive constant υ:

$$\frac{\Delta l}{\sqrt{\Delta t}} \to \upsilon \neq 0 \;. \tag{17.7}$$

Now we can define the densities of *TFT* and *AD* as continuous functions of space and time, denoted by $n_T(x, t)$ and $n_D(x, t)$. Let m be the total density $n_T + n_D$. We have $p_T = n_T/m$, $p_D = n_D/m$; thus, the *TFT* reproductive rate is

$$\frac{n_T}{m}R + \frac{n_D}{m}[wP + (1-w)S] \;. \tag{17.8}$$

Likewise, the *AD* reproductive rate is

$$\frac{n_D}{m}P + \frac{n_T}{m}[wP + (1-w)T] \;. \tag{17.9}$$

The payoff matrix of the replicator–diffusion game equations (17.2) follows readily:

$$A = \begin{bmatrix} R & wP + (1-w)S \\ wP + (1-w)T & P \end{bmatrix} , \tag{17.10}$$

with w given by

$$w = \frac{2\upsilon}{\sqrt{\pi}} \frac{\mu_T \mu_D}{(\mu_T + \mu_D)^{5/2}} \;. \tag{17.11}$$

Notice that for consistency with the assumption made above – that the interaction length is constant across space – the total density should vary very slowly in time and smoothly across space. Using numerical integration of Equation (17.2) with A given by Equations (17.10) and (17.11), we have found this requirement to be fulfilled when μ_T and μ_D were not vastly different.

A spatial version of Hamilton's rule

To analyze the replicator–diffusion model of the IPD by means of the theory developed in Section 17.2, we must first consider the nonspatial version of the system and investigate its equilibria. With this aim in view, it is convenient to introduce a cost–benefit parameterization of the IPD payoffs (Brown *et al.* 1982). Assume that a cooperator exhibits some behavior that benefits the fitness of its partner, the recipient, by an amount b which is larger than 0. The benefit is independent of the recipient's behavior. By providing its partner with the benefit b, the cooperator incurs a cost $-c$, $c > 0$. Again, this cost is independent of the recipient's behavior. If the effects on fitnesses are additive, with a baseline value taken to be 1, one obtains the following parameterization: $T = 1 + b$, $R = 1 + b - c$, $P = 1$, $S = 1 - c$.

Using this parameterization, we see that the replicator equation of the game admits two stable equilibria (corresponding to each pure strategy) whenever

$$w \geq \frac{c}{b} \, , \tag{17.12}$$

which is the condition found by Brown *et al.* (1982) for *TFT* and *AD* to simultaneously be ESSs in the standard, nonspatial game. Then Proposition 17.2 asserts that there exists a traveling wave replacing *AD* with *TFT* if

$$\frac{w}{1 + (1 - w)F(\mu_T/\mu_D)} > \frac{c}{b} \, . \tag{17.13}$$

This inequality provides a Hamilton rule (Hamilton 1964) for the increase of cooperation in a nonsocial, spatial environment. The left-hand side (hereafter denoted by H) generalizes the coefficient of reciprocation defined for the nonspatial IPD (Brown *et al.* 1982), which gives the probability that an individual's cooperative act is returned via reciprocation from other *TFT*. The right-hand side of the inequality is the cost–benefit ratio of cooperation. This spatial Hamilton rule can be further extended to include a cost to mobility (Ferrière and Michod 1996).

Inequality (17.13) defines a set of mobility rates μ_T and μ_D that cause an invasion of defectors *AD* by *TFT*: a traveling wave replaces *AD* with *TFT*. All other parameters being fixed, this set is delineated by the c/b isoclines drawn on the surface $H(\mu_T, \mu_D)$ (see Figure 17.1). [Notice that if a pair μ_T, μ_D satisfies inequality (17.13), then it automatically meets inequality (17.12).] We find that a range of mobility rates exists for which

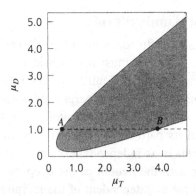

Figure 17.1 Mobility rates leading to an invasion of an established *AD* population by a wave of *TFT* players. The model assumes time-limited memory and is given by the system of replicator–diffusion equations (17.2). The payoff matrix *A* is specified by Equations (17.10) and (17.11) with $v = 1$. The shaded area contains all pairs of mobility rates μ_T, μ_D such that the coefficient of reciprocation $H(\mu_T, \mu_D)$ is larger than the cost–benefit ratio c/b (fixed at 0.22). For given defectors' mobility μ_D larger than a minimum value ($\approx$0.11) – for example, $\mu_D = 1.0$ (dashed line) – there is an interval (A, B) of μ_T mobility rates over which a *TFT* invading wave displaces a resident *AD* population.

TFT can invade provided that *AD* mobility exceeds a minimum threshold. In general, this range includes the mobility rate of resident defectors, μ_D, but it is skewed around μ_D so that *TFT* players may be much more mobile than defectors and yet successfully displace them.

As a consequence of the particular form taken by the function *F* in Equation (17.13) [$F(u) \approx u^{-0.61}$, see Proposition 17.2], a condition for the invasion by rare *AD*s of a *TFT* population is obtained by reversing inequality (17.13). (Note that the particular form of *F* is only an approximation. Dealing with the exact function would call for further investigation.) This condition determines the stability of *TFT* once established. It turns out that *TFT* is jeopardized by *AD* endowed with either high or very low mobility (see Figure 17.1). Also, *TFT* is immune to invasion for a much wider range of *AD* mobility rates when its own rate of mobility increases. Thus, by moving at higher rates, cooperators find more efficient protection against reinvasion by *AD*.

Why mobility can favor *Tit For Tat*

To answer this question and to give some intuitive understanding of the above results, Ferrière and Michod (1995) have developed an auxiliary model focusing on the stochastic motion of the players. Heuristically, the growth of an initially small cluster relies on two conditions: first, that the cluster can spread outward from the edge, and, second, that its core is not destroyed by *AD* intruders (Axelrod 1981; Eshel and Cavalli-Sforza 1982; Wilson *et al.* 1992). The first condition is ensured whenever *TFT*s can make safe moves toward the front of the invasion, that is, whenever *TFT* pioneers can avoid being suckered as they move outward. A *TFT* pioneering to the front of an invasion will not be suckered there if it can get assorted with another *TFT* also on a pioneering move or if it moves together with a known *AD* (in which case it will retaliate). The auxiliary model set up by Ferrière and Michod (1995) shows that both conditions are more likely to be met for high (but not too high) mobility in *TFT* and *AD*. Likewise, the second condition is met if a defector entering the core of a *TFT* cluster gets assorted there with another *AD* or if it undergoes retaliation by a *TFT* also moving back to the core. Again, the likelihood that either case will be realized is maximized at high *TFT* and *AD* mobility rates. To summarize, the following events are crucial for the emergence of *TFT* and are enhanced by significant mobility of the players: assortative meetings of *TFT*s at the front of an invasion or of *AD*s in the core of the cluster, and tracking of *AD*s by *TFT*s toward the front or toward the core.

17.4 Invasion of *Tit For Tat* in Games with Space-limited Memory

There are two important assumptions underlying the IPD replicator–diffusion equations investigated in the previous section. First, the memory is "space-extended" but "time-limited." That is, a player can recognize its opponent wherever they meet, but the player's memory is limited to the last round of the game. Second, the local density of each strategy is assumed to vary very slowly, in agreement with the assumption made in the microscopic description of the population that the spatial axis can be divided into contiguous cells of constant length, each cell containing two individuals. Hutson and Vickers (1995) have developed a different reaction–diffusion model of the spatial IPD where these assumptions have been modified or relaxed. In the Hutson–Vickers model, memory is not restricted to the last interaction but instead is space-limited: a player can remember any of its previous opponents provided that neither has moved out of the cell where they first met. Furthermore, a cell may now contain a variable number of players. The goal of this section is to present some important results drawn from their approach after examining structural differences between this model and the previous game-diffusion equations.

Model description

The Hutson–Vickers model is fully expounded in Chapter 22 (Section 22.2, A model for invasion of *Tit For Tat*). Here, we content ourselves with highlighting the specificities of this model.

- *Local interactions.* The spatial axis is still divided into contiguous cells of constant length l, but now each cell may contain many individuals. Opponents of any player in a given cell are drawn randomly within that cell.
- *Repeated interactions.* The Hutson–Vickers model relaxes the assumption that local population size varies on a slow time scale. Consequently, the probability w of players meeting is no longer a constant. In their model, Hutson and Vickers (1995) recast w into a dynamic "getting-to-know" function [denoted by $g(x, t)$] that gives the proportion of *AD* (or *TFT*) players within a cell that a typical *TFT* (or *AD*) player has already met. They further define $G = g n_T n_D$ as the number density of *TFT–AD* pairs within a cell that have already met.
- *Memory.* Memory is not limited to the last round. A *TFT* player recognizes an opponent on a second or subsequent occasion provided that neither has left the cell where the encounter occurred. This is in contrast with the game-diffusion model, where recognition may occur wherever the encounter takes place, but only on the next interaction.
- *Population regulation.* The per capita death rate is made density dependent. Therefore, it varies in space and time (but, as before, it is not influenced by the outcome of the game).

Main properties of the model

The analysis of the Hutson–Vickers model stems from ideas similar to those underlying Proposition 17.2. First, one has to determine the possible population equilibria assuming that player densities are spatially homogeneous. One may then turn to the effect of locally perturbing the stable equilibria, thereby mimicking the effect of an invasion attempt. When players have spatially homogeneous distributions, their densities and the getting-to-know function depend only on t. Then the model reduces to a system of ordinary differential equations [set $\partial^2 u / \partial x^2 = 0$ and $\partial^2 v / \partial x^2 = 0$ in Equations (22.17) and replace all partial derivatives with respect to t with ordinary derivatives in Equations (22.17) and (22.19) of Chapter 22]. Standard techniques of local stability analysis can now be used. The system turns out to have one of three simple structures:

1. There are stable equilibria with only *TFT*-players and only *AD*-players, and an unstable coexistence equilibrium.
2. There is a stable pure *AD* state and the only other equilibrium, that with just *TFT* players, is unstable.
3. In addition to the equilibria of structure (2), there are two coexistence states (one stable and one unstable) of *TFT* and *AD*.

The next step aims at determining whether a pure *TFT* state may evolve starting from initial conditions where *TFT* players are localized within an established *AD* population. This requires that the homogeneous *TFT* state must be stable, which actually happens with structure (1). Thus, we must deal with a situation similar to that handled by Proposition 17.2: two stable states and traveling waves that may "connect" them. However, the Hutson–Vickers model is not written as a system of replicator–diffusion equations (hence Proposition 17.2 does not apply), and a theoretical treatment presents rather formidable difficulties. A computational study suffices, however, to demonstrate the remarkable richness of the model's behavior. The most noteworthy point, as illustrated by the numerical example presented hereafter, is that large or small players' mobilities cannot be claimed to be unambiguously good or bad for the evolution of cooperation.

Existence of invading waves of *Tit For Tat*

In contrast with the case of replicator–diffusion models, no general theorem is available to guarantee the existence of traveling wave solutions to Hutson–Vickers equations. Yet numerical procedures do provide evidence that invasion dynamics develop wave patterns. As in Proposition 17.2, the sign of the wave speed determines whether the wave replaces *AD* with *TFT*, or *TFT* with *AD*. Figure 17.2 shows how for a specific set of parameter values the mobility rates influence the outcome of the game. To a certain extent, the results confirm those of the previous section. For small or large μ_T the cooperators are defeated, whereas for medium μ_T *TFT* successfully invades.

Hutson and Vickers (1995) gave the following interpretation of this result. Consider what happens as μ_T is reduced at points A and B in Figure 17.2. These points indicate stalemates, that is, traveling waves with zero speed. When μ_T is reduced, the wave front of the *TFT* players steepens so that density is reduced in the leading edge of the wave. The key factor is that at A the number of encounters between any given two players is determined by the death rate, whereas at B it is determined by mobility. At B, the getting-to-know function g and the number of *AD*–*TFT* pairs that

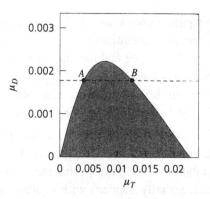

Figure 17.2 Mobility rates leading to an invasion of an established *AD* population by a wave of *TFT* players. The model assumes space-limited memory and can be written as a system of reaction–diffusion equations [see Chapter 22, Equations (22.17) and (22.19)]. The shaded area contains all pairs of mobility rates μ_T, μ_D for which there exists a traveling wave replacing *AD* with *TFT*. For given defectors' mobility μ_D smaller than a maximum value (≈ 0.0021) – for example, $\mu_D = 0.0018$ (dashed line) – there is an interval (A, B) of μ_T mobility rates over which invasion by *TFT* occurs. Model parameters: $\alpha = 1.3725$, $\beta = 0.9$, $\gamma = 1.45$, $\delta = 1$, $k = 4$, $b = 0 = d$, $\sigma = 1$, $\theta = 100(\mu_T + \mu_D)$.

have already met G increase significantly, causing the per capita birth rate in *TFT* to increase. Thus the *TFT* players have the advantage and a wave develops in which they advance. By contrast, at A there is little change in functions g and G so that the per capita birth rate of *TFT* is mainly influenced by the decrease in *TFT* density, which gives *AD* the advantage. The same type of argument suggests that reducing μ_D is always bad for *AD*, which is consistent with Figure 17.2 but contrasts with conclusions drawn from the replicator–diffusion model.

17.5   Concluding Comments

The importance of spatial structure for the IPD has long been realized (see, e.g., Axelrod 1984). Intuitively, individual mobility in the IPD is expected to raise an insurmountable obstacle to the spread of cooperation by allowing egoists to exploit cooperativeness and escape retaliation (Houston 1993). Dugatkin and Wilson (1991) and Enquist and Leimar (1993) addressed the issue, but their models had several limitations: only *AD* players were mobile; mobility was represented implicitly through some traveling cost; and only the question of the stability of *TFT* against *AD* was considered.

Reaction–diffusion models offer a natural framework to incorporate temporal and spatial effects in games. These models represent players that move in space in a random manner at a rate controlled by specific parameters. Their development is rooted in the Taylor–Jonker replicator equations (17.1). Players' mobility is included through the standard diffusion approximation of spatial motion, which yields second-order derivatives with respect to the spatial variable in Equations (17.2). We call the resulting system a "replicator–diffusion model." The reaction term can be modified further to allow for population limitation through density-dependent payoffs (Cressman and Dash 1987).

Once the reaction–diffusion model has been set up, one can address the central question in game theory: can an established population of one or several strategies be invaded by an initial spatially limited distribution of individuals playing an alternative strategy? Propositions 17.1 to 17.3 provide some insights into this problem in the context of replicator–diffusion models. The spatial dimension does not affect the stability of an internal strategy mix (i.e., all strategies are represented), which is an ESS in the standard game (Proposition 17.1). When there are only two strategies and both are ESSs in the nonspatial game, space dramatically alters the picture by allowing one strategy to displace the other (Proposition 17.2). Finally, in games with three (or more) strategies, spatial patterns (that is, spatially heterogeneous but temporally "frozen" distributions of coexisting strategies) develop when the replicator equation possesses a stable internal equilibrium that is not an ESS (Proposition 17.3). These results have been extended to spatial games including logistic population regulation (Cressman and Vickers 1997).

From the point of view of finding explicit, tractable invasibility criteria, two-strategy replicator–diffusion models are quite remarkable. If there is only one pure ESS in the standard game or if there is a mixed ESS, the stability property carries over nicely to the spatial game. A difficulty arises when both pure strategies are ESSs in the nonspatial game. In the spatial setting, the mathematical theory (Hutson and Vickers 1992) offers three statements that constitute the core of Proposition 17.2: one strategy invades and replaces the other (no coexistence); the invasion dynamics develop as a traveling front; there is a clear-cut invasibility criterion based on the sign of the speed of the traveling wave. On the basis of numerical simulations, the same invasibility rule proves to also apply to the more involved Hutson–Vickers model. Therefore, in these models it is the emergence of traveling waves that determines the evolutionary fate of individuals. The wave acts as a "vehicle" for population conflict (which mainly occurs around the fringe of the wave). In a sense, selection operates "at the level of the wave," although the wave itself is not a self-reproducing unit, just an expanding one. Obviously, the properties of waves are not in the definition of the system, instead they are derived from the individuals' behavioral and demographic traits. A similar phenomenon has been observed in individual-based models of host–parasitoid interactions where the formation of spiral waves determine the invasion success of mutant parasitoids (Boerlijst *et al.* 1993).

Other versions of the spatial IPD, designed as cellular automata, have recently been issued (Lindgren and Nordahl 1994; Nakamaru *et al.* 1997; see Chapter 13). Differences between these models and the reaction–diffusion approach lie in various (biological) assumptions about individual mobility and the effect of the game on individual life histories. The game pay-offs translate into a transmission rate (i.e., the probability of invading a neighboring site) in the model designed by Lindgren and Nordahl (1994), whereas they determine mortality rates in the framework by Nakamaru *et al.* (1997). The former model was analyzed though computer simulations; the latter received an analytical treatment by means of pair-approximation techniques (see Chapters 13 and 18). In both models, mobility is restricted to the dispersal of one offspring into a vacant neighboring site. Consequently, neither model allows connections to be drawn between the outcome of the game and different levels of individual mobility. Van Baalen and Rand (1998) have also developed a pair-approximation model of competition between altruists and non-altruists in a viscous population, in which they incorporated a rate of mobility (the same for both types of individuals). Although their system is not an iterated game, there is an interesting parallel between its behavior and that of the replicator–diffusion model. Again, invasion appears to be governed by a "spatially extended" Hamilton rule, where the coefficient of relatedness is recast into a coefficient of reciprocation depending on the birth, death, and mobility rates – much like the left-hand side of Equation (17.13). Also, the unit of selection becomes a "characteristic cluster" whose structure is described by a stable distribution of pairs of neighboring site occupancies, altruist–altruist, altruist–selfish, altruist–empty (a distribution that can be calculated from the model parameters). Van Baalen and Rand's model predicts that altruism can invade a selfish population background provided that the individual mobility rate is close to some optimum, intermediate value. As in the reaction–diffusion models, this ensures that the "scale of dispersal" is larger than the "scale of interaction." In other words, dispersal should be limited to guarantee a sufficient proportion of altruist–altruist pairings, but strong enough to ensure that altruists can "export" themselves and propagate through the environment.

The issue of invasion in spatial games arises from the study of a fascinating biological enigma – the origin and maintenance of cooperation – and yields profound mathematical challenges. The key relation between the existence of a traveling wave and invasion from a localized cluster is widely accepted on the basis of overwhelming numerical simulations; however, it has yet to be proved mathematically (see Chapter 22). The most urgent issue might be to further probe how the local mean-field description of spatial games based on reaction–diffusion models departs from the dynamics of the underlying discrete system of interacting individuals. Individual models cannot reach a sufficient level of generality, nor do they succeed at pointing out details at the individual level that are critical for understanding the macroscopic dynamics. Intermediate descriptions – for example, through moment or correlation equations (see Chapters 18 to 21) – have yet to be improved with respect to dealing with the initial stages of invasion processes, when the invading population is limited to a small area in space. In the meantime, we believe that the models of spatial games described in this chapter represent a significant improvement over previous mathematical attempts to describe the IPD and explain the evolution of cooperation.

Part D

Simplifying Spatial Complexity: Techniques

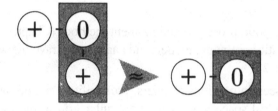

Introduction to Part D

Which systematic techniques can we apply for simplifying spatially explicit models in ecology? In the following chapters, we have assembled a menu of tools tailored to capturing and reducing spatial complexity. Chapters in this part are tutorials, each introducing or extending one method. They should help you to acquire the necessary skills for applying these techniques to ecological problems of your own choice.

The tools described in this part cover three complementary types of models in spatial ecology:

- Probabilistic cellular automata and pair approximations (Chapters 18 and 19)
- Dynamical point processes and moment methods (Chapters 20 and 21)
- Deterministic continuum models and analyses of traveling waves (Chapters 22 and 23)

As we have seen in previous chapters, probabilistic cellular automata discretize ecological space into sites on a regular grid or on a more general contact network. On such a geometry, ecological dynamics unfold as sites change their states depending on states of adjacent sites. In contrast, individuals in dynamical point processes occupy locations in continuous space and thus are not restricted to any given set of sites; individuals may multiply, move, or disappear in response to other individuals present in their neighborhood. Ecological models of cellular automata and point processes are typically stochastic: two realizations of the same ecological process are unlikely to be identical. Continuum models are different: while individuals here are situated in continuous space, they are assumed to be so abundant that their spatial density distributions can be regarded as continuous and their dynamics as deterministic.

For each such modeling framework, techniques have been developed to reduce its spatial complexity. Methods of choice for cellular automata are pair approximations and their refinements; point processes can be simplified by way of moment methods; and the dynamics of continuum models are often best understood by investigating the existence and propagation of traveling density waves. The various examples presented in Part C have introduced these different tools: each has its domain of utility and thus ought to be chosen according to the ecological question at stake. We believe that you will find it worthwhile to learn about these alternatives: such an overview allows you to decide which tools are best suited to the specific ecological scenario you wish to investigate.

The first two chapters in this part introduce pair approximations. Chapter 18 reviews the fundamental concepts of this technique for grid-based models. While mean-field approaches keep track only of global spatial densities (and therefore are restricted to spatially uniform systems), pair approximations step beyond this simplified view by tracking through time probabilities of states of pairs of adjacent sites. Satō and Iwasa also present so-called improved pair approximations that allow specific *a priori* assumptions about the density of triplet configurations to be incorporated into analyses. This can be done either in a static way (constant discounting) or dynamically (variable discounting). Both refinements are designed to enhance the accuracy of pair approximations.

Chapter 19 shows how pair approximations can exploit knowledge about underlying contact structures. Triangular and square lattices are studied in detail, and other two-dimensional contact networks, resulting from local randomization of lattice links, are investigated as well. To represent ecological processes on grids, it is often helpful to consider simultaneous changes in the states of two adjacent sites (think, for example, of the consumption of a prey individual and the resulting satiation of a predator); van Baalen allows for this generalization. The chapter also emphasizes the importance of local fluctuations for refining predictions of pair approximations.

In Chapter 20, the pair-approximation method is extended to continuous space. Individuals are represented by their actual location, avoiding an artificial discretization of the habitat. With each point in a pattern corresponding to a single individual, ecological change can then be envisaged as the stochastic dynamics of point processes. Bolker, Pacala, and Levin analyze how spatial densities of pairs of individuals depend on the distance between the individuals and how these pair densities are expected to change over time. The resulting moment methods for simplifying spatial complexity successfully predict outcomes of intra- and interspecific competition, and provide analytical insight into intricacies of spatial interactions.

Chapter 21 elaborates on the method of moments in continuous space and presents a three-tiered procedure for its derivation. Ecological processes at the level of individuals are described by a spatially explicit, stochastic process. Expected dynamics of this process are then expressed in terms of correlation densities, with global densities and pair densities as special cases. A closure assumption finally yields self-contained dynamics. Dieckmann and Law discuss their method within the broader context of relaxation projections, evaluate the relative performance of different closure assumptions, and stress the importance of fluctuation corrections complementing the correlation corrections derived from pair dynamics.

Chapter 22 gives an introduction to reaction–diffusion models – the most popular type of deterministic models in spatial ecology to date. Assuming continuous spatial density distributions, ecological interactions at each location of a habitat are described by deterministic population dynamics, defining the reaction part of the model. The model's diffusion part allows for the coupling of different locations that results from movement processes. Hutson and Vickers explain the use of comparison methods and review analytical techniques for understanding the formation of Turing patterns and the dynamics of traveling waves. They underline the flexibility of reaction–diffusion modeling by presenting applications to mutation-selection processes and to systems with memory.

Invasion waves are empirically important and are investigated in Chapter 23. Concentrating on this specific class of spatio-temporal patterns, Metz, Mollison, and van den Bosch can incorporate more realistic movement processes and life-history details than is possible for reaction–diffusion models. While invasion waves and foci expansions occur in many different settings – ranging from epidemiological to evolutionary – their main characteristics are often determined solely by the speed of the resulting wave front. The chapter explains ways to predict this speed and discusses complications that arise when space is inhomogeneous, individuals interact (directly or indirectly), or movement of individuals occurs across different spatial scales.

Part D thus provides introductions to three major techniques for simplifying spatial complexity. The utility of these tools has been demonstrated in Part C; here, systematic treatments are given.

18

Pair Approximations for Lattice-based Ecological Models

Kazunori Satō and Yoh Iwasa

18.1 Introduction

In recent years, the effects of spatial configuration on population and evolutionary processes have been the subject of intensive research efforts in ecology and evolutionary biology. One reason for this surge of interest is the observation of intricate, natural spatio-temporal structures. Such structures are exemplified, for instance, by wave regeneration of fir forests, in which trees show a large-scale, wavelike pattern of regeneration with many stripes of dieback zones moving slowly downwind at a constant rate (Iwasa *et al.* 1991; Satō and Iwasa 1993; Jeltsch and Wissel 1994). Another example of spatio-temporal structure studied in forest ecology is the formation and closure of canopy gaps (Kubo *et al.* 1996; see Chapter 13). It is now acknowledged that considering spatio-temporal structures spontaneously formed by demographic processes and ecological interactions is sometimes essential for understanding population and evolutionary dynamics, and that traditional modeling in theoretical ecology assuming complete spatial mixing often fails to capture these dynamics.

A simple and useful method for modeling population and evolutionary dynamics in a spatially explicit way is to use a lattice, or cellular automaton, model. Typically, one considers a large regular lattice, which may be linear, square, hexagonal, or triangular, in which an individual occupies one vertex, or lattice site. We assume that each individual interacts directly and strongly only with its nearest neighbors. Most ecological interactions between individuals – for example, competition for resources, disease transmission, cooperative interaction such as attracting common pollinators, and reproduction – occur at spatial scales much smaller than that of the whole population. If the mobility and dispersal ability of individuals are rather limited, ecological and demographic processes will result in characteristic

341

spatial patterns in which individuals of the same type tend to form clumps. These clumps cause population phenomena that are qualitatively different from those expected in well-mixed systems and can affect species coexistence, invasion, evolution, and species or genetic diversity. The evolution of a pathogen's virulence and of the host's resistance may critically depend on the spatial structure of the habitat, and the conditions for the evolution of altruism or spite in completely mixed models can be very different from those in lattice models. These qualitative departures are relatively insensitive to the particular choice of lattice – whether the lattice is linear, square, hexagonal, or triangular, for example, or whether or not interaction is restricted to nearest neighbors. A more flexible type of model in which individuals are represented as points distributed over two-dimensional continuous space interacting with nearby points shows results similar to those from lattice models (see Chapters 14, 20, and 21). What matters is that interactions between individuals are local, an assumption satisfied in most cases of population biology. This effect is especially important for sedentary organisms, such as terrestrial plants, but it also has important implications for territorial animals (Gordon 1997).

An advantage of lattice models over traditional population dynamic models that are merely based on population densities or abundances is the ease and flexibility with which ecological interactions occurring between nearby individuals can be incorporated. In this respect, lattice models are closely related to a group of models called "individual-based models" (Judson 1994) or "agent-based models" (Axelrod 1997). In these models, the state of the system is described by a group of "individuals" that interact in a relatively simple manner. Complicated dynamical properties of large biological systems comprising many individuals have become more tractable due to recent advances in the power of computing facilities. In fact, lattice models are often used in complex systems theory to illustrate how a number of individual agents interacting only locally in a rather simple way can sometimes create very complex behaviors that are difficult to predict without actually running the simulation (e.g., Forrest and Jones 1994). In the field of applied mathematics and theoretical physics, lattice models are called interacting particle systems (Liggett 1985; Durrett and Levin 1994b). Finally, lattice models have the advantage of explicitly considering stochasticity caused by finite numbers of individuals, called demographic stochasticity in population ecology and random drift in population genetics. This stochasticity is often neglected in traditional models of differential equations in theoretical ecology.

A major problem with these spatially explicit models, whether lattice models or models based on continuous space, is that analysis is often restricted to direct computer simulation. If the model includes a number of parameters, it is extremely time consuming and often simply infeasible to examine the parameter dependence of the model's behavior in detail. Moreover, due to inherent stochasticity, these models require the consideration of many replicates for the same set of parameters to determine salient aspects of the model's behavior. Because simulation models can often be studied only for a limited number of parameter combinations (usually those that look plausible to the modeler), analytical tools for studying lattice models are needed.

Mean-field approximation, which neglects spatial structures by assuming complete mixing of individuals, often shows behavior that is qualitatively different from that observed in direct computer simulations of lattice models, as is demonstrated by many chapters of this book (see Chapters 6 to 11).

Fortunately, there is a useful method for analyzing lattice models. The method, pair approximation, was introduced by Matsuda in an analysis of the evolution of social interaction in lattice-structured populations (Matsuda 1987; Matsuda *et al.* 1987). In this method, one constructs a closed dynamical system comprising two components: first, the overall densities of lattice sites in specified states and, second, the conditional probability that a randomly sampled nearest neighbor of a site is in a specified state. Pair approximation is a type of decoupling approximation, developed in statistical physics, and is also referred to as a "moment closure method" (see Chapters 14, 19, 20, and 21).

The importance of considering correlations between interacting individuals has already been pointed out, for example, in the case of insect population dynamics (e.g., Ives and May 1985), where a spatially clumped distribution creates a conditional density in the neighborhood of an individual much higher than the density from random samples. Pair approximation is different because it treats the degrees of spatial clumping as a separate dynamic variable rather than as a given parameter, since the degree of clumping and nearest-neighbor correlation are also the result of demographic processes and ecological interactions (Harada and Iwasa 1994).

Whereas Chapter 13 provides several examples of pair approximation applied to ecological issues, this chapter focuses on the technique as such.

18.2 Pair Approximation

Let us consider the simplest example of a lattice-structured population, the "lattice logistic model," also referred to as the "basic contact process." This model is based on the following assumptions.

■ *Uniform lattice.* The system describes, for example, a large population of sedentary plants growing in a lattice-structured habitat. Each individual occupies one site of the lattice, and, as a result of competition (or self-thinning) among plants within the site, each site can maintain at most one individual. Thus, each lattice site is either occupied by a plant (+) or empty (0). The lattice can be one-dimensional (linear) or two-dimensional (square, triangular, or hexagonal). We assume that the lattice space is infinitely large, homogeneous, and isotropic. All lattice sites have the same number of nearest neighbors z.

■ *Death.* Each individual plant dies at a constant rate, which is defined as the expected number of death events in a unit time interval resulting in a transition from + to 0. The death rate at a site is independent of the state of neighboring sites.

■ *Birth.* Each individual attempts to reproduce vegetatively into its nearest-neighbor sites at a given constant rate, which is defined as the expected number of birth events by an individual in a unit time interval. Newborn individuals are sent to nearest-neighbor sites with equal probability but fail to become established there if the site is already occupied by another individual.

Because what matters is the ratio of the death and birth rates, we can reduce the number of parameters to one by rescaling the time unit. Thus without a loss of generality we set the death rate to 1 and the birth rate for each individual to b/z. Accordingly, b is the maximum reproduction rate of plants, which is realized when all nearest-neighbor sites are empty. An empty site will be filled by a plant at a rate proportional to the number of nearest-neighbor plants.

We illustrate these processes as follows:

$$\text{Death:} \quad \oplus \xrightarrow{1} \textcircled{0}, \tag{18.1}$$

$$\text{Birth:} \quad \oplus\text{-}\textcircled{0} \xrightarrow{b/z} \oplus\text{-}\oplus, \tag{18.2}$$

where $\bigcirc\text{-}\bigcirc$ indicates a pair of nearest neighbors in which the large circle represents the site undergoing transition. Note that an individual cannot reproduce unless a nearest-neighbor site is empty. This constraint causes density-dependent reproduction, where the birth rate is proportional to the availability of empty sites.

When trying to describe the dynamics of the densities of the two states (or, equivalently, the fraction of lattice sites occupied by plants), we immediately recognize that the birth rate depends on the availability of vacant sites among the nearest neighbors of a randomly chosen occupied site rather than on the overall average availability of empty sites. Thus it is difficult to construct a simple model of population dynamics as an ordinary differential equation based only on the overall density of the plant population. Instead, we have to distinguish between two notions, global (or singlet) densities and local (or environ) densities (Matsuda *et al.* 1992). The state of a site is denoted by σ, which is either $+$ or 0. Global densities ρ_σ, with $\sigma \in \{+, 0\}$, are the probabilities that a randomly chosen lattice site is in state σ. Local densities $q_{\sigma/\sigma'}$, with $\sigma, \sigma' \in \{+, 0\}$, are the conditional probabilities that a randomly chosen nearest neighbor of a site in state σ' is in state σ. The difference between ρ_σ and $q_{\sigma/\sigma'}$ indicates nearest-neighbor correlation. These local densities can be expressed in terms of doublet densities $\rho_{\sigma\sigma'}$, which are the probabilities that a randomly chosen pair of nearest-neighbor sites are in state $\sigma\sigma'$. Doublet densities can be expressed as the products of global densities and local densities: $\rho_{\sigma\sigma'} = \rho_{\sigma'\sigma} = \rho_\sigma q_{\sigma'/\sigma} = \rho_{\sigma'} q_{\sigma/\sigma'}$. Higher-order densities, such as triplet densities or quartet densities, can be defined similarly. By definition, global and local densities satisfy

$$\sum_{\sigma\in\{+,0\}} \rho_\sigma = 1 \, , \tag{18.3}$$

$$\sum_{\sigma\in\{+,0\}} q_{\sigma/\sigma'} = 1 \quad \text{for any } \sigma' \in \{+, 0\} \, . \tag{18.4}$$

In all the models discussed in this chapter, the initial configuration is uniformly distributed and the processes are spatially homogeneous and isotropic. Hence probabilities ρ_σ and $\rho_{\sigma\sigma'}$ remain independent of location.

These probabilities change over time t. According to processes (18.1) and (18.2), global density develops over time as follows:

$$\frac{d\rho_+}{dt} = -\rho_+ + bq_{0/+}\,\rho_+ \, . \tag{18.5}$$

The first and the second terms on the right-hand side of the equation correspond to the death and birth processes, respectively. Because the nearest-neighbor sites must be empty for a plant to reproduce vegetatively, the second term includes local density of vacant sites $q_{0/+}$ instead of global density ρ_0.

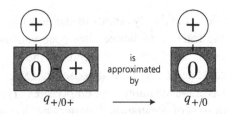

Figure 18.1 The key assumption of pair approximation. The average density of + sites adjacent to 0+ pairs, $q_{+/0+}$, is approximated by $q_{+/0}$. Neglecting the effect of the third site also implies that no distinction is made between different possible spatial arrangements of the three adjacent sites (see Figure 18.2a).

Similarly, the doublet density changes according to the dynamics

$$\frac{d\rho_{++}}{dt} = -2\rho_{++} + 2\frac{b}{z}\rho_{+0} + 2\frac{b}{z}(z-1)q_{+/0+}\rho_{+0} \ . \tag{18.6}$$

This equation includes the death term (the first term on the right-hand side) and the birth terms (the second and third terms). The first term describes transitions of pairs in state ++ to either +0 or 0+. Both transitions occur at a rate of 1 (the death rate) and thus give rise to the factor 2. The factor 2 is needed in the second and the third terms because we do not assume any asymmetry in the interaction between sites, which means $\rho_{+0} = \rho_{0+}$. The second term indicates the case where + reproduces into the nearest-neighbor 0 site in this pair. The third term corresponds to events in which an empty site becomes occupied as a result of vegetative reproduction of an adjacent + site. Thus, in the two-dimensional square lattice ($z = 4$) we must consider three sites (in general, $z - 1$ sites) in the dynamics of local densities. Here we neglect differences in the configuration of these three connected sites on the two-dimensional square lattice space (see Figure 18.1). This assumption is fulfilled exactly for models on Cayley-tree-structured lattices (or Bethe lattices) that include no loops (see Figure 18.2b).

Even when assuming that $q_{+/0+}$, the average density of + sites adjacent to 0+ pairs, is the same for all configurations of such +0+ triplets, Equation (18.6) is incomplete without knowledge of $q_{+/0+}$. Although for this purpose we can formulate the dynamics of higher-order densities, such as triplet densities $\rho_{\sigma\sigma'\sigma''} = \rho_{\sigma'\sigma''}q_{\sigma/\sigma'\sigma''}$, these equations include densities of still higher orders and we can never arrive at a closed system of differential equations. To overcome this difficulty, we adopt an approximation, called the decoupling method (or moment closure), which neglects higher-order correlations. The particular approximation we discuss in this chapter is called pair approximation because we trace pair correlations but neglect

(a)   (b)

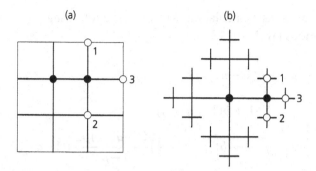

Figure 18.2 Different spatial arrangements of three adjacent sites. Pair approximation treats as equivalent the three triplet configurations indicated by the pair of filled circles together with one of the open circles marked 1 to 3. (a) Square lattice. Here, the neighborhood structure of the two right-angled configurations is different from that of the straight configuration. (b) Cayley tree. On this lattice, the three triplets have the same neighborhood structure.

three-site correlations (Matsuda *et al.* 1992; Satō *et al.* 1994). For example, consider a site A that is adjacent to a site B, which in turn is adjacent to another site C. If A and C are not nearest neighbors, in pair approximation we neglect C's effect on the probability of finding the pair AB in a specified state (see Figure 18.1). In pair approximation we therefore focus only on nearest-neighbor correlation; correlations between non-nearest neighbors are approximately reconstructed from nearest-neighbor correlations. In general, $q_{\sigma/\sigma'\sigma''}$ is replaced by $q_{\sigma/\sigma'}$. Underlying this assumption is the idea that the correlation between lattice sites decreases monotonically with the distance between them. Thus a site is affected less by distant neighbors than by intermediate neighbors. The higher-order conditional probabilities in Equation (18.4) thus can be reduced as follows:

$$q_{+/0+} \approx q_{+/0} = \frac{\rho_{+0}}{\rho_0} = \frac{(1 - q_{+/+})\rho_+}{1 - \rho_+} \, . \tag{18.7}$$

Because doublet densities are the products of global and local densities, we can express the local densities as the ratio of doublet densities to singlet (global) densities. Hence the dynamics of local density can be written as follows:

$$\frac{dq_{+/+}}{dt} = \frac{d(\rho_{++}/\rho_+)}{dt} = -\frac{\rho_{++}}{\rho_+^2}\frac{d\rho_+}{dt} + \frac{1}{\rho_+}\frac{d\rho_{++}}{dt} \, . \tag{18.8}$$

After straightforward calculation, we obtain the following set of equations
from Equations (18.5) to (18.8):

$$\frac{d\rho_+}{dt} = -\rho_+ + b(1 - q_{+/+})\,\rho_+ \,, \qquad\qquad (18.9)$$

$$\frac{dq_{+/+}}{dt} = -q_{+/+}[-1 + b(1 - q_{+/+})]$$
$$+ 2\left\{-q_{+/+} + b\left[\tfrac{1}{z} + \left(1 - \tfrac{1}{z}\right)\frac{(1 - q_{+/+})\rho_+}{1 - \rho_+}\right](1 - q_{+/+})\right\}\,. \qquad (18.10)$$

We then can calculate the internal equilibrium of the closed dynamical system given by Equations (18.9) and (18.10) with two variables ρ_+ and $q_{+/+}$
as follows:

$$\hat{\rho}_+ = 1 - \frac{z - 1}{(z - 1)b - 1}\,, \qquad \hat{q}_{+/+} = 1 - 1/b\,. \qquad (18.11)$$

In contrast, if we neglect the spatial structure by adopting the mean-field
approximation, in which local density $q_{+/+}$ is assumed to be equal to global
density ρ_+ in Equation (18.9), we obtain

$$\frac{d\rho_+}{dt} = -\rho_+ + b(1 - \rho_+)\rho_+\,. \qquad\qquad (18.12)$$

At equilibrium this yields

$$\hat{\rho}_+ = 1 - 1/b\,. \qquad\qquad (18.13)$$

Figure 18.3 shows that the equilibrium abundance given by the pair approximation [Equation (18.11)] is closer to the values from computer simulations than the equilibrium abundance predicted by the mean-field approximation [Equation (18.13)]. The pair-approximation method gives a
better quantitative prediction of the equilibrium abundance of plants (i.e.,
the global density) than the dynamics of mean-field approximation, but it
is not perfect. It still overestimates the abundance of individuals in the
equilibrium lattice populations (Figure 18.3), which suggests the importance of correlations between non-nearest neighbors. There are methods to
improve the approximation by considering stronger clumping, such as the
improved pair-approximation method developed by Satō *et al.* (1994) and
Harada *et al.* (1995). In the following section, we introduce this method by
investigating a more complex three-state model.

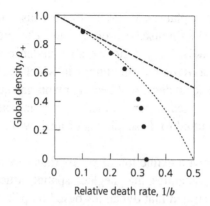

Figure 18.3 Global densities in the lattice logistic model. Results of Monte Carlo simulations are shown as filled circles; predictions by pair approximation, as the dotted curve; and predictions by mean-field approximation, as the dashed line.

18.3 Improved Pair Approximation

Here we consider an epidemic model for sedentary hosts living in a lattice-structured habitat (Satō *et al.* 1994). Specifically, we examine a population of terrestrial plants that reproduce only to neighboring sites. The infectious disease attacking the plants is transmitted only between nearest neighbors. Furthermore, we assume that the diseased plants may lose the ability to reproduce but enjoy fairly normal viability. This assumption is quite plausible from the pathogen's viewpoint, because killing the infected host is detrimental to the pathogen's transmission.

The corresponding model without a lattice structure – that is, the completely mixed epidemic model – is the standard susceptible–infected–recovered (SIR) model, in which a healthy host increases logistically in the absence of disease and the population size is controlled by the availability of vacant sites into which plants can reproduce. In these models, attempts by diseased plants to invade an equilibrium population of healthy plants can result in only two outcomes: the invasion either fails or it succeeds, leading to an endemic state with diseased and healthy plants stably coexisting.

Surprisingly, the computer simulations of the lattice model reveal a completely different outcome: the lattice epidemic model often exhibits a temporal spread of the infectious disease followed by the extinction of both diseased and healthy plants. This "pathogen-driven extinction" occurs especially when the disease transmission rate is high. Such extinctions are not possible in traditional epidemic models that assume complete mixing. In these models, a threshold density of healthy plants exists below which the number of infected plants starts to decline because there are not enough healthy plants to maintain disease propagation. The situation is quite different for lattice models. Even if the overall abundance of healthy plants is very small, those that remain form clumps in which each healthy host has neighbors. Thus the local density of healthy plants can be quite high even if their global density is nearly zero. Consequently, infected individuals that happen to be transmitted to these islands of healthy plants spread and kill all the plants.

This pathogen-driven extinction of plants thus cannot be explained by simple mean-field dynamics that neglect spatial structure. Satō *et al.* (1994) have demonstrated that dynamics based on pair approximation can in fact predict such pathogen-driven host extinction and that the predicted parameter range for this outcome is quite accurate. In doing so, they modified the pair-approximation method since they found that ordinary pair approximation fails to capture the strong clumping of remaining healthy plants as they become rare. This modified method is called improved pair approximation.

For illustration, we consider lattice dynamics with three states: a site occupied by a healthy plant ($+$); a site occupied by an infected plant ($-$); and an empty site (0). We consider a continuous-time Markov process, in which state transition occurs asynchronously so that within a very short time interval at most a single transition can occur. The dynamics of the whole lattice population can be specified by the following processes:

$$\text{Death:} \quad \oplus \text{ or } \ominus \xrightarrow{1} \textcircled{0} \,, \tag{18.14}$$

$$\text{Birth:} \quad \oplus\text{-}\textcircled{0} \xrightarrow{m_+/z} \oplus\text{-}\oplus \,, \tag{18.15}$$

$$\text{Infection:} \quad \ominus\text{-}\oplus \xrightarrow{m_-/z} \ominus\text{-}\ominus \,. \tag{18.16}$$

Here, process (18.14) describes the death of an individual plant; the death rate is the same for healthy and infected plants.

For the birth rate given in (18.15), an offspring is assumed to settle in an empty site neighboring its parent, and m_+ is a constant parameter representing maximum fecundity of a healthy plant, which is realized when all nearest-neighbor sites are empty; z is again the number of nearest neighbors for each site.

The infection rate in process (18.16) represents the transmission of a pathogen from an infected plant to a healthy plant in a nearest-neighbor site. The maximum transmission rate of a pathogen m_- is realized when all its nearest-neighbor sites are occupied by healthy hosts. We assume that an infected plant cannot reproduce but has the same chance of survival as a healthy plant.

From the dynamics of the disease propagation model given by processes (18.14) to (18.16), the following set of equations of doublet densities is obtained (for details see Satō *et al.* 1994):

$$\frac{d\rho_{++}}{dt} = -2\left[1 + \left(1 - \tfrac{1}{z}\right)m_- q_{-/++}\right]\rho_{++} \\ + 2m_+ \left[\tfrac{1}{z} + \left(1 - \tfrac{1}{z}\right)q_{+/0+}\right]\rho_{+0}, \tag{18.17}$$

$$\frac{d\rho_{--}}{dt} = -2\rho_{--} + 2m_- \left[\tfrac{1}{z} + \left(1 - \tfrac{1}{z}\right)q_{-/+-}\right]\rho_{+-}, \tag{18.18}$$

$$\frac{d\rho_{+-}}{dt} = -\left\{2 + m_- \left[\tfrac{1}{z} + \left(1 - \tfrac{1}{z}\right)q_{-/+-}\right]\right\}\rho_{+-} \\ + \left(1 - \tfrac{1}{z}\right)m_+ q_{+/0-}\rho_{-0} + \left(1 - \tfrac{1}{z}\right)m_- q_{-/++}\rho_{++}, \tag{18.19}$$

$$\frac{d\rho_{+0}}{dt} = -\left\{1 + \left(1 - \tfrac{1}{z}\right)m_- q_{-/+0} + m_+ \left[\tfrac{1}{z} + \left(1 - \tfrac{1}{z}\right)q_{+/0+}\right]\right\}\rho_{+0} \\ + \left(1 - \tfrac{1}{z}\right)m_+ q_{+/00}\rho_{00} + \rho_{++} + \rho_{+-}, \tag{18.20}$$

$$\frac{d\rho_{-0}}{dt} = -\left[1 + \left(1 - \tfrac{1}{z}\right)m_+ q_{+/0-}\right]\rho_{-0} \\ + \left(1 - \tfrac{1}{z}\right)m_- q_{-/+0}\rho_{+0} + \rho_{+-} + \rho_{--}, \tag{18.21}$$

$$\frac{d\rho_{00}}{dt} = -2\left(1 - \tfrac{1}{z}\right)m_+ q_{+/00}\rho_{00} + 2\rho_{+0} + 2\rho_{-0}. \tag{18.22}$$

The resulting changes in the singlet densities $\rho_\sigma = \sum_{\sigma' \in \{+,-,0\}} \rho_{\sigma\sigma'}$ are

$$\frac{d\rho_+}{dt} = \left(-1 - m_- q_{-/+} + m_+ q_{0/+}\right)\rho_+, \tag{18.23}$$

$$\frac{d\rho_-}{dt} = \left(-1 + m_- q_{+/-}\right)\rho_-. \tag{18.24}$$

Equations (18.17) to (18.22) for doublet densities are not yet closed because the right-hand sides contain triplet densities, $q_{+/0\sigma}$ and $q_{-/+\sigma}$.

In pair approximation, $q_{\sigma/\sigma'\sigma''}$ is replaced with $q_{\sigma/\sigma'}$ to close the system of equations. But here we introduce an improved version of pair approximation that considers the tendency for stronger spatial clumping when the population is close to extinction than is suggested by ordinary pair approximation. Because reproduction of host and pathogen takes place only into nearest-neighbor sites, sites that are in the same state tend to become aggregated in space, forming a strong positive correlation of state between nearest neighbors. Because the remaining rare + sites form a clump that includes some + and 0 sites, a 0 site neighboring a + site is likely to be in such a clump. Consequently, $q_{+/0+}$ tends to be larger than $q_{+/0}$, and $q_{+/00}$ tends to be smaller than $q_{+/0}$. Yet, it is less inaccurate to approximate $q_{+/0-}$ using $q_{+/0}$ because relevant states (+ and −) tend to be segregated. We therefore assume that

$$q_{+/0-} = q_{+/0} , \tag{18.25}$$

$$q_{+/00} = \varepsilon q_{+/0} , \tag{18.26}$$

where ε is a positive constant less than 1 that is called the "discounting factor" (Harada *et al.* 1995). For the case of very small ρ_+, the discounting factor ε is estimated to be the critical value for ergodicity in the contact processes: $\varepsilon = 0.8093$ for $z = 4$ (Katori and Konno 1990; Konno and Katori 1990). Combining Equations (18.25) and (18.26) with the relationship between conditional probabilities,

$$q_{+/0+}q_{+/0} + q_{+/00}q_{0/0} + q_{+/0-}q_{-/0} = q_{+/0} , \tag{18.27}$$

we obtain

$$q_{+/0+} = 1 - q_{-/0} - \varepsilon q_{0/0} \geq q_{+/0} . \tag{18.28}$$

On the verge of population extinction, $q_{-/+\sigma}$ is expected to remain positive because of the high probability that the pathogen will exist next to the host. This expectation is supported by extensive Monte Carlo simulations. In contrast, $q_{+/0}$ approaches zero while $q_{+/0+}$ remains positive (similarly, $q_{-/+}$ approaches zero and $q_{-/+-}$ remains positive around the transition from the endemic parameter region to the disease-free region; once we determine which state is doomed to extinction, we often find the same state in the local neighborhood). Therefore, although $q_{-/+-}$ tends to be larger than $q_{-/+}$, it is expected that they have the same order of magnitude and that it is acceptable to ignore the difference between them:

$$q_{-/+\sigma} = q_{-/+} \quad \text{for } \sigma \in \{+, -, 0\} . \tag{18.29}$$

Satō *et al.* (1994) call the set of approximations given by Equations (18.25), (18.26), (18.28), and (18.29) improved pair approximation. Note that if we set ε equal to 1, the approximation is reduced to ordinary pair approximation.

Satō *et al.* (1994) suggested that on a one-dimensional linear lattice a pathogen can never invade a host population. This is because of the different way disease propagates in the one-dimensional linear lattice compared with the two-dimensional square lattice space. Later this was proved mathematically by Schinazi (1996).

We now concentrate on the two-dimensional square lattice. Using improved pair-approximation analysis, Satō *et al.* (1994) found that the equilibrium state of the lattice model can fall into four different domains: (1) extinction due to excess mortality; (2) disease-free equilibrium; (3) endemic equilibrium; and (4) pathogen-driven extinction. Figure 18.4 illustrates the phase diagram on the parameter space of the host mortality $1/m_+$ (horizontal axis) and the relative fecundity of the pathogen m_-/m_+ (vertical axis). When m_+ is smaller than m_+^{sus}, which we call the "sustenance threshold," the host cannot survive even in the absence of the pathogen. This critical host fecundity is higher (i.e., the resulting $1/m_+$ is lower) than in the corresponding mean-field model (see Figure 18.5) since the clumping of healthy individuals makes reproduction less efficient than under complete mixing. As mentioned before, this is because of the reduced availability of vacant nearest-neighbor sites in lattice models. For a given host fecundity m_+ greater than m_+^{sus}, the population resists the pathogen invasion when pathogen fecundity m_- is below a certain threshold m_-^{end}, which we call the "endemic threshold." As m_- increases beyond this threshold, the population dynamics attain an endemic equilibrium in which both the host and pathogen populations coexist. The behaviors of the model shown by these three parameter regions are observable in the mean-field model as well. However, if m_- increases further and exceeds a second threshold m_-^{ext}, which we call the "extinction threshold," the population is driven to extinction by invasion of infected hosts, an outcome that is not possible in models that neglect spatial structure (Figure 18.5). Extinction occurs despite the fact that before infection the healthy host population had maintained a stable equilibrium with high abundance.

We also have to consider the dependence of the phase diagram on the discounting factor ε. We find that as long as $\varepsilon < 1$ ($\varepsilon = 1$ indicates the case of ordinary pair approximation) we have pathogen-driven extinction,

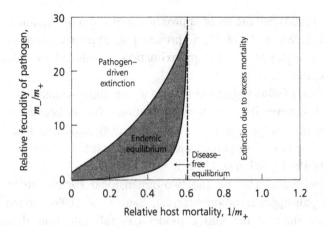

Figure 18.4 Phase diagram of the lattice epidemic model as predicted by pair approximation. The resulting four regions, each delimiting sets of parameters leading to qualitatively different outcomes, correspond to extinctions due to excess mortality ($m_+ < m_+^{sus}$), disease-free equilibria ($m_+ > m_+^{sus}$ and $m_- < m_-^{end}$), endemic equilibria ($m_+ > m_+^{sus}$, $m_- > m_-^{end}$, and $m_- < m_-^{ext}$), and pathogen-driven extinctions ($m_+ > m_+^{sus}$ and $m_- > m_-^{ext}$).

while for $\varepsilon < 8/9$, extinction occurs at any value of m_+. (For the derivation of the threshold value, see Satō *et al.* 1994.)

In the traditional SIR epidemic model, which is similar to the Lotka–Volterra model, there is no parametric region for pathogen-driven extinction. Indeed, mean-field approximation yields the following dynamical system, which can be obtained by replacing local densities $q_{-/+}$, $q_{0/+}$, and $q_{+/-}$ with the global densities ρ_-, ρ_0, and ρ_+, respectively, in Equations (18.23) and (18.24):

$$\frac{d\rho_+}{dt} = (-1 - m_-\rho_- + m_+\rho_0)\rho_+ , \tag{18.30}$$

$$\frac{d\rho_-}{dt} = (-1 + m_-\rho_+)\rho_- . \tag{18.31}$$

The phase diagram for the mean-field approximation is shown in Figure 18.5. The absence of pathogen-driven extinction in the lattice epidemic model demonstrates a sharp contrast between mean-field approximation and pair approximation, and underlines the importance of considering spatial structure. Monte Carlo simulations support the parameter regions for pathogen-driven extinction predicted by pair approximation (Satō *et al.* 1994).

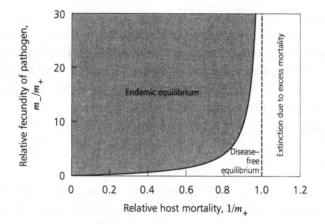

Figure 18.5 Phase diagram of the lattice epidemic model as predicted by mean-field approximation. The resulting three parameter regions correspond to extinctions due to excess mortality, disease-free equilibria, and endemic equilibria. Notice that a region of pathogen-driven extinctions (predicted by pair approximation and actually occurring in Monte Carlo simulations) is absent.

18.4   Improved Pair Approximation with Variable Discounting

In this section, we discuss the application of improved pair approximation to a two-state lattice model studied by Harada *et al.* (1995). They considered a lattice logistic model, where individuals engage in "social" interaction by modifying the mortality of their neighbors. Harada *et al.* (1995) also allowed for the migration of individuals, modeled as a random exchange of states between nearest neighbors. They applied improved pair approximation using two different variants, improved pair approximation with constant discounting (IPAC), and improved pair approximation with variable discounting (IPAV). As described in the previous section, the parameter ε is constant in IPAC; in contrast, IPAV treats ε as a function of the migration rate.

Their model is defined by the following processes:

Death: $\oplus\text{-}\oplus \xrightarrow{\ d\ } \oplus\text{-}\textcircled{0}$,   (18.32)

Birth: $\oplus\text{-}\textcircled{0} \xrightarrow{\ b/z\ } \oplus\text{-}\oplus$,   (18.33)

Migration: $\oplus\text{-}\textcircled{0} \xrightarrow{\ m/z\ } \textcircled{0}\text{-}\oplus$.   (18.34)

Process (18.32) represents the death of an individual at rate d. Unlike in previous sections, the death rate is not a constant and may be modified by social interaction with neighboring individuals. If social interactions are cooperative, the death rate of an individual is lower when it is surrounded by many neighbors than when it has fewer neighbors. If, on the other hand, social interactions are competitive, a larger number of neighbors increases the death rate. This can result from egoistic social acts, resource competition, pathogen transmission, and local environmental pollution. Specifically, we assume

$$d = d_0 - \beta \frac{n_+}{z} , \qquad\qquad (18.35)$$

where d_0 is a positive constant denoting an intrinsic death rate and β is a constant specifying the type and strength of social interaction. A positive β implies cooperative interaction; a negative β indicates competitive interaction between neighbors. The quantity n_+ stands for the number of $+$ sites among nearest neighbors and satisfies $0 \leq n_+ \leq z$. In Equation (18.35), the effect of social interaction is assumed to be proportional to the number of nearest neighbors n_+. Because mortality cannot be negative, even for $n_+ = z$, we assume $d_0 \geq \beta$.

Process (18.33) describes birth. This is the same process as that in (18.2). Reproduction is possible only into empty nearest-neighbor sites; hence reproduction decreases if individuals are tightly clumped.

Process (18.34) represents the migration of an individual. It amounts to the exchange of a $+$ and a 0 site that are adjacent to each other. Because individuals are uniform in their characteristics (i.e., they have no heritable differences in behavior), the exchange of individuals between two neighboring occupied sites does not change the spatial pattern of the population and therefore need not be taken into account. The maximum rate of migration for each individual is denoted by m.

The model is the same as that described in the previous section if there is no migration ($w = m/b = 0$) and if there is no social interaction ($\beta = 0$), which fits well to improved pair approximation with discounting factor ε smaller than 1 ($\varepsilon = 0.8093$). However in the limit of strong mixing ($w \rightarrow \infty$), this approximation is not valid and we must set ε equal to 1 to be consistent with mean-field approximation. To connect these two situations smoothly, Harada *et al.* (1995) introduced the following assumption:

$$\varepsilon = \frac{\alpha w + \varepsilon_0}{\alpha w + 1} , \qquad\qquad (18.36)$$

where $\varepsilon_0 = 0.8093$. Equation (18.36) agrees with the mean-field approximation in the case of an infinite migration rate $w \rightarrow \infty$, and with improved pair approximation with constant discounting in the case of $w = 0$. To capture the effects of b and β, ε_0 is determined by a fitting procedure as

$$\varepsilon_0 = 1.0 + 0.14u - 0.60u^2 + 0.24uv \quad (r^2 = 0.83) , \qquad (18.37)$$

where we set $u = d_0/b$ and $v = \beta/b$.

The appropriateness of choosing ε between 0 and 1 has been shown mathematically for two-state lattice models by Belitsky *et al.* (1997). Harada *et al.* (1995) have shown that IPAV [Equations (18.36) and (18.37)] is quite accurate in predicting the results of direct computer simulations.

18.5 Concluding Comments

In this chapter, we have introduced pair-approximation methods by applying them to several lattice models. Whereas Chapter 13 focuses on the application of two-state lattice models to ecological systems (e.g., the population dynamics of plants with both vegetative reproduction and seed production, and the gap dynamics in tropical seasonal forests), here we have concentrated on three-state lattice models.

Here, we briefly review some additional applications of pair-approximation methods. Satō and Konno (1995) extended the lattice epidemic model of Satō *et al.* (1994) by introducing an additional process that allows the reproduction of diseased individuals into empty nearest-neighbor sites. They assume that all offspring of infected mothers are infected (that is, perfect efficiency of vertical transmission) and consider a direct transition from 0 to −:

$$\ominus\text{-}⓪ \xrightarrow{\alpha m_-/z} \ominus\text{-}\ominus . \qquad (18.38)$$

This model is equivalent to that described in Section 18.3 when $\alpha = 0$, and corresponds to a model for the successional dynamics of grass, shrubs, and trees by Durrett (1995b) when $\alpha = 1$ (see also Durrett and Swindle 1991; Durrett and Schinazi 1993; and references in Satō and Konno 1995). For the model with $\alpha = 1$, Durrett (1995b)

- concluded that coexistence cannot occur in a one-dimensional linear lattice with nearest neighbors;
- suggested that coexistence can occur for some parameter regions except on a one-dimensional linear lattice with nearest neighbors;
- derived the condition of coexistence and non-coexistence for a large range of interactions predicted by mean-field approximation.

Satō and Konno (1995) demonstrated the existence of an additional range of parameter space, where equilibrium populations comprise only diseased hosts and no healthy hosts. This behavior was predicted by both mean-field approximation and pair approximation. Another three-state lattice model analyzed by pair approximation and described in detail in Chapter 13 is based on competition between two strains of bacteria. One of the strains produces a toxin (colicin) that kills its neighbors; the other is sensitive to the toxin but has the higher reproductive rate in its absence (Iwasa *et al.* 1998). The difference between lattice models and complete mixing models had already been demonstrated experimentally: the competitive outcomes between liquid culture (complete mixing) and agar culture (spatial structure) are known to be very different.

Tainaka (1994) and Satulovsky and Tomé (1994) independently analyzed the same predator–prey lattice model. Tainaka (1994) showed both that the equilibrium prey population level decreases with an increase in prey reproduction rate and that this counterintuitive result of the lattice models can be explained by pair approximation but not by mean-field approximation. The findings of Satulovsky and Tomé suggest that pair-approximations can sometimes give rise to limit cycles. These cycles, however, do not occur readily owing to the strong dependence on the initial configuration and the demographic stochasticity, which occurs even on large lattices.

Although in recent years mathematical insight into the dynamical properties of lattice models has been accumulating quickly, exact critical values for population sustenance, even for basic contact processes, are not available. Computer simulations, on the other hand, are time consuming and suffer from other limitations (dependencies on parameters are difficult to study in detail, and the noise resulting from demographic stochasticity often obscures results). We therefore believe that, with a growing appreciation of the consequences of spatial structure for understanding population dynamics and evolutionary dynamics, pair approximation provides a useful analytical method for predicting dynamical phenomena in heterogeneous biological systems.

19

Pair Approximations for Different Spatial Geometries

Minus van Baalen

19.1 Introduction

The standard assumption underlying the formulation of models for population dynamics (such as the logistic growth equation, the Lotka–Volterra predator–prey model, and the Kermack and McKendrick epidemiological equations, to name a few) is that populations spread homogeneously through space and that individuals mix rapidly. It is not a new insight that spatial structure is often an essential component of the ecological (and evolutionary) dynamics of populations, and there have been many approaches to understanding the various consequences of spatial structure. In this chapter I address one of the more recently developed techniques for modeling spatial population dynamics.

The oldest approach is to assume that populations are subdivided into different discrete subpopulations that are linked through migration (the "metapopulation" approach). This may be a reasonable assumption for certain systems (groups of parasites living in different hosts, for example), but space often has a more a continuous aspect. For example, a forest may be highly structured without having clear boundaries between subpopulations. Such situations are often modeled using a diffusion formalism, but this approach has its shortcomings as well. In particular, when one considers spatial spread of a population (or gene), individuality (discreteness) and its associated stochasticity may be important (Durrett and Levin 1994b). In a diffusion model, the rate of population growth is determined by the spread of "nano-individuals" at the wave front, whereas in reality it is often determined by the more erratic process of dispersal and subsequent successful settlement of individuals. Not only might this give quantitatively wrong estimates [e.g., the conditions under which an epidemic can arise; see Chapter 6 and Jeltsch *et al.* (1997)], it can also yield qualitatively wrong predictions (the term "atto-foxes" has been used to describe some spurious results from diffusion-based population models; see Mollison 1991).

By their very nature, diffusion models do not incorporate individuality. There exists, however, a suite of powerful mathematical techniques to deal with models that are based on stochastic, discrete events (Durrett and Levin 1994b). In some variants, individuals are represented as points and inhabit a continuous spatial domain (see Chapters 20 and 21; Pacala and Tilman 1994; Dieckmann *et al.* 1997); in others, individuals inhabit a discretized spatial domain, that is, a lattice of sites. This chapter focuses on the latter of these.

Even if localized, discrete and stochastic events make it difficult to study the exact dynamics of a system. However, the rates of change of certain average quantities (macroscopic spatial statistics) can be predicted with some accuracy. The fundamental approach is to derive the expected rate of change of an average quantity f (such as the proportion or number of sites in a particular state) by averaging all possible events over the entire lattice; that is,

$$\frac{dE(f)}{dt} = E\left(\frac{df}{dt}\right) = \sum_{\text{All sites } x} \sum_{\text{All events } e_x} r(e_x)(f_{e_x} - f) , \qquad (19.1)$$

where $r(e_x)$ gives the probability per unit time that an event e occurs at location x changing the average from f to f_{e_x}. The main technical problem that we need to address in this chapter originates with the fact that rates $r(e_x)$ usually depend on the full spatial configuration.

When f stands for a quantity such as the proportion of sites occupied by a given species, classical (nonspatial) models can be obtained. However, f can just as easily stand for a configuration statistic involving more than one site. In that case, Equation (19.1) describes the dynamics of this configuration statistic and allows us, in principle, to work out how spatial structure changes over time. The simplest spatial "configuration" to which to apply this technique is pairs of nearest neighbors. In other words, instead of having as average quantity the proportion of sites in a given state (which yields the "density" of that state), the formalism is applied to the states of pairs of neighboring sites. The density concept is thus extended to pairs of nearest neighbors (or pairs for short).

The main advantage of knowing pair densities is that they provide information about the spatial distribution of states on the lattice. I will explain how they do so in some detail in this chapter. Pair-dynamics models originated in theoretical physics and were introduced into theoretical biology by Matsuda *et al.* (1992) to analyze spatial dynamics of predator–prey systems. They subsequently have been applied to host–parasite systems (Satō *et al.* 1994; Keeling 1995), models for plant competition (Harada and Iwasa 1994), and the evolution of altruism (Matsuda *et al.* 1992; Harada *et al.* 1995; Nakamaru *et al.* 1997; van Baalen and Rand 1998).

A minor note for the connoisseur of spatial models may be appropriate here. In the simplest pair-dynamics models, events change the state of only a single site at a time. For some applications this may be sufficient, but in many ecologically interesting systems, events change two sites at a time. Maybe the most important example of such a two-site event is movement of an individual from one site to another: an occupied site is vacated as another becomes occupied. In addition, other types of events cause simultaneous changes in a pair of neighboring sites. For example, in a predator–prey model, a predation event is modeled as one in which a "prey"–"hungry predator" pair becomes a "satiated predator"–"empty site" pair (de Roos *et al.* 1991). Models that allow simultaneous changes in neighboring sites are called "artificial ecologies" by Rand *et al.* (1995). To account for processes such as movement of individuals, in this chapter I use this formalism to allow events that change *pairs* of neighboring sites, instead of just single sites. Because movement is allowed, the method can be used to describe "viscous" populations, a term introduced by Hamilton (1964) to characterize populations that do not exhibit panmixis, but do not have a sharply subdivided spatial structure either.

A major incentive for developing pair-dynamics models has been the limited usefulness of probabilistic cellular automata, an increasingly popular way of studying spatial dynamics. It is relatively straightforward to model a spatial ecological system by setting up a lattice of sites and defining a set of rules that change the state of sites depending on their state and that of their environment. The advantage of this approach is that spatial phenomena are explicitly included. A disadvantage, however, is that simulating these models is rather time consuming. More seriously, the results are sometimes difficult to interpret and are not easily generalized. One is effectively limited to observing what happens in the simulation. Often, for example, it is difficult to explain *why* some species persist in spatial simulations and others do not. Pair-dynamics models provide analytical insight into this question. Used this way, pair-dynamics models are tools for capturing the essence of more-detailed, explicitly spatial models.

(a)   (b)

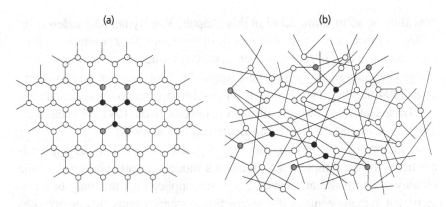

Figure 19.1 Examples of (a) a regular two-dimensional lattice and (b) a random network, both with a neighborhood size of three. In both graphs, a focal site (black) and its neighbors up to two links away (dark and light gray) are indicated.

Pair-dynamics models sometimes cannot account for all features of the full spatial dynamics, particularly when the dynamics give rise to large-scale spatial structures (these cannot be accurately described on the basis of nearest-neighbor correlations alone). Nonetheless, the analytical insight pair-dynamics models provide is valuable. In this chapter I demonstrate that, even if pair-dynamics models do not describe the exact dynamics of a full system, they can provide good approximations for invasion and equilibrium conditions. Moreover, pair-dynamics models provide analytical insight into the relationship between lattice structure and population dynamics.

Computer simulations are usually defined on square grids where every site is connected to either four or eight neighbors. This is of practical convenience, as such lattices can easily be implemented in a program and displayed on a screen. However, it should be realized that such square lattices are quite special; even with a fixed number of connections per site, sites can be arranged in many ways. For example, with four connections per site, the sites can be laid out geometrically in a flat lattice (the usual case), but also in a lattice (based on tetrahedrons) that fills three-dimensional space, and of course in many other, less regular lattices.

That the structure of a lattice has consequences for population dynamics becomes apparent once one realizes that growing clusters of individuals have different overall shapes on different lattices: expanding foci are roughly circular on two-dimensional lattices such as the lattice depicted in Figure 19.1a, roughly spherical on three-dimensional lattices; and more tree-like in random lattices (Figure 19.1b). Obviously, for a model of plant population dynamics the choice of a two-dimensional lattice can be justified. However, a case can be made for more random lattices to describe other systems, such as parasites that transmit themselves across the social network of their host. Such a network does not necessarily correspond to the host individuals' spatial arrangement and may resemble a random network such as shown in Figure 19.1b (see also Watts and Strogatz 1998).

The aim of this chapter is twofold. First, I outline how to derive and use pair-dynamics equations, that is, differential equations that describe the rate of change of the density of pairs of neighbors (instead of only the densities of singlets, as is usually done). I then indicate how pair dynamics depend on the geometric structure of the lattice.

Working out the derivation in some detail is useful because it makes explicit the types of assumptions that must be made to "close" the set of equations. This term refers to the fact that the differential equations depend on quantities outside their scope (the densities of "triplets" and more complicated configurations). "Closure" means adopting an assumption that allows the differential equations to be completely expressed in terms of the quantities whose dynamics are described by the differential equations (in our case, pair densities). This is an important step, because pair dynamics turn out to depend on the densities of triplets and even larger configurations. The standard approach is to approximate triplets using pairs. However, as I show, knowledge of the geometrical structure of the lattice can be used to provide better estimates. A limitation of the approach adopted here is that it applies only to lattices with a fixed number of neighbors. Morris (1997) and Rand (1999) provide discussions of possible approaches for lattices that have variable numbers of connections per site.

Because the first part of this chapter is quite technical, it is accompanied by a parallel series of boxes in which a pair-dynamics model is derived and analyzed for a simple example. The example is the spatial equivalent of the well-known logistic growth model, where in addition to the birth and death of individuals, movement is also included. I compare the pair-dynamics model with explicit simulations and outline how invasion conditions can be derived. (Such invasion conditions give valuable insight into the conditions for persistence of a given population, because persistence requires that a population must bounce back when it is brought to low densities.) In this analysis, particular attention again is paid to how the spatial dynamics depend on the geometrical structure of the lattice.

19.2   The Dynamics of Pair Events

The set of sites that belong to the lattice (which is assumed to be finite but large) is denoted by S. Every site is assumed to have n neighbors. Let $L \subset S \times S$ represent the set of connections (or *pairs* of sites) on the lattice. That is, two sites x and y form a pair if $xy \in L$.

Every site $x \in S$ is in a state σ_x, where Ω is the set of all possible states. The state of the entire lattice is denoted by σ; the state of a pair xy is denoted by $\sigma_x \sigma_y$.

In an artificial ecology, the state of the lattice σ changes over time because events change the state of sites or pairs of sites. Here, all events are defined in terms of pairs:

$$\sigma_x \sigma_y \rightarrow \sigma'_x \sigma'_y , \tag{19.2}$$

where the state of the lattice changes from σ_x to σ'_x at site x and from σ_y to σ'_y at site y simultaneously. (Notice that this formalism also includes all "single-site" events. Single-site events can be analyzed separately, but for the moment we consider a single-site event a special pair event.)

Any event $\sigma_x \sigma_y \rightarrow \sigma'_x \sigma'_y$ has an associated rate (probability per unit time)

$$r_\sigma(\sigma_x \sigma_y \rightarrow \sigma'_x \sigma'_y) . \tag{19.3}$$

Usually, it is assumed that all interactions are local: therefore, rates affecting a pair xy depend only on the state of the pair's immediate environment E_{xy}:

$$r_\sigma(\sigma_x \sigma_y \rightarrow \sigma'_x \sigma'_y) = r(\sigma_x \sigma_y \rightarrow \sigma'_x \sigma'_y | \sigma_{E_{xy}}) , \tag{19.4}$$

where E_{xy} is a list of all pairs in the local environment of the pair at xy (see Figure 19.2).

A specific example of a simple artificial ecology is given in Box 19.1.

Pair densities

The initial state of the lattice, $\sigma(0)$, together with the event rates specify a stochastic dynamical system: the state of the lattice follows a stochastic trajectory $\sigma(t)$. However, if the lattice is large enough, some quantities (such as the average number of pairs in a particular state) change almost deterministically. Here, I explain how differential equations can be found that (approximately) describe the expected dynamics of these quantities.

One can derive the expected rate of change in p_a, the probability that a given site is in state a. This is basically the classical "density" concept, and indeed the differential equations that result are of the type commonly used in population biology. The disadvantage is that all information with

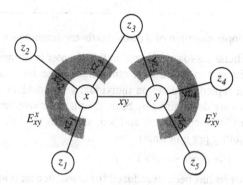

Figure 19.2 An example of a pair at xy and its "left" ($xz_i \in E^x_{xy}$) and "right" ($yz_i \in E^y_{xy}$) environments. The combination of the two specifies the entire environment E_{xy} of the pair.

regard to spatial structure is lost: one is forced to assume that populations are "well mixed."

Some spatial information is retained if the states of *pairs* of nearest neighbors are traced instead of the states of single sites. One can work out the expected rate of change in the "pair densities" p_{ij}, where p_{ij} gives the probability of finding a given pair in state ij. Knowing the pair densities permits computation of the conditional probabilities

$$q_{j|i} = p_{ij}/p_i . \qquad (19.5)$$

These conditional probabilities give the density of j sites as "experienced" by sites in state i: effectively, $q_{j|i}$ is the *local* density of j in the environment of i [Matsuda *et al.* (1992), use the term "environs densities" for these quantities].

Pair dynamics of simple birth–death–movement process

Consider a particular pair combination ab. Obviously, its density p_{ab} will change when pair events directly create or destroy both partners of ab pairs. In other words, p_{ab} increases with all $ij \to ab$ events (see Figure 19.3a) and p_{ab} decreases with all $ab \to kl$ events. In addition, the density of ab pairs will be affected by events that occur in ab's neighborhood; for example, ab pairs will be created if $ij \to bl$ events occur next to an ai pair (such an event indirectly changes the ai into an ab pair; see Figure 19.3b). Throughout this chapter, I use the symbols a and b to refer to the pair combination of states whose dynamics are in focus. The indices i, j, k, and l are used to sum over other states.

The contribution of direct events to the rate of change of p_{ab} is straightforward. For example, the creation of ab pairs from ij pairs occurs at rate $p_{ij}\bar{r}_\sigma(ij \to ab)$, which is just the density of ij pairs multiplied by the

Box 19.1 A simple example of a birth–death–movement process

The simple artificial ecology analyzed in this chapter is defined on a lattice where every site is connected to n other sites. The sites may be either empty (o) or occupied by a single individual (×), so $\Omega = \{o, \times\}$. The spatial dynamics are determined by birth, death, and movement.

"Birth events" that change o× and ×o pairs into ×× pairs occur with rate (i.e., probability per unit time)

$$r(\times o \rightarrow \times\times) = r(o\times \rightarrow \times\times) = \phi b .$$

The factor $\phi = 1/n$ has been introduced for convenience: it allows b to be interpreted as a *per capita* rate instead of a *per neighbor-pair* rate. Note that "mirror image" events (like the two above) always have the same rates.

"Death events" change × sites into o sites, or in terms of pairs, $\times j$ pairs into oj pairs (where $j \in \Omega$). They occur with rate

$$r(\times j \rightarrow oj) = r(j\times \rightarrow jo) = \phi d .$$

"Movement events" swap × and o sites in ×o and o× pairs (the individual moves to the empty neighbor site). They occur with rate

$$r(\times o \rightarrow o\times) = r(o\times \rightarrow \times o) = \phi m .$$

Notice that in this model the rates do not depend on the environment of the pairs. Such a dependency would arise if, for example, an individual's rate of reproduction were dependent on how many occupied neighbors it has. In that case, b would not be constant but would be given by a function $b(n_{\times\times})$, where $n_{\times\times}$ is the number of × neighbors of the × in the ×o pair. Such environment-dependent rates result, for example, from altruistic behavior, where altruistic individuals help their neighbors at their own cost (Matsuda *et al.* 1992; van Baalen and Rand 1998). Intra- and interspecific competition also lead to environment-dependent rates.

average event rate. We must sum over all possible source pairs ij to compute the total rate of direct ab creation. [Calculating average rates such as $\bar{r}_\sigma(ij \rightarrow ab)$ is discussed below.]

The contribution of indirect events is a bit more complicated. First, the contribution of, say, $ij \rightarrow bl$ events depends on how many j neighbors an ai pair has on average. This average is given by $(n - 1)q_{j|ia}$: the i in an ai pair has $n - 1$ neighbors, and the likelihood of finding any one of them in state j is given by the conditional probability $q_{j|ia}$. Note that this conditional probability depends on *triplet* densities: $q_{j|ia} = p_{aij}/p_{ia}$. Thus the set of differential equations for pairs depends on quantities outside its scope. (I return to this problem in Section 19.4.) The average rate of neighborhood events is given by $\bar{r}_\sigma(ij \rightarrow bl|aij)$; note that this is not the

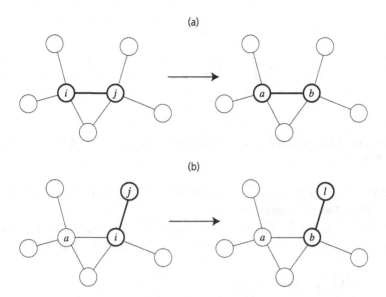

Figure 19.3 The bookkeeping of pairs can depend both directly and indirectly on pair events. For example, (a) ab pairs are created directly from $ij \to ab$ events, whereas (b) they are created indirectly when an $ij \to bl$ event changes the i of an ai pair into a b.

average rate of these events over the entire lattice, but the average rate of these events in ij pairs in the neighborhood of ai pairs. If the event rates are constant, conditional and unconditional averages are the same; if they are dependent on the environment, these averages are different.

Putting together all direct and indirect events that create as well as destroy ab pairs yields the following "master equation":

$$\frac{dp_{ab}}{dt} = -p_{ab}\sum_{kl}\left[\bar{r}_\sigma(ab \to kl) + \sum_i (n-1)q_{i|ab}\bar{r}_\sigma(ia \to kl|iab)\right.$$

$$+ \sum_j (n-1)q_{j|ba}\bar{r}_\sigma(bj \to kl|abj)\bigg]$$

$$+ \sum_{ij}\left[p_{ij}\bar{r}_\sigma(ij \to ab) + p_{jb}\sum_k (n-1)q_{i|jb}\bar{r}_\sigma(ij \to ka|ijb)\right.$$

$$\left. + p_{ai}\sum_l (n-1)q_{j|ia}\bar{r}_\sigma(ij \to bl|aij)\right].$$

$$\tag{19.6}$$

Note that this equation itself does not contain any specific references to the actual locations of ab pairs on this lattice (there is no summation over all pairs xy in the lattice). In other words, Equation (19.6) does not explicitly depend on space. It depends, however, on average event rates, and these

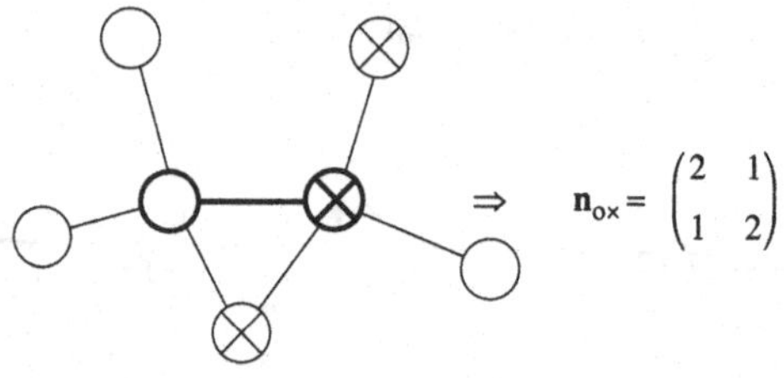

Figure 19.4 Example of how to characterize the state $n_{o\times}$ of the local environment of a pair in state $o\times$.

average rates depend on the state of the lattice. For this reason I have given the average rates a subscript σ.

A specific example of how to apply Equation (19.6) to the simple birth–death–movement process introduced in Box 19.1 is given in Box 19.2.

19.3 Average Event Rates

In the simplest models, pair event rates are constants. Often, however, such rates depend on the local environment of a pair. For example, if a is an altruistic individual and b is an empty site, the probability of an $ab \rightarrow aa$ (birth) event may depend on the number of other as in a's neighborhood. The mean rate of $ab \rightarrow aa$ events therefore depends on the mean number of as next to ab pairs. Of course, more complicated schemes are possible, such as in hypercycle dynamics where a helps the reproduction of b, b helps the reproduction of c, and so on, until eventually there is a type that helps a (Boerlijst *et al.* 1993; see also Chapters 9 and 10). The approach I present here allows all such schemes; it even allows for nonlinear effects (when the effect of two is is not equal to twice the effect of a single i).

Consider a pair event $ab \rightarrow kl$. When the event rate depends on the environment of the pair (say it increases with every neighbor in state i in a's neighborhood), the rates must be averaged over all possible configurations (i.e., over all configurations with zero is, with one i, with two is, etc.).

To be as general as possible, I represent the entire neighborhood of a given pair ab at xy using two vectors,

$$n_{ab}^a = \begin{pmatrix} \vdots \\ n_{ia} \\ \vdots \end{pmatrix} \quad \text{and} \quad n_{ab}^b = \begin{pmatrix} \vdots \\ n_{jb} \\ \vdots \end{pmatrix}, \tag{19.7}$$

with $i, j \in \Omega$. The vector n_{ab}^a simply counts the different types of neighbors around a in the ab pair, while n_{ab}^b does the same around b. These two vectors together give a two-column matrix,

$$n_{ab} = \begin{pmatrix} n_{ab}^a & n_{ab}^b \end{pmatrix} ,$$ (19.8)

which therefore fully characterizes the state of the environment n_{ab} of the ab pair (see Figure 19.4). While the environment E_{xy} itself simply specifies the geometrical *connections* around two sites x and y, the neighborhood n_{ab} specifies the *state* of this environment around a pair of sites in states a and b. The assumption that all interactions occur only among neighbors and are not dependent on the actual configuration (i.e., whether a neighbor is located to the north or to the south, or to any other direction, is irrelevant) implies that this configuration completely determines the rate of events $ab \rightarrow kl$.

In principle, we could calculate the average event rate $\bar{r}_\sigma(ij \rightarrow kl)$ by averaging over the environments of all ij pairs on the lattice. An equivalent and conceptually advantageous alternative, however, is to work out the frequency distribution of all possible neighborhoods n_{ij} on the lattice. If the proportion of a given neighborhood n_{ij} (relative to all possible configurations) is denoted by $F_\sigma(n_{ij})$, the average rate is given by

$$\bar{r}_\sigma(ij \rightarrow kl) = \sum_{n_{ij}} F_\sigma(n_{ij})\, r(ij \rightarrow kl|n_{ij}) .$$ (19.9)

The only component on the right-hand side that still depends on the state of the lattice is $F_\sigma(n_{ij})$. This quantity can be interpreted as a (local) density, this time not of a simple pair but of a larger spatial configuration. Thus, the pair equations may depend not only on the densities of triplets (which are needed to calculate local densities $q_{i|jk}$), but also on the densities of more complicated local configurations.

Conditional average rates are calculated in a similar fashion. The complicating factor here is that they depend on one of the pair's neighbors (they are the mean rate over triplets, not pairs). Writing down the expression is straightforward:

$$\bar{r}_\sigma(bj \rightarrow kl|abj) = \sum_{n_{abj}} F_\sigma(n_{abj})\, r(bj \rightarrow kl|n_{abj}) ,$$ (19.10)

where $F_\sigma(n_{abj})$ is the frequency distribution of the neighborhoods n_{abj} of abj triplets (n_{abj} are three-column matrices that count the number of neighbors of a, b, and j similar to the way n_{ab} describes the neighborhood of an ab pair). Note that in the case of environment-independent pair event rates, calculating the densities of the larger local configurations is not necessary.

Box 19.2 Pair dynamics of the simple birth–death–movement process

In the artificial ecology defined in Box 19.1, there are four different pair combinations: oo, ×o, o×, and ××. As an example of the dynamics of their densities, the differential equation for $dp_{\times o}/dt$ is worked out in detail. Setting up the bookkeeping, using all direct and indirect (neighbor) pair events that affect the density of ×o pairs, yields the transitions schematically shown below.

Contributions of direct and indirect events affecting the density of ×o pairs. To calculate net transition rates, the rates indicated in the diagram have to be multiplied by the density of the source pairs.

Summing all terms in this figure and simplifying the ensuing expression using $q_{o|ij} + q_{\times|ij} = 1$, $(n-1)\phi = (n-1)/n = 1-\phi$, and $p_{o\times} = p_{\times o}$ yields

$$
\begin{aligned}
\frac{dp_{\times o}}{dt} = & - p_{\times o} \left[\phi b + d + (1-\phi)q_{o|\times o}m + (1-\phi)q_{\times|o\times}(b+m) \right] \\
& + p_{oo}\,(1-\phi)q_{\times|oo}(b+m) \\
& + p_{\times\times}\left[d + (1-\phi)q_{o|\times\times}m \right].
\end{aligned}
$$

The first term, incorporating all events that destroy ×o pairs, has four components. The first component represents the × individual giving birth into the o site, the second represents the death of the × individual, the third represents the departure of the × individual through movement (note that this depends on the proportion of empty sites surrounding it), and the fourth represents the arrival of another × individual at the empty site (which may happen if a neighboring × individual reproduces or moves).

Similar considerations for dp_{oo}/dt and $dp_{\times\times}/dt$ yield the other two differential equations that describe the pair dynamics of this system:

$$
\begin{aligned}
\frac{dp_{oo}}{dt} = & - p_{oo}\,2(1-\phi)q_{\times|oo}(b+m) \\
& + p_{\times o}\,2\left[d + (1-\phi)q_{o|\times o}m \right],
\end{aligned}
$$

continued

Box 19.2 *continued*

$$\frac{dp_{\times\times}}{dt} = + p_{\times 0}\, 2\big[\phi b + (1 - \phi)q_{\times|0\times}(b + m)\big]$$
$$- p_{\times\times}\, 2\big[d + (1 - \phi)q_{0|\times\times}m\big]\,.$$

The factor 2 arises in these expressions because oo and $\times\times$ are symmetric, so that all events can happen "on both sides."

Thus we have obtained a set of three differential equations that describe the pair dynamics of the artificial ecology defined in Box 19.1. Note that because the sum of the p_{ij} equals 1, one of the differential equations is actually redundant: we could do away with, for example, the differential equation for p_{00} and substitute $p_{00} = 1 - 2p_{\times 0} - p_{\times\times}$ in the remaining two equations for $p_{\times 0}$ and $p_{\times\times}$.

Because we must analyze the dynamics without keeping track of the entire lattice, the densities of triplets (and in the case of density-dependent rates, the densities of the larger local configurations) have to be estimated from the distribution of pairs. This estimation "closes" the system of differential equations because they are now entirely defined in terms of pairs. How to close the system is dealt with in Section 19.4.

Here we summarize our understanding developed so far: given

- a lattice L,
- a set of states Ω, and
- a list of possible events $ij \to kl$ and their rates $r(ij \to kl|n_{ij})$,

differential equations can be constructed for the expected rates of change in the frequency p_{ab} of all pair combinations ab, as a function of all

- pair frequencies p_{ab},
- conditional probabilities $q_{i|ab}$, and
- frequencies $F_\sigma(n_{ab})$ and $F_\sigma(n_{iab})$ of configurations surrounding pairs and triplets, respectively.

Of these, only the pair frequencies are known, because the differential equations keep track of the numbers of pair combinations. The conditional probabilities $q_{k|ij}$ and the frequencies of larger configurations either have to be calculated from the explicit state of the lattice (which would require explicit simulations) or have to be worked out by other means.

In principle, differential equations can be derived for the dynamics of triplets (and of more complex configurations) analogously to the derivation of the differential equations for pairs. However, there are various reasons for not pursuing this avenue. First, the bookkeeping is much more complicated, as all transitions from one configuration to another must be incorporated. Second, the dynamics of configurations will depend on "configurations of configurations," which means that the problem has only been carried to the next level. We therefore estimate the frequencies of more complex configurations from the simpler ones (pairs); this is the essence of a solution to what is often called the "moment-closure" problem.

19.4 Pair Approximations for Special Geometries

The pair equations that we have derived so far are exact (on infinitely large lattices); no simplifying assumptions have been made. The problem is that they depend on the density of configurations that are outside their scope. To avoid a cascade of dependency on ever more complex configurations, the system of differential equations has to be "closed." That is, if the aim is to describe the dynamics of pairs, everything has to be expressed in terms of configurations no more complex than pairs. This implies that the frequencies of all configurations larger than pairs have to be approximated.

Consider the conditional probability $q_{i|ab}$. This gives the probability that a site next to the a of an ab pair is in state i. The most straightforward approximation for the conditional probability $q_{i|ab}$ is based on the simple heuristics that the more distant site of the pair (in this case b) might not influence the probability of finding an i next to the a; that is,

$$q_{i|ab} \approx q_{i|a} \tag{19.11}$$

(Matsuda *et al.* 1992; see also Chapters 13 and 18). The error this assumption introduces may be considerable. For example, if b is rare globally, $q_{b|a}$ is likely to be small for $a \neq b$. However, if b is clustered on the lattice, $q_{b|ab}$ may be much larger, because the ab pair is likely to be picked from within such a cluster and then more bs are likely to be nearby.

Any approximation introduces errors and information is inevitably lost. In this section I discuss how knowledge of the geometrical structure of the lattice can be used to derive improved "closure" assumptions (i.e., expressions for $q_{i|ab}$ in terms of pair densities). The basic method is outlined first for "random" lattices – lattices in which every site is connected to n other sites, but with no overall spatial structure, such as depicted in Figure 19.1b. I then discuss how to correct for lattices that are more regular and that do have an overall spatial structure – such as the flat lattice in Figure 19.1a.

Random lattices

On a random lattice like that depicted in Figure 19.1b, sites are randomly connected to n other sites. Consequently, if the number of sites is large, the probability that the members of a pair have neighbors in common is negligible. Thus, for the probability that a given triplet is in state iab one can write

$$p_{iab} = p_i p_a p_b C_{ia} C_{ab} T_{iab} , \qquad (19.12)$$

where p_i denotes the probability of finding a site in state i,

$$p_i = \sum_j p_{ij} , \qquad (19.13)$$

C_{ij} denotes the *pair correlation* between i and j sites,

$$C_{ij} = \frac{p_{ij}}{p_i p_j} , \qquad (19.14)$$

and T_{ijk} denotes the *triple correlation* of ijk chains, which is basically the error in the estimate based only on pairs. Notice that there is no correlation factor C_{ib}; the only way b can "influence" the probability distribution of a's other neighbors is through the triple correlation.

The values of triple correlations T_{iab} are determined by the full spatial dynamics of the system under consideration. Unless they are estimated from full stochastic simulations, however, their values are unknown. Therefore, to arrive at a closed set of differential expressions for pairs, assumptions have to be made with respect to the triple correlations. The simplest approach is to assume that they are constant. That is, we substitute an estimate τ_{ijk} for every triple correlation T_{ijk}. In fact, the standard pair approximation follows from the assumption that all $T_{iab} = 1$. One then obtains

$$q_{i|ab} = \frac{p_{iab}}{p_{ab}} = \frac{p_i p_a p_b C_{ia} C_{ab}}{p_a p_b C_{ab}} = p_i C_{ia} = q_{i|a} . \qquad (19.15)$$

Thus for chain-like triplets the simplest estimate for $q_{i|ab}$ is indeed simply $q_{i|a}$; the fact that the a has a b neighbor becomes irrelevant.

Some authors (Harada *et al.* 1995; Keeling 1995) have analyzed improved pair approximations that are based on the assumption that these triple correlations have values not equal to 1, particularly for *bab*-type triplets (see Chapter 18). The reason for doing so is best understood by considering a biological example. Let b stand for a site that is occupied by a member of a rare population and let a stand for an empty site. What is the likelihood that in an ab pair the a becomes occupied and becomes a b? There are two ways for this to happen: the b in the pair reproduces or another b neighboring a reproduces. The probability of the latter happening is proportional to $q_{b|ab}$. However, under the classical pair approximation, this would be approximated by $q_{b|a}$, which is very small when b is rare. Analysis of simulations have shown that if b is rare and tends to form clusters, *within* such clusters $q_{b|a}$ is not small at all. By setting τ_{bab} to a value larger than 1, we assume that, if an a has one b neighbor, it is likely to have more. In other words, increasing the triple correlations increases the degree of crowding on the lattice, which may have various consequences.

It should be noted that the estimates τ_{iab} for the triple correlations are not independent, as they have to satisfy the consistency condition

$$\sum_i q_{i|ab} = \sum_i q_{i|a}\tau_{iab} = 1 . \tag{19.16}$$

If all τ_{iab} are equal to 1, this condition is satisfied. But if a value not equal to 1 is chosen for one τ_{iab}, the others have to be corrected such that Equation (19.16) holds for all $q_{i|a}$.

Triangular lattices

Most cellular automata assume square lattices, but the way to improve on the classical pair approximation is most easily understood by first considering a triangular lattice – that is, a lattice in which every site has six neighbors arranged in a hexagon.

On such a lattice, a triplet can be in one of two different configurations, "chain-like" (or *open*) or "triangular" (or *closed*). In fact, there is a 2/5 chance that a randomly picked triplet is in a closed configuration and a 3/5 chance that it is open (see Figure 19.5). We can take this information into account when calculating conditional probabilities.

For open triplets, denoted by $\angle iab$, we can still write

$$p_{\angle iab} = p_i p_a p_j C_{ia} C_{ab} T_{\angle iab} , \tag{19.17}$$

but for closed triplets we must take into account an extra correlation factor C_{bi}:

$$p_{\triangle iab} = p_i p_a p_j C_{ia} C_{ab} C_{bi} T_{\triangle iab} . \tag{19.18}$$

Then, if θ denotes the probability of finding the triplet in closed form ($\theta = 2/5$), one obtains

$$q_{i|ab} = q_{i|a}\left((1 - \theta)T_{\angle iab} + \theta C_{ib} T_{\triangle iab}\right) . \tag{19.19}$$

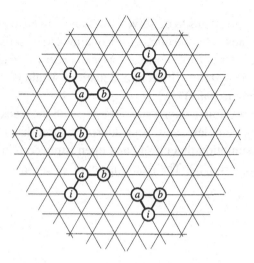

Figure 19.5 The five different configurations for an *iab* triplet on a triangular lattice.

On a triangular lattice there are *two* kinds of triple correlation, $T_{\angle iab}$ for open triangles and $T_{\triangle iab}$ for closed triangles. There is no *a priori* reason these should be the same. Let $\tau_{\angle iab}$ and $\tau_{\triangle iab}$ be our estimates for the triple correlations $T_{\angle iab}$ and $T_{\triangle iab}$. The consistency condition then becomes

$$\sum_i q_{i|ab} = \sum_i q_{i|a}\left((1-\theta)\tau_{\angle iab} + \theta C_{ib}\tau_{\triangle iab}\right) = 1 \ . \tag{19.20}$$

With $\tau_{\angle iab} = \tau_{\triangle iab} = 1$, this holds (for all $q_{i|a}$ and $\theta > 0$) *only* if all $C_{ib} = 1$, in which case the pairs are entirely uncorrelated. Therefore, we cannot simply set $\tau_{\angle iab} = \tau_{\triangle iab} = 1$. The simplest assumption that satisfies condition (19.20) is that all $\tau_{\angle iab} = 1$ and all $\tau_{\triangle iab} = 1/C_{ib}$. Note, however, that this implies that we have "uncorrected" the correlation for closed triplets (because we divide by the correlation among the far ends of the triplet). In fact, we have recovered the classical pair approximation, because now the proportion of triangles θ disappears from our estimate.

To arrive at more sophisticated approximations, one must take into account a number of facts. The first is that open triple correlations should be symmetric (i.e., $\tau_{\angle iab} = \tau_{\angle bai}$); the second is that all closed triple correlations should also be rotationally symmetric (i.e., $\tau_{\triangle iab} = \tau_{\triangle abi} = \tau_{\triangle bia} = \tau_{\triangle bai} = \tau_{\triangle aib} = \tau_{\triangle iba}$). Thus our estimates for $\tau_{\triangle iab}$ have to satisfy the set of Equations (19.20) under these symmetry conditions: the same triple correlation estimate $\tau_{\triangle iab}$ may occur in up to three different equations (i.e., in those for $\sum_i q_{i|ab}$, $\sum_a q_{a|bi}$, and $\sum_b q_{b|ai}$). It is always possible to find solutions, that is, approximations for the τs. Actually, there are many more triple correlation estimates to be solved for than there are equations: if N is the number of states in Ω, the number of equations to be satisfied is of order N^2 while the number of triple correlations is of order N^3. (Incidentally, this also shows that many more differential equations would be needed to track triplets in addition to pairs).

I have found no simple expressions for $\tau_{\triangle iab}$ on the basis of the scheme $\tau_{\angle iab} = 1$. The simplest consistent approximation that I have found is based on the assumption that closed and open triple correlations are equal (i.e., $\tau_{\triangle iab} = \tau_{\angle iab} = \tau_{iab}$). The resulting set of equations

$$\sum_i q_{i|a}\big((1 - \theta) + \theta C_{ib}\big)\tau_{iab} = 1 \tag{19.21}$$

is satisfied by

$$\tau_{iab} = \begin{cases} 1 & \text{if } i \neq b \\ \frac{1}{q_{b|a}}\big[1 - \sum_{i \neq b} q_{i|a}\big((1 - \theta) + \theta C_{ib}\big)\big] & \text{if } i = b \end{cases}, \tag{19.22}$$

which leads to the simple estimate

$$q_{i|ab} \approx \begin{cases} q_{i|a}\big((1 - \theta) + \theta C_{ib}\big) & \text{if } i \neq b \\ 1 - \sum_{j \neq b} q_{j|ab} & \text{if } i = b \end{cases}. \tag{19.23}$$

In a later section, this approximation is compared with explicit simulations of the artificial ecology described in Box 19.1. Keeling *et al.* (1997a) and Rand (1999) also provide examples of pair-dynamics models that are based on this approach. (Notice, however, that these authors did not apply the correction terms τ_{iab} to their estimates, which may have caused a small error in their results.)

Square lattices

Configurations other than triangles can be taken into account in a similar manner. Take, for example, an iab triplet on a square lattice ($n = 4$). There is no *direct* connection between the b and the i, but there is an indirect one through an intermediate site. In other words, the iab triplet may form part of an $iabj$ square quadruplet. Given the square configuration, the probability of finding it in state $iabj$ is

$$p_{\square iabj} = p_i p_a p_b p_j\, C_{ia} C_{ab} C_{bj} C_{ij}\, T_{\square iabj} . \tag{19.24}$$

One can arrive at an estimate for $q_{i|ab}$ in a fashion similar to that for triangular lattices. Averaging over all j configurations and using the fact that on a square lattice 1/3 of the triplets are straight and 2/3 form part of a square, one arrives at the estimate

$$q_{i|ab} \approx q_{i|a}\left(\tfrac{1}{3}\tau_{\angle iab} + \tfrac{2}{3}\sum_j q_{j|b}C_{ij}\tau_{\square iabj}\right), \tag{19.25}$$

where the estimates $\tau_{\angle iab}$ and $\tau_{\square iabj}$ again have to be chosen such that the conditional probabilities sum to 1.

In this formulation, the b from the example above can be thought of as modifying the probability distribution of the intermediate site, which in turn modifies the probability distribution of a's other neighbor. This way, the influence of b "percolates" through the intermediate site and so modifies $q_{i|a}$.

However, the influence of b may percolate through many more routes. This suggests that we need not confine ourselves to triangles or squares. Actually, every closed chain that goes through iab will contribute to the conditional probability $q_{i|ab}$. It should therefore be possible to find increasingly sophisticated estimates for $q_{i|ab}$ by including closed loops (and other closed structures) of greater length while still using only pairs as building blocks.

Notice, however, that this approach still assumes that higher-order correlations ($T_{\square ijkl}$ etc.) are all either fixed or expressed in terms of pair frequencies. There is no *a priori* reason these correlations should not have dynamics of their own, and it may well be that for an adequate description of configurations, higher-order correlations *must* be included. Then either specific assumptions can be made as to the size of some of these correlations, or the analysis should be extended to include the dynamics of more complex configurations. As an example of the first approach, one can set $T_{\angle ijk}$ to a value not equal to 1, and work out the consequences. This is basically the underlying strategy of the improved pair approximations (see Chapter 18; Satō *et al.* 1994; Harada *et al.* 1995; Keeling 1995). If this scheme does not work, the higher-order correlations have to be derived from the dynamics of the system. Thus differential equations for configurations more complex than pairs have to be derived and analyzed; not surprisingly, this is a difficult undertaking (see Morris 1997 for an example).

Toward higher-order approximations

At this point we have expressed the conditional probabilities $q_{i|ab}$ in terms of pair densities. If the model that is studied has constant pair event rates, the differential equations are now fully closed. However, if event rates are density dependent, another approximation step is necessary.

As we have seen, with density-dependent event rates, the differential equation for ab depends on averages over neighborhoods surrounding the ab pairs (n_{ab} and n_{iab}). In a spatially explicit simulation, these frequency distributions [$F_\sigma(n_{ab})$ and $F_\sigma(n_{iab})$] can be measured. Because we want to avoid simulations of the full lattice and develop a model purely in terms of pairs, we have to approximate these frequency distributions in terms of pairs. Thus, to calculate the average event rates, estimates for the likelihood of larger configurations [i.e., $\Pr(n_{ab})$ for $F_\sigma(n_{ab})$ and $\Pr(n_{abj})$ for $F_\sigma(n_{iab})$] must be formulated.

The average rate $\bar{r}_\sigma(ab \to kl)$ depends on the frequency distribution of the configurations $\Pr(n_{ab})$. This can be approximated in much the same way we approximate triplet frequencies, using pair correlations as building blocks. The main difficulty is that many more correlations have to be taken into account.

As a start, consider a random lattice, so that members of a pair are unlikely to have common neighbors. The assumption then is that the neighbors of a and b are independently distributed; that is,

$$\Pr(n_{ab}) = \Pr(n_{ab}^a)\Pr(n_{ab}^b) . \tag{19.26}$$

Recall that a neighborhood n_{ab} consists of two vectors giving the number of ia and bj pairs. Having no neighbors in common means that the pairs in the left and right environments do not share sites and can be treated independently. A simple assumption then is that probabilities follow a multinomial distribution

$$\Pr(n_{ab}^a) = (n-1)! \prod_i \frac{(q_{i|a})^{n_{ia}}}{n_{ia}!} , \tag{19.27}$$

where $\sum_{ia} n_{ia} = n - 1$ (there are $n - 1$ neighbors to be distributed, because one of the neighbors, namely, the other member of the pair, is already given). If a and b share neighbors, or if there are chains of connections between the neighbors of each, this method introduces an error, because a's neighbors and b's neighbors can no longer be considered independent.

For example, if a pair forms part of only one triangle, both sites have $n - 2$ "independent" neighbors and one shared neighbor. The probability distribution of the independent neighbors can still be represented by a multinomial, except that the n_{ia} (or n_{bj}) should sum to $n - 2$. If it is assumed that open triplets are uncorrelated, the probability distribution of the shared neighbor can be estimated by

$$\Pr(\text{common neighbor of type } i|ab) = q_{i|a}C_{ib}\tau_{\triangle iab} , \tag{19.28}$$

as we have seen in the section on estimating the frequencies of closed triplets. Although it would involve some cumbersome combinatorics, in principle both probability distributions can be calculated to find the probability distribution of the entire configuration around the pair.

However, all these calculations become simpler if it is assumed that event rates are linear functions of their environment, that is, if the effect of one neighbor is independent of that of the other neighbors. In this case, the rate averaged over the configurations equals the rate for the mean configuration. That is,

$$
\bar{r}_\sigma (ab \rightarrow kl) = \sum_{n_{ab}} F_\sigma(n_{ab}) \, r(ab \rightarrow kl | n_{ab})
$$

$$
= r(ab \rightarrow kl, \overline{n_{ab}}) \,, \tag{19.29}
$$

where

$$
\overline{n_{ab}} = \sum_{n_{ab}} F_\sigma(n_{ab}) n_{ab} \,. \tag{19.30}
$$

I do not attempt to further discuss the complexities that arise in trying to incorporate density-dependent event rates in pair-dynamics models. In particular, overlap among the neighbors of a pair introduces all sorts of extra correlations to be taken into account. Most likely, the only feasible approach is to assume multinomial distributions [Equation (19.27)] and accept the fact that some error will remain if the system under study is defined on a regular lattice.

19.5 Pair Approximations versus Explicit Simulations

We have seen in the previous section that it is possible to improve on classical pair approximation by introducing new correlations to incorporate knowledge of the geometrical structure of the lattice. The resulting expressions are cumbersome, however, and substituting them into the differential equations leads to complicated sets of expressions that are still rather difficult to analyze.

More important, at this stage it is not known to what extent these refinements actually improve the accuracy of the differential equations for pairs. When is classical pair approximation sufficient? When do we need to incorporate extra correlations? Is it really worth the trouble? In other words, some estimate of the errors associated with the various approximations is needed. Error analysis can be used to assess the accuracy of the approximation (Morris 1997), but as this approach is worthy of a chapter in itself, in this section the accuracy of pair-dynamics models is assessed merely by comparing their results with explicit spatial simulations.

Box 19.3 Singlet dynamics and the mean-field equation

To calculate local densities $q_{j|i}$, the singlet densities should be known. These can be derived from the differential equations for the dynamics of pairs. Because it is assumed that all sites have the same number of neighbors, we have $p_i = \sum_j p_{ij}$ (which is a standard relationship from probability theory). Thus, from the pair equations derived in Box 19.2, we can derive the equation for $dp_\times/dt$ by summing $dp_{\times o}/dt$ and $dp_{\times\times}/dt$. This yields

$$\frac{dp_\times}{dt} = (bq_{o|\times} - d)p_\times .$$ (a)

Ignoring spatial structure implies assuming $q_{o|\times} = p_o = 1 - p_\times$, which then leads to

$$\frac{dp_\times}{dt} = \big(b(1 - p_\times) - d\big)p_\times .$$ (b)

This is the well-known model for logistic population growth. It can be verified that this model results from the spatially explicit one under mean-field conditions, that is, if the population is well connected (n is large) and/or well mixed (the rate of movement m is very large). If $b > d$, a small population will grow logistically toward a carrying capacity $\bar{p}_\times = 1 - d/b$.

(A note of caution: if the number of neighbors is *not* constant across the lattice, $p_\times = p_{\times o} + p_{\times\times}$ does not hold, and the differential equation for singlets has to be derived separately.)

Consider the system introduced in Boxes 19.1 to 19.3. This system models a population of individuals inhabiting a lattice (so the sites may be either empty or occupied, $\Omega = \{o, \times\}$). The events that change the distribution are birth, death, and movement (see Box 19.1). Notice that the rates are constant (i.e., they do not depend on the environment of a pair), so that we do not have to average over configurations. The resulting pair-dynamics equations are given in Box 19.2.

Stochastic, event-based simulations of the artificial ecology have been run for two types of lattices, a random lattice (2500 sites with six neighbors per site) and a triangular lattice (also 2500 sites with six neighbors per cell and with periodic boundary conditions), starting with a low number of randomly distributed individuals [expected initial density $p_\times(0) = 0.001$]. The parameters chosen for the simulations are as follows: per capita birth rate $b = 2$; per capita mortality rate $d = 1$; and per capita movement rate $m = 1$. These runs can be compared with results obtained by numerical integration of the pair-approximation model.

Figure 19.6 shows a run for the random lattice case and the corresponding pair-approximation model (classical pair approximation

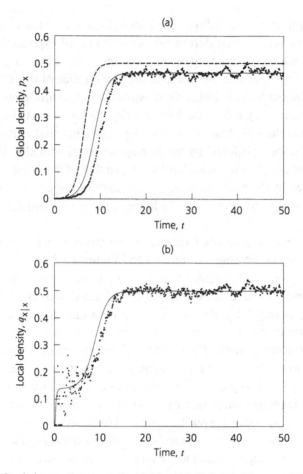

Figure 19.6 Logistic growth on a random lattice. (a) Global density $p_\times$ of the simulation (dots), the trajectory predicted by the pair approximation (continuous line), and the trajectory predicted by the equivalent mean-field model (nonspatial logistic growth; dashed line). (b) Local densities $q_{\times|\times}$ in the simulation (dots) and as predicted by the pair-approximation model (continuous line). Notice that the local density $q_{\times|\times}$ equilibrates much faster than the global density $p_\times$; this fact is very useful for deriving invasion conditions.

with $\theta = 0$). Because in the simulation the lattice is finite, the population density fluctuates at first due to demographic stochasticity. Such demographic stochasticity is inevitable. However large the lattice, if the initial population consists of only a few individuals, it will be a matter of chance whether they reproduce before they die. There is always a probability that the population goes extinct even if its expected rate of growth is positive. Also note that the simulation lags a bit behind the prediction made by the pair-dynamics model. This lag is a direct consequence of demographic stochasticity, since at low overall densities random events can have a considerable effect, either accelerating or decelerating population growth.

In the trajectory that follows the initial fluctuations, the fit is striking. What can be seen is that during the initial phase of exponential growth, the local density $q_{x|x}$ converges to a constant value (much larger than the global density p_x at this initial stage; the significance of this effect is discussed by Matsuda *et al.* 1992). Only when the lattice fills up and different clusters start to mingle does the local density rise again. Both the exponential phase and the end phase, where the population has settled at its carrying capacity, are well predicted by the pair-approximation model. Notice that this carrying capacity is somewhat lower than that of the nonspatial model; this is caused by the nonhomogeneous distribution of the population, which causes individuals to "experience" a higher density of conspecifics than exists globally.

For the regular, triangular lattice, the situation is a bit different. As can be seen by comparing Figure 19.6 and Figure 19.7, population growth is much slower on the regular lattice. This is no surprise because on a two-dimensional lattice, growth of a focus of individuals is confined to its boundary (consequently, the area covered by a cluster increases roughly in proportion to t^2). What can also be seen is that the pair approximation does not perform as well. The classical pair approximation (with $\theta = 0$) is widely off the mark. The pair approximation that was derived earlier (with $\theta = 2/5 = 0.4$) predicts local densities $q_{i|x}$ fairly well, but its global density p_x increases much faster than in the simulation. It may come as a surprise that increasing θ even further (to 0.6) produces an approximation that is reasonably accurate. There is no *a priori* reason to assume a high value of θ, but doing so apparently captures the consequences of clustering quite well, particularly in the early phase of cluster formation (where $q_{x|x}$ equilibrates while p_x increases) and for the final equilibrium. Only during the intermediate phase where the lattice fills up does the approximation fail to perform as well.

19.6   Invasion Dynamics

Studies based on a probabilistic cellular automaton framework have shown that results of classical game theory (which is based on random encounters between individuals) can be significantly affected by spatial structure (see Chapter 8; Axelrod 1984; Nowak and May 1992; Boerlijst *et al.* 1993). Pair approximation provides a tool for understanding these effects.

One of the basic concepts in biological game theory (as in any branch of evolutionary theory) is that of fitness. Fitness should be defined as the invasion capacity of a rare mutant (Metz *et al.* 1992; Rand *et al.* 1994). In a well-mixed, nonspatial system this poses no theoretical problems, but

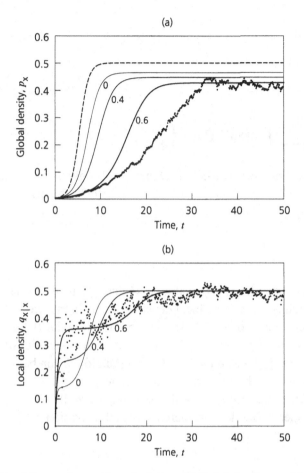

Figure 19.7 Logistic growth on a triangular lattice. (a) Global density $p_\times$ of the simulation (dots); trajectories (continuous lines) as predicted by the pair approximation based on different values of θ (indicated in the graph), the proportion of triplets that are in closed form; and trajectory (dashed line) predicted by the equivalent mean-field model (nonspatial logistic population growth). (b) Local densities $q_{\times|\times}$ in the simulation (dots) and as predicted by the pair-approximation models (continuous lines).

deriving the "invasion exponent" of a rare mutant in an explicitly spatial system is no trivial matter. This problem has been investigated in more detail by van Baalen and Rand (1998); here, the basic approach is illustrated using the simpler problem of a population's invasion of an "empty" world. Apart from evolutionary considerations, this gives insight into the factors that determine a population's persistence (since extinction is basically the reverse of invasion).

Using $q_{\times|oo}\,p_{oo} = p_{\times oo} = q_{o|o\times}\,p_{o\times}$, the dynamics of an invading population can be written in matrix form:

$$\frac{dP_\times}{dt} = M(Q_\times)P_\times \,, \tag{19.31}$$

where

$$P_\times = \begin{pmatrix} p_{o\times} \\ p_{\times\times} \end{pmatrix} \quad \text{and} \quad Q_\times = \begin{pmatrix} q_{o|\times} \\ q_{\times|\times} \end{pmatrix} . \tag{19.32}$$

Because the "invasion matrix" $M(Q_\times)$ depends only on local densities that equilibrate relatively quickly (Matsuda *et al.* 1992), the invasion dynamics are given by

$$\frac{dP_\times}{dt} = c\tilde{Q}_\times e^{\lambda t} \,, \tag{19.33}$$

where λ is the dominant eigenvalue of M and $\tilde{Q}_\times$ is its normalized eigenvector (which is a vector of local densities $q_{i|\times}$). In this case, the invasion condition becomes obvious: λ should be positive. Van Baalen and Rand (1998) argue that by extending this approach to the dynamics of a rare mutant in a lattice dominated by a resident population, λ can be interpreted as a fitness measure, while $\tilde{Q}_\times$ gives information about the associated "unit of selection" (i.e., the entity whose fitness is maximized).

For the specific model considered above, the invasion matrix is

$$M(Q_x) =$$

$$\begin{pmatrix} (b+m)(1-\phi)q_{o|o\times} - b\phi \\ -(b+m)(1-\phi)q_{\times|o\times} & 2b\phi + 2(b+m)(1-\phi)q_{\times|o\times} \\ -d - (1-\phi)q_{o|\times o} \\[2mm] d + (1-\phi)mq_{o|\times\times} & -2d + 2(1-\phi)mq_{o|\times\times} \end{pmatrix} \tag{19.34}$$

From this, the condition for invasion [$M(Q_\times)$ has a positive dominant eigenvalue] can be calculated:

$$b(1-\phi)\tilde{q}_{o|o\times} - d + m(1-\phi)(\tilde{q}_{o|o\times} - \tilde{q}_{o|\times o}) > 0 \,. \tag{19.35}$$

Using improved pair approximation, Equation (19.23), this yields

$$b(1-\phi) - d - m(1-\phi)\tilde{q}_{\times|\times} \\ > \theta\tilde{q}_{\times|\times}[(b+m)(1-\phi) + m(1-\phi)\tilde{q}_{\times|\times}] \,, \tag{19.36}$$

where $\tilde{q}_{\times|\times} = 1 - \tilde{q}_{0|\times} = 1 - d/b$. That $\tilde{q}_{0|\times}$ equals d/b follows readily from the singlet equation (a) in Box 19.3.

On a random lattice, θ equals zero, in which case the right-hand side of inequality (19.35) vanishes. Thus, a non-moving population can invade a random lattice if $b(1 - \phi) > d$. Invasion is thus more difficult the lower the number of connections per site (because this reduces the factor $1 - \phi = 1 - 1/n$). This result is in accordance with the conclusion of Matsuda *et al.* (1992) that, for a focus to grow on a lattice, the birth rate must exceed the death rate by a certain amount. It can also be concluded that movement facilitates invasion: if m becomes very large, the invasion condition will become $b > d$, which is the invasion condition for the nonspatial model.

Triangular lattices are even more difficult to invade, because the right-hand side of the invasion condition will be positive, so that higher values of b or m are necessary. The ecological reason for this is that competition for space is more intense on regular lattices than on random lattices. Even if a lattice is totally empty, members of the invading population will crowd together, effectively competing for space with each other (which is indicated by $q_{\times|0\times} > 0$). On a random lattice this effect is almost absent because there an expanding focus of individuals has many more sites to grow into.

19.7 Concluding Comments

Simulations of probabilistic cellular automata are excellent for developing intuition regarding spatial processes (see, for example, Chapters 6 to 9; Crawley and May 1987; Boerlijst *et al.* 1993; Claessen and de Roos 1995). The drawback, however, is that they take much computer time to simulate (transient behavior persists) and they are difficult to analyze and compare with classical models for population dynamics (de Roos *et al.* 1991; Claessen and de Roos 1995). Pair-dynamics models fill the gap between unwieldy spatially explicit models and nonspatial models that are easy to analyze but fail to capture the spatial effects.

Even though a fairly simple pair-dynamics model may still involve a substantial number of differential equations, there are far fewer than in an equivalent cellular automaton model (whose dynamical dimension equals the number of sites on the lattice). The first advantage of pair-dynamics models is therefore purely practical: even if the model turns out to be too complex to obtain analytical results, within the same amount of computer time a much greater region of parameter space can be explored with a pair-dynamics model than with a probabilistic cellular automaton. In addition, pair-dynamics models do not "suffer" from demographic stochasticity, but whether this is actually an advantage will depend on the system under consideration – demographic stochasticity may be the dominant process on smaller lattices. A third, more important advantage is that pair-dynamics models allow direct insight into the effect of space, because classical population-dynamical models result in the limiting case of high movement rates. Finally, analytical insight often *is* possible. For example, in this chapter, explicit persistence conditions are derived that indicate how persistence of an artificial ecology depends on individual-based properties. Related to the persistence problem is the problem of fitness in "viscous populations": both require an understanding of invasion conditions (van Baalen and Rand 1998).

Although pair-dynamics models incorporate an essential aspect of spatial structure, they ignore other aspects: in particular, the standard "closure assumption" of pair approximation underestimates the consequences of spatial clustering. The assumption that triple correlations are simply absent (i.e., that the probability of encountering a particular triplet configuration is fully given by pair densities) fails to incorporate certain aspects of population clustering. For example, the simulations presented in this chapter show that local competition for space reduces the rate of population growth on a two-dimensional lattice. A number of improved approximations have been published (see Chapter 18; Satō *et al.* 1994; Harada *et al.* 1995; Keeling 1995) that incorporate this effect. They presuppose that certain kinds of triplets (in particular *bab*-type triplets) are more common than expected on the basis of pairs. This increase in triple correlations implies that if *a* has one rare neighbor *b*, it is likely to have more (which would not be the prediction of classical pair approximation). For example, if *b* stands for an infected host and *a* for a susceptible host, this would lead to increased clustering and hence increased competition for hosts among the infecting parasites.

Here, an improved approximation is proposed that is not based on such *a priori* assumptions of triple correlations of *certain* types of triplets, but rather on an evaluation of how triple correlations may arise as a consequence of the lattice structure itself. In contrast to random lattices, on regular lattices (for example, the triangular lattice considered in this chapter) the members of a pair often have common neighbors. This introduces extra correlations, and when these are taken into account one can *predict* that, for example, *bab*-type triple correlations may be larger than 1.

An important limitation of the analysis presented here is that the number of neighbors per site should be constant. If this assumption is relaxed, and there are good reasons for wanting to do so – for example, to model a network of social relations in which some individuals have more contacts than others – the analysis becomes more complex (Morris 1997; Keeling *et al.* 1997a; Rand 1999). Not only does one have to introduce additional equations for the dynamics of single sites (which can no longer be derived from the pairs), one may also have to make assumptions about many more higher-order correlations.

The extent to which pair-dynamics models are satisfactory depends on the goal of the modeler. As we have seen, these models do not capture all of the phenomena that can be observed in simulations of fully spatial probabilistic cellular automata. Basically, the approximation fails whenever spatial structures arise that are difficult to "describe" using pairs alone. More technically, the method fails whenever significant higher-order correlations arise – that is, whenever the frequency of particular triplets (or triangles, squares, or all sorts of star-like configurations) starts to diverge from what one would expect on the basis of pair densities. Thus, pair-dynamics models satisfactorily describe probabilistic cellular automata in which only "small-scale" patterns arise. Larger, "meso-scale" patterns such as spirals are difficult to capture using this method.

However, probabilistic cellular automata occupy only one end of the spectrum; models for classical nonspatial population dynamics are at the other end, and it is there, in particular, that pair-dynamics models can be valuable tools. In the first place, pair-dynamics models can be used to test the assumptions underlying mean-field models. If a pair-dynamics model does not behave significantly differently from the equivalent mean-field model, it is probably not worth bothering about space. Second, and more important, because they are more open to mathematical analysis, pair-dynamics models may give real insight into *why* spatial models behave differently from nonspatial mean-field models. This way, pair-dynamics models can be used "to add space" to a well-understood but nonspatial model without having to resort to analyzing explicitly spatial models.

Acknowledgments Discussions with David Rand and Andrew Morris have helped me greatly to understand the subject and to avoid many pitfalls. I would like to thank Vincent Jansen and Matthew Keeling for their comments on this chapter. My research is currently supported by the French Centre National de la Recherche Scientifique (CNRS), but part of this research was carried out at the University of Warwick in cooperation with Professor D.A. Rand and Dr. A. Morris, and was supported by the Applied Nonlinear Mathematics Initiative of the UK Engineering and Physical Science Research Council (EPSRC) and by the UK National Environmental Research Council (NERC).

20

Moment Methods for Ecological Processes in Continuous Space

Benjamin M. Bolker, Stephen W. Pacala, and Simon A. Levin

20.1 Introduction

Spatial dynamics of populations have long been of interest to ecologists (Skellam 1951; Levin and Paine 1974; Andow *et al.* 1990), but recent advances in data collection and in computational power have put these concepts within the reach of many ecologists for the first time. Computational models (Wilson *et al.* 1993, 1995b; McCauley *et al.* 1993; Pacala and Deutschman 1995) suggest important and previously unexplored effects of space and discrete individuals on population dynamics. Analytic approaches that capture these effects are emerging, building on methods developed in other contexts. This chapter presents a general method for deriving approximate equations for spatial dynamics in continuous space and time that has advantages over classical and many modern approaches.

We are interested in spatial pattern formation in plant communities and in the effects of pattern on plant competition. Our goal is to find general methods for exploring this problem that are

- analytically tractable, so that we can gain insight into qualitative behaviors of the system and analyze how they depend on the parameters;
- sufficiently general, so that some of the same tools can be applied to answer a range of different questions about spatial dynamics in ecology; and
- close enough to the characteristics of real populations that we can eventually fit the models to field data on individual behavior.

We focus on *spatial point processes* (Diggle 1983; Gandhi *et al.* 1998), continuous-time dynamical systems for discrete individuals interacting in a continuous habitat. Closely related *lattice models* such as interacting particle systems, where individuals interact on a square grid, have been more intensively studied (Satō *et al.* 1994; Durrett and Levin 1994b). Our equations, which provide approximations of spatial point process dynamics when individuals interact with many neighbors, are complementary to lattice models and are more useful in some ecological contexts.

Our approach is to derive equations for the expected densities and spatial structure of different kinds (e.g., species or genotypes) of individuals given a particular set of ecological interactions (e.g., competition, predation, or disease transmission). Using these *moment equations* for a given ecological interaction, we apply a standard set of analytic tools – including finding equilibria and invasion eigenvectors as functions of the parameters – to draw ecological conclusions about the determinants of spatial dynamics.

20.2 Moment Methods

Consider a large set of isolated subpopulations, each starting from the same initial conditions and each following the same ecological rules – a statistical *ensemble*. Sampling all of the populations at a given time measures the *probability distribution* of the possible states – for example, the population sizes – of the populations. Similarly, sampling each subpopulation at different spatial points within its range gives a *spatial distribution* of the numbers of individuals found at each location. Probability distributions and spatial distributions are closely related; in particular, for *metapopulations*, where space is divided into discrete patches with global dispersal among patches, each patch is nearly an isolated subpopulation and the probability and spatial distributions can be treated similarly.

Classical ecological models such as the Lotka–Volterra competition model assume both large populations, so they can ignore stochastic correlations measured by the probability distribution, and well-mixed populations, so they can ignore spatial fluctuations measured by the spatial distribution. These are first-order models in the sense that they track only the expected mean densities of the populations, which are the first moments of both the probability distribution and the spatial distribution.

Second-order models go beyond first moments by retaining information about population variances, which are the second moments of the probability distribution, but they still ignore third and higher moments. We can write dynamic equations for an arbitrary number of moments (subject only to the increasing complexity of the equations), but at some point we must cut off the chain of moments and make some assumption about the values of the higher moments in order to obtain analytical results.

For example, some authors have parameterized clumped distributions of hosts, or clumped distributions of parasites within hosts, in host–parasite models by assuming that they follow a negative-binomial distribution with a static shape parameter (Hochberg *et al.* 1990; Hassell *et al.* 1991b). Other modelers retain the *dynamics* of the variance term by assuming that all probability distributions are Gaussian, with changing means and variances, but still setting the third and higher central moments to zero (Grenfell *et al.* 1995b). For distributions with only nonnegative values (such as population densities), one can instead assume that the moments decrease in a geometric series (Dushoff 1997). Population geneticists also use moment approximations, characterizing non-independent distributions of alleles at different loci (linkage disequilibria) by their second and higher moments (Nagylaki 1992; Turelli and Barton 1994).

In metapopulation dynamics (Levin 1974; Slatkin 1974) the within-patch *covariances*, measuring dependence between population sizes of different species in a patch, become important. Levin (1974) pointed out that negative covariance – the tendency for two species not to inhabit the same patch – would be important in metapopulation models for plant competition (Horn and MacArthur 1972). He suggested, though, that covariance could be ignored when a species was invading, because it would initially be randomly distributed among patches. Our preliminary work with second-order moment equations for metapopulations shows that even during an invasion the covariances grow rapidly enough to affect population dynamics. In this chapter we explore exactly this kind of process in the more general context of fully spatial models.

In fully spatial models, where individuals can live at different distances from each other (either in continuous space or as n-step neighbors in a lattice or network), and where spatial patterns and interactions occur at different scales, moment equations can follow correlations at a single distance or at many different distances. Most of the current models for ecological spatial systems with discrete individuals use second-order moment equations that track only correlations between nearest neighbors on a lattice or network. Moment equations for correlations at all possible distances in continuous space are well-established in the physics literature and have been used in studies of chemical dynamics (Keizer 1985; Molski and Keizer 1995); however, to the best of our knowledge, they have never been studied in an ecological context.

20.3 A Spatial Logistic Model

This section introduces all of the ideas and machinery needed to analyze simple spatial ecological models. We define a model in terms of individual birth and death rates; we show how to summarize spatial simulation results with spatial statistics, and how these statistics form the basis of a set of second-order moment equations; and we analyze the equations, checking the conclusions against simulation output.

Model description

As the simplest example of the spatial moment equations for a stochastic point process, we present a spatial analogue of the well-known logistic growth model (Murray 1990). Consider a population of discrete, identical individuals inhabiting a continuous, homogeneous habitat. These individuals die at a constant expected rate μ and reproduce at a constant expected rate f; all processes are stochastic and Markovian, so the waiting times between births and deaths are exponentially distributed. Newborns instantaneously disperse from their parents in a random direction to a distance governed by a *dispersal kernel* $D(r)$ (see Box 20.1). When they arrive, they are able to establish with a probability determined by the *local density* $d(x)$ around their arrival point, which is the density weighted according to a *competition kernel* $U(r)$:

$$d(x) = \sum_{\text{All plants } i} U(|y_i - x|) , \tag{20.1}$$

where y_i is the location of the ith plant. We chose the establishment probability to be a linearly decreasing function of local density,

$$E(x) = \left[1 - \left(\frac{\alpha}{f}\right) d(x)\right]_+ . \tag{20.2}$$

The competition coefficient α is scaled by total fecundity so that the expected fecundity for a plant producing f seeds is $f - \alpha\bar{d}$, where $\bar{d}$ is the average local density experienced by dispersing offspring; the subscript $+$ indicates that if the local density is so high that it would drive the establishment probability negative, the probability is set to zero instead. It would be more natural to make the establishment probability a decreasing exponential, but nonlinear interactions would add a step to the derivation (a Taylor expansion of the competition function around the expected local density). We try to stay away from regions in parameter and state space where the local density frequently pushes the establishment probability all the way down to zero.

Box 20.1 Dispersal and competition kernels

Spatial interactions in our models are governed by spatial *kernels*, symmetric, normalized, decreasing functions of distance that determine the probability or relative strength of interactions. Kernels have a functional form and a *scale parameter* (λ) that sets the spatial scale of interaction; large values of λ mean strongly localized, short-range interactions.

Dispersal kernels give the probability of an organism or propagule moving a particular distance. Dispersal kernels appear in invasion ecology (Skellam 1951), community dynamics (Ribbens *et al.* 1994), and plant epidemiology (van den Bosch *et al.* 1990a). Common functional forms are exponential, generalized normal, or Bessel functions (see, e.g., Abramowitz and Stegun 1965). The normalized form of the modified Bessel function of second kind and order 0, $\lambda^2 K_0(\lambda r)/(2\pi)$, is a monotonically decreasing function like the exponential distribution (see Figure 16.14 and Box 23.1). It is the expected distribution of individuals random-walking from a central point and stopping with constant probability per unit time (Renshaw 1991).

Competition kernels determine the strength of competition (for example, by resource preemption) between two individuals located a given distance apart. Mechanistic competition kernels could, for instance, calculate root overlap as a function of distance, or use canopy geometry to calculate shading (see Chapter 2). For analytical convenience, our models use phenomenological Bessel functions; other functional forms are also possible.

Competition could decrease per capita fecundity or establishment, or increase per capita mortality (or some combination of the three). Pure density-dependent establishment is simple and makes ecological sense as a germination or establishment probability – for example, the probability of a tree seedling surviving to age five is a function of its light environment, which is directly affected by the local density of canopy trees. Density-dependent establishment is also a good approximation to the form of density dependence typical of lattice models (where individuals dispersing into an occupied square die), although the density dependence embodied in lattice models is more discontinuous, acting only in occupied cells.

Simulation and summary statistics

What are the dynamics and equilibrium behavior of this population? We can simulate the model by keeping track of individuals' positions in continuous space and in small discrete time steps, picking random variates to decide whether they die or reproduce, then calculating the local density

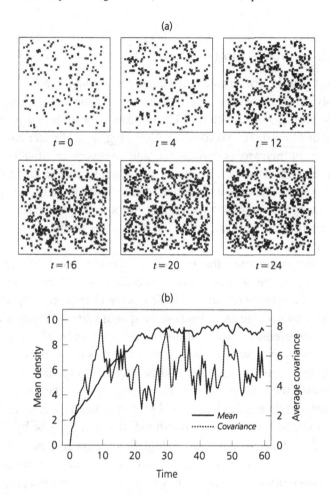

Figure 20.1 Stochastic simulation results. (a) Snapshots of the state of the system at various times. (b) Summary measures of the simulation: mean density (left axis) and average covariance (right axis) versus time. Parameters: $f = 0.8$, $\mu = 0.4$, $\alpha = 0.02$; U and D are Bessel functions (see Box 20.1) with scale parameters $\lambda_U = \lambda_D = 1.0$; length = 10, $\bar{n}(0) = 2$.

around each offspring's location and picking another random variate to determine whether it establishes successfully (Figure 20.1a shows the output of such a process).

We can also calculate summary statistics to show how the population develops. The *mean density* $\bar{n} = \langle n(x) \rangle$ of individuals per unit area (averaging over space) is an easy measure, but it eliminates all information about spatial structure. There are many possible choices of summary statistics of spatial structures (Diggle 1983; Cressie 1991). The *spatial covariance* is defined as

$$c(|\Delta x|) = \langle (n(x) - \bar{n})(n(x + \Delta x) - \bar{n}) \rangle \ . \tag{20.3}$$

We assume that the covariance between any two points x and $x + \Delta x$ depends only on their distance apart, $|\Delta x|$, and not on their absolute locations. The average in Equation (20.3) is taken over all points x and over all angles of the distance vector Δx. Assuming *second-order stationarity* and *isotropy* in this way is convenient, and common, in situations where we are interested in statistical aspects of patterns rather than in the spatial configurations themselves (Cressie 1991). Throughout this chapter, we also assume spatial ergodicity, such that averages across space and averages across different realizations of the system are identical and can both be denoted by $\langle \ldots \rangle$.

We further summarize the *average covariance* as $\bar{c} = 2\pi \int (U * D)(r)c(r) \, dr$, where $*$ represents a two-dimensional *convolution* (see Box 20.2). The average covariance indicates the change in competition experienced by an average plant because of spatial pattern. Positive average covariance represents crowding at the spatial scale set by $U * D$ (aggregation), while negative average covariance represents even spacing (regularity); $\bar{c}/\bar{n}$ is the excess in neighborhood density due to spatial structure, and $\bar{c}/\bar{n}^2$ measures the relative excess. For example, once the system in Figure 20.1 reaches its equilibrium ($t > 24$), $\bar{c} \approx 5$ and $\bar{n} \approx 9$, so on average crowding increases absolute neighborhood densities by $\frac{5}{9}$ or by a proportion of $1+\frac{5}{81}$. These statistics are analogous to "crowding indices," which are defined in terms of the means and variances of quadrat samples. For example, the coefficient of variation, which also measures the proportional effects of crowding, is defined as variance/mean^2, similar to our statistic $\bar{c}/\bar{n}^2$ (Pielou 1977). Figure 20.1b shows the time-dependent mean and average covariance of the population depicted in Figure 20.1a; the mean density starts small and rises logistically, while the average covariance starts from about zero (indicating a random spatial distribution) and then rises, ending up with strong fluctuations around a positive equilibrium value.

Mean dynamics

We can analyze the dynamics of the mean density by finding the expected rate of change of the number of plants at a point (see Appendix 20.A). The result is

$$\frac{d\bar{n}}{dt} = (f - \mu - \alpha\bar{n})\bar{n} - 2\pi\alpha \int (U * D)(r)c(r) \, dr \ . \tag{20.4}$$

Box 20.2 Convolutions

A one-dimensional convolution between two functions A and B is defined as $(A * B)(z) = \int A(x)B(x - z)\, dx$. It combines two spatial variables or kernels by taking the integral of their product for every value of the *lag* z. Convolution is also equivalent to smoothing one kernel (using the second kernel as a "moving window"), or finding the overlap between two distributions as a function of the distance between their centers.

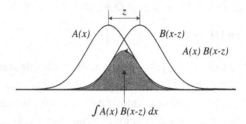

A two-dimensional convolution is defined in polar coordinates as $(A * B)(r) = \iint A(r')B(R(r, r', \theta))r'\, dr'd\theta$, where $R(r, r', \theta) = (r^2 + r'^2 - 2rr' \cos \theta)^{1/2}$ is the length of the third side of a triangle with sides of length r and r' forming an angle θ. Notation and computation of the two-dimensional convolution are ugly, but the meaning of convolutions remains the same in any dimension.

Convolutions arise in moment equations as a way of averaging the strength of interactions at all possible distances. In the case of the average covariance, for example, the term $(U * D)(r)$ appears because the seedling's competitive neighborhood, $U(r)$, is averaged over all possible locations in the dispersal neighborhood of its parent, $D(r)$.

The first term on the right-hand side is simply the usual logistic term [with exponential growth rate $f - \mu$ and carrying capacity $(f - \mu)/\alpha$]. The second term is $\alpha \bar{c}$; combining the two terms containing α gives $\bar{n} + \bar{c}/\bar{n}$ as the effective number of competitive neighbors. Knowing the spatial structure of the population, we could predict the equilibrium mean $\bar{n}^*$ as a function of the average covariance, setting the left-hand side of Equation (20.4) to zero and rearranging to obtain

$$(\bar{n}^*)^2 - \left(\frac{f - \mu}{\alpha}\right)\bar{n}^* + \bar{c} = 0 \; ; \tag{20.5}$$

neighborhood models (Pacala and Silander 1985, 1987) use this approach.

Covariance dynamics

Equations (20.4) and (20.5) give the dynamics and equilibrium of the mean density in terms of the average covariance. The dynamics of the covariance come from the expected change in the product of the densities in two places a given distance apart (see Appendix 20.B). The result contains a *spatial third moment* term, which we set to zero, leaving

$$\frac{1}{2}\frac{dc(r)}{dt} = -\mu c(r) + (f - \alpha\bar{n})\left[(D * c)(r) + \bar{n}\,D(r)\right]$$
$$- \alpha\bar{n}\left[(U * c)(r) + \bar{n}\,U(r)\right] . \tag{20.6}$$

Even without detailed analysis, this equation illuminates the processes determining spatial pattern. The first term on the right-hand side shows that mortality reduces pattern (either clustering or spacing) toward randomness; the second term shows that local dispersal generates clustering; and the last term shows that competition generates even spacing.

The expressions proportional to $\bar{n}$ in the second and third terms on the right-hand side of Equation (20.6) are important; they represent the inevitable growth of spatial pattern around an individual plant. If we derive the standard "continuum" limit of the spatial logistic model where we keep space explicit but allow all stochastic fluctuations and all local spatial correlations to disappear, then we can show that the spatially homogeneous equilibrium [$n(x)$ set to carrying capacity everywhere] is stable with respect to any small perturbation (Murray 1990). In contrast, the equivalent state for the second-order moment equations, where individuals are randomly distributed in space [$\bar{n} = K, c(r) = 0$ for all r], is not even an equilibrium because of the $\bar{n}$ terms in Equation (20.6); if we start with a random distribution of individuals at their nonspatial carrying capacity, they immediately begin to form spatial patterns.

Model accuracy

Does the second-moment approximation work for realistic scenarios where plants have ecologically reasonable growth rates, life-history strategies, and competition and dispersal scales?

Figure 20.2 compares predicted equilibrium mean densities from the moment equations and from a simulator (for simplicity, we set the competition and dispersal scales equal, $\varepsilon = 1$; see Box 20.3). We show time averages of the mean after the system has reached equilibrium; because the system is *ergodic*, sampling over time is equivalent to sampling across space or across ensembles, provided that we use long enough scales in time

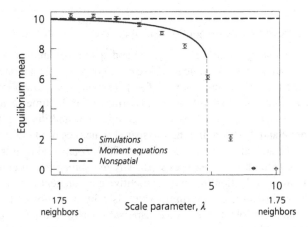

Figure 20.2 Comparison of simulation results and results from moment equations for a range of scale parameters (with equal competition and dispersal kernels). Error bars indicate simulation results (mean ± standard deviation for 100 samples taken at unit time steps after a transient). Parameters: $f = 0.6$, $\mu = 0.4$, $\alpha = 0.02$, $\lambda_U = \lambda_D = \lambda$, length = 30.

and space to wash out any correlations. The horizontal axis of Figure 20.2 shows the spatial scale parameter λ (which is large for local interactions and small for global interactions; see Box 20.1) as well as the effective number of neighbors. For a uniform neighborhood, where individuals compete equally with all neighbors within distance ρ, the effective number of neighbors is the carrying capacity K times $\pi \rho^2$; for the Bessel kernel, the equivalent calculation gives $4\pi K / \lambda^2$.

The total strength of competition is fixed by α, so the strength of competition per competitor varies inversely with the effective number of neighbors. The effective number of neighbors determines the magnitude of spatial fluctuations; as the neighborhood grows large, individuals interact with more and more individuals in the population, so the spatial variances vanish. Above about 40 neighbors the means in Equation (20.2) are statistically indistinguishable from the nonspatial establishment probabilities; below about four neighbors the moment equations collapse, predicting that the equilibrium mean goes to zero – which does happen in the simulator, but not in the way predicted by the moment equations. Luckily, the range of 4–40 interacting neighbors is consistent with the population densities and interaction scales of many ecological systems.

Examining the system in more detail, Figure 20.3 compares the spatial pattern (covariance) at equilibrium and the phase-plane dynamics for a "shady" (shade-tolerant) and a "weedy" species (see Box 20.3). The shady species has a clustered pattern at equilibrium (positive average covariance),

---

Box 20.3 Plant life-history and spatial strategies

How do life-history parameters (f, μ, α) and spatial parameters (λ_D, λ_U) relate to plant competitive strategies? We can understand plant strategies in terms of the logistic parameters $r = f - \mu$ (exponential growth rate) and $K = r/\alpha$ (carrying capacity); $R = f/\mu$ (maximum lifetime fecundity); and $\varepsilon = \lambda_U/\lambda_D$ (the relative scale of competition and dispersal). The parameters r and K determine the nonspatial dynamics of plants; R and ε determine the spatial dynamics (see Section 20.3, Analytical results).

Lifetime fecundity R measures *sensitivity to competition*; it equals the ratio of total reproduction in a noncompetitive environment to total reproduction at competitive equilibrium (which is 1 by definition). The scale ratio ε is large for plants with long dispersal relative to the size of their competitive neighborhoods (remember that scale parameters λ are large for local interactions).

"Weedy" plant species are sensitive to competition and have long dispersal relative to competition $(R, \varepsilon$ are large); *shade-tolerant*, or "shady," species are insensitive to competition and have short dispersal $(R, \varepsilon$ are small). In general, weedy species are early-successional, while shady species are late-successional. Section 20.3 analyzes and contrasts the spatial dynamics of species with these two types of strategies.

---

while the weedy species has an evenly spaced pattern at equilibrium (negative average covariance), and the moment equations predict simulation results very well (Figure 20.3a). Figure 20.3b shows the dynamics of each species growing from a random initial distribution to its equilibrium mean and covariance. The full dynamics of the system are infinite-dimensional (because covariance is a function of distance), but the plane shows a useful projection of the dynamics through the mean and the average covariance. The figure shows discrepancies between the simulation and the moment equation predictions, especially during the transient dynamics of the shady species; this means that non-vanishing third moments suppress the transient level of the average covariance.

Analytical results

We analyze the *quasi-equilibrium* covariance (describing the equilibrium spatial structure of the population when the total population size is held constant) by setting the left-hand side of Equation (20.6) to zero and $\bar{n}$ to a constant $\hat{n}$. Finding the quasi-equilibrium is the first step toward a full-equilibrium solution; it will also help us to analyze the invasion and coexistence properties of a two-species competitive system (see Section 20.4).

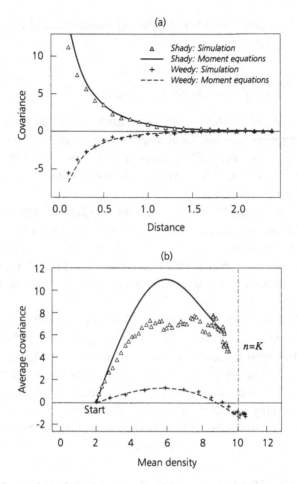

(a)

(b)

Figure 20.3 Comparison of simulation results with moment equation predictions for weedy and shady species (see Box 20.3). (a) Equilibrium spatial covariances (points indicate simulation; lines, moment equations). (b) Phase plane dynamics (mean density versus average covariance). Vertical line represents nonspatial carrying capacity (K); points, simulation; lines, moment equations. Parameters: shady species $f = 1.2$, $\mu = 0.4$, $\alpha = 0.08$; weedy species $f = 0.6$, $\mu = 0.4$, $\alpha = 0.02$; scale parameters $\lambda_D = \lambda_U = \pi$ for both species, length = 30.

To simplify the quasi-equilibrium we need the *Fourier transform*, or rather the *Hankel transform*, which is the two-dimensional, radially symmetric version of the Fourier transform (James 1995): we write the Hankel transform of $c(r)$ as $\tilde{c}(q)$. Hankel transforms are linear, so we can interchange them with sums, integrals, or products with scalars. They turn convolutions into products, so that $A * B$ becomes $\tilde{A}\tilde{B}$ – this useful property is actually the only reason we need Hankel transforms.

Solving Equation (20.6) for the Hankel-transformed quasi-equilibrium covariance $\tilde{c}^*(q)$ yields

$$\tilde{c}^*(q) = \hat{n}\frac{(f - \alpha\hat{n})\tilde{D}(q) - \alpha\hat{n}\tilde{U}(q)}{\mu - (f - \alpha\hat{n})\tilde{D}(q) + \alpha\hat{n}\tilde{U}(q)} \ . \tag{20.7}$$

As in Equation (20.6), this equilibrium value shows that clustering is increased by reproduction and local dispersal, and is decreased by mortality and by competition. From $\tilde{c}^*$ in Equation (20.7) we can numerically obtain the back-transform c^* for given values of the parameters (f, μ, α) and shapes of the transformed kernels ($\tilde{D}$, $\tilde{U}$).

At this point we exploit the fact that the competition and dispersal kernels D and U are normalized Bessel functions (see Box 20.1). When dispersal and competition scales are relatively large ($\lambda \ll 1$), for the equilibrium mean and covariance we can use the Hankel transform to simplify the quasi-equilibrium covariance into a sum of Bessel functions (see Appendix 20.C).

Figure 20.4 shows the overall effect of parameters on the scaled equilibrium covariance $\tilde{c}^*/\bar{n}^*$. Shady species, falling in the lower left-hand corner of Figure 20.4, tend to cluster at equilibrium, while weedy species tend to be evenly spaced at equilibrium. The figure shows only the equilibrium spatial structure; note that the transient dynamics of all species undergo clustering (roughly speaking, the effects of competition take time to build up), and weedy species in the field may appear clustered because they are in such a transient phase of their dynamics.

The ecological conclusion of this analysis is that, for realistic parameter values, spatial dynamics lead to even spacing in monocultures. This is because plants typically have values of R, maximum lifetime fecundity, on the order of hundreds or thousands (Harper and White 1974), placing them at or beyond the right-hand edge of Figure 20.4. To account for spatial heterogeneity in monocultures we must invoke spatial processes beyond the scope of this chapter, such as abiotic environmental heterogeneities or disturbance.

20.4 A Spatial Competition Model

To illustrate a case where endogenous spatial structure does have important effects, we now turn to the spatial equivalent of Lotka–Volterra competition: individuals of two species competing (phenomenologically) in a homogeneous arena by inhibiting each other's (and their own) establishment. We give only a rough outline of the methods (which are very similar to

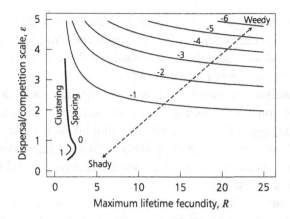

Figure 20.4 Spatial structure $\bar{c}^*/\bar{n}^*$ as a function of lifetime fecundity R and scale ratio ε (see Box 20.3). Positive $\bar{c}^*/\bar{n}^*$ represents clustering and negative $\bar{c}^*/\bar{n}^*$ represents spacing; "Shady" and "Weedy" show the locations of the shady and weedy species tested in Figure 20.3. Parameters: $\lambda = 1$.

those described in Section 20.3 for the spatial logistic equation), concentrating instead on ecological results. More technical details are given in Appendix 20.D.

Our approach is to derive equations for the densities $(\bar{n}_1, \bar{n}_2)$ of the species and their three covariances [c_{11}, c_{12}, c_{22}, where $c_{ij}(r)$ is the covariance between species i and j] using the techniques shown in Appendices 20.A and 20.B. We can then find the quasi-equilibrium covariances (the equilibrium values of the c_{ij} for fixed $\bar{n}_i$) and derive *invasion criteria*, inequalities governing whether one species can invade the other, by fixing one species at its monoculture equilibrium mean and covariance and making the mean density of the other species very small (see Appendix 20.D).

In general terms, the invasion criterion is

$$\underbrace{\eta_{12}}_{\substack{\text{Invasion}\\\text{rate}}} = \underbrace{\sigma_{12}}_{\substack{\text{Nonspatial}\\\text{invasion rate}}} - \underbrace{\alpha_{11}\frac{\bar{c}_{11}}{\bar{n}_1}}_{\substack{\text{Invader}\\\text{clustering}}} + \underbrace{\alpha_{12}\frac{\bar{c}_{22}}{\bar{n}_2}}_{\substack{\text{Resident}\\\text{clustering}}} - \underbrace{\alpha_{12}\frac{\bar{c}_{12}}{\bar{n}_1}}_{\substack{\text{Spatial}\\\text{segregation}}} > 0 \ . \qquad (20.8)$$

The first term, σ_{12}, gives the nonspatial invasion rate – this would be the invading species' growth rate if both species had completely global interactions. The spatial terms describe how the spatial structure of the population at quasi-equilibrium modifies the invader's competitive ability. When the invader clusters with itself ($\bar{c}_{11} > 0$), it increases self-competition, weakening its invasion; when the resident clusters with itself ($\bar{c}_{22} > 0$), it reduces its own density and encourages invasion. Spatial segregation ($\bar{c}_{12} < 0$) weakens interspecific competition and encourages invasion. Life-history and spatial strategies determine the balance of these terms, which in turn determines the outcome of invasion and competition.

We can analyze the most general invasion criterion – with different spatial scales of competition and dispersal for each species – numerically or with symbolic algebra software, but the solutions do not provide analytic insight (we have to solve a cubic equation and a 4×4 linear system). Solving for various special cases, though, contributes to our general understanding of spatial competition.

In particular, here we illustrate one process that is *not* consistent with the explanations produced by existing models. Suppose species 2 has global competition and dispersal, and species 1 has local competition and global dispersal. In this case, the only terms that remain in Equation (20.8), besides the nonspatial terms, are "invader clustering" and "spatial segregation" when species 1 invades species 2, and "resident clustering" when species 2 invades species 1. In this case, increasing the sensitivity to competition of species 1 (R_1) benefits species 1 by decreasing its clustering (when it is either the resident or the invader); increasing either species' sensitivity to competition (R_1 or R_2) benefits species 1 by increasing segregation.

If the balance of these terms is right, which happens over a broad range of parameters, then the two species can invade each other and coexist even for parameters where they cannot coexist nonspatially. Suppose species 1 is a competitively inferior but "gap-filling" species that benefits from spatial segregation because it has local dispersal and competition. In contrast species 2, which has global competition and dispersal, cannot benefit from spatial segregation. Longer scales give it more, rather than fewer, opportunities. If species 1 competes and disperses at longer scales, it loses its gap-filling advantage and goes extinct. This spatial strategy is the complete opposite of the usual colonization strategy discussed in the ecological literature, where one species overcomes a competitive handicap by dispersing farther to reach gaps (Skellam 1951; Shmida and Ellner 1984; Tilman 1994; Holmes and Wilson 1998). In this case, species 1 (still the competitive loser, but now a long disperser) benefits by settling space that species 2, a short disperser, leaves empty. If species 1 competes and disperses at even longer scales, it now gains, rather than loses, opportunities to exploit favorable conditions.

Figure 20.5 shows the dynamics of the moment equations, the nonspatial dynamics, and the dynamics of one realization of the full stochastic

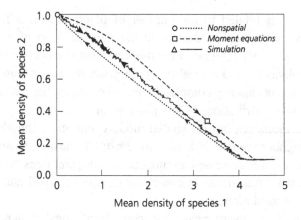

Figure 20.5 Phase-plane dynamics showing coexistence of a gap-filling species with a superior competitor in spatial competition. Dashed line represents moment equations (equilibrium at □); continuous line, full stochastic system (equilibrium at △); dotted line, nonspatial dynamics (equilibrium at ○).

system for such a scenario where short-dispersing species 1 goes extinct in the nonspatial case (with global competition and dispersal) but coexists in spatial competition. The moment equations predict the qualitative dynamics correctly (coexistence at a stable internal equilibrium), although the locations of the moment-equation and true equilibria differ.

20.5 Extensions and Related Work

We now consider spatial moment equations in a wider context, both in terms of their applicability to the dynamics of experimental and natural ecological communities and in terms of their connections to other classes of models for spatial dynamics.

Third moments

The application of second-order moment equations to natural systems in ecology, as well as in other fields such as population genetics, depends critically on the assumption that central third moments are small. In the limit of large movement scales, we can show formally that central third moment terms converge to zero faster than covariances, but how well will the third moment approximation work for practical problems?

For these problems, which involve measurement and statistics rather than formal mathematical arguments, it will be useful to gain some intuition about the third spatial moment. The competition-weighted integral of the covariance, $\bar{c}$, is related to the number of plants around a focal plant, or to the mean population density in the neighborhood of a plant; the competition-weighted integral of the third moment is related to the number of pairs of plants around a focal plant. In patch models, where the distribution of patch occupancy completely describes the spatial pattern, we can easily construct distributions with a given mean, variance, and skew (non-spatial third moment), but for spatial models with more interdependence between neighborhoods it is harder to construct consistent patterns with given moments. Large-scale inhomogeneities in plant density, and possibly sharp spatial transitions in density (e.g., checkerboard patterns), will generate large third moments.

We are working toward measuring spatial third moments in natural populations and are trying to improve our intuition as to what essential ecological information is represented by third spatial moments and under which conditions the dynamics of a system will naturally lead to large (or small) third spatial moments.

Connections with other models

As mentioned in Section 20.1, the methods given here closely parallel other approaches to spatial statistical models. What are the differences, and what are the advantages and disadvantages of moment equations in continuous space and time?

Reaction–diffusion equations (Fife 1979; Murray 1990; see also Chapters 16, 17, 22, and 23) are the classical modeling approach to population dynamics in continuous space. The moment equations presented here differ in three ways: they assume expected densities are spatially homogeneous, they allow for nonlocal movement, and they take local spatial correlations into account. The first two differences are minimal: moment equations can be made explicitly spatial (Lewis and Pacala, in press), and deterministic models can incorporate nonlocal movement (van den Bosch *et al.* 1988c; van den Bosch *et al.* 1990a). The last difference is fundamental: none of the clustering phenomena described in this paper occur in the reaction–diffusion system. For competitive systems, local spatial correlations are important; we hope to explore other systems such as aggregation models (Grünbaum 1994) in the future to see how reaction–diffusion and moment equations compare in those cases.

Lattice models (see Chapters 3, 6, 7, 8, 10, 11, 12, and 15) cover most existing analytic models of stochastic processes in continuous space. For sparsely occupied lattices, where nearest neighbors are always many grid cells apart but still interact with each other over long distances, the distances between individuals become roughly continuous. In this limit, lattice models begin to resemble point processes, and the two types of models should give qualitatively similar results. Many of our findings should be echoed by analogous results from the literature on lattice models; where the models differ, we argue that our moment equations incorporate fewer artificial assumptions.

Pair approximations (Satō *et al.* 1994; Durrettn and Levin 1994b; see also Chapters 13, 18, and 19) use exactly the same conceptual approach as continuous-space moment equations, tracking the occupancy of nearest neighbors in a lattice conditional on occupancy of a focal cell. Ellner *et al.* (1998) have recently extended pair approximation to track more distant relationships. Satō *et al.* (1994) have developed an improved pair approximation, which uses a single static parameter for third-moment or triple correlations. An alternative approach, *local structure functions*, extends pair approximation by doing a bookkeeping on the probabilities of various local configurations beyond pairs of nearest neighbors, effectively tracking higher-order and more distant correlations (Hiebeler 1997).

As discussed above, there are limits in which lattice models converge to point processes; point processes converge to reaction–diffusion equations when competition and dispersal are both local, but dispersal scales are longer than competitive scales. Other limiting processes go directly from lattice models to reaction–diffusion equations (Durrett and Neuhauser 1994). Reaction–diffusion equations have no dependence on local clustering, whereas lattice equations, with their "all-or-nothing" form of density-dependent establishment, have too strong a dependence. Moment equations for point processes represent a happy medium.

20.6 Concluding Comments

Stochastic spatial pattern formation occurs in ecological populations because of the discrete nature of individuals and the local nature of dispersal. The clustering feeds back on itself, and on population dynamics, because of the local nature of competition. As we have shown here, the effects of stochastic pattern formation and spatial segregation in particular can quantitatively change the densities of monocultures and qualitatively change the outcome of two-species competition. We have used second-order moment equations, an analytic framework that approximates stochastic processes in continuous space, to understand the balance of competition and dispersal in determining the growth and effects of spatial pattern. Our moment equations are related to deterministic continuous-space models, on the one hand, and stochastic lattice models, on the other, but they have advantages of their own – not least their close connection to the measurable, individual-level demographic processes occurring in ecological populations.

Acknowledgments We would like to thank Jonathan Dushoff, George Hurtt, Mark Lewis, and Paul Moorcroft for helping with the development of these ideas. We are pleased to acknowledge the support of the National Aeronautics and Space Administration through grants NAGW-4688 and NAG56422 and the Andrew W. Mellon Foundation to Simon Levin and Stephen Pacala. Simon Levin's work is supported in part by the University Research Initiative Program of the Office of Naval Research through grant number ONR-URIP N00014-92-J-1527 to the Woods Hole Oceanographic Institution.

Appendix 20.A   Mean Equation

In this appendix we start from the rules described in the body of the chapter and go through the steps required to derive the equation for the mean density, complete with a spatial (covariance) correction term.

We set fecundity to f, mortality to μ, and establishment probability to $1 - (\alpha/f) \int n(y) U(|x - y|) \, dy$ (the change in establishment probability is scaled by $1/f$ to make the total change in effective fecundity equal to α times the local density). We temporarily discretize the system into small boxes (of size h) with no more than one individual per box, writing the expected change in box occupancy over a small time Δt (averaged across different realizations of the system)

$$\Delta N(x) = \left[\underbrace{-\mu N(x)}_{\text{Death}} + \underbrace{\left(\sum_z f N(z) \, h D(|x - z|) \right)}_{\text{Reproduction}} \right.$$

$$\left. \times \underbrace{\left(1 - \frac{\alpha}{f} \sum_y N(y) \, U(|x - y|) \right) \eta(x)}_{\text{Establishment}} \right] \Delta t \; . \tag{20.9}$$

Note that the sums are over all space; x, y, and z are the (vector) locations of boxes; and $|x - y|$ gives the distance between locations. The kernels D and U are normalized functions of distance (see Box 20.1). Here, $\eta(x) = 1 - N(x)$ is a term that ensures that plants can never establish in a box that is already occupied. This term is not necessary to our derivation – it disappears as we scale h to zero – but it incorporates the strong density dependence typically found in lattice models (see Chapter 9), and so it allows us to scale from single-occupancy lattice models to continuous point processes with one equation.

Next we rearrange terms and take spatial averages on both sides:

$$\langle \Delta N(x) \rangle = \Big[-\mu \langle N(x) \rangle + f \sum_z \eta(x) \langle N(z) \rangle h D(|x - z|)$$

$$- \alpha \sum_z \sum_y \eta(x) U(|x - y|) D(|x - z|) h \langle N(y) N(z) \rangle \Big] \Delta t . \tag{20.10}$$

We use the definitions $\langle N(x) \rangle = N$ and $\langle N(x)N(z) \rangle = N^2 + C'(|x - z|)$ and assume second-order stationarity and isotropy so that N is independent of x and C' depends only on $|x - z|$ (so $\eta = 1 - N$ is also independent of x). We also move the origin to $x = 0$ without loss of generality and use the fact that D and U are normalized (sum to 1) to obtain

$$\frac{\Delta N}{\Delta t} = -\mu N$$

$$+ \eta \left(fN - \frac{\alpha}{h} N^2 - \alpha \sum_z \sum_y U(|y|) D(|z|) h C'(|y - z|) \right) . \tag{20.11}$$

Dividing both sides of the equation by h and scaling to continuous space and time ($\Delta t, h \to 0$), with $\bar{n} = \lim_{h \to 0} N/h$, $c' = \lim_{h \to 0} C'/h^2$, and $\eta \to 1$, we have

$$\frac{d\bar{n}}{dt} = -\mu \bar{n} + f \bar{n} - \alpha \bar{n}^2 - \alpha \iint U(|y|) D(|z|) c'(|y - z|) \, dy \, dz$$

$$= -\mu \bar{n} + f \bar{n} - \alpha \bar{n}^2 - 2\pi \alpha \int (U * D)(r) c'(r) \, dr$$

$$= -\mu \bar{n} + f \bar{n} - \alpha \bar{n}^2 - \alpha \bar{n}(U * D)(0) - 2\pi \alpha \int (U * D)(r) c(r) \, dr$$

$$= (f' - \mu - \alpha \bar{n}) \bar{n} - \alpha \bar{c} . \tag{20.12}$$

In the last two lines of Equation (20.12), we have substituted $c(r) + \bar{n}\delta(r)$ for $c'(r)$ and then incorporated the extra term by modifying the effective fecundity $f' = f - \alpha(U * D)(0)$. The Dirac δ function at zero in the covariance is a continuous spike with infinite height but with area equal to 1; it follows in the continuous limit from our definition of covariance. As specified when we discretized the system, h is small enough that there is never more than one individual per box. Thus $\langle N(x)^2 \rangle = \langle N(x) \rangle$ and $c'(0) = \lim_{h \to 0}(N - N^2)/h^2 = \lim_{h \to 0} \bar{n}/h$, which produces a δ function. This term is the nonspatial variance of the distribution; even a completely random spatial pattern with no spatial autocovariance has an associated sampling distribution, which in this case is a Poisson distribution and so has a variance equal to the mean density. In Equation (20.12) this term appears as parent–offspring competition, which we can incorporate into the fecundity (any field measurement of fecundity contains parental competition by default); in the covariance equation, however, it has important consequences.

Appendix 20.B   Covariance Equation

To derive the equations for the covariance, we follow essentially the same procedure of taking expectations and scaling, although the expressions involved are more complicated. Recalling the definition of covariance, we start with the expected change in $\langle N(w)N(x) \rangle$:

$$\Delta \langle N(w)N(x) \rangle = N(w)\Delta N(x) + N(x)\Delta N(w) + \Delta N(x)\Delta N(w)$$
$$= 2N(w)\Delta N(x) \ . \tag{20.13}$$

We drop the $\Delta N(x)\Delta N(w)$ term by assuming that Δt is too small for more than one event to occur (both birth and death change the occupancy of only one box at a time). Because we are assuming second-order stationarity, we can interchange the roles of any two points, so we set $N(w)\Delta N(x) = N(x)\Delta N(w)$. Next we substitute in the same expression for $\Delta N(x)$ as given in Equation (20.10), rearranging and moving expectations inside sums as we go:

$$\Delta \langle N(w)N(x) \rangle = 2\Big[-\mu \langle N(w)N(x) \rangle + f \sum_z \eta(x) \langle N(w)N(z) \rangle h D(|x-z|)$$
$$\tag{20.14}$$
$$- \alpha \sum_z \sum_y \eta(x) U(|x-y|) D(|x-z|) h \langle N(w)N(y)N(z) \rangle \Big] \Delta t \ .$$

Expanding $\langle N(w)N(x) \rangle$ and $\langle N(w)N(y)N(z) \rangle$ in terms of moments using $\langle N(w)N(y)N(z) \rangle = N^3 + N \left(C'(|w-y|) + C'(|w-z|) + C'(|y-z|) \right) + M_3(w, y, z)$, where M_3 is the so-called central third moment, we get

$$\Delta \langle N(w)N(x) \rangle = 2\Big[- \mu \big(N^2 + C'(|w-x|)\big)$$
$$+ f \sum_z \eta(x) \big(N^2 + C'(|w-z|)\big) \rangle h D(|x-z|)$$
$$- \alpha \sum_z \sum_y \eta(x) U(|x-y|) D(|x-z|) \tag{20.15}$$
$$\times h\Big(N^3 + N \big(C'(|w-y|) + C'(|w-z|) + C'(|y-z|)\big)\Big)$$
$$+ M_3(w, y, z)\Big] \Delta t \ .$$

Now we subtract $2N\langle \Delta N \rangle$ [the small-Δt limit of $\langle \Delta N \rangle^2$ by the same argument as given in Equation (20.13)], drop M_3 and $\eta(x)$, divide both sides by $2h$, and let h and Δt go to zero:

$$\frac{1}{2}\frac{dc'(|w-x|)}{dt} = -\mu c'(|w-x|) + f \int D(|x-z|) c'(|w-z|)\, dz$$
$$\tag{20.16}$$
$$- \alpha \iint U(|x-y|) D(|x-z|) \big[\bar{n}(c'(|w-z|) + c'(|w-y|))\big]\, dy\, dz \ .$$

If we recognize the δ functions in c' $[c'(r) = c(r) + \bar{n}\delta(r)]$ and take them out of the integrals as in Appendix 20.A, and rearrange the integrals a little bit, we arrive at the dynamical equation for the covariance given as Equation (20.6).

Appendix 20.C Analyzing the One-species System

If we substitute normalized Bessel functions for the kernels in Equation (20.7), using the fact that the Hankel transform of the normalized Bessel function equals $\lambda^2/(\lambda^2 + q^2)$, we can separate the expression into a sum of (transformed) Bessel kernels by partial fractions; rational functions of transformed kernels (i.e., sums of kernels divided by other sums of kernels) decompose very neatly into sums of kernels by partial fractions. We denote the normalized Bessel kernel, $\lambda^2 K_0(\lambda r)/(2\pi)$, as $\mathcal{K}(\lambda)$.

Separating the kernels by partial fractions and applying the inverse Hankel transform, we get

$$c^*(r) = \bar{n}\big(g_+\mathcal{K}(\lambda_D\phi_+) + g_-\mathcal{K}(\lambda_D\phi_-)\big) \tag{20.17}$$

with

$$g_\pm = \phi_\pm^2 \frac{(\phi_\pm^2 - \varepsilon^2)(\phi_\pm^2 - 1)}{\phi_\mp^2 - \phi_\pm^2} , \tag{20.18}$$

where $\varepsilon = \lambda_U/\lambda_D$. The equilibrium covariance is the sum of two Bessel functions: the values of $\phi_\pm$ indicate how the scales of these functions differ from the original dispersal scale. For example, if $\phi_\pm < 1$, then the spatial scale of the covariance will be longer than the competition and dispersal scales. The values of $\phi_\pm^2$ are the roots of a quadratic equation derived from the numerator of Equation (20.7):

$$\phi_\pm^2 = \frac{1}{2\mu}\bigg(-\big[f - (\varepsilon^2 + 1)(\mu + \alpha\bar{n})\big]$$
$$\pm \sqrt{\big[f - (\varepsilon^2 + 1)(\mu + \alpha\bar{n})\big]^2 - 4\mu\varepsilon^2(\mu - f + 2\alpha\bar{n})} \bigg) . \tag{20.19}$$

Given this value for the quasi-equilibrium covariance, we can also integrate $(D * U)c^*$, getting

$$\bar{c}^* = \frac{\bar{n}\lambda^2\varepsilon^2}{4\pi(\varepsilon^2 - 1)} \sum_{i=+,-} \frac{\phi_i^2(1 - \varepsilon^2)\log(\phi_i^2) + \varepsilon^2(\phi_i^2 - 1)\log(\varepsilon^2)}{(\phi_i^2 - 1)(\phi_i^2 - \varepsilon^2)} , \tag{20.20}$$

which is ugly but straightforward.

Now that we have the quasi-equilibrium value, we can numerically solve the system of Equations (20.5) and (20.20). For more analytic insight into the determinants of the shape and magnitude of the equilibrium covariance, we can restrict ourselves to the case where λ is small and $\bar{n} = (f - \mu)/\alpha + O(\lambda)$. Then, since $\bar{c}$ itself is proportional to λ, we can solve for the (approximate) equilibrium covariance by assuming $\bar{n} = K$ inside Equation (20.19), which simplifies to

$$\phi_\pm^2 = \frac{1}{2}\left(-\varepsilon^2 R \pm \varepsilon\sqrt{\varepsilon^2 R^2 - 4(R - 1)}\right) , \tag{20.21}$$

where $R = f/\mu$ is the maximum net lifetime fecundity of an individual. If $\varepsilon^2 < 4(R - 1)/R^2$, then the quantity inside the square root sign is negative and $\phi_\pm$ and $g_\pm$ become complex, but all imaginary parts cancel out in the calculation of $\bar{c}$ so that Equation (20.20) is still correct.

Appendix 20.D   Analyzing the Two-species System

Analyzing the two-species system follows exactly the same procedures as given in Appendices 20.A and 20.B, although the equations are more complicated. The general expression is as follows:

$$
\begin{aligned}
\frac{dc_{ij}(r)}{dt} = {}& -(\mu_i + \mu_j)c_{ij}(r) + f_i(D_i * c_{ij})(r) + f_j(D_j * c_{ij})(r) \\
& + 2\delta_{ij} f_i \bar{n}_i D_i(r) - \sum_k \big[\alpha_{ik} \bar{n}_k (D_i * c_{ij})(r) \\
& + \alpha_{jk} \bar{n}_k (D_j * c_{ij})(r) + 2\alpha_{ik}\delta_{ij} D_i(r)\bar{n}_i\bar{n}_k \\
& + \alpha_{ik}\bar{n}_i (U_{ik} * c_{jk})(r) + \alpha_{jk}\bar{n}_j (U_{jk} * c_{ik})(r) \big] \\
& - (\alpha_{ij} U_{ij}(r) + \alpha_{ji} U_{ji}(r))\bar{n}_i\bar{n}_j \ .
\end{aligned}
\tag{20.22}
$$

At equilibrium, we can set the left-hand side to zero, apply the Hankel transform, and rearrange to get $\tilde{c}_{ij}^*$:

$$
\tilde{c}_{11}^* = \frac{\bar{n}_1 (m_1 \tilde{D}_1 - \alpha_{11}\bar{n}_1 \tilde{U}_{11} - \alpha_{12}\tilde{U}_{12}\tilde{c}_{12}^*)}{\mu_1 - m_1 \tilde{D}_1 + \alpha_{11}\bar{n}_1 \tilde{U}_{11}}
\tag{20.23}
$$

and

$$
\tilde{c}_{12}^* = \frac{-(\alpha_{12}\tilde{U}_{12} + \alpha_{21}\tilde{U}_{21})\bar{n}_1\bar{n}_2 - \alpha_{21}\bar{n}_2\tilde{U}_{21}\tilde{c}_{11}^* - \alpha_{12}\bar{n}_1\tilde{U}_{12}\tilde{c}_{22}^*}{(\mu_1 + \mu_2) - m_1\tilde{D}_1 - m_2\tilde{D}_2 + \alpha_{11}\bar{n}_1\tilde{U}_{11} + \alpha_{22}\bar{n}_2\tilde{U}_{22}} \ ,
\tag{20.24}
$$

where $m_i = f_i - \alpha_{ii}\bar{n}_i - \alpha_{ij}\bar{n}_j$ with $j \neq i$. The equation for $\tilde{c}_{22}^*$ is obtained from Equation (20.23) by swapping indices 1 and 2.

If we suppose that species 1 invades species 2, so that $\bar{n}_1 \ll 1$ (but $\tilde{c}_{11}^*/\bar{n}_1$ and $\tilde{c}_{12}^*/\bar{n}_1$ are of order unity), then $\tilde{c}_{11}^*/\bar{n}_1$ reduces to

$$
\frac{\tilde{c}_{11}^*}{\bar{n}_1} = \frac{m_1\tilde{D}_1}{\mu_1 - m_1\tilde{D}_1} \ .
\tag{20.25}
$$

In this linearized case, the spatial pattern of the invading species is solely determined by its dispersal properties and not by competition. Similarly, we obtain

$$
\frac{\tilde{c}_{12}^*}{\bar{n}_1} = \frac{-(\alpha_{12}\tilde{U}_{12} + \alpha_{21}\tilde{U}_{21})\bar{n}_2 - \alpha_{21}\bar{n}_2\tilde{U}_{21}(\tilde{c}_{11}^*/\bar{n}_1) - \alpha_{12}\tilde{U}_{12}\tilde{c}_{22}^*}{(\mu_1 + \mu_2) - m_1\tilde{D}_1 - m_2\tilde{D}_2 + \alpha_{22}\bar{n}_2\tilde{U}_{22}} \ .
\tag{20.26}
$$

We *could* separate this expression by partial fractions, but in the general case where kernels are not identical the answer would involve first solving a cubic equation to find the scale parameters and then substituting those solutions into a 4 × 4 linear system to find the coefficients – too complicated to be of much use analytically.

We can gain insight, however, into the special case where species 2 is "global," with infinite dispersal and competition scales, so that the auto-covariance of species 2 as well as its competition and dispersal kernels become negligible. Such a choice corresponds to the extreme case of a plant that is much larger than its competitor, and so senses the environment on a much larger scale. Now Equation (20.26) simplifies considerably:

$$\frac{\tilde{c}_{12}^*}{\bar{n}_1} = \frac{-\alpha_{12}\tilde{U}_{12}\bar{n}_2}{(\mu_1 + \mu_2) - m_1\tilde{D}_1} . \tag{20.27}$$

We could solve Equation (20.27) directly by partial fractions, but we will simplify still further and assume that species 1 disperses far beyond its competitive neighborhood, so that its competition kernel is effectively a δ function, or (after applying the Hankel transform) $\tilde{U}_{12} = 1$. Then we get

$$\frac{\tilde{c}_{12}^*}{\bar{n}_1} = \frac{-\alpha_{12}\bar{n}_2^*}{\mu_1 + \mu_2}\left(1 + \frac{m_1}{\mu_1 + \mu_2 - m_1}\mathcal{K}(\lambda\phi)\right) , \tag{20.28}$$

with

$$\phi = \sqrt{1 - \frac{m_1}{\mu_1 + \mu_2}} . \tag{20.29}$$

Segregation decreases with increasing density-independent mortality (which tends to randomize the spatial pattern); in addition, if $m_1 > 0$ (i.e., $f_1 > \alpha_{12}K_2$), which is necessary for the establishment probability of the invader to be positive, then segregation increases with increasingly localized dispersal.

As a last step, we specialize still further to the case where $0 < m_1 \ll 1$, and the effects of the spatially local competition function [which appear as the constant term in Equation (20.28)] dominate. Then only the spatial segregation term remains, and it is easy to analyze when coexistence occurs. When we translate from the (f, μ, α) parameters to r, K, and R (see Box 20.3), we find that both R_1 and R_2 (sensitivity to competition) increase segregation.

21

Relaxation Projections and the Method of Moments

Ulf Dieckmann and Richard Law

21.1 Introduction

Theory in spatial ecology has to steer a narrow and challenging course between the Scylla of oversimplification and the Charybdis of intractability. Until about 15 years ago, most of theoretical ecology was based on the mean-field paradigm, thus targeting well-mixed ecological systems. Although the underlying assumption of spatial homogeneity is violated for many, if not most, ecological populations and communities in the field, mean-field approaches appeared to be the only way forward. They even took center stage in certain areas, as in epidemiological systems (Bailey 1975; Anderson and May 1991), today recognized as typical examples of ecological processes for which space matters. It was only with the advent and ready availability of modern computer technology that explorations into critical effects of spatial heterogeneities became feasible (Levin 1974, 1976; Weiner and Conte 1981; Weiner 1982; Pacala and Silander 1985; Holsinger and Roughgarden 1985; Pacala 1986; Hogeweg 1988). Today, computer screens and journals abound with images of spatially extended simulations that have convincingly demonstrated that many predictions of classical ecological theory are inappropriate in the presence of spatially structured habitats or short-range ecological interactions.

Despite their value as counterexamples to mean-field predictions and their usefulness in exploring the emergence of macroscopic effects resulting from microscopic ecological mechanisms, simulation studies often remain inconclusive. Are the reported phenomena robust under changed ecological parameters? Where, among the noisy dynamics of individual-based and stochastic models, is the ecological signal? How many (usually time-consuming) spatial simulations have to be run before reliable conclusions can be drawn? These questions remind us that only part of our ecological understanding is based on description: on top of this, we look for mechanistic explanations and for reliable generalizations from observations. Specifically, we are interested in qualitative rules, and for this reason we would like heterogeneous ecological processes to be amenable to tools that allow robust conclusions to be drawn. For many systems, such sound qualitative insight can only be derived from careful quantitative analyses (well-known examples from classical ecological theory are many inequality conditions and results of bifurcation analyses).

Is it realistic to hope for a middle ground between oversimplified mean-field models and intractable computer simulations? The answer depends on how many *essential* degrees of freedom there are in spatial ecological systems. A degree of freedom here is a quantitative piece of information needed to specify the current state of and the expected change in a given system. How many degrees of freedom are considered essential often depends on the purpose of an investigation. Think of the trajectory traced by a stone thrown into the air. A detailed description of the flying stone would account for the state of all its atoms. For practical predictions of the stone's expected path, however, most of these details are utterly irrelevant: only the position and velocity of its center of mass are essential, and even the stone's orientation and rotational speed can be neglected. Another illustration is provided by milk being poured into a cup of coffee. To specify the initially intricate pattern of milk and coffee mixing, very many variables are needed. After a short while, however, the milk concentration becomes uniform and can be specified by a single number. Likewise, in ecological systems we can often ignore most of the physiological or biochemical details of individuals, provided that our interest rests at the population level. Lewontin has introduced the term *dynamic sufficiency* for distinguishing between essential and dispensable degrees of freedom: a subset of variables is called dynamically sufficient if sufficiently accurate predictions of future dynamics can be based on these variables alone (Lewontin 1974).

Sometimes the dynamically sufficient number of variables is small from the outset (the thrown stone), and sometimes it quickly decreases as a consequence of internal processes (the milk drop). The rapid "destruction" of degrees of freedom is also typical for many spatially extended systems: often, a small set of variables is dynamically sufficient for adequately capturing the system's state. In a grassland, for instance, observing a few spatial statistics in a few square meters may provide most of the information required to characterize the whole area. We do not need to know the precise position of every single shoot in this area to predict the system's expected

Box 21.1 Relaxation projections in spatial ecology

Spatio-temporal processes usually possess some degrees of freedom operating on a fast time scale: any initial pattern p, taken from the set P of all possible patterns, will quickly converge to a more limited set P_r of patterns. A smaller number of variables – called essential degrees of freedom – is required to characterize patterns within P_r than within P.

Spatial statistics are variables for describing the state of a spatio-temporal process at any point in time. If S denotes a set of spatial statistics sufficient to characterize patterns in P, and S_r contains those needed for P_r, the statistics in $S' = S \setminus S_r$ are only needed to distinguish between patterns in $P' = P \setminus P_r$ and hence are no longer essential after the system's fast degrees of freedom have decayed.

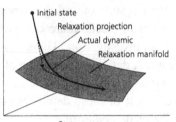

State space

Illustration of the rapid decay of initial conditions toward a relaxation manifold in a three-dimensional dynamic. The relaxation projection defines a relaxation manifold and a projection of initial conditions onto this manifold; both are supposed to approximate the actual dynamics' first phase of rapid decay.

As S' thus retains no essential information about the state of the process after relaxation, all spatial statistics of resulting patterns can be expressed approximately as functions of those in S_r, $S' = f(S_r)$. Hence, the functions f simplify the complexity of describing the *state* of a spatial process. In addition, the functions f simplify the *dynamics* of the process: in the space S of statistics, the functions f define

■ a projection of arbitrary initial conditions $s = (s_r, s')$ in S onto a relaxation manifold, $(s_r, s') \to (s_r, f(s_r))$, and

■ the shape of the relaxation manifold itself, which is the subspace of all states s in S invariant under that projection, thus satisfying $(s_r, s') = (s_r, f(s_r))$.

The projection mimics the fast relaxation of the spatio-temporal process. The relaxation manifold can be used for constructing a simplified dynamic on that manifold: starting from a state s_r on the relaxation manifold, the missing statistics s' can be reconstructed as $s' = f(s_r)$. Now the change $\frac{d}{dt}s$ of the state $s = (s_r, s')$ follows from the dynamics of the spatio-temporal process and implies a change $\frac{d}{dt}s_r$ along the relaxation manifold.

continued

Box 21.1 *continued*

We call projections of this kind *relaxation projections*. In general, projections are non-invertible mappings, and objects projected consequently carry a diminished amount of information. In particular, relaxation projections remove the dynamically nonessential information from a spatial pattern. The art of constructing relaxation projections amounts to finding suitable small sets of statistics S and simple functions f so that the actual relaxation process as well as the resulting relaxation manifold are well approximated and the relaxation projection itself is sufficiently simple to allow for the derivation of tractable dynamics.

---

development at the population level: specifying a much smaller number of essential degrees of freedom will suffice. But only in systems that eventually become completely homogeneous, like the milk in the cup of coffee, does the number of essential degrees of freedom become minimal. In other cases, a certain amount of heterogeneity is preserved in the system since certain mechanisms counteract full mixing. In ecological systems, these mechanisms arise from the local interactions between individuals, their restricted dispersal, and their dependence on a potentially heterogeneous habitat; we therefore cannot simply expect all variables apart from mean densities to become superfluous. Instead, a balance between forces of mixing and ordering (Watt 1947) will lead to states with more essential degrees of freedom than in mean-field approximations, but still with only a fraction of the vast number of degrees of freedom possible.

This general expectation is corroborated by observations and explicit simulations of spatial ecological dynamics that start from arbitrarily complicated initial configurations and whose internal dynamics, after a while, reliably sustain only a small class of patterns. Imagine gardeners setting up a plot with ornate patterns of plants, like those in the baroque gardens at Versailles. Without continuous examination and lots of effort on the part of the gardeners, that intricate distribution will soon relax into a more natural configuration, characterized by considerably fewer essential degrees of freedom than were initially present. Interactions between plants tend to operate at a local scale; therefore, in the absence of external heterogeneities, any long-range correlations between plants present in the gardeners' original setup cannot persist – leaving the Versailles gardens unattended for a century would eradicate most of the artful patterns originally devised. In our quest for a tractable but still dynamically sufficient description of ecological systems in space and time, we can capitalize on this tendency

Box 21.2 The Dirac delta function in spatial ecology

Dirac's delta function is a particularly useful tool for characterizing the distribution of individuals across continuous spaces. Imagine a plot of plants, each occupying a specific location. One way of describing their spatial distribution relies on quadrat counts: after imposing a square grid on the ecological space, the number of plant individuals within each square is counted. The resulting information can be represented as a histogram, where the volume of each column is proportional to the number of individuals present in the underlying square (Figure a).

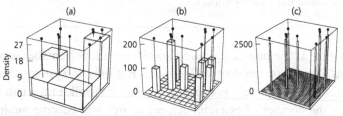

Histograms of increasing resolution, constructed for the same spatial pattern of individuals distributed over the unit square (depicted as points on the top plane). The high-resolution limit yields a Dirac delta function at the location of each individual.

The information captured by quadrat counts, however, is incomplete. *Within* each square of the grid, no information is available about the distribution of individuals. The situation can be improved by choosing a grid of higher resolution in which each square is of a smaller size. For a sufficiently fine grid, most of the squares will contain either zero or one individual, with squares containing two or more individuals being the exception. If the volume of histogram columns is still to correspond to the number of individuals, the height of columns has to be increased as their base squares are shrunk (Figure b). To assess the number of individuals within a certain area, we can then just add the volume of all histogram columns over that area.

Plants and animals, of course, possess a spatial extension as well as a spatial location. It is often convenient to distinguish these two aspects of spatial structure by thinking of individuals as being centered at some location, while including information about their extension, if necessary, in the state of individuals. Although on finer grids accuracy is improved, information regarding the precise location of individuals within squares is still discarded.

For theoretical considerations, then, it is natural to let the size of squares shrink to zero, so that full information about the location of individuals is retained. Keeping our interpretation of the volumes of histogram columns requires letting their height go to infinity. From this limit we obtain a Dirac

continued

Box 21.2 *continued*

delta function at the location of each individual. These functions are zero everywhere, except at one specific location. There, their height is infinite, such that their integral is 1 (for illustration of this point, a very fine grid is shown in Figure c). By integrating over all delta functions within a given area, we thus obtain the abundance of individuals occupying that area.

In summary, a spatial pattern of individuals can be described by a sum of delta functions, each peaked at the location of a single individual, thus describing its contribution to the population's density function.

---

of natural systems to "destroy" degrees of freedom by relaxation. Hence, for models of ecological heterogeneity, we have to cast the set of all possible configurations onto that much smaller subset toward which the system will rapidly relax; we call this general procedure a *relaxation projection* (see Box 21.1). Relaxation projections help establish the middle ground between oversimplification and intractability. By culling the dynamically irrelevant degrees of freedom initially present in many spatially heterogeneous systems, we reduce the number of variables to manageable proportions. Obviously, this culling must not go too far: retaining only mean densities would take us back to mean-field descriptions.

These considerations lead to the following question: if we are to capture the essential aspects of spatial heterogeneity in a dynamically sufficient way, which variables should we use to complement mean densities? To address that central issue in detail, this chapter uses a widely applicable class of stochastic models for individual-based and spatially extended ecological systems, described in Section 21.2. In Section 21.3, we introduce and analyze a flexible set of spatial statistics, called correlation densities or spatial moments, which are candidates for providing the extra spatial information needed on top of mean densities. Section 21.4 explains how relaxation projections can be applied to the dynamics of spatial moments. We examine alternative projection schemes and show that some of these provide powerful descriptions of spatially heterogeneous ecological change. Section 21.5 considers extensions of the framework and examines consequences of the stochastic effects that inevitably arise when individuals interact with finite numbers of neighbors. We conclude that the novel class of ecological models developed in Sections 21.3 and 21.4 allows for robust and generalizable insights into the inner workings of spatially structured ecological populations and communities.

21.2   Individual-based Dynamics in Continuous Space

In this section we introduce a class of spatially explicit, individual-based, stochastic birth–death–movement processes that, on the one hand, can be calibrated to reflect ecological dynamics as observed in the field and, on the other hand, are suitable targets for applying the technique of relaxation projections (Sections 21.3 and 21.4). Such individual-based stochastic processes in continuous space also underlie the analyses in Chapter 20, and it is helpful to make them explicit.

Patterns of individuals in continuous space

A natural starting point for specifying ecological details of a spatially explicit model is the demographic events experienced by individuals. As a step toward the real world, we allow birth and death rates of individuals to depend on their local environment. In the class of models below, individuals can disperse (at birth) and relocate (during their lifetime) within a given habitat. We think it is preferable to envisage individuals at locations in continuous space rather than at discrete sites on a grid. Avoiding the discretization of space into regular cells – which, for most ecological systems, is artificial – offers a more faithful and direct correspondence between model parameters and model dynamics and those quantities and processes that can be measured in the field. This modeling framework has the additional advantage of adequately reflecting the gradual effects of increased physical distance on the interaction strength between individuals, and it allows for continuous changes of dispersal and relocation probabilities with distance traveled.

Individuals can belong to different species i, with $i = 1, \ldots, n$, where n is the number of species in the community. All individuals inhabit a space of locations x; the habitat can have one, two, or three spatial dimensions. Most applications in plant ecology, of course, focus on planar habitats. The spatial extension of the habitat is measured by A, denoting the habitat's length, area, or volume, respectively. Individuals l in species i are situated at spatial locations x_{il}, with $l = 1, \ldots, AN_i$, where N_i is the mean population density of species i. The distribution of individuals in each species i is described by *spatial density functions* $p_i(x) = \sum_l \delta_{x_{il}}(x)$, where $\delta_{x_{il}}$ denotes Dirac's delta function peaked at location x_{il} (for an explanation of this representation and the motivation behind it, see Boxes 21.2 and 21.3). The density function $p_i(x)$ is thus peaked at all locations occupied by individuals of species i and is zero elsewhere. At any moment in time, the spatial pattern within the community is given by collecting the density functions of all species into a vector $p(x) = (p_1(x), \ldots, p_n(x))$.

Box 21.3 Properties of the Dirac delta function

The delta function was introduced by the English physicist Paul A.M. Dirac (Dirac 1926, 1958), and relates to previous work by G. Kirchhoff and O. Heaviside (Jammer 1966). Dirac's idea was to construct a strictly localized function on the real numbers: $\delta(x)$ is zero for any x, except for $x = 0$, where it is peaked.

To make this notion precise, the delta function is defined by

$$\int_{-\infty}^{+\infty} f(x')\delta(x' - x)\, dx' = f(x)$$

for any continuous function f. This is called the *sifting property of the delta function*: since the term $\delta(x' - x)$ is 0 except at $x' = x$ (where the delta function's argument vanishes), only at that point can the value of f contribute to the integral. This can also be expressed by $f(x')\delta(x' - x) = f(x)\delta(x' - x)$.

From this definition we can derive other useful properties of the delta function:

- The delta function is symmetric: $\delta(-x) = \delta(x)$.
- Its integral equals 1: $\int_{-\infty}^{+\infty} \delta(x')\, dx' = 1$.
- One or more zeros in the delta function's argument contribute according to their inverse slope: $\delta(g(x)) = \sum_i \delta(x - x_i)/|g'(x_i)|$, summing over all zeros x_i of g.
- Its primitive function is the unit-step or Heaviside function: $\int_{-\infty}^{x} \delta(x')\, dx' = \theta(x)$, where $\theta(x)$ equals 0 for negative and 1 for positive x.
- The Fourier and Laplace transforms of the delta function are equal to 1.

The Dirac delta function can also be envisaged as the limit of a series of functions. Setting $h(x)$ to $1/\varepsilon$ for $|x| < \varepsilon/2$ and to 0 elsewhere, we can write $\delta(x) = \lim_{\varepsilon \to 0} h(x)$. Compare this with the figures in Box 21.2: while the width of such a function shrinks to zero, its height goes to infinity. Setting h to a normal distribution with mean 0 and standard deviation ε has the same effect. All properties of the delta function can alternatively be derived from such limit representations (theory of distributions; Schwartz 1950).

In more than one dimension, the delta function is defined as the product of one-dimensional delta functions: $\delta(x) = \delta(x_1)\delta(x_2)\ldots\delta(x_n)$ for $x = (x_1, x_2, \ldots, x_n)$. For convenience, the location of a delta function's peak is often given as a subscript: $\delta(x' - x) = \delta_x(x')$.

Birth, death, and movement events

The per capita death rates of individuals can depend on the presence or absence of other individuals in their local environment. We denote by $w_{ij}^{(d)}(x - x')$ the strength with which an individual of species j at location x' affects the mortality of an individual of species i at location x. The functions $w_{ij}^{(d)}$ are called *interaction kernels* for density-dependent death and are scaled so that they integrate to 1, $\int w_{ij}^{(d)}(\xi)\, d\xi = 1$. (In this chapter we denote absolute spatial locations by xs and relative locations by ξs.) The reason for calling these functions kernels becomes evident when we realize that, in the death rate of an individual of species i at location x in a community with spatial pattern p,

$$D_i(x, p) = d_i + \sum_j d'_{ij} \int w_{ij}^{(d)}(x' - x)\big[p_j(x') - \delta_{ij}\delta_x(x')\big]\, dx' \, , \quad (21.1a)$$

$w_{ij}^{(d)}$ occurs as the kernel of a convolution integral (see also Boxes 20.1 and 20.2). Here, d_i denotes the intrinsic per capita death rate of species i, and d'_{ij} weighs the density-dependent effect of species j on the mortality of individuals in species i. The integral in Equation (21.1a) collects the contributions from all locations x' according to their interaction strength $w_{ij}^{(d)}(x' - x)$ and local density $p_j(x')$. Obviously, individuals do not compete with themselves; therefore the contribution of the focal individual at location x, given by $\delta_x(x')$, is removed from the integration when the summation over j reaches the focal species i (the Kronecker symbol δ_{ij} equals 1 for $i = j$ and is 0 otherwise).

The rate of movement of an individual in species i at location x to another location x' is given by

$$M_i(x, x', p) = m_i(x' - x) \, . \quad\quad\quad (21.1b)$$

Equation (21.1b) implies that rates of movement events are homogeneous in space; they only depend on the distance moved $x' - x$, and not on absolute location x or on spatial pattern p. Nevertheless, the *movement kernel* m_i can differ from species to species. It is convenient to keep the movement kernel unnormalized: its integral $|m_i| = \int m_i(\xi)\, d\xi$ measures the expected total per capita rate of movement.

Like death events, we allow the per capita reproduction rate for each individual to depend on its local environment. Unlike death events, however, an offspring takes, by means of dispersal, a spatial location different from that of its parent. The rate of reproduction of an individual in species i at location x, giving rise to a new individual at location x', is given by

$$B_i\left(x, x', p\right) = \left[b_i + \sum_j b'_{ij} \int w_{ij}^{(b)}(x'' - x)\right.$$
$$\left. \times \left[p_j(x'') - \delta_{ij}\delta_x(x'')\right]dx''\right]m_i^{(b)}(x' - x) . \tag{21.1c}$$

Here b_i and b'_{ij} denote the density-independent and density-dependent components of the per capita birth rate, respectively; $w_{ij}^{(b)}$ is the interaction kernel for density-dependent birth and $m_i^{(b)}$ is the *dispersal kernel*. Like the kernels of interaction, the dispersal kernel is normalized to 1. In Equation (21.1c), competitive interactions occur for $b'_{ij} < 0$; these will be the most natural choice for many ecological systems. Yet, mixtures of competition with neutral ($b'_{ij} = 0$) or mutualistic ($b'_{ij} > 0$) interactions are also possible and can be readily incorporated. Equations (21.1a) and (21.1c) describe the spatial analogues of Lotka–Volterra competition and provide a natural starting point for exploring interaction effects in spatially heterogeneous systems. Alternative assumptions, allowing for nonlinear dependencies of D_i and B_i on p, are discussed in Section 21.5.

Pattern dynamics

Starting from an initial spatial configuration, we can now investigate how a pattern changes with the ecological events described above. According to Equations (21.1), the dynamics of patterns are stochastic (any two runs from the same starting patterns are expected to result in different patterns) and Markovian (rates of change at any moment in time depend only on the current pattern). Such processes are characterized by so-called *master equations* (van Kampen 1992). Denoting by $P(p)$ the probability density for observing pattern p, the rate of change in this probability density is given by

$$\frac{d}{dt}P(p) = \int \left[w(p|p')P(p') - w(p'|p)P(p)\right]dp' , \tag{21.2a}$$

where $w(p'|p)$ is the probability density per unit time that any event will turn a pattern p into another pattern p'. Hence the first term on the right-hand side corresponds to an increase in the probability density of pattern p due to events that lead to p and originate from a different pattern p', while the second term captures the decrease of the probability density at p resulting from events that change pattern p into a different pattern p'.

Box 21.4 The generalized delta function

The generalized delta function Δ extends the functionality of the Kronecker symbol and the Dirac delta function from the realm of integers and real numbers, respectively, to that of functions.

Markovian jump processes on integers, reals, or functions are described by master equations (for an introduction, see, e.g., van Kampen 1992). This class of equations is based on transition rates and characterizes the flow of probability into and out of states of a stochastic process, as in Equation (21.2a). Transitions between states often result from different types of event, each of which changes the state of the process in a particular manner. Switch functions are used to link descriptions of events and their rates to transition rates of the process; these functions are "on" only for a particular transition. An important role for all three delta functions is to serve as such switches.

State variable	Master equation / transition rates / switch function
Integer i	$\frac{d}{dt} P_i = \sum_{i'} [w_{ii'} P_{i'} - w_{i'i} P_i]$
	$w_{i'i} = \sum_{\Delta i} E_{\Delta i, i}\, \delta_{i+\Delta i, i'}$
	$\sum_{i'} F_{i'}\, \delta_{i'i} = F_i$ (Kronecker symbol)
Real number r	$\frac{d}{dt} P(r) = \int [w(r\mid r') P(r') - w(r'\mid r) P(r)]\, dr'$
	$w(r'\mid r) = \sum_{\Delta r} E(\Delta r, r)\, \delta(r + \Delta r - r')$
	$\int F(r')\, \delta(r' - r)\, dr' = F(r)$ (Dirac delta function)
Function f	$\frac{d}{dt} P(f) = \int [w(f\mid f') P(f') - w(f'\mid f) P(f)]\, df'$
	$w(f'\mid f) = \sum_{\Delta f} E(\Delta f, f)\, \Delta(f + \Delta f - f')$
	$\int F(f')\, \Delta(f' - f)\, df' = F(f)$ (Generalized delta function)

While the sums in the equations for transition rates w extend over all possible jumps Δi, Δr, or Δf, only permitted jumps with non-vanishing event rates E contribute to the transition rates w. Substituting the transition rates into the corresponding master equation, the switch functions select only those states i', r', or f' that can be reached by permitted jumps.

We can envisage the generalized delta function Δ as an infinitely narrow and infinitely high peak in the space of functions f. Such a heuristic notion, together with the functional integration occurring in the master equation for functions, is made exact by defining the generalized delta function so as to collapse the functional integration in $\int F(f')\, \Delta(f' - f)\, df'$ by yielding the functional F's value at the location of Δ's peak, $F(f)$. This definition is analogous to that of Dirac's

continued

Box 21.4 *continued*

delta function (Box 21.3). After the functional transition rates are thus combined with the functional master equation, neither a functional integration nor a generalized delta function remain in the end result. Or, in the words of Dirac (1958): "The use of delta functions thus does not involve any lack of rigour in the theory, but is merely a convenient notation, enabling us to express in a concise form certain relations which we could, if necessary, rewrite in a form not involving delta functions, but only in a cumbersome way which would tend to obscure the argument."

The functions $w(p'|p)$ are also known as *transition rates* and for our model are simply given by summing and integrating over all possible birth, death, and movement events that can turn a pattern p into a pattern p',

$$
\begin{aligned}
w(p'|p) = \sum_i \iint & B_i(x, x', p) p_i(x) \Delta(p + u_i \delta_{x'} - p') \, dx dx' \\
+ \sum_i \int & D_i(x, p) p_i(x) \Delta(p - u_i \delta_x - p') \, dx \qquad \text{(21.2b)} \\
+ \sum_i \iint & M_i(x, x', p) p_i(x) \Delta(p - u_i \delta_x + u_i \delta_{x'} - p') \, dx dx' .
\end{aligned}
$$

The unit vector u_i for species i has 1 as its ith element and zeros elsewhere. We refer to Δ as the generalized delta function (Dieckmann 1994; Dieckmann *et al.* 1997, in press), see Box 21.4. Like Dirac's delta function δ (Boxes 21.2 and 21.3), it is peaked where its argument is zero, and is zero elsewhere. Unlike Dirac's delta function, however, generalized delta functions take functions as arguments. All three sums on the right-hand side of Equation (21.2b) comprise terms of the form $\Delta(p_{\text{event}} - p')$. They therefore only contribute to the transition rate $w(p'|p)$ when $p_{\text{event}} = p'$, in other words, only if the considered event can turn the current pattern p into the new pattern p'. For example, the expression $\Delta(p + u_i \delta_{x'} - p')$ is "switched on" when $p + u_i \delta_{x'} = p'$ (that is, only if a birth event in species i, producing a new individual at location x', changes pattern p into pattern p').

Models of the kind described in this section are flexible tools for studying the dynamics of spatially extended ecological systems. Their parameterization gets closer than mean-field models to individual-based processes in the field and should permit incorporation of empirical measurements. Moreover, their intrinsic mixture of randomness and determinism

(a)

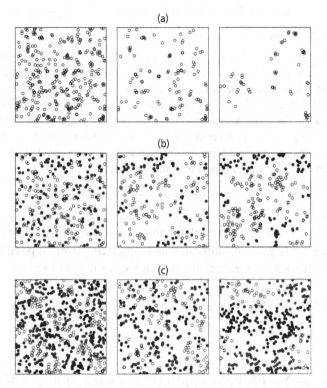

(b)

(c)

Figure 21.1 Examples of individual-based birth–death–movement processes. (a) A single species with dispersal and local logistic competition. Owing to spatial aggregation, the asymptotic mean density is significantly below the mean-field prediction. (b) Two dispersing species with competitive interactions. Under mean-field conditions, this pair of species cannot coexist. (c) Three dispersing species with competitive interactions. Notice how quickly any regularity of the initial pattern is destroyed. For all three systems, stochastic realizations are shown at three different moments in time: initial pattern (left column), intermediate pattern (center column), and a pattern from the asymptotically stable distribution (right column). Parameters for all three systems are given in Table 21.1.

corresponds well to processes in natural populations. Models of this kind are also easily implemented as computer simulations (see Figures 21.1, 14.1, and 14.5). Thus individual realizations of the stochastic processes defined above are readily available; their properties can be investigated in detail without introducing any further simplifying assumptions. At the same time, these complicated interactive dynamics are ideal candidates for demonstrating how to simplify spatial complexity. As we will see in the following two sections, considerable insight can be gained by subjecting these dynamics to relaxation projections and to the resulting method of moments.

Table 21.1 Parameter values for the three systems in Figure 21.1.

System	Times[a]	Parameters[b]
Figure 21.1a	0, 10, 30	$N_1(0) = 200$ $b_1 = 0.4, d_1 = 0.2, d'_{11} = 0.001$ s.d. $w_{11}^{(d)} = 0.04$, s.d. $m_1^{(b)} = 0.02$
Figure 21.1b	0, 5, 15	$N_1(0) = N_2(0) = 150$ $b_1 = b_2 = 0.6, d_1 = d_2 = 0.2$ $d'_{11} = d'_{22} = 0.001, d'_{12} = 0.004, d'_{21} = 0.005$ s.d. $w_{11}^{(d)} =$ s.d. $w_{12}^{(d)} =$ s.d. $m_1^{(b)} =$ s.d. $m_2^{(b)} = 0.06$ s.d. $w_{21}^{(d)} =$ s.d. $w_{22}^{(d)} = 0.05$
Figure 21.1c	0, 0.2, 2	$N_1(0) = N_2(0) = N_3(0) = 200$ $b_1 = b_2 = b_3 = 0.8, d_1 = d_2 = d_3 = 0.2$ $d'_{11} = d'_{22} = d'_{33} = 0.001, d'_{12} = 0.003, d'_{13} = 0.004$ $d'_{21} = 0.005, d'_{23} = 0.003, d'_{31} = 0.002, d'_{32} = 0.006$ s.d. $w_{11}^{(d)} =$ s.d. $w_{12}^{(d)} =$ s.d. $w_{13}^{(d)} =$ s.d. $w_{21}^{(d)} =$ s.d. $w_{22}^{(d)} =$ s.d. $w_{23}^{(d)} =$ s.d. $w_{31}^{(d)} =$ s.d. $w_{32}^{(d)} =$ s.d. $w_{33}^{(d)} = 0.06$ s.d. $m_1^{(b)} = 0.06$, s.d. $m_2^{(b)} = 0.07$, s.d. $m_3^{(b)} = 0.05$

[a] Times at which the three snapshots are taken. Each snapshot depicts the unit square with periodic boundary conditions.
[b] Interaction and dispersal kernels are Gaussian with standard deviations (s.d.) as indicated. Parameters not mentioned are zero.

21.3 Dynamics of Correlation Densities

With the general class of birth–death–movement models in place (Section 21.2), we consider how to simplify the spatial complexity of these models by applying a relaxation projection. As a first step in this process, we have to decide which are the essential degrees of freedom in the spatial patterns under consideration.

Scales of spatial heterogeneity

Spatial heterogeneity can occur at various scales, and therefore a basic distinction is helpful here:

- *either* patterns are small scale relative to the extension of the habitat, in which case the whole habitat contains many varied instances of a unit pattern,
- *or* we are dealing with large-scale patterns, for which the habitat is not large enough to comprise many replicates of any unit pattern.

The term "unit pattern" here is not meant in the sense of a spatial tiling, but rather refers to the totality of spatial configurations at the scale above which the pattern becomes repetitive and conveys no further information on essential degrees of freedom (for a way of estimating this scale, see Chapter 12). Think, for example, of an ecological habitat that is divided into two disjunct spatial domains, each of which is occupied by a single species. Here, the pattern of spatial segregation spans the entire habitat and thus occurs at a large scale relative to this habitat. In contrast, imagine that local clumps are formed by individuals of one species, and that these are gradually invaded by individuals of a second species, while at the same time new clumps of the former are re-established elsewhere. In this case, a not-too-small habitat will comprise many clumps and we therefore refer to the pattern as being small scale (with the various shapes of clumps playing the role of "unit patterns"). The dichotomy is important because essential degrees of freedom for large-scale patterns vary from pattern to pattern. This means that for each type of large-scale pattern, it would be necessary to evaluate which essential degrees of freedom best capture the system's state. For small-scale patterns, however, matters are simpler: here, average information on the local environments of individuals often can capture the essentials of the spatial pattern. Although relaxation projections can also be devised for large-scale patterns (Ellner *et al.* 1998), in this chapter we focus attention on techniques suitable for understanding the implications of small-scale spatial heterogeneity.

Correlation densities

Given that interactions between individuals are local, the fate of each individual is determined by its local environment. To understand the ensuing dynamics, we must take a "plant's-eye view" (Turkington and Harper 1979) of spatial heterogeneity. If patterns are sufficiently small scale, individuals within each species experience similar "views," and essential degrees of freedom are thus given by descriptions of their average local environment. For example, Mahdi and Law (1987) measured average local environments in a community of grassland species by determining the expected densities of several other species at various radial distances around individuals of a focal species (see Figure 21.2a). We can do the same for the spatial patterns simulated in Figure 21.1b; results are shown in Figure 21.2b.

The quantities illustrated in Figure 21.2 are special cases of a general class of spatial statistics. These statistics are based on the densities at which certain spatial configurations appear in a given spatial pattern.

■ The simplest spatial configuration is a singlet; the density of single individuals in a species across a given habitat simply corresponds to the mean density of the considered species.

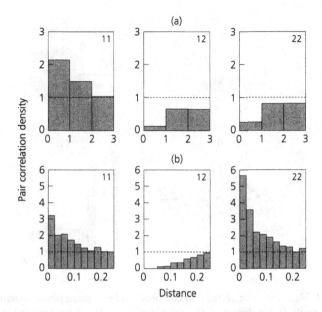

Figure 21.2 Pair correlation densities for field data and simulated pattern. Pair densities are normalized so that mean-field expectations are 1 (shown as dashed lines). (a) Measurements by Mahdi and Law (1987). Species 1 is *Carex caryophyllea*; species 2 is *Carex flacca*; distances are measured in cm. At short distances, small cross-correlations indicate heterospecific segregation. (b) Pair densities for the spatial pattern in Figure 21.1b at time $t = 15$. As can be verified by examining the original pattern, large auto-correlations and small cross-correlations suggest the existence of conspecific aggregations at spatial scales of 0.1 to 0.2.

- Now consider the spatial density of pairs comprising an individual of species i and one of species j (where j can be equal to i) that are a vectorial distance ξ apart (see Figures 21.3a and 21.3b). Pair densities thus give information about the average local environments of individuals in each species. Local environments often are isotropic, that is, there are no special directions in space along which the pattern is aligned. In such cases (as in Figure 21.2), it is convenient to construct angular averages of pair densities, or, equivalently, to determine the spatial density of pairs that are a given length $r = |\xi|$ apart (see Figures 21.3c and 21.3d).
- Apart from pairs, we can also examine spatial configurations comprising three individuals; for instance, individuals of species j and k that are at distances ξ and ξ', respectively, from an individual of species i. In an isotropic setting, these triangular configurations are characterized by three radii, r, r', and $r'' = |\xi' - \xi|$, corresponding to the three edges of the spanned triangle.

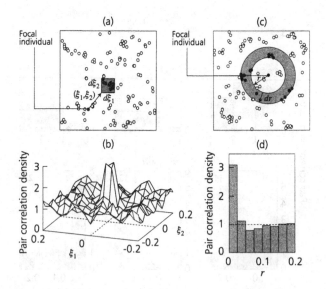

Figure 21.3 Measuring pair correlation densities. (a) For a given planar pattern of area A and mean density N, the density of pairs at distance $\xi = (\xi_1, \xi_2)$ is assessed by counting, for a focal individual at location x, the number of paired individuals within the square $(x_1 + \xi_1, x_1 + \xi_1 + d\xi_1) \times (x_2 + \xi_2, x_2 + \xi_2 + d\xi_2)$, using a sufficiently small spatial resolution $d\xi$. Repeating this procedure with each individual in turn being the focal one and dividing the total count by A and by $d\xi_1 d\xi_2$ yields the pair density at distance ξ. Figure (b) illustrates how the resulting pair densities depend on ξ. (c) For isotropic patterns, it is useful to average pair densities over all angles of ξ. Equivalently, paired individuals are counted in rings of radius $r = |\xi|$ and width dr, centered on a focal individual. Total counts over all focal individuals are divided by A and by $2\pi r\, dr$. Figure (d) shows the resulting radial pair density. Pair densities in (b) and (d) are normalized by division through N^2 so that mean-field expectations are 1.

For any given pattern p we can thus define correlation densities of order m by

$$C_{i_1 \ldots i_m}(\xi_1, \ldots, \xi_{m-1}, p) = \frac{1}{A} \int p_{i_1}(x) \prod_{j=2}^{m} p_{i_j}(x + \xi_{j-1})\, dx , \qquad (21.3a)$$

where i_1 to i_m specify the species of the m individuals constituting the spatial configuration, and ξ_1 to ξ_{m-1} denote the distances of individuals 2 to m, respectively, from individual 1. The integration extends over all locations x in the area A, which is assumed to be large enough that effects resulting from its boundary are negligible. As described above, special cases of correlation densities are given by mean densities (also referred to as global or singlet densities),

$$C_i(p) = \frac{1}{A} \int p_i(x)\, dx , \qquad (21.3b)$$

by pair densities (also referred to as local, environs, or doublet densities),

$$C_{ij}(\xi, p) = \frac{1}{A} \int p_i(x) p_j(x + \xi) \, dx \, , \qquad (21.3c)$$

and by triplet densities,

$$C_{ijk}(\xi, \xi', p) = \frac{1}{A} \int p_i(x) p_j(x + \xi) p_k(x + \xi') \, dx \, . \qquad (21.3d)$$

All correlation densities are given as integrals over products of spatial density functions and are therefore also known as spatial moments. Pair densities (second moments) C_{ij} are also called auto-correlations for $i = j$ and cross-correlations for $i \neq j$. For mnemonic convenience, we denote mean densities (first moments) by $N_i = C_i$ and triplet densities (third moments) by $T_{ijk} = C_{ijk}$.

The spatial pattern p of an ecological community comprising m individuals is completely characterized (apart from its absolute location and orientation) by a correlation density of order m. This is because for two individuals the pair correlation density has a single peak at a location given by the vectorial distance between the two individuals. For three individuals, the triplet correlation function would have such a single peak at a location determined by the distances between individuals in the triangular configuration. Analogously, for m individuals, information about all relative locations of individuals is available from the mth-order correlation density. Although it is difficult to measure correlation densities of high order, it is evident that information about a spatial pattern can be represented either in location-based form (by specifying a pattern p) or in correlation-based form (by specifying a correlation density C of sufficient order). These two representations contain exactly the same information about the relative position of all individuals. We can regard this equivalence as a coordinate transformation, expressing the same information in two different ways.

What, then, is the advantage of describing the dynamics of spatial patterns in correlation-based form? There are several reasons for doing so; here we give a preview of features that will be developed in this chapter.

■ Quantitative descriptions of spatial patterns ought to be based on suitable summary statistics; correlation densities define such statistics. In particular, this class of statistics includes both mean densities and pair densities, two statistics that are natural for describing small-scale patterns and are readily applied to field data (Ripley 1981; Diggle 1983). Therefore, the expected dynamics of low-order correlation densities provide valuable insight into the main characteristics of a developing spatial process.

- Because first-order correlations correspond to mean densities, mean-field models can be envisaged as a subset of correlation-based models. Correlation densities thus offer a systematic way of gradually extending and refining the set of summary statistics by successive integration of higher correlation orders. Such a task would be more difficult if based on sets of disparate types of spatial statistics.
- High-order correlations often relax much faster than low-order correlations (see Figure 21.4); consequently, essential degrees of freedom are captured by the low-order end of the correlation spectrum. It is for this reason that, for many spatio-temporal processes, relaxation manifolds (see Box 21.1) take a simple and tractable form in the space of correlation densities, whereas these manifolds tend to have a complicated topology in the space of spatial patterns or when expressed in terms of other spatial statistics. This feature of correlation densities greatly facilitates the application of relaxation projections.
- Pair densities naturally arise when assessing population-level consequences of pairwise interactions under spatial heterogeneity. In fact, the integrals over pair densities in Equations (21.1a) and (21.1c) extend the principle of "mass action" to spatially heterogeneous settings: formally speaking, individuals respond to linear functionals of spatial distributions.
- Whereas essential degrees of freedom in large-scale patterns can vary widely from process to process, correlation densities offer a universal representation of these essential degrees of freedom for small-scale patterns. A general theory of small-scale heterogeneity in ecological processes can therefore be built on the basis of correlation densities.

Corrected correlation densities

A simple modification of the correlation densities defined above is often helpful. Whenever a spatial configuration of order $m > 1$ is considered, the same individual may figure twice in the description of the configuration. Whereas formally it is correct to count pairs that individuals form with themselves, from an ecological point of view such pairs are meaningless and misleading. For this reason, it is useful to construct corrected correlation densities, from which so-called self-pairs and similar repetitive configurations are removed. Corrected pair densities are marked by a tilde and, corresponding to Equation (21.3c) for uncorrected pair densities, are defined by

$$\tilde{C}_{ij}(\xi, p) = \frac{1}{A} \int p_i(x) \left[p_j(x + \xi) - \delta_{ij}\delta_x(x + \xi) \right] dx . \qquad (21.4a)$$

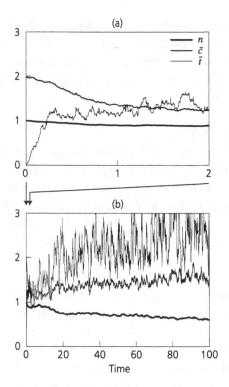

Figure 21.4 Relaxation time scales of correlation densities. Already for a single-species system, dynamics of pair and triplet densities cannot be depicted in full. We thus use integral statistics $\bar{c} = N_1^{-2} \int w_{11}^{(d)}(\xi)\tilde{C}_{11}(\xi)\,d\xi$ and $\bar{t} = N_1^{-3} \iint w_{11}^{(d)}(\xi)w_{11}^{(d)}(\xi')$ $\times \tilde{T}_{111}(\xi, \xi')\,d\xi\,d\xi'$ instead and, for convenience, also normalize mean densities to $n = N_1/N_1(0)$. As an initial condition, we choose a distribution of individuals, designed to satisfy $\bar{t} = 0$. Even in the reduced state space $(n, \bar{c}, \bar{t})$, the system's response to such an artificial state is evident from the figure: (a) fast relaxation at a very short time scale is followed by (b) subsequent changes occurring over a much longer time scale. To diminish the amount of noise, time series are averaged over 10 realizations. Parameters are the same as in Figure 21.1a, except for s.d. $w_{11}^{(d)} = 0.015$ and s.d. $m_1^{(b)} = 0.05$.

With this correction, the focal individual of the first density function at location x is removed from the second density function evaluated at location $x + \xi$. Of course, such removal is only necessary if the two density functions describe the same species; otherwise, individuals contributing to these functions belong to different species and therefore can never be identical; this is taken care of by the Kronecker symbol δ_{ij}. Similarly, we can remove self-pairs and self-triplets from the third-order correlation densities,

$$\tilde{T}_{ijk}(\xi, \xi', p) = \frac{1}{A}\int p_i(x)\left[p_j(x + \xi) - \delta_{ij}\delta_x(x + \xi)\right] \qquad (21.4b)$$
$$\times \left[p_k(x + \xi') - \delta_{ik}\delta_x(x + \xi') - \delta_{jk}\delta_{x+\xi}(x + \xi')\right]dx \, ,$$

where, for the third density function, we have to subtract contributions of the focal individuals of both the first and second density functions.

With this additional refinement in place, we can now turn to describing the dynamics of spatial patterns in terms of the corresponding dynamics of corrected correlation densities.

Correlation dynamics

Whenever a stochastic process of pattern dynamics is run, it creates a different realization (see, for example, Figure 14.2). This is in keeping with field observations which show that, although we may try to generate very similar initial spatial setups in, for example, plots of plants intended as replicates, the actual processes in each plot usually result in a wide variety of spatial patterns at later points in time. To extract the deterministic signal of a spatial ecological process from among all accompanying stochastic noise, averages over replicates are needed. If, however, we were to average spatial density functions across the many different patterns that can arise from a single initial condition, we would merely obtain an uninformative and rather unstructured "average pattern": in the long run, high abundances at some location in one replicate would cancel out low abundances at the same location in other replicates. It is therefore important to average over spatial statistics, rather than across spatial patterns. For example, the mean density of a particular species averaged across realizations is a meaningful ecological quantity.

In this vein, we define expected values of corrected correlation functions as

$$\tilde{C}_{i_1...i_m}(\xi_1, \ldots, \xi_{m-1}) = \int \tilde{C}_{i_1...i_m}(\xi_1, \ldots, \xi_{m-1}, p) P(p) \, dp \, , \qquad (21.5)$$

that is, by weighting the correlation densities for a pattern p with the probability density $P(p)$ for that pattern to occur. We can now study the dynamics of expected correlation densities [for notational convenience we omit the arguments $(\xi_1, \ldots, \xi_{m-1})$ of $\tilde{C}_{i_1...i_m}$ in the next three equations],

$$\frac{d}{dt}\tilde{C}_{i_1...i_m} = \int \tilde{C}_{i_1...i_m}(p) \frac{d}{dt} P(p) \, dp \, . \qquad (21.6a)$$

Using Equation (21.2a), this gives

$$\frac{d}{dt}\tilde{C}_{i_1...i_m} = \int \tilde{C}_{i_1...i_m}(p) \int \left[w(p|p') P(p') - w(p'|p) P(p) \right] dp' dp \, , \quad (21.6b)$$

which, by separating the difference into two integrals, swapping the integration variables p and p' in the first of them, and then joining the two integrals again, yields

$$\frac{d}{dt}\tilde{C}_{i_1...i_m} = \int \left\{ \int [\tilde{C}_{i_1...i_m}(p') - \tilde{C}_{i_1...i_m}(p)]w(p'|p)dp' \right\} P(p)dp . \quad (21.6c)$$

The term in curly braces is known as the first jump moment (van Kampen 1992) of the mth-order correlation density; it describes the correlation density's expected rate of change around a given pattern p, and we denote it by $a_{i_1...i_m}(\xi_1, \ldots, \xi_{m-1}, p)$.

Dynamics of mean densities

We can now evaluate the expected rate of change in mean densities, or first spatial moments, by using Equations (21.2b) and (21.3b) in Equation (21.6c). According to Equation (21.2b), $a_i(p)$, the first jump moment of the first spatial moment in species i, can be decomposed into three contributions coming from birth, death, and movement events, respectively,

$$a_i(p) = a_i^{(b)}(p) + a_i^{(d)}(p) + a_i^{(m)}(p) . \quad (21.7)$$

The birth contribution is

$$a_i^{(b)}(p) = \int [N_i(p') - N_i(p)] \quad (21.8a)$$

$$\times \sum_j \iint B_j(x, x', p)p_j(x)\Delta(p + u_j\delta_{x'} - p')\,dx\,dx'\,dp' ,$$

which, after collapsing the integration over p' using the definition of the generalized delta function (see Box 21.4), simplifies to

$$a_i^{(b)}(p) = \sum_j \iint [N_i(p + u_j\delta_{x'}) - N_i(p)]B_j(x, x', p)p_j(x)\,dx\,dx' . \quad (21.8b)$$

For all $j \neq i$, the term in square brackets is 0, while for $j = i$ it is $\frac{1}{A}$. We see this by considering $N_i(p + u_j\delta_{x'}) = \frac{1}{A}\int (p + u_j\delta_{x'})_i(x)\,dx = \frac{1}{A}\int p_i(x)\,dx + \frac{1}{A}\delta_{ij}\int \delta_{x'}(x)\,dx = N_i(p) + \frac{1}{A}\delta_{ij}$. This gives

$$a_i^{(b)}(p) = \frac{1}{A}\iint B_i(x, x', p)\,p_i(x)\,dx\,dx' , \quad (21.8c)$$

or, by using Equation (21.1c) and exploiting that $m_i^{(b)}$ is the probability density for dispersal and thus integrates to 1,

$$a_i^{(b)}(p) = \frac{1}{A}\int \left[b_i + \sum_j b'_{ij} \int w_{ij}^{(b)}(x'' - x) \right.$$

$$\left. \times [p_j(x'') - \delta_{ij}\delta_x(x'')]\,dx'' \right] p_i(x)\,dx , \quad (21.8d)$$

or, when separating terms for density-independent and density-dependent birth,

$$a_i^{(b)}(p) = b_i \frac{1}{A} \int p_i(x)\, dx + \sum_j b'_{ij} \frac{1}{A} \iint w_{ij}^{(b)}(x'' - x)$$
$$\times p_i(x) \left[p_j(x'') - \delta_{ij}\delta_x(x'') \right] dx\, dx'' \ . \tag{21.8e}$$

We now replace the integration over x with one over $\xi = x'' - x$, use the definitions of $N_i(p)$ and $\tilde{C}_{ij}(\xi, p)$, and thus obtain

$$a_i^{(b)}(p) = b_i N_i(p) + \sum_j b'_{ij} \int w_{ij}^{(b)}(\xi)\tilde{C}_{ij}(\xi, p)\, d\xi \ . \tag{21.8f}$$

For death events, we have $N_i(p - u_j\delta_x) - N_i(p) = -\frac{1}{A}\delta_{ij}$, which yields

$$a_i^{(d)}(p) = -d_i N_i(p) - \sum_j d'_{ij} \int w_{ij}^{(d)}(\xi)\tilde{C}_{ij}(\xi, p)\, d\xi \ , \tag{21.8g}$$

while $N_i(p - u_j\delta_x + u_j\delta_{x'}) - N_i(p) = 0$ for movement events gives

$$a_i^{(m)}(p) = 0 \ . \tag{21.8h}$$

Collecting all three results together, and using Equations (21.6c) and (21.7), gives the expected rate of change in the mean density of species i

$$\frac{d}{dt} N_i = + b_i N_i + \sum_j b'_{ij} \int w_{ij}^{(b)}(\xi)\tilde{C}_{ij}(\xi)\, d\xi$$
$$- d_i N_i - \sum_j d'_{ij} \int w_{ij}^{(d)}(\xi)\tilde{C}_{ij}(\xi)\, d\xi \ . \tag{21.9}$$

This result is readily interpreted. Movement does not contribute to the dynamics of mean densities at all: moving individuals around does not alter their abundance. Notice that, according to Equation (21.9), for both birth and death events the interaction kernels $w_{ij}^{(b)}(\xi)$ and $w_{ij}^{(d)}(\xi)$ of individuals of species j at distance ξ from individuals in species i, are weighted with the density $\tilde{C}_{ij}(\xi)$ at which individuals of species j are expected to be found at a distance ξ around individuals of species i. Effective local densities – that is, the integrals in Equation (21.9) – are then simply the sum of all such contributions. In this way, the mean densities within a spatial pattern change according to the "individual's-eye view" of its local environment, as described by the pair densities. If spatial heterogeneity were removed by setting $\tilde{C}_{ij}(\xi) = N_i N_j$ for all i, j, and ξ, we would recover the familiar Lotka–Volterra dynamics of mean-field theory, $\frac{d}{dt} N_i = (b_i - d_i)N_i + \sum_j (b'_{ij} - d'_{ij})N_i N_j$. For spatially heterogeneous systems, however, this simple description is incorrect and the dynamics of mean densities become contingent on those of pair densities.

Dynamics of pair densities

To provide the information on corrected pair densities $\tilde{C}_{ij}(\xi)$ as required in Equation (21.9), we need to work out their expected rates of change. For this purpose, we follow the same sequence of steps as was used for deriving the dynamics of mean densities and thereby gain the following general result:

$$\frac{d}{dt}\tilde{C}_{ij}(\xi) =$$

$$
\begin{array}{ll}
\text{Birth} \left\{
\begin{array}{l}
+ \delta_{ij} b_i m_i^{(b)}(-\xi) N_i \\[2mm]
+ b_i \displaystyle\int m_i^{(b)}(\xi')\tilde{C}_{ij}(\xi+\xi')\,d\xi' \\[2mm]
+ \delta_{ij} m_i^{(b)}(-\xi) \displaystyle\sum_k b'_{ik} \int w_{ik}^{(b)}(\xi'')\tilde{C}_{ik}(\xi'')\,d\xi'' \\[2mm]
+ \displaystyle\sum_k b'_{ik} \iint m_i^{(b)}(\xi')w_{ik}^{(b)}(\xi'')\tilde{T}_{ijk}(\xi+\xi',\xi'')\,d\xi'd\xi'' \\[2mm]
+ b'_{ij} \displaystyle\int m_i^{(b)}(\xi')w_{ij}^{(b)}(\xi+\xi')\tilde{C}_{ij}(\xi+\xi')\,d\xi'
\end{array}
\right. & \\[2mm]
\text{Death} \left\{
\begin{array}{l}
- d_i \tilde{C}_{ij}(\xi) \\[2mm]
- \displaystyle\sum_k d'_{ik} \int w_{ik}^{(d)}(\xi'')\tilde{T}_{ijk}(\xi,\xi'')\,d\xi'' \\[2mm]
- d'_{ij} w_{ij}^{(d)}(\xi)\tilde{C}_{ij}(\xi)
\end{array}
\right. & \\[2mm]
\text{Movement} \left\{
\begin{array}{l}
- |m_i| \tilde{C}_{ij}(\xi) \\[2mm]
+ \displaystyle\int m_i(\xi')\tilde{C}_{ij}(\xi+\xi')\,d\xi'
\end{array}
\right. &
\end{array}
$$

$$+ \langle i,j,\xi \to j,i,-\xi \rangle \,. \tag{21.10}$$

The derivation of this equation is provided in Appendix 21.A. All terms on the right-hand side have a precise interpretation, given below. We first focus on the i individual of the ij pair; analogous considerations apply to events undergone by the j individual, as discussed at the end of the section.

The first five terms on the right-hand side describe the increase in the density of ij pairs at vectorial distance ξ resulting from birth events in species i.

■ A new i individual, arising from density-independent birth, generates a parent–offspring pair of type ii with its parent. This is accounted for by the first term, which therefore only contributes to the expected dynamics of auto-correlation densities; this is reflected by the Kronecker delta δ_{ij}. Multiplying the mean density of individuals N_i by the density-independent per capita birth rate b_i gives the rate for such events creating a pair at any distance ξ; different distances occur at spatial density $m_i^{(b)}(-\xi)$ of parents around dispersed offspring.

■ The second term also accounts for density-independent birth, but focuses on the new pair that the i offspring forms with the j neighbor of its parent. If the i individual of an ij pair at distance $\xi + \xi'$ gives birth to a new individual at distance ξ' from its parent, a new ij pair at distance ξ is generated. The per capita rate for density-independent birth is b_i, the density of ij pairs at distance $\xi + \xi'$ is $\tilde{C}_{ij}(\xi + \xi')$, and the spatial density of offspring settling around the i parent is $m_i^{(b)}(\xi')$. Multiplying these three factors and integrating over all possible distances ξ' of offspring dispersal yields the second term.

■ The third term describes how, under density-dependent birth, a new parent–offspring pair changes the density of ii pairs. Fecundity of an i individual can be modified by the presence of a k individual at distance ξ''. The expected density of k individuals around the i individual is $\tilde{C}_{ik}(\xi'')$, the interaction kernel for the ik pair yields $w_{ik}^{(b)}(\xi'')$, so that integrating over all interaction distances ξ'' gives the third term's integral. We then have to sum over all species k that exert such an effect: sign and weight for the contribution of each species are determined by coefficients b_{ik}'. The resulting density-dependent per capita birth rate is multiplied by δ_{ij}, since this term only affects the dynamics of auto-correlations, and by $m_i^{(b)}(-\xi)$, the density of parent individuals as seen from the i offspring.

■ The fourth term describes the effect of new offspring–neighbor pairs of type ij arising from density-dependent birth. As in the previous term, the fecundity of an i individual in an ij pair at distance $\xi + \xi'$ can be modified by the presence of a k individual at distance ξ'' relative to the i individual. Under these circumstances, an i offspring, located at distance ξ' from its parent, generates a new ij pair at distance ξ. The density of the original ijk triplet configuration is $\tilde{T}_{ijk}(\xi + \xi', \xi'')$, the interaction kernel for the ik pair yields $w_{ik}^{(b)}(\xi'')$, and the spatial density of offspring around parent individuals is $m_i^{(b)}(\xi')$. Multiplying these factors and integrating over all possible dispersal distances ξ' and interaction distances

ξ'' weights the effect of species k in the fourth term's sum. The resulting integrals are summed over all species k that exert such an effect on i's fecundity: signs and weights for these contributions are again determined by the interaction coefficients b'_{ik}.

■ The fifth term, the last to account for birth effects, describes the consequences of density-dependence of birth rates *within* the ij pair. The density of such pairs at distance $\xi + \xi'$ is $\tilde{C}_{ij}(\xi + \xi')$, and the effect that interaction with j has on i's birth rate is given by $b'_{ij} w^{(b)}_{ij}(\xi + \xi')$. An i offspring at distance ξ' from the i parent, arising at density $m^{(b)}_i(\xi')$, creates a new ij pair at distance ξ. Integrating these three factors over all dispersal distances ξ' yields the fifth term.

The next three terms account for the decrease in the density of ij pairs at distance ξ due to death events in species i.

■ Density-independent death of the i individual of the ij pair occurs at per capita rate d_i. Multiplying d_i by the density $\tilde{C}_{ij}(\xi)$ of ij pairs at distance ξ in the sixth term gives the resulting decrease in that density.

■ Density-dependent death of the i individual of the ij pair can result from the presence of a k individual at distance ξ''. Such triplet configurations occur at density $\tilde{T}_{ijk}(\xi, \xi'')$, and the cumulative effect of individuals of species k, situated at different distances from the i individual, is obtained by weighting the triplet density with $w^{(d)}_{ik}(\xi'')$, given by the interaction kernel for density-dependent death, and integrating the product over all interaction distances ξ''. Multiplying this contribution of species k with the corresponding interaction coefficient d'_{ik} and summing over all species k gives the seventh term.

■ Like birth events, the density-dependent component of death rates can also be affected by interaction of the focal i individual with its j partner *within* the ij pair. The interaction coefficient for this interaction is d'_{ij}, the effect of distance ξ on the strength of interaction is measured by $w^{(d)}_{ij}(\xi)$, and the pair configuration occurs at density $\tilde{C}_{ij}(\xi)$. The product of these three factors gives the eighth term.

The next two terms describe how the density of ij pairs at distance ξ changes because of movement of the i individual.

■ When the i individual moves, the original ij pair at distance ξ is destroyed. The ninth term reflects this: here, the per capita rate $|m_i|$ for i's movement is multiplied by the density $\tilde{C}_{ij}(\xi)$ of the original pair configuration.

■ On the other hand, a new ij pair at distance ξ is created when the original pair distance is $\xi + \xi'$ and the i individual moves a distance ξ'. This effect is captured by the tenth term, which weights the density $\tilde{C}_{ij}(\xi + \xi')$ of the original pair configuration with the movement kernel $m_i(\xi)$ for species i and integrates over all possible movement distances ξ'.

In all events described above we have focused on the i individual. Of course, analogous events can occur to the j individual of the ij pair. When swapping the roles of the i and j individuals, the distance vector ξ must also be reversed. The last term $\langle i, j, \xi \rightarrow j, i, -\xi \rangle$ accounts for this: it is shorthand for all preceding terms after changing i to j, j to i, and ξ to $-\xi$.

21.4 Moment Closures and their Performance

A critical step toward simplifying the spatial complexity of ecological birth–death–movement models is to close the correlation dynamics derived in Equations (21.9) and (21.10). In this section, we introduce alternative closures and investigate their relative performance.

Moment hierarchies and moment closures

While the right-hand side of the mean densities' dynamics, Equation (21.9), depends on pair densities, that of the pair densities' dynamics, Equation (21.10), contains triplet densities. This observation can be generalized: when interactions are pairwise, for any order m of correlation densities, dynamics are contingent on correlation densities of order $m + 1$. The resulting sequence of equations is called a *moment hierarchy*, the head of which is given by Equations (21.9) and (21.10). This hierarchy precludes exploiting the dynamics of N unless we have information on $\tilde{C}$; this information comes from the dynamics of $\tilde{C}$, for which we need information on $\tilde{T}$, and so on.

To escape from this cascade of dependencies, the moment hierarchy has to be truncated. In other words, we have to express the correlation densities of order $m + 1$ in terms of those of order m and below. Such expressions are called *moment closures*. In the previous section, we saw that inserting

$$\tilde{C}_{ij}(\xi) = N_i N_j \tag{21.11}$$

into Equation (21.9) yields a closed dynamical system that is the mean-field approximation of the considered birth–death–movement process. This comes as no surprise: assuming $\tilde{C}_{ij}(\xi) = N_i N_j$ for all ξ implies pair densities that are independent of distance or, equivalently, the absence of salient spatial structure. These are precisely the circumstances under which the mean-field assumption is valid. We therefore conclude that using the moment closure given in Equation (21.11) to truncate the moment hierarchy after the first order recovers the mean-field approximation.

To improve on this result we go beyond the first order of the hierarchy: the natural next step is to truncate the hierarchy after its second order. For this purpose, we have to express triplet densities, or third moments $\tilde{T}$, in terms of first and second moments N and $\tilde{C}$. Notice that in the derivation of Equations (21.9) and (21.10) no approximations have been introduced; consequently, these equations are exact. The price that has to be paid for a moment closure is that the resulting description is an approximation. In fact, the performance of such an approximation may vary with the moment closure chosen. So what criteria can guide the choice of moment closures to give a good approximation and the desired truncation of the moment hierarchy at the second order?

Conditions for moment closures

Moment closures are not uniquely determined and thus alternative versions can be chosen. Yet, some conditions narrow the range of options.

- Condition (C1). In the absence of any pair correlations, no information on spatial structure is available to closures for triplet densities, and therefore individuals in triplets must also be assumed to be uncorrelated. In other words, if $\tilde{C}_{ij}(\xi) = N_i N_j$ for all i, j, and ξ, any consistent moment closure ought to yield $\tilde{T}_{ijk}(\xi, \xi') = N_i N_j N_k$ for all i, j, k, ξ, and ξ'.

- Condition (C2). Because attention is focused on small-scale spatial structure, pairs of individuals separated by large distances are assumed to be uncorrelated. Therefore, $\lim_{|\xi| \to \infty} \tilde{C}_{ij}(\xi) = N_i N_j$ and $\lim_{|\xi| \to \infty} \frac{d}{dt} \tilde{C}_{ij}(\xi) = \frac{d}{dt}(N_i N_j) = N_i \frac{d}{dt} N_j + N_j \frac{d}{dt} N_i$ for all i and j, hold in general: any potentially developing structure is supposed not to affect pair densities at large distances. When evaluating Equation (21.10) in this limit, consistency with Equation (21.9) requires that we assume (i) $\lim_{|\xi| \to \infty} \int w_{ik}^{(d)}(\xi'') \tilde{T}_{ijk}(\xi, \xi'') \, d\xi'' = N_j \int w_{ik}^{(d)}(\xi'') \tilde{C}_{ik}(\xi'') \, d\xi''$ and (ii) $\lim_{|\xi| \to \infty} \iint m_i^{(b)}(\xi') w_{ik}^{(b)}(\xi'') \tilde{T}_{ijk}(\xi + \xi', \xi'') \, d\xi' d\xi'' = N_j \int w_{ij}^{(b)}(\xi'') \tilde{C}_{ij}(\xi'') \, d\xi''$ for all i, j, and k, and for all kernels $w_{ik}^{(d)}$, $w_{ik}^{(b)}$, and $m_i^{(b)}$. Conditions (i) and (ii) are fulfilled if, and only if, $\lim_{|\xi| \to \infty} \tilde{T}_{ijk}(\xi, \xi'') = N_j \tilde{C}_{ik}(\xi'')$ holds for all i, j, k, and ξ''.

The two conditions above provide criteria for *valid* moment closures. We now have to consider how to characterize *good* moment closures.

Moment closures as relaxation projections

To distinguish better moment closures from less suitable ones, the notion of relaxation projections (see Box 21.1) becomes critical. Usually, not all degrees of freedom in a system change on the same time scale. Some degrees of freedom are fast and decay quickly, while others are slower and thus remain essential for a longer time. In particular, triplet densities often have a much faster pace of change than both pair and mean densities. Figure 21.4 gives an example illustrating this feature.

After a system's fast degrees of freedom have decayed, they are no longer independent variables and instead become functions of the slower degrees of freedom. For example, when pair densities $\tilde{C}$ have decayed, it must become possible to express them in terms of mean densities N. Such a relation is provided by Equation (21.11); it truncates the moment hierarchy at first order. The mean-field approximation therefore can also be interpreted as a relaxation projection. Likewise, after triplet densities $\tilde{T}$ have decayed, they lose their role as essential degrees of freedom and can be expressed as functions of pair densities $\tilde{C}$ and mean densities N. Such relations are the moment closures we are seeking for truncating the moment hierarchy at second order. As described in Box 21.1, good moment closures match a system's relaxation manifold and define a projection onto this manifold that resembles the system's actual relaxation dynamics.

Candidate moment closures

To express triplet densities $\tilde{T}$ as a function of pair densities $\tilde{C}$ and mean densities N, different assumptions for the relaxation manifold can be made. Here we introduce and investigate four possible candidates:

$$\tilde{T}_{ijk}(\xi, \xi') = \tilde{C}_{ij}(\xi)N_k + \tilde{C}_{ik}(\xi')N_j$$
$$+ \tilde{C}_{jk}(\xi' - \xi)N_i - 2N_iN_jN_k \,, \tag{21.12a}$$

$$\tilde{T}_{ijk}(\xi, \xi') = \tilde{C}_{ij}(\xi)\tilde{C}_{ik}(\xi')/N_i \,, \tag{21.12b}$$

$$\tilde{T}_{ijk}(\xi, \xi') = \tfrac{1}{2}\big[\tilde{C}_{ij}(\xi)\tilde{C}_{ik}(\xi')/N_i$$
$$+ \tilde{C}_{ij}(\xi)\tilde{C}_{jk}(\xi' - \xi)/N_j \tag{21.12c}$$
$$+ \tilde{C}_{ik}(\xi')\tilde{C}_{jk}(\xi' - \xi)/N_k - N_iN_jN_k\big] \,,$$

$$\tilde{T}_{ijk}(\xi, \xi') = \tilde{C}_{ij}(\xi)\tilde{C}_{ik}(\xi')\tilde{C}_{jk}(\xi' - \xi)/(N_iN_jN_k) \,. \tag{21.12d}$$

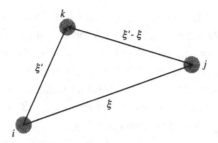

Figure 21.5 Triplet configuration for moment closure. Individuals of species i, j, and k are separated by vectors ξ, ξ', and $\xi' - \xi$.

All these candidate closures satisfy Conditions (C1) and (C2) as defined above. As a counterexample, consider an alternative candidate, $\tilde{T}_{ijk}(\xi, \xi') = \frac{1}{3}[\tilde{C}_{ij}(\xi)N_k + \tilde{C}_{ik}(\xi')N_j + \tilde{C}_{jk}(\xi' - \xi)N_i]$, for which Condition (C1) holds but (C2) is violated.

The salient ingredients for these closures are the pair correlations along the three sides of the triangular configuration of the i, j, and k individuals: $\tilde{C}_{ij}(\xi)$, $\tilde{C}_{ik}(\xi')$, and $\tilde{C}_{jk}(\xi' - \xi)$ (see Figure 21.5). The closures above can be constructed according to the following recipe.

- First, choose the power of the closure. This is the number of pair densities that are multiplied by each other and determines the "building blocks" of the closure: for a power-1 closure, each building block is a pair density; for a power-2 closure it is the product of two pair densities; and for a power-3 closure, the product of three pair densities.
- Second, construct all building blocks arising from permutations among the three corners of the triangle.
- Third, modify each building block by multiplying or dividing by mean densities such that it satisfies Condition (C1).
- Fourth, add all blocks.
- Fifth, modify the sum by adding or subtracting $N_i N_j N_k$ blocks and divide by the resulting *net* number of blocks so that Condition (C2) is met.

Closures in Equations (21.12a), (21.12c), and (21.12d) simply follow this recipe for powers 1, 2, and 3, respectively. Higher powers are not considered since they would require repeating pair densities in building blocks.

The closure in Equation (21.12b) is obtained from the same recipe by modifying the second step. This is motivated by the asymmetric way triplet densities enter into Equation (21.10): triplet densities there characterize the density of j and k individuals around a focal i individual (or of i and k individuals around a focal j individual, see below), whose birth and death rates are affected by its local surroundings. The i individual therefore has a special role in the triplet; this is reflected in the closure in Equation (21.12b), which concentrates on the pair densities along those two edges of the ijk triangle that are connected to the corner occupied by the i individual (i.e., the ij and ik edges). For a focal j individual, the same consideration applies: now the triplet configuration in Equation (21.10) is jik, and in Equation (21.12b) the special role is assumed by the j individual. The closure in Equation (21.12c), in contrast, gives equal weight to all three pairs of edges of the triangle; the resulting expression for $\tilde{T}_{ijk}$ is thus completely symmetric under permutations of the i, j, and k individuals. Closures that reflect the special role of the one focal individual cannot be constructed for powers 1 and 3; thus, the asymmetric power-2 closure, Equation (21.12b), is the only asymmetric closure included in the considered set of candidates.

The power-1 moment closure in Equation (21.12a) can be motivated by defining so-called *central* third moments, $T_{ijk}^{(c)}(\xi, \xi', p) = \frac{1}{A} \int [p_i(x) - N_i][p_j(x + \xi) - N_j][p_k(x + \xi') - N_k] dx$, and is based on the assumption that these are vanishing (see Chapter 20; Bolker and Pacala 1997). In contrast, the motivation for power-2 closures comes from a probabilistic argument. When assessing the probability density of a triangular configuration based on pair correlations along edges, we notice that only two of the triangle's three edge vectors can be chosen independently; the third edge vector directly follows from choosing the other two. Envisaging the probability density for a particular triangle as the joint probability density for two of its edge vectors and assuming that the contribution of both edges is statistically independent naturally leads to moment closures that use products of two pair densities as building blocks (see Chapter 14; Dieckmann *et al.* 1997, in press). The power-3 closure in Equation (21.12d) has a long tradition in the literature of theoretical physics, where it is known as the Kirkwood superposition approximation (Kirkwood 1935; Ziman 1979).

Testing the candidate closures

We now test the performance of the four candidate closures by comparing the results of individual-based simulations with predictions derived from these moment closures.

The first test considers triplet densities directly. This requires measuring, at one moment in time, the distribution of triplets in the three-dimensional space spanned by the edge lengths (r, r', r'') of all triangles formed by triplets of individuals. For a triplet configuration with a pair (ξ, ξ') of edge vectors, we have $r = |\xi|$, $r' = |\xi'|$, and $r'' = |\xi' - \xi|$ (see Figure 21.5). Contour surfaces resulting from such a measurement of triplet densities are

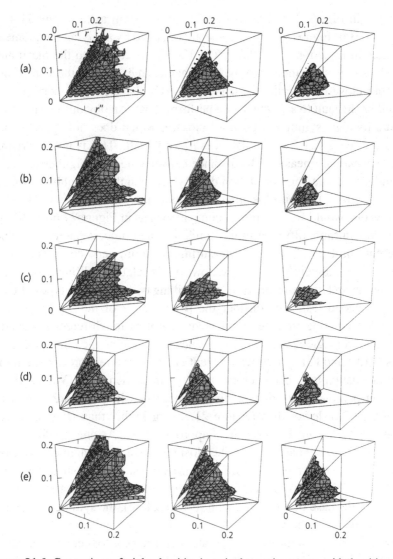

Figure 21.6 Comparison of triplet densities in a single-species system with densities predicted by candidate moment closures. (a) Triplet densities with radial distances (r, r', r'') realized at time $t = 100$ in the system described in Figure 21.4 are shown as contour surfaces. Triplet densities can be envisaged as clouds of varying intensity, and contour surfaces of increasing value are nested inside one another; the surfaces shown correspond to triplet densities of 2 (left column), 2.5 (center column), and 3 (right column), all measured relative to a mean-field expectation of 1. Notice that the three main diagonals of the boundary planes through the origin span a pyramid, outside of which triplet densities are zero – a confinement resulting from the three triangle inequalities for (r, r', r''). Realized singlet and pair densities at time $t = 100$ are combined with one of the four moment closures in Equations (21.12a) to (21.12d) and predicted triplet densities are shown in panels (b) to (e), respectively. To diminish the amount of noise, densities are averaged over 100 realizations.

depicted in Figure 21.6a, based on the same system as in Figure 21.4 at time $t = 100$. Figures 21.6b to 21.6e are obtained by applying the moment closures in Equations (21.12a) to (21.12d), respectively, to the mean and pair densities of the individual-based dynamics measured at the same time. All four closures describe the triplet density's shape roughly correctly; it would be difficult to say which one performs best. An obvious discrepancy occurs for the asymmetric power-2 closure, which does not approximate the triplet density for small distances r'' very well. This is not surprising as, in a triplet, r'' measures the distance between the two neighbors of a focal individual; the pair correlation density between these neighbors (which can differ from 1 when r'' is small) does not enter into Equation (21.12b). Ideally, one would want to repeat comparisons as in Figure 21.6 for different points in time, different initial conditions and process parameters, and different numbers of species. Measuring, depicting, and comparing densities in three-dimensional spaces, however, is relatively difficult and time consuming and we therefore turn to describing two alternative types of test.

In a second test, we consider the performance of the candidate closures in predicting time series of mean densities for different parameter settings of a single-species system. When used in conjunction with Equations (21.9) and (21.10), to what extent can the closures forecast the transient and asymptotic behavior exhibited by mean densities? We use three different parameter settings, leading to mean densities that are similar to (Figure 21.7a), lower than (Figure 21.7b), or higher than (Figure 21.7c) those expected from the mean-field approximation. As one would expect, there is little to choose between the closures when the dynamics are close to those of the mean field (Figure 21.7a). Under strong spatial aggregation (Figure 21.7b), the asymmetric power-2 closure performs better than the others. In the case of overdispersion (Figure 21.7c), the symmetric power-2 closure performs best.

We extend the comparison of closure performances to predict population dynamics in a two-species system. In a third test, we investigate phase portraits of trajectories for the mean densities of two competing species, with each trajectory starting from a different initial condition. Results of individual-based dynamics are shown in Figure 21.8a. Moment closures in Equations (21.12a) to (21.12d) are used to describe the dynamics of auto-correlation and cross-correlation pair densities, Equation (21.10), which affect the dynamics of mean densities, Equation (21.9). The phase portraits predicted by those moment closures are displayed in Figures 21.8b to 21.8e. In this test, the power-2 closures clearly perform best and yield accurate

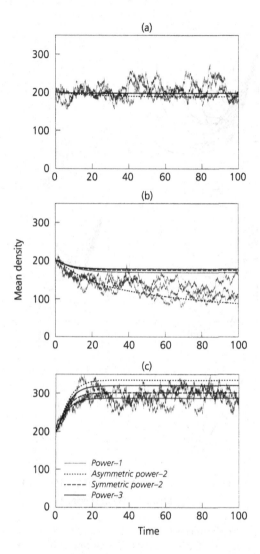

Figure 21.7 Comparison of mean-density dynamics of a single-species system with predictions by candidate moment closures. For a spatial logistic model, three different parameter combinations result in asymptotic mean densities (a) similar to, (b) lower than, or (c) higher than the mean-field equilibrium of $N_1 = 200$. In each case, time series of mean densities for three realizations are shown in gray. Dynamics predicted by pair correlation dynamics in conjunction with one of the moment closures in Equations (21.12a) to (21.12d) are superimposed in black. Parameters are the same as in Figure 21.1a, except for (a) s.d. $w_{11}^{(d)} = 0.15$, s.d. $m_1^{(b)} = 0.15$; (b) s.d. $w_{11}^{(d)} = 0.05$, s.d. $m_1^{(b)} = 0.05$; (c) s.d. $w_{11}^{(d)} = 0.015$, s.d. $m_1^{(b)} = 0.15$.

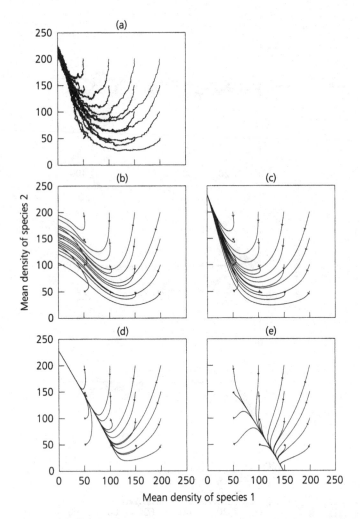

Figure 21.8 Comparison of phase portraits of a two-species system with predictions by candidate moment closures. Each panel shows trajectories of mean densities (N_1, N_2), starting from a grid of 16 initial conditions. (a) Mean paths of the stochastic process, averaged over 20 realizations. (b to e) Phase portraits predicted by pair correlation dynamics in conjunction with one of the moment closures in Equations (21.12a) to (21.12d): (b) power-1 closure; (c) asymmetric power-2 closure; (d) symmetric power-2 closure; (e) power-3 closure. Notice that trajectories are projections from higher-dimensional dynamics and may thus intersect one another; see also Figure 14.3. Parameters: $b_1 = b_2 = 0.4$, $d_1 = d_2 = 0.2$, $d'_{11} = d'_{22} = 0.001$, $d'_{12} = 0.005$, $d'_{21} = 0.002$, s.d. $w^{(d)}_{11}$ = s.d. $w^{(d)}_{12}$ = s.d. $w^{(d)}_{21}$ = s.d. $w^{(d)}_{22}$ = s.d. $m^{(b)}_1$ = 0.03, s.d. $m^{(b)}_2$ = 0.2.

portraits of the actual dynamics. In particular, only the asymmetric power-2 closure captures the fine details of the phase portrait. Performance of the power-1 closure is compromised by an instability at small densities, and the power-3 closure even results in prediction of a wrong attractor.

The first test above (summarized in Figure 21.6) is independent of the correlation dynamics and therefore directly probes the accuracy of the moment closures. In contrast, the last two tests (recorded in Figures 21.7 and 21.8) rely on evaluations of the *joint* performance of closures and correlation dynamics. While a benefit of the first test is its independence of additional theory, more discriminating results and more direct relevance to predictive quality are advantages of the latter pair. On the basis of these tests, we have chosen the asymmetric power-2 closure for exploring ecological processes in Chapter 14. There is, however, much to be learned about moment closures, and we do not intend to suggest that this particular closure performs best under all circumstances. To gain further insight, there appears to be no alternative to empirical tests for particular ecological settings, in the manner of Figures 21.7 and 21.8.

21.5 Further Developments and Extensions

In this chapter we have derived a closed dynamical system for describing spatially heterogeneous change in general ecological communities of interacting species. Instead of considering the simultaneous dynamics of mean and pair densities, Equations (21.9) and (21.10), we can further simplify the description by using Equation (21.10) alone. This is possible because Condition (C2), see Section 21.4, which is satisfied for small-scale spatial patterns, implies $\lim_{|\xi| \to \infty} \tilde{C}_{ii}(\xi) = N_i^2$: consequently, information about mean densities is implicit in the description of pair dynamics. After inserting one of the moment closures from Equations (21.12), Equation (21.10) represents the central result of this chapter and provides a powerful tool for reducing the complexity of individual-based models. We have explained how this result is based on the *method of moments*: expressing the essential degrees of freedom for a developing spatial pattern in terms of spatial moments and truncating the resulting moment hierarchy using a suitable moment closure yields an approximation for the expected dynamics of moments. The utility of such moment dynamics for describing and understanding ecological processes under spatial heterogeneity is examined in Section 21.4; a variety of successful applications are also presented in Chapter 14.

To conclude this chapter, we briefly review several directions for further elaborations on the method of moments. These extensions remove some remaining restrictions of the results presented and broaden the scope of ecological systems amenable to simplifying spatial complexity.

External heterogeneity and internal states

Moment dynamics can account for small-scale spatial heterogeneity in the external environment of populations. As discussed in Section 4.3, which variables are considered external to a system depends on how the system's boundaries are defined. On a short time scale, for example, the concentration of soil nutrients in a spatially extended plant community may be fixed in time and thus can be considered external to the system. In the long run, however, we expect feedback between the dynamics of plants and nutrients. Another example is the effect that landscape patterns have on the dispersal behavior of animals. Some butterflies, for instance, inhabit meadows that arise in gaps created by forest fires, and while the meadows are easily traversed, stretches of forest act as dispersal barriers. In this case, there is no significant feedback between the dynamics of butterflies and forest trees. With or without such feedback, spatial heterogeneity of environmental conditions encountered by a focal population can be formally treated on par with the heterogeneity arising from the presence of extra populations. Straightforward generalizations of Equations (21.10) therefore account for a population's unfolding against the backdrop of a heterogeneous landscape. Salient state variables are the auto-correlations of the focal population, complemented by the auto-correlations of environmental conditions (these may be variable or fixed) as well as the cross-correlations between the population and its environment. Moment dynamics of this kind can provide insights into complex interactions between internal and external mechanisms of ecological pattern formation.

Another extension of moment dynamics allows the incorporation of details about the state of individuals. For example, instead of characterizing a population of trees just by the locations on which they are centered, the size of each tree (height or crown diameter) can be accounted for as well. Implications of asymmetric competition for light, for instance, can thus be studied in spatially heterogeneous settings. Also, a plea for dynamic neighborhoods (see Section 2.5) that vary with, for example, plant size can be met in this way. Other such internal states by which location-based information can be augmented are the age of individuals, their physiological status, or their phenotype. In each case, pair correlation densities take two extra arguments, measuring the internal state of paired individuals in addition to their distance. Merging physiologically structured and spatially structured population models in such a way is not easy; yet, such integration would bring models closer to the reality of field systems and is therefore an important path to follow in developing ecological theory.

Fluctuation and correlation corrections

The importance of taking an "individual's-eye view" in determining the response of a population to its environment has already been emphasized. In this context, we must realize that, for populations of finite density, any bounded neighborhood around an individual will comprise only a finite number of individuals. When numbers of neighbors are large, resulting sampling variation is negligible. But for many systems, neighbors around each individual are few enough for the latter to experience substantially varying local environments. We thus have to ask whether and how such fluctuations feed through to population dynamics and their description by moment equations. The answer comes in three steps.

We observe, first, that local fluctuations do not bear on the expected dynamics of moments if individuals respond linearly to the local densities of individuals in their surroundings. Equations (21.1a) and (21.1c) are based on this assumption – per capita birth and death rates in these spatial analogues of Lotka–Volterra competition depend linearly on the pattern of individuals – and therefore no fluctuation-related terms arise in Equations (21.9) or (21.10). Under such circumstances, the mean response of a population to a distribution of environments equals the response predicted for the distribution mean; consequently, local fluctuations have no effect.

Second, even in the absence of spatial structure (i.e., without pair correlations), the behavior of a well-mixed population of individuals that respond nonlinearly to relatively small numbers of neighbors is different from that of a population under mean-field conditions. A typical result is explained in Box 21.5.

Third, we can investigate the implications of nonlinear responses of individuals to local fluctuations in the presence of pair correlations. Let us consider a single species for which per capita birth and death rates are nonlinearly dependent on local densities, so that, analogous to Equations (21.1a) and (21.1c), we have $D(x, p) = d(n(x, p))$ and $B(x, x', p) = b(n(x, p)) m^{(b)}(x' - x)$, with local densities $n(x, p) = \int w(x'' - x)[p(x'') - \delta_x(x'')]dx''$. Under mean-field conditions, the global mean density simply changes according to $\frac{d}{dt}N = f(N)N$, with a per

Box 21.5 Fluctuation corrections in well-mixed systems

Even in the absence of spatial structure, fluctuations in the local environments of individuals can cause departures from mean-field predictions if individuals respond nonlinearly to their environment.

As an example (R. Ferrière, personal communication; Leitner, unpublished), consider a completely randomly distributed population of individuals, whose recruitment from one season to the next is affected by the number of competing neighbors.

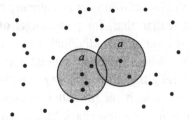

For a mean density N and a uniform weighting of competitors within a neighborhood area a, the average number of neighbors equals aN. Around this average, neighbor numbers vary according to a Poisson distribution: the probability of finding n neighbors is $P_n = \frac{1}{n!}(aN)^n e^{-aN}$. If we assume that recruitment in the absence of any competition is R_0 and that it decreases by a factor c with each competing neighbor, then for n neighbors we have $R_n = R_0 c^n$. The expected per capita offspring number is thus given by $\langle R \rangle = \sum_{n=0}^{\infty} R_n P_n = R_0 e^{-aN} \sum_{n=0}^{\infty} \frac{1}{n!}(acN)^n = R_0 e^{-aN} e^{acN} = e^{\ln R_0 - a(1-c)N}$. If the between-season dynamics depend entirely on recruitment, the next season's mean density is $N' = N\langle R \rangle = Ne^{r(1-N/K)}$, with $r = \ln R_0$ and $K = \frac{\ln R_0}{a(1-c)}$. This is the famous Ricker equation (Ricker 1954) and is quite different from $N' = N R_0 c^{aN}$, the result expected in the absence of fluctuations.

Results of this type are also available for more complicated interactions between individuals (Durrett and Levin 1994b; Czárán 1997) and are always important if individuals respond nonlinearly to small numbers of neighbors. Notice that even though the systems described by these models are assumed to be well mixed, the discreteness of individuals and the finiteness of population densities can result in substantial departures from mean-field predictions. To distinguish such corrections of mean-field results from those arising from spatial structure, we refer to the former as *fluctuation corrections* and to the latter as *correlation corrections*.

capita growth rate $f = b - d$. With local fluctuations and correlations, however, reasoning analogous to that in Section 21.3 shows that the mean-field result ought to be replaced with $\frac{d}{dt}N = [f(\bar{n}) + \frac{1}{2}\sigma^2 f''(\bar{n})]N$, with an average $\bar{n} = N^{-1}\int w(\xi)\tilde{C}(\xi)d\xi$ and a variance σ^2 of local densities. Comparing both dynamics for the global mean density, we observe two corrections relative to the mean-field result. First, a *correlation correction* replaces the global mean density N with the average local density $\bar{n}$ in the argument of the response function f. This reflects the fact that, in the presence of correlations, the average environment around individuals is different from the environment averaged over the entire habitat. Without pair correlations, we have $\tilde{C}(\xi) = N^2$ [see Equation (21.11)], which yields $\bar{n} = N$, thus recovering the mean-field result. Second, a *fluctuation correction* adds the extra term $\frac{1}{2}\sigma^2 f''(\bar{n})$ to the population's response function. Notice that the sign of the term and thus the direction of this correction depends on the response function's curvature at the average local density. We also see that this term vanishes if there are no fluctuations in local densities ($\sigma^2 = 0$) or if the response function is linear ($f'' = 0$).

Other spatial statistics and further reductions of complexity

Low-order correlation densities often capture the essential degrees of freedom in spatial ecological processes. For isotropic systems, spatial complexity can be simplified even further by considering angular averages of pair densities, as explained in Section 21.3 and Figure 21.2. Such a transformation of Equation (21.10) then describes pair densities as a function of radial distance instead of vectorial distance. One further step in removing non-essential degrees of freedom can be accomplished by a short-range expansion of Equation (21.10), approximating pair dynamics by the dynamics of intercept, slope, and curvature of pair densities at distance zero. That slope in particular is often a good measure of a pattern's correlation length, characterizing how rapidly pair correlations decay with distance. Correlation lengths can change in the course of ecological dynamics and thus carry key information about patterns as well as processes.

These extensions emphasize that alternative kinds of spatial statistics may be most appropriate for describing a given ecological system. In fact, correlation functions are not the only choice of statistics suitable for simplifying spatial complexity – for particular systems, other relaxation projections may offer superior performance. Chapter 15 provides an example by constructing a patch-based approximation of a spatial host–parasite system. Instead of tracing through time the probabilistic cellular-automaton dynamics of that system, only the expected average abundances of species in patches of different types are considered, thus yielding a deterministic description of substantially reduced complexity. Other examples are the so-called local structure approximations of Hiebeler (1997) and the pair-edge approximation of Ellner *et al.* (1998). The latter approach is especially suited to a class of large-scale spatial patterns where two types of populations live in adjacent domains and a traveling invasion wave develops at their interface (see also Chapters 16, 17, 22, and 23); for such patterns, the position of the invasion front is the central degree of freedom.

This chapter has introduced a general theory of small-scale heterogeneity in ecological processes that is based on correlation densities. Eventually, a tool kit of successful relaxation projections and associated spatial statistics should become available to help ecologists reduce various kinds of complex spatio-temporal models to manageable approximations.

Appendix 21.A   Derivation of Pair Dynamics

In this appendix we explain how to derive the dynamics of corrected pair densities, presented in Equation (21.10).

The expected rates of change in corrected pair densities $\tilde{C}_{ij}(\xi)$ are determined by the first jump moments of these densities, see Equation (21.6c),

$$\frac{d}{dt}\tilde{C}_{ij}(\xi) = \int a_{ij}(\xi, p) P(p)\, dp \,, \tag{21.13a}$$

with contributions coming from birth, death, and movement events,

$$a_{ij}(\xi, p) = a_{ij}^{(b)}(\xi, p) + a_{ij}^{(d)}(\xi, p) + a_{ij}^{(m)}(\xi, p) \,. \tag{21.13b}$$

Let us focus on the birth contribution first. From Equation (21.6c) and Equation (21.2b), we have

$$a_{ij}^{(b)}(\xi, p) = \int \left[\tilde{C}_{ij}(\xi, p') - \tilde{C}_{ij}(\xi, p) \right]$$
$$\times \sum_k \iint B_k(x, x', p) p_k(x) \Delta(p + u_k \delta_{x'} - p')\, dx\, dx'\, dp' \,, \tag{21.14a}$$

which, after collapsing the integration over p' using the definition of the generalized delta function (see Box 21.4), simplifies to

$$a_{ij}^{(b)}(\xi, p) = \sum_k \iint \left[\tilde{C}_{ij}(\xi, p + u_k \delta_{x'}) - \tilde{C}_{ij}(\xi, p) \right]$$
$$\times B_k(x, x', p) p_k(x)\, dx\, dx' \,. \tag{21.14b}$$

With $\tilde{C}_{ij}(\xi, p + u_k\delta_{x'}) = \frac{1}{A}\int(p + u_k\delta_{x'})_i(x)[(p + u_k\delta_{x'})_j(x + \xi) - \delta_{ij}\delta_x(x + \xi)]\,dx = \frac{1}{A}\int[p_i(x) + \delta_{ik}\delta_{x'}(x)][p_j(x + \xi) + \delta_{jk}\delta_{x'}(x + \xi) - \delta_{ij}\delta(\xi)]\,dx = \tilde{C}_{ij}(\xi, p) + \frac{1}{A}[\delta_{jk}p_i(x'-\xi) + \delta_{ik}p_j(x'+\xi) + \delta_{ik}\delta_{jk}\delta(\xi) - \delta_{ik}\delta_{ij}\delta(\xi)] = \tilde{C}_{ij}(\xi, p) + \frac{1}{A}[\delta_{jk}p_i(x' - \xi) + \delta_{ik}p_j(x' + \xi)]$, we obtain

$$
\begin{aligned}
a_{ij}^{(b)}(\xi, p) &= \\
&= \sum_k \frac{1}{A}\iint \left[\delta_{jk}p_i(x'-\xi) + \delta_{ik}p_j(x'+\xi)\right]B_k(x,x',p)p_k(x)\,dx\,dx' \\
&= \frac{1}{A}\iint \left[p_i(x'-\xi)B_j(x,x',p)p_j(x) + p_j(x' + \xi)B_i(x,x',p)p_i(x)\right]dx\,dx' \\
&= \frac{1}{A}\iint p_i(x)p_j(x'+\xi)B_i(x,x',p)\,dx\,dx' + \langle i,j,\xi \to j,i,-\xi\rangle \ . \quad (21.14c)
\end{aligned}
$$

As in Equation (21.10), the expression $\langle i, j, \xi \to j, i, -\xi\rangle$ is shorthand and stands for all its preceding terms after changing i to j, j to i, and ξ to $-\xi$. Inserting B_i according to Equation (21.1c) yields

$$
\begin{aligned}
a_{ij}^{(b)}(\xi, p) = \frac{1}{A}\iint p_i(x)p_j(x' + \xi)\Big[b_i + \sum_k b'_{ik}\int w_{ik}^{(b)}(x'' - x) \\
\times \left[p_k(x'') - \delta_{ik}\delta_x(x'')\right]dx''\Big]m_i^{(b)}(x' - x)\,dx\,dx' + \langle i,j,\xi \to j,i,-\xi\rangle \ ,
\end{aligned}
\quad (21.14d)
$$

or, when separating terms for density-independent and density-dependent birth,

$$
\begin{aligned}
a_{ij}^{(b)}(\xi, p) = &+ b_i \frac{1}{A}\iint p_i(x)p_j(x' + \xi)m_i^{(b)}(x' - x)\,dx\,dx' \\
&+ \sum_k b'_{ik}\frac{1}{A}\iiint p_i(x)p_j(x' + \xi) \\
&\times \left[p_k(x'') - \delta_{ik}\delta_x(x'')\right]m_i^{(b)}(x' - x)w_{ik}^{(b)}(x'' - x)\,dx\,dx'\,dx'' \\
&+ \langle i,j,\xi \to j,i,-\xi\rangle \ .
\end{aligned}
\quad (21.14e)
$$

The x-integration of $p_i p_j$ in the first term and of $p_i p_j p_k$ in the second resembles the definitions of $\tilde{C}_{ij}$ and $\tilde{T}_{ijk}$ in Equation (21.4a) and (21.4b), respectively. For full compatibility with integrands in these definitions, we must subtract (and then add in again) one self-pair term for $\tilde{C}_{ij}$, $p_i(x)p_j(x' + \xi) = p_i(x)[p_j(x' + \xi) - \delta_{ij}\delta_x(x' + \xi)] + p_i(x)\delta_{ij}\delta_x(x' + \xi)$, and two such terms for $\tilde{T}_{ijk}$, $p_i(x)p_j(x' + \xi)[p_k(x'') - \delta_{ik}\delta_x(x'')] = p_i(x)[p_j(x' + \xi) - \delta_{ij}\delta_x(x' + \xi)][p_k(x'') - \delta_{ik}\delta_x(x'') - \delta_{jk}\delta_{x'+\xi}(x'')] + p_i(x)\delta_{ij}\delta_x(x' + \xi)[p_k(x'') - \delta_{jk}\delta_{x'+\xi}(x'')] + p_i(x)[p_j(x' + \xi) - \delta_{ij}\delta_x(x' + \xi)]\delta_{jk}\delta_{x'+\xi}(x'')$. With $\xi' = x' - x$ and $\xi'' = x'' - x$, we thus obtain

$$a_{ij}^{(b)}(\xi, p) = + b_i \int m_i^{(b)}(\xi') \tilde{C}_{ij}(\xi + \xi', p) \, d\xi'$$

$$+ \delta_{ij} b_i m_i^{(b)}(-\xi) N_i(p)$$

$$+ \sum_k b'_{ik} \iint m_i^{(b)}(\xi') w_{ik}^{(b)}(\xi'') \tilde{T}_{ijk}(\xi + \xi', \xi'', p) \, d\xi' d\xi''$$

$$+ \delta_{ij} m_i^{(b)}(-\xi) \sum_k b'_{ik} \int w_{ik}^{(b)}(\xi'') \tilde{C}_{ik}(\xi'', p) \, d\xi''$$

$$+ b'_{ij} \int m_i^{(b)}(\xi') w_{ij}^{(b)}(\xi + \xi') \tilde{C}_{ij}(\xi + \xi', p) \, d\xi'$$

$$+ \langle i, j, \xi \to j, i, -\xi \rangle .$$

(21.14f)

We now turn to calculating the death contribution,

$$a_{ij}^{(d)}(\xi, p) = \sum_k \int \left[\tilde{C}_{ij}(\xi, p - u_k \delta_x) - \tilde{C}_{ij}(\xi, p) \right] D_k(x, p) p_k(x) \, dx .$$ (21.15a)

With $\tilde{C}_{ij}(\xi, p - u_k \delta_x) = \frac{1}{A} \int [p_i(x') - \delta_{ik} \delta_x(x')][p_j(x' + \xi) - \delta_{jk} \delta_x(x' + \xi) - \delta_{ij} \delta(\xi)] \, dx' = \tilde{C}_{ij}(\xi, p) + \frac{1}{A}[-\delta_{jk} p_i(x - \xi) - \delta_{ik} p_j(x + \xi) + \delta_{ik} \delta_{jk} \delta(\xi) + \delta_{ik} \delta_{ij} \delta(\xi)] = \tilde{C}_{ij}(\xi, p) - \frac{1}{A}\{\delta_{ik}[p_j(x + \xi) - \delta_{jk} \delta(\xi)] + \delta_{jk}[p_i(x - \xi) - \delta_{ik} \delta(\xi)]\}$, we obtain

$$a_{ij}^{(d)}(\xi, p) = -\frac{1}{A} \int p_i(x) \left[p_j(x + \xi) - \delta_{ij} \delta(\xi) \right] D_i(x, p) \, dx$$

$$+ \langle i, j, \xi \to j, i, -\xi \rangle .$$

(21.15b)

After inserting D_i according to Equation (21.1a), this yields

$$a_{ij}^{(d)}(\xi, p) = - d_i \frac{1}{A} \int p_i(x) \left[p_j(x + \xi) - \delta_{ij} \delta(\xi) \right] dx$$

$$- \sum_k d'_{ik} \frac{1}{A} \iint p_i(x) \left[p_j(x + \xi) - \delta_{ij} \delta(\xi) \right]$$

$$\times \left[p_k(x'') - \delta_{ik} \delta_x(x'') \right] w_{ik}^{(d)}(x'' - x) \, dx dx''$$

$$+ \langle i, j, \xi \to j, i, -\xi \rangle .$$

(21.15c)

Notice that the integral in the first term matches the definition of $\tilde{C}_{ij}$. To achieve full compatibility with the definition of $\tilde{T}_{ijk}$ in the second term, we subtract (and then add in again) a self-pair term, as we did in calculating the birth contribution: $p_i(x)[p_j(x + \xi) - \delta_{ij} \delta(\xi)][p_k(x'') - \delta_{ik} \delta_x(x'')] = p_i(x)[p_j(x + \xi) - \delta_{ij} \delta_x(x + \xi)][p_k(x'') - \delta_{ik} \delta_x(x'') - \delta_{jk} \delta_{x+\xi}(x'')] + p_i(x)[p_j(x + \xi) - \delta_{ij} \delta_x(x + \xi)] \delta_{jk} \delta_{x+\xi}(x'')$. With $\xi'' = x'' - x$, we arrive at

$$a_{ij}^{(d)}(\xi, p) = - d_i \tilde{C}_{ij}(\xi, p)$$

$$- \sum_k d'_{ik} \int w_{ik}^{(d)}(\xi'') \tilde{T}_{ijk}(\xi, \xi'', p) \, d\xi''$$

$$- d'_{ij} w_{ij}^{(d)}(\xi) \tilde{C}_{ij}(\xi, p)$$

$$+ \langle i, j, \xi \to j, i, -\xi \rangle .$$

(21.15d)

Finally, we calculate the contribution of movement events to the dynamics of corrected pair densities,

$$a_{ij}^{(m)}(\xi, p) = \sum_k \iint \left[\tilde{C}_{ij}(\xi, p - u_k \delta_x + u_k \delta_{x'}) - \tilde{C}_{ij}(\xi, p) \right]$$
$$\times M_k(x, x', p) p_k(x) \, dx dx' \,. \tag{21.16a}$$

With $\tilde{C}_{ij}(\xi, p - u_k \delta_x + u_k \delta_{x'}) = \frac{1}{A} \int [p_i(x'') - \delta_{ik}\delta_x(x'') + \delta_{ik}\delta_{x'}(x'')][p_j(x'' + \xi) - \delta_{jk}\delta_x(x''+\xi) + \delta_{jk}\delta_{x'}(x''+\xi) - \delta_{ij}\delta(\xi)] \, dx'' = \tilde{C}_{ij}(\xi, p) + \frac{1}{A}[-\delta_{jk}p_i(x-\xi) + \delta_{jk}p_i(x'-\xi) - \delta_{ik}p_j(x+\xi) + \delta_{ik}\delta_{jk}\delta(\xi) - \delta_{ik}\delta_{jk}\delta_x(x+\xi) + \delta_{ik}\delta_{ij}\delta(\xi) + \delta_{ik}p_j(x'+\xi) - \delta_{ik}\delta_{jk}\delta_x(x'+\xi) + \delta_{ik}\delta_{jk}\delta(\xi) - \delta_{ik}\delta_{ij}\delta(\xi)] = \tilde{C}_{ij}(\xi, p) - \frac{1}{A}\{\delta_{ik}[p_j(x'+\xi) - p_j(x+\xi) - \delta_{jk}\delta_x(x'+\xi) + \delta_{jk}\delta(\xi)] + \delta_{jk}[p_i(x'-\xi) - p_i(x-\xi) - \delta_{ik}\delta_x(x'-\xi) + \delta_{ik}\delta(\xi)]\}$, and with $M_i(x, x', p) = m_i(x'-x)$ from Equation (21.1b), we get

$$a_{ij}^{(m)}(\xi, p) = +\frac{1}{A} \iint p_i(x)\big[p_j(x'+\xi) - p_j(x+\xi) $$
$$ - \delta_{ij}\delta_x(x'+\xi) + \delta_{ij}\delta(\xi) \big] m_i(x'-x) \, dx dx' \tag{21.16b}$$
$$+ \langle i, j, \xi \to j, i, -\xi \rangle \,.$$

For matching the definition of $\tilde{C}_{ij}$, we rewrite this equation by splitting the integral in two,

$$a_{ij}^{(m)}(\xi, p) = +\frac{1}{A} \iint p_i(x) \big[p_j(x'+\xi) - \delta_{ij}\delta_x(x'+\xi) \big] m_i(x'-x) \, dx dx'$$
$$-\frac{1}{A} \iint p_i(x) \big[p_j(x+\xi) - \delta_{ij}\delta(\xi) \big] m_i(x'-x) \, dx dx' \tag{21.16c}$$
$$+ \langle i, j, \xi \to j, i, -\xi \rangle \,.$$

With $\xi' = x' - x$, this yields

$$a_{ij}^{(m)}(\xi, p) = +\int m_i(\xi') \tilde{C}_{ij}(\xi + \xi', p) \, d\xi'$$
$$- \tilde{C}_{ij}(\xi, p) \int m_i(\xi') \, d\xi' \tag{21.16d}$$
$$+ \langle i, j, \xi \to j, i, -\xi \rangle \,.$$

Inserting the three results from Equations (21.14f), (21.15d), and (21.16d) into Equations (21.13) recovers the dynamics of corrected pair densities, Equation (21.10).

22

Methods for Reaction–Diffusion Models

Vivian Hutson and Glenn T. Vickers

22.1 Introduction

In the classical literature of theoretical biology, most models were based on the assumption that the effects of physical space could be removed by taking spatial averages. This approach has led to various classes of models, of which the most common are ordinary differential equations (ODEs) and difference equations. Yet, probably the most obvious observation is that in nature the distribution of animal and plant species varies greatly in space, even when the spatial environment is fairly uniform. Early investigators were, of course, aware of this fact, but including spatial structure presented problems of considerable technical difficulty. Progress in mathematical techniques means that at least some important questions can be treated theoretically, and others through a combination of theory and computations on the powerful computers now available. A range of techniques, some of them only recently applied to biology, are discussed in this volume. In this chapter we consider the oldest of the major models, for which a large body of theory and well-developed, powerful, computational methods exist. The model is based on a system of so-called reaction–diffusion equations and is one of a class of what may best be described as continuous models.

Continuous models have long been used to treat discrete systems such as gas dynamics. There the behavior of huge numbers of molecules is described by differential equations. Reactions of chemical mixtures have often been modeled using reaction–diffusion equations – that is, systems of nonlinear (technically, parabolic) partial differential equations. As pointed out by Fife (1979), there are obvious analogues with various processes in living organisms – for example, in the Hodgkin–Huxley model for the propagation of nerve impulses where the electric charge diffuses and interacts with chemicals. Some of the applications in the present volume concern interactions between groups consisting of large numbers of individuals – for example, animals or plants, and bacteria or cells. The key idea is that the system has something in common with those mentioned before; that is, it deals with large numbers and in some sense (although by no means an obvious one) with high densities. Thus, it seems, the idea is to ignore spatial interactions at very small scales. If modeling these small-scale interactions is crucial, then one needs an "individual-based" model, as described in many chapters in this volume. Hence, one can expect that many interesting phenomena cannot be captured by a reaction–diffusion model. Yet such a model may be excellent in many situations and it is a great deal more tractable than individual-based models. In fact, in many situations the two classes of models may lead to results that are qualitatively the same (see Sherratt 1996; Sherratt *et al.* 1997). A vast mathematical theory for reaction–diffusion equations and a considerable body of literature on its application to biological models now exist. It is not our intention to review the theory, rather it is our aim to provide the non-specialist with an overview of some of the techniques that are particularly useful in the biological context and to discuss applications that illustrate the techniques described. In choosing the material we have considered three key issues.

First, we discuss areas where including spatial diffusion gives results significantly different from those of ODE models based on spatial averaging. Intuition might suggest that, because diffusion is a smoothing process, in spatially homogeneous environments the addition of a spatial structure will not have a significant effect. This is not, however, always the case, a fact perhaps first recognized (at least in the biological context) by Turing (1952), some of whose ideas are described in Section 22.3.

Second, a frequently occurring situation in biology concerns so-called bistable models. Suppose that initially only one type of individual is present and a small quantity of another type is introduced, say by mutation. In the bistable situation, the first type is at an asymptotically stable equilibrium, but the system has another stable equilibrium, typically consisting of a population in which only the second type is present. If the mutant has a low density everywhere, it will be eliminated. Thus for a spatially averaged model, the mutant will only invade if initially its density is considerable everywhere. If, however, spatial diffusion is included, a spatially localized "blob" of the second type may invade. The problem is deciding when this will happen and describing how large an initial distribution is required. How the blob may have been formed is a separate issue. Several mechanisms may contribute to its formation, such as localized mating, and often an individual-based model may be needed to explain it. This seems to be a common situation, particularly in genetics and evolutionary game theory.

The third theme concerns another approach to the issue of invasion. In general it may be extremely difficult to carry out the analysis required for the bistable situation just described. A simplification that circumvents problems associated with the boundary of the spatial region and has other technical advantages is based on the following idea. Suppose the spatial region is large enough to be reasonably considered infinite and one-dimensional (the x-axis). Imagine that the two types of individuals initially exist on their own in distinct regions, say $x < 0$ and $x > 0$, separated from each other by an "iron curtain" (which may be a sea). If the curtain is removed (or a land-bridge is formed), it is intuitively reasonable that a "traveling wave front" will develop. Although this approach is one-dimensional, it is a reasonable approximation to the situation in two- or three-dimensional space (for large times) and provides considerable insight – in particular, the direction of the wave front indicates which species will invade. Of special interest here is the role of diffusion in determining this direction.

This chapter has the following structure. Simple derivations of the basic equations in standard situations are described in Section 22.2. We then discuss a model in which the two types of individuals are required to have a "memory," showing that a reaction–diffusion equation can, with care, be derived in this nonstandard situation. This, we believe, is a convincing demonstration of the flexibility of reaction–diffusion equations in modeling some rather complicated biological situations. In Section 22.3, a local technique is described and applied to a Turing-type instability, a classical and beautiful idea that has been applied to a host of models of biological phenomena. Section 22.4 discusses how comparison methods may be used; Section 22.5 outlines the traveling-wave approach to investigating invasion and competition. Finally, in Section 22.6, we present a more technical approach to a problem of the very interesting mutation–selection class, which uses some of the ideas discussed in the other sections.

We emphasize that our account is not intended for mathematicians who are expert in reaction–diffusion equations. Rather it is for mathematically inclined biologists and mathematicians from other fields who require a brief introduction to the topic. For those wishing to investigate the topic more deeply, the best starting point is, perhaps, the excellent book by Fife (1979), which discusses both the derivation of the equations and some of the standard techniques. Britton (1986) and Grindrod (1991) provide other useful accounts. Two recent, highly technical references for the theory are by Leung (1989) and Pao (1992). Henry (1981) has written an essential classic. Murray (1989) contains a mixture of wide-ranging discussions of numerous models and mathematical techniques and is highly recommended. The recent book by Shigesada and Kawasaki (1997) deals specifically with the question of invasion.

Finally, we note that there are well-known methods for the numerical solution of reaction–diffusion equations. For example, the Numerical Algorithms Group (NAG) library contains efficient subroutines for this purpose.

22.2   Continuous Models

In this section we outline a derivation of the basic equations governing the reaction–diffusion model. This derivation is based on the assumption that we are dealing with high population densities and large regions. It is to be expected that a continuous model (i.e., one involving spatial and temporal derivatives) will often provide a good approximation. However, some caution is necessary when interpreting the results of any model to ensure that the underlying assumptions have not been violated.

Random walks and diffusion

Suppose that the individuals of a population move randomly. For simplicity, the spatial region is assumed to be one-dimensional; later it will become clear how the equations should be modified to allow for two- (or even three-) dimensional regions. The derivation follows that of Hoppensteadt (1982). Divide the environment (the x-axis) into sites labeled by an integer m covering the entire x-axis. Let $u_{m,n}$ be the number of individuals in site m at time step n. During one time step, the individuals are assumed to move by at most one site. The probability that they move at all is k; if they do move, they are equally likely to move to the right or left. Thus

$$u_{m,n+1} = \tfrac{k}{2} u_{m+1,n} + (1-k)u_{m,n} + \tfrac{k}{2} u_{m-1,n} \tag{22.1}$$

is the number of individuals that are at site m at the next time step. We now assume that we can write

$$u_{m,n} = u(m\Delta x, n\Delta t) , \tag{22.2}$$

where $u(x, t)$ is a smooth function (in fact twice differentiable in x and once in t). Hence Equation (22.1) can be rewritten as

$$u(m\Delta x, (n+1)\Delta t) - u(m\Delta x, n\Delta t) = \tag{22.3}$$

$$\tfrac{1}{2}k \left[u((m+1)\Delta x, n\Delta t) - 2u(m\Delta x, n\Delta t) + u((m-1)\Delta x, n\Delta t) \right] .$$

Here Δx is the size of a site and Δt is the interval between successive time steps. Intuitively, the existence of u is more likely if Δx and Δt are small.

However, they are not independent in size because of the assumption that an individual can only move to an adjacent site during time Δt. Taylor expansions of u now give the following (where $x = m\Delta x$ and $t = n\Delta t$):

$$u_{m+1,n} = u(x + \Delta x, t)$$

$$= u(x, t) + \Delta x \frac{\partial u}{\partial x} + \frac{(\Delta x)^2}{2} \frac{\partial^2 u}{\partial x^2} + \cdots, \qquad (22.4a)$$

$$u_{m-1,n} = u(x - \Delta x, t)$$

$$= u(x, t) - \Delta x \frac{\partial u}{\partial x} + \frac{(\Delta x)^2}{2} \frac{\partial^2 u}{\partial x^2} - \cdots. \qquad (22.4b)$$

Thus Equation (22.3) implies that

$$\frac{\partial u}{\partial t} = \mu \frac{\partial^2 u}{\partial x^2}, \qquad \text{where} \quad \mu = \frac{k(\Delta x)^2}{2\Delta t}. \qquad (22.5)$$

This is the diffusion equation and μ is called the diffusion coefficient (or diffusivity). The equation is also known as the heat conduction equation.

The above derivation is very cavalier and many serious objections can be raised. [There are discussions by Skellam (1973), Okubo (1980), and Fife (1979) for those requiring a more rigorous approach.] Murray (1989) provides an alternative argument. He shows that, for a random walk, the probability of finding a particular individual in the spatial interval $(x, x + \Delta x)$ and in the time interval $(t, t+\Delta t)$ approaches the Gaussian distribution

$$\frac{1}{\sqrt{4\pi \mu t}} \exp\left(-\frac{x^2}{4\mu t}\right). \qquad (22.6)$$

Remarkably, this approximation is accurate even when the number of sites and time steps is quite modest – Murray suggests six is sufficient. Now the Gaussian distribution is the solution of the diffusion equation when the initial distribution is $\delta(x)$, a Dirac delta function, or an "atom of probability" at $x = 0$. It is certainly encouraging that the limiting form of the solution to the random-walk problem satisfies the diffusion equation, but this derivation leaves the precise relationship unclear.

The classic approach to diffusion is through Fick's law, which states that the flux J of a quantity (which may be the concentration of a chemical, the number density of a certain species of animal, or heat per unit volume) is proportional to the gradient of the concentration u of that quantity. Thus

$$J = -\mu \frac{\partial u}{\partial x}, \qquad (22.7)$$

where μ is the diffusion coefficient. Assuming that the quantity being considered is conserved, it can be shown that

$$\frac{\partial u}{\partial t} = \frac{\partial}{\partial x}\left(\mu \frac{\partial u}{\partial x}\right),$$

(22.8)

which is the diffusion equation again. The advantage of this last approach is that it indicates how one should treat situations in which the diffusion coefficient is not a constant but instead depends on x or u. There is evidence that, for some insect dispersal problems, μ should increase with the population density. Okubo (1986) presents a review of various models and their properties.

All models have limitations and the dispersal model above is no exception. The issue usually is whether to use a cellular model (discrete in space and/or time) instead of the continuous model derived above (see Chapter 9). Admittedly, there is no known simple answer. A point of criticism of the continuous model is that it predicts waves of infinite speed. This would seem to be a fatal flaw, but actually the effect is quite small. The "Gaussian tail" might propagate unreasonably, but $\exp(-x^2)$ decays remarkably quickly. Even granted that its propagation is spurious, its effect upon computations is truly negligible. As indicated above, the Gaussian distribution only requires very modest values of x and t to be accurate (it mirrors the amazing accuracy of Stirling's formula for $n!$). It is certainly true that cellular models have a richness of behavior that continuous models lack. This does not prove the inadequacy of continuous models; rather, the proponents of cellular models must demonstrate that the exotic behavior that they are predicting does actually occur in the systems that they are purporting to model. Finally, in at least some cases the qualitative predictions of the two approaches are found to be quite similar [see Sherratt (1996) and Sherratt *et al.* (1997), who present careful comparisons of the two approaches].

If the spatial region is not one-dimensional, then the above argument must be modified. In the present context it would be inappropriate to go into details, and we do not consider regions of higher dimensions. However, for those familiar with vector notation, we note that the form of the equations may be very easily written down. The general equation for the effect of diffusion alone is

$$\frac{\partial u}{\partial t} = \nabla \cdot (\mu \nabla u)$$

$$= \mu \nabla^2 u \quad \text{(if } \mu \text{ is constant)} .$$

(22.9)

In the mathematical literature, Δ is usually used for the diffusion operator rather than ∇^2. The final equation for modeling a single species is obtained by simply adding a term that models self-interaction. This term is known as the reaction term, and the corresponding system of ODEs is known as the reaction system. Thus if the growth rate of the population density u of a species is $f(u, t)$ in the absence of spatial effects (i.e., when u is constant in space), then we posit the partial differential equation (PDE)

$$\frac{\partial u}{\partial t} = f(u, t) + \mu \frac{\partial^2 u}{\partial x^2} \qquad (22.10)$$

when the spatial region is one-dimensional and the diffusion coefficient is a constant. The case of many species/strategies simply means that we interpret u as a vector.

In biology, interest centers on a finite spatial region. Sometimes, if this region is very large, the boundary has only a local effect on the solution and its principal features can be well approximated by assuming that the region is infinite. However, sometimes this approximation may not be valid, and then boundary conditions must be prescribed. The most common assumption is that there is no migration across the boundary, that is, it is impermeable. This implies, in mathematical terminology, "zero Neumann boundary conditions." With one-dimensional space, if the region is $[0, l]$ then the corresponding conditions are

$$\frac{\partial u}{\partial x}(0) = 0 = \frac{\partial u}{\partial x}(l) \quad \text{for all } t . \qquad (22.11)$$

More general diffusion processes

A model that is based on an increase in diffusion due to population pressure has the nonlinear diffusion term

$$\frac{\partial}{\partial x}\left(D(u)\frac{\partial u}{\partial x}\right) , \qquad (22.12)$$

where D is an increasing function. A common assumption is that $D(u) = u^m$, where $m > 1$. However, this assumption is probably biologically unrealistic because it predicts that the diffusion approaches zero as the population density u approaches zero. A "good" feature of this type of model is that disturbances propagate with finite speed (see Grindrod 1991). A "bad" feature is that the disturbances are much more difficult to handle mathematically. [For a sample paper in this area, see Grindrod and Sleeman (1987) and the references therein.]

It is probable that in many situations dispersal is not a local effect. One obvious scenario is the wide dispersal of seeds by wind. This is an important area of research that has not received a great deal of attention. The following model is continuous in space but discrete in time:

$$N_{t+1}(x) = \int_{-\infty}^{\infty} k(x, y) \, f[N_t(y)] \, dy \, , \qquad (22.13)$$

where N_t is the density at time step t. [For a discussion of this equation see Kot *et al.* (1996) and Chapters 16 and 23.]

A model for invasion of *Tit For Tat*

This section demonstrates how a model can be constructed for a specific problem: the evolution of cooperation via the iterated Prisoner's Dilemma (IPD). Some of the features of this problem are likely to be encountered in almost every ecological situation, such as birth/death rates and population limitation effects; others may be novel, such as the incorporation of a memory effect. It is not appropriate here to give a review of the vast IPD literature or even a very detailed account of the model (see Hutson and Vickers 1995). Rather, the aim is to show the versatility of continuous models, even within the classical framework of reaction–diffusion equations. Some groundwork is required, however, to make the features of the model intelligible.

Consider a population in which whenever a pair of individuals meet there is a contest. In this contest each player must choose either to play C (cooperate) or D (defect). The payoffs for the various types of plays are as follows:

		Player 2 plays:	
		C	D
Player 1 plays:	C	α	β
	D	γ	δ

Here, the payoff to each player is measured in per capita growth rate and the entries are ordered so that

$$\gamma > \alpha > \delta > \beta \quad \text{and} \quad 2\alpha > \beta + \gamma \, . \qquad (22.14)$$

If the players only meet once, then the only rational strategy is for each to defect. However, the situation changes if the same two players compete repeatedly, leading to the IPD. There are now more options available to the players because each may decide to play C or D in a round using any rule of their choice. The rules adopted here are D (which now means defect in every round) and T, the *Tit For Tat* (*TFT*) strategy. A T-player always starts by cooperating and then repeats his or her opponent's last play. Now consider a population of individuals, each of whom is either a T- or a D-player. These individuals are able to move around in space and so meet different opponents. At each meeting a single round of the game is played. For the T strategy to make sense, T-players must be able to recognize a past adversary. During a first encounter – that is, with an unrecognized opponent – a T-player always cooperates. If a T-player meets someone on a second (or later) occasion, he or she instantly recognizes that opponent, remembers what the opponent played last time, and plays the same.

Suppose that the environment, taken as the x-axis, is divided into contiguous cells or sites of size l. We discretize time into units Δt, a time interval in which a fraction k of the individuals in a cell leave that cell. Players may also leave the cell by dying or by moving to an adjacent cell; new players may enter by being born or by diffusing from an adjacent cell. Simple clonal reproduction – that is, like begets like – is assumed. A T-player will recognize an opponent on subsequent occasions provided that neither has left the cell where the encounter occurred.

Suppose that a typical cell contains lu T-players and lv D-players, so that u and v are the line densities of the two player types. Let the number of D-players that a typical T-player has already met (and so will recognize the next time they meet) be gvl. This T-player can meet any of three types of individuals and the fraction of time that will be occupied by each type of meeting is as follows:

T-player	Known D-player	Unknown D-player
$\dfrac{u}{u+v}$	$\dfrac{gv}{u+v}$	$\dfrac{(1-g)\,v}{u+v}$

The corresponding payoffs are α, δ, and β, implying that the rate of increase of the T-players is

$$u\left[\frac{u\alpha}{u+v} + \frac{gv\delta}{u+v} + \frac{(1-g)v\beta}{u+v}\right] = \frac{u}{u+v}(\alpha u + \beta v) + \frac{G(\delta-\beta)}{u+v} , \quad (22.15a)$$

where $G = guv$. Similarly the rate of increase of D-players due to interactions is

$$v\left[\frac{v\delta}{u+v} + \frac{gu\delta}{u+v} + \frac{(1-g)u\gamma}{u+v}\right] = \frac{v}{u+v}(\gamma u + \delta v) - \frac{G(\gamma-\delta)}{u+v} , \quad (22.15b)$$

because a typical D-player will have already met gul of the T-players in its cell and so only receives δ when it plays them again. These growth rates must be modified by population limitation terms, which we take to have the forms $uF(u+v)$ and $vF(u+v)$ for the T- and D-players, respectively. In the discussion that follows, it is assumed that logistic growth is appropriate, but other limitation terms would be equally possible. Thus we have

$$F(U) = b - (d + \sigma U) ,$$

(22.16)

where b and d are the intrinsic birth and death rates, respectively, and σ measures the importance of overcrowding. The final effect to be considered is diffusion, which is modeled by a random walk. If k_T is the probability that a T-player moves out of a cell and into an adjacent cell in time Δt, then the diffusion rate of the T-players is $\mu_T = k_T l^2/(2\Delta t)$. Similarly, the diffusion rate of the D-players can be written as $\mu_D = k_D l^2/(2\Delta t)$. The final form of the equations for the growth of the populations is

$$\frac{\partial u}{\partial t} = \frac{u}{u+v}(\alpha u + \beta v) + \frac{G}{u+v}(\delta - \beta)$$
$$+ uF(u+v) + \mu_T \frac{\partial^2 u}{\partial x^2} ,$$

(22.17a)

$$\frac{\partial v}{\partial t} = \frac{v}{u+v}(\gamma u + \delta v) - \frac{G}{u+v}(\gamma - \delta)$$
$$+ vF(u+v) + \mu_D \frac{\partial^2 v}{\partial x^2} .$$

(22.17b)

The possible variation of G in time and space now must be considered. Recall that $G = guv$ is the number density of $T - D$ pairs within a cell that have already met. This density will be affected by deaths, emigration, and encounters. The first two of these will change G but not g. Thus, if Δu, Δv are the changes in u, v due to the combined effects of death and emigration in the time interval Δt, then the corresponding change in G is

$$g(u\Delta v + v\Delta u) = G\left(\frac{\Delta u}{u} + \frac{\Delta v}{v}\right)$$
$$= -G\left(2d + 2\sigma(u+v) + k_T \frac{T}{\Delta t} + k_D \frac{D}{\Delta t}\right)\Delta t .$$

(22.18)

The number of encounters per unit time in a cell is taken to be proportional to the number of pairings within that cell. The encounter rate m is used for the constant of proportionality. It follows that the number of encounters that each individual has per time unit is proportional to the local value of $u + v$, and also that the change in G due to meetings in the time interval Δt is $(uv - G) \, m\Delta t$. Thus we have

$$\frac{\partial G}{\partial t} = muv - G \left(m + 2d + 2\sigma (u + v) + \theta \right) , \qquad (22.19a)$$

where

$$\theta = \frac{2}{l^2} \left(\mu_T + \mu_D \right) . \qquad (22.19b)$$

The full system, given by Equations (22.16), (22.17), and (22.19), provides a continuous model for the spatial IPD game. This model is analyzed extensively in Hutson and Vickers (1995). For a review of this and other models, see Chapter 17.

22.3 Linearized Stability and the Turing Bifurcation

For systems of ODEs, one of the simplest, best-known, and most useful techniques for examining the stability of an equilibrium point is linearization. The results from this analysis are somewhat limited in that they only predict the behavior in the neighborhood of the equilibrium point; nonetheless, they often provide valuable, if tentative, insight into the behavior of the system. An analogous type of analysis, albeit often more complicated, holds for reaction–diffusion systems, and this is outlined below. We will then be able to consider the mechanism of a Turing bifurcation by which spatial patterns may be produced.

To introduce the technique for the treatment of the linearized stability, we consider first a single species with density u and suppose that the spatial domain is inhomogeneous but one-dimensional, say $[0, 1]$. Recall that the boundary is assumed to be impermeable. For logistic self-interaction, the governing equations are as follows:

$$\frac{\partial u}{\partial t} = u[a(x) - u] + \mu \frac{\partial^2 u}{\partial x^2} , \quad \text{with} \quad \frac{\partial u}{\partial x} = 0 \quad (x = 0, 1) . \quad (22.20)$$

Here the function a represents the spatial environment. An important question from the biological point of view is, If a small distribution of the species is introduced into a region where there are initially none, will the species become established or will it die out? Intuition may suggest that the outcome depends on whether the spatial average $\hat{a} = \int_0^1 a(x) \, dx$ is positive or negative, but we shall see that this guess is quite misleading.

In preparation for answering this question, recall that for the system of ODEs

$$\frac{dv}{dt} = g(v) , \tag{22.21}$$

where v is an n-vector and $g(0) = 0$, the stability of the equilibrium point $v = 0$ is found by considering the system

$$\frac{dv}{dt} = Av , \quad A = \left(\frac{\partial g_i}{\partial v_j}\right)_0 , \tag{22.22}$$

obtained by linearizing Equation (22.21) about 0. The substitution $v = we^{\lambda t}$ (where w is a constant vector) leads to

$$\lambda w = Aw . \tag{22.23}$$

Thus the solution of the linear system can be written as

$$v = \sum_{i=0}^{n-1} C_i \, w_i \, e^{\lambda_i t} , \tag{22.24}$$

where λ_i and w_i are the eigenvalues and (normalized) eigenvectors of A, respectively, and C_i are constants. If any eigenvalue has a positive real part, then 0 will be unstable and v will increase (from initially small values). Clearly, $u = 0$ is a solution of Equation (22.20), and we follow a similar argument. Consider, therefore, the linearization of Equation (22.20) about $u = 0$. This leads to the equation

$$\frac{\partial u}{\partial t} = a(x)u + \mu \frac{\partial^2 u}{\partial x^2} . \tag{22.25}$$

The substitution $u = e^{\lambda t}\phi(x)$ gives

$$\mu \frac{d^2\phi}{dx^2} + a(x)\phi = \lambda\phi , \tag{22.26}$$

with the boundary conditions

$$\frac{d\phi}{dx} = 0 \quad \text{at } x = 0, 1 . \tag{22.27}$$

Equation (22.23) has nontrivial solutions only if λ has certain special values and Equations (22.26) and (22.27) have exactly the same property. In the case of Equations (22.26) and (22.27), these special values, which are still called eigenvalues, are known to be real and so may be ordered starting with the largest (which will be finite). Thus we have a sequence λ_n of eigenvalues with $\lambda_n \to -\infty$ as $n \to \infty$ and associated (normalized) eigenfunctions ϕ_n (taking the place of eigenvectors in the algebraic problem). Furthermore the "principal" eigenfunction ϕ_0, corresponding to the largest eigenvalue λ_0, can be shown to be positive everywhere. The general solution to Equation (22.20) can now be written as

$$u = \sum_{i=0}^{\infty} C_i \, \phi_i \, (x) e^{\lambda_i t} \, . \tag{22.28}$$

It follows that $u = 0$ is stable if $\lambda_0 < 0$ and unstable if $\lambda_0 > 0$. Furthermore, the theory ensures that the linearized system has the same stability properties as the full nonlinear system Equation (22.20). Although further arguments are needed to fully justify the assertion, this observation suggests (as is indeed the case) that whether or not the species becomes established depends on the sign of the principal eigenvalue λ_0. Comparing this conclusion with the ODE theory for a system, we see that we have exchanged the problem of calculating the largest eigenvalue of a matrix for the same problem for a linear, ordinary differential operator. If a is a constant, then, of course, $\lambda_0 = a$. Otherwise the second problem is considerably more difficult; indeed if the environment is (more realistically) two-dimensional, the analogue of Equations (22.26) and (22.27) is an eigenvalue problem for a linear elliptic, partial differential operator, which is even more difficult to handle. It is, however, a very well-known problem in many contexts and there are well-established techniques for dealing with it. One such technique is the Rayleigh–Ritz variational principle, which makes it quite easy to prove the following. Assuming that $a(x) > 0$ for some x (but not necessarily that a is positive everywhere), λ_0 is a decreasing function of μ and $\lambda_0 \to \hat{a}$ as $\mu \to \infty$, while $\lambda_0 \to \max a(x)$ if $\mu \to 0$. Referring back to the intuition previously discussed, we see that this is quite correct for large diffusion rates: if $\hat{a} > 0$ then $\lambda_0 > 0$ and the species will become established. However, if μ is small, it is enough that $a(x) > 0$ somewhere (and hence $\lambda_0 > 0$) for the species to become established – a conclusion quite different from that suggested. It is clear that the diffusion rates play an important role and that conclusions using spatial averages and an ODE model must be treated with suspicion. A final remark: it can be shown that if $\lambda_0 > 0$, then there is a nonconstant, positive, stationary (i.e., time-independent) solution that is globally stable in the sense that for any initial condition that is not identically zero, the solution approaches this stationary state.

Consider next the application of similar ideas for systems. In general, the linearization leads to a linear eigenvalue problem for a system of elliptic equations, which is a great deal more difficult to handle than Equations (22.26) and (22.27). However, in a limited but important class of examples, a great deal of progress is possible. We illustrate such a situation by outlining an argument that yields an extremely interesting biological conclusion, the existence of spatial patterns. For simplicity, we restrict the analysis to a pair of equations in one spatial dimension that is assumed to be infinite, and we assume that the environment is spatially homogeneous (so that it does not enter explicitly into the reaction terms).

Suppose that $\bar{u} = (u_1, u_2)$ is a constant equilibrium point and that it is asymptotically stable for the reaction system. Diffusion is generally thought of as a smoothing or stabilizing process, and most people's intuition would suggest that if spatial diffusion were included, then $\bar{u}$ would also be stable for the full reaction–diffusion system. With remarkable insight, Turing observed that, under certain circumstances, diffusion can *destabilize* a stable equilibrium and a pattern can be formed (here "pattern" has the conventional meaning of a stable, stationary, spatially inhomogeneous solution). Turing proposed this in a contribution (Turing 1952) regarded by Murray (1989) to be one of the most important papers in theoretical biology of the twentieth century.

Consider the pair of equations

$$\frac{\partial u_i}{\partial t} = u_i f_i(u) + \mu_i \frac{\partial^2 u_i}{\partial x^2} \quad (i = 1, 2) , \tag{22.29}$$

where the spatial domain is the x-axis. Linearizing them about $\bar{u}$ gives the system

$$\frac{\partial u_i}{\partial t} = (Au)_i + \mu_i \frac{\partial^2 u_i}{\partial x^2} , \tag{22.30}$$

where A is a constant 2×2 matrix. The following conditions are imposed on the system.

(1) $\bar{u}$ is asymptotically stable for the reaction system. Thus the eigenvalues of A have negative real parts and

$$\text{trace}(A) = a_{11} + a_{22} < 0 , \tag{22.31a}$$

$$\det A = a_{11}a_{22} - a_{12}a_{21} > 0 . \tag{22.31b}$$

(2) $a_{11} > 0, a_{22} < 0$ and $a_{12} < 0, a_{21} > 0$.

To test whether $\bar{u}$ remains stable when diffusion is included, $u = \alpha e^{\lambda t} \cos kx$ is substituted into Equation (22.29); it is found that there is a nontrivial solution if and only if

$$\begin{vmatrix} \lambda - a_{11} + k^2 \mu_1 & -a_{12} \\ -a_{21} & \lambda - a_{22} + k^2 \mu_2 \end{vmatrix} = 0 . \tag{22.32}$$

The solutions λ have negative real parts if and only if both

$$a_{11} + a_{22} - (\mu_1 + \mu_2) k^2 < 0 , \tag{22.33a}$$

and

$$H(k) \overset{\text{def}}{=} \mu_1 \mu_2 k^4 - (\mu_1 a_{22} + \mu_2 a_{11}) k^2 + \det A > 0 . \tag{22.33b}$$

Because Equation (22.31a) obviously implies Equation (22.33a), for an instability we require $H(k) < 0$, which is possible if

$$\mu_1 a_{22} + \mu_2 a_{11} > 2(\mu_1 \mu_2 \det A)^{1/2} . \tag{22.34}$$

We first note that Equation (22.34) cannot hold when $\mu_1 = \mu_2$, since this would require $a_{11} + a_{22} > 0$, which contradicts Equation (22.31a). However, with $\mu_2 = 1$ and a small enough μ_1, Equation (22.34) will be satisfied. For appropriate k there will then be an eigenvalue λ with positive real part and the solution of Equation (22.29) will grow. The analysis can be extended to a finite space domain by choosing k so that $\cos kx$ satisfies the boundary conditions given by Equation (22.11). The analysis can also be extended to rectangular and cuboid regions in two and three dimensions, respectively, without great difficulty. Extension to general regions is possible, but the calculations are considerably more difficult.

We have described the basic mechanism of a Turing instability, which is caused by having "very unequal" diffusion rates. A key assumption is (2) above, which ensures that the pair of species is of the "activator–inhibitor" type, with species 1 being the activator and species 2 the inhibitor. It is not easy to see why this assumption leads to an instability and thus to patterns. An explanation is offered by Edelstein-Keshet (1988). The discovery of this phenomenon required considerable intuition on the part of Turing.

This mechanism has been much exploited. A particularly interesting application is to the modeling of animal-coat patterns (see Murray 1988). The work of Meinhardt (1982) and Murray (1989) is strongly recommended for background information; Edelstein-Keshet (1988) also provides an attractive, introductory account. More mathematical details are presented by Fife

(1979) and Britton (1986). An application, in the somewhat novel context of game theory, to the spatial version of the *TFT* game (see Section 22.2) and the evolution of cooperation is described by Hutson and Vickers (1995).

22.4 Comparison Methods

The analysis described in Section 22.3 outlines a method for tackling the behavior of a system in a small neighborhood of an equilibrium point and thus is local in character. Of course, most equations in biology require the analysis of solutions that are not so severely restricted and are thus of a global nature. The global analysis of the corresponding reaction system may often itself be highly nontrivial, and the addition of a spatial structure adds greatly to the difficulty. One of the most powerful techniques available is based on the idea of bracketing the solution between certain comparison functions; these functions are not solutions but satisfy weaker conditions. The method is well known in the classical theory of ODEs (see, e.g., Miller and Michael 1982) and has been intensively developed in the past 30 to 40 years in the reaction–diffusion context. There is, of course, not enough space here to review the topic (see Leung 1989; Pao 1992); our aim is to give an idea of the analysis by concentrating on one particular, but representative, technique.

Consider the scalar equation

$$\frac{\partial u}{\partial t} = f(u) + \frac{\partial^2 u}{\partial x^2} \tag{22.35}$$

and assume for simplicity that the spatial domain is the x-axis.

Definition: $\underline{u}$ is called a subsolution of Equation (22.35) if it is smooth and satisfies

$$\frac{\partial \underline{u}}{\partial t} \leq f(\underline{u}) + \frac{\partial^2 \underline{u}}{\partial x^2} . \tag{22.36}$$

A supersolution, $\bar{u}$, satisfies the opposite inequality.

Notice that a solution is both a sub- and supersolution. The idea, of course, is that it is often much easier to find sub- and supersolutions than to solve Equation (22.35).

Theorem 22.1 Let $\underline{u}, \bar{u}$ be sub-/supersolutions, respectively, with $\underline{u}(x, 0) \leq \bar{u}(x, 0)$. Then one of the following alternatives must hold for $t > 0$: (1) $\underline{u} < \bar{u}$, or (2) $\underline{u} = \bar{u}$.

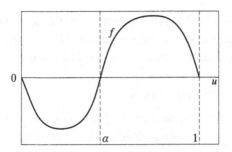

Figure 22.1 This form of the reaction term leads to bistable dynamics.

Theorem 22.2 Let $\underline{u}(x)$ be a stationary subsolution and u a solution satisfying $u(x, 0) = \underline{u}(x)$. Then either $u = \bar{u}$ or u is strictly increasing in t for each fixed x.

There are extensions of these results to finite spatial domains with impermeable boundaries, but they are a little more complicated to state. We now illustrate the use of these theorems by means of two examples.

Suppose the density u of a certain species satisfies the equation

$$\frac{\partial u}{\partial t} = u(1 - u) + \mu \nabla^2 u \,, \tag{22.37}$$

and $M \geq u(x, 0) \geq m > 0$ for all x. Let $\underline{u}(t)$, $\bar{u}(t)$ be solutions of the following initial value ODE problems:

$$\frac{d\underline{u}}{dt} = \underline{u}\,(1 - \underline{u}) \,, \qquad \underline{u}(0) = m \,, \tag{22.38a}$$

$$\frac{d\bar{u}}{dt} = \bar{u}\,(1 - \bar{u}) \,, \qquad \bar{u}(0) = M \,. \tag{22.38b}$$

Of course, we know that $\underline{u}(t)$ and $\bar{u}(t) \to 1$ as $t \to \infty$. From Theorem 22.1 (used twice) we have $\underline{u}(t) \leq u(x, t) \leq \bar{u}(t)$ for $t > 0$. Hence $u(x, t) \to 1$. Although this is a somewhat trivial example, it nicely demonstrates the idea of bracketing the solution between easily found sub- and supersolutions.

As a second example, we consider the equation

$$\frac{\partial u}{\partial t} = f(u) + \mu \nabla^2 u \,, \tag{22.39}$$

where f is as in Figure 22.1. This is a bistable example, where $u = 0$ and $u = 1$ are attractors for the reaction system. It is easy to show that if $u(x, 0)$ is near 0 (respectively near 1), then $u(x, t) \to 0$ (respectively 1). An interesting question is to decide under what circumstances an

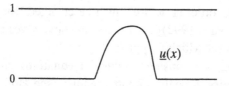

Figure 22.2 A subsolution of compact support.

initial distribution of the species will become established. To tackle this, an extension of the idea of a smooth subsolution is needed. This is quite easy to obtain, although we do not describe the technicalities here (for details, see Fife 1979). The argument leads to the following result. Suppose that $\int_0^1 f(u)\,du > 0$. Then there is a function $\underline{u}(x)$ of compact support (that is, zero except on a finite interval), such that if $u(x, 0) \geq \underline{u}(x)$, then $u(x, t) \rightarrow 1$ as $t \rightarrow \infty$ for each x. Figure 22.2 illustrates the meaning of this result. If the initial value of u is such that it exceeds some critical "blob," then u will spread and approach its saturation level everywhere. Notice that this result is of quite a different character from that for the reaction system where one would need $u(x, 0) > \alpha$ for all x. Fixation is much easier in the spatial model. Thus 1 is an attractor in a very strong sense. This has led to the use of the term "dominant attractor" (see Fife 1979).

Let us now consider the extent to which these results may be extended to systems. For illustration, consider a pair of equations:

$$\frac{\partial u}{\partial t} = f(u, v) + \mu \nabla^2 u , \qquad \frac{\partial v}{\partial t} = g(u, v) + \nu \nabla^2 v . \qquad (22.40a)$$

A subsolution pair $(\underline{u}, \underline{v})$ is required to satisfy

$$\frac{\partial \underline{u}}{\partial t} \leq f(\underline{u}, \underline{v}) + \mu \nabla^2 \underline{u} , \qquad \frac{\partial \underline{v}}{\partial t} \leq g(\underline{u}, \underline{v}) + \nu \nabla^2 \underline{v} . \qquad (22.40b)$$

The system is said to be quasi-monotone increasing if

$$\frac{\partial f}{\partial v} \geq 0 , \qquad \frac{\partial g}{\partial u} \geq 0 . \qquad (22.41)$$

Theorem 22.3 Theorems 22.1 and 22.2 are also valid for systems that are quasi-monotone increasing.

From the viewpoint of tackling biological problems, the extra condition required presents a serious difficulty. In effect it means that the system is mutualistic. There is one important exception. For two species, if $\partial f / \partial v \leq 0$, $\partial g / \partial u \leq 0$ (so that the system is competitive), then there is an analogous result. In fact the "order structure" must be changed [for more details, see Leung (1989), or Pao (1992)]. The theorem above gives much information for those systems for which it applies.

For illustration, we consider a class of equations that arise in evolutionary game theory, namely, bimatrix games (see Hofbauer *et al.* 1997; Hofbauer 1998). Another example involving a pair of obligate mutualists is discussed in Hutson (1986). Bimatrix games have been used to model asymmetric contests both in biology (see Hofbauer and Sigmund 1988, p. 139) and in economics (see Harsanyi and Selten 1988). The background in a simple case is as follows.

There are two types of players, A and B, each with two options. A-players choose to play option i with probability p_i and B-players choose option j with probability q_j. The payoff to an A-player when it plays option i against an opponent playing option j is a_{ij}, and the payoff to the opposing B-player is b_{ji}. Thus the expected payoff to an A-player when it plays option i is $(Aq)_i$, where $A = (a_{ij})$. We write

$$p_1 = u , \quad p_2 = 1 - u , \quad q_1 = v , \quad q_2 = 1 - v . \tag{22.42}$$

Adding an evolutionary-dynamic perspective to the game leads to the replicator dynamic (see Hofbauer and Sigmund 1988, p. 141). In an economics context, it is often reasonable to fix the total number of players, which yields the following pair of reaction–diffusion equations to model spatial structure and diffusion:

$$\frac{\partial u}{\partial t} = u(1-u)[(a_1 + a_2)v - a_1] + \mu \frac{\partial^2 u}{\partial x^2} , \tag{22.43a}$$

$$\frac{\partial v}{\partial t} = v(1-v)[(b_1 + b_2)u - b_1] + \nu \frac{\partial^2 v}{\partial x^2} , \tag{22.43b}$$

where $a_1 = a_{22} - a_{12}$, $a_2 = a_{11} - a_{21}$, $b_1 = b_{22} - b_{12}$, $b_2 = b_{11} - b_{21}$. In a biological context the number of players (i.e., the population size) should also be determined by the model. In this case modifications are necessary, but we do not pursue this here. Both the strategy pairs $(0, 0)$ and $(1, 1)$ are stable equilibria. The key question is which is "stronger." We take over the idea of "dominance" for the one-species model and inquire under what conditions there is a subsolution of compact support – in which case $(1, 1)$ is reasonably described as dominant. The system is clearly quasi-monotone increasing and the comparison theorems may be used. We obtain the following result (for details see Hutson 1986). Let f and g denote the reaction terms in the u and v equations, respectively. Define

$$h(u) = \min[\mu^{-1} f(u, u), \nu^{-1} g(u, u)] . \tag{22.44}$$

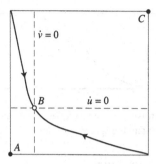

Figure 22.3 The phase plane for the reaction system corresponding to Equations (22.43). A and C are asymptotically stable and B is a saddle. The stable manifold of B is also shown.

If $\int_0^1 h(s)\,ds > 1$, there is a subsolution pair $(\underline{u}, \underline{v})$ of compact support. Further, if $u(x, 0) \geq \underline{u}(x)$ and $v(x, 0) \geq \underline{v}(x)$, then $u(x, t)$ and $v(x, t)$ converge to 1 uniformly on compact sets.

This last example shows, in a less trivial context, behavior similar to that of the single species problem described earlier. The phase portrait of the reaction system, that is, the ODE model, is sketched in Figure 22.3. Orbits to the right of the stable manifold of B tend to C; those to the left tend to A. Thus for strategy $(1, 1)$ to win, the initial distribution must be fairly large at all spatial points. For the reaction–diffusion model, a distribution that is initially spatially limited may successfully invade.

We close by remarking that a considerable body of theory has been built up over the past 15 years or so concerning the dynamics of quasi-monotone systems [largely due to Hirsch and Smith; for a complete review, see Smith (1995)]. The theory is both attractive and powerful. However, even in the ODE case, it suffers from the severe restriction (from the biological point of view) that its applicability is limited by the quasi-monotone condition. Systems that do not satisfy this condition are more difficult to treat, although there is a considerable literature dealing with them.

22.5   Traveling Waves

Continuing with the plan of investigating the question of invasion and competition between types, we outline another approach that can be very effective. It is assumed that the spatial region is large so that the effect of the boundary can be ignored and space is one-dimensional. This idea was introduced into biology in the classic work of Fisher (1937) in the study of the spatial spread of genes (see also Aronson and Weinberger 1975), and has been much extended (see Murray 1989, p. 276). For illustration, consider

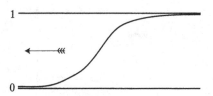

Figure 22.4 The shape of a traveling wave at a particular moment in time. If $c < 0$ then the wave is traveling to the left and the species invades.

the scalar reaction–diffusion equation

$$\frac{\partial u}{\partial t} = f(u) + \frac{\partial^2 u}{\partial x^2} , \tag{22.45}$$

where the diffusion parameter μ has been removed by a rescaling of the spatial variable x by $\sqrt{\mu}$. Suppose that, as in Figure 22.1,

$$f(u) = u(1 - u)(u - \alpha) . \tag{22.46}$$

Then the reaction system on $[0, 1]$ has two stable equilibria, 0 and 1, and α is unstable. We look for solutions of the form $u(x - ct)$, and with $z = x - ct$, require that $u(z) \to 0$ as $z \to -\infty$ and $u(z) \to 1$ as $z \to \infty$. In this particular case the solution may be found explicitly and is shown in Figure 22.4. It is

$$u(z) = \frac{1}{1 + e^{-z/\sqrt{2}}} , \quad z = x - ct , \quad c = \sqrt{2}(\alpha - 1/2) . \tag{22.47}$$

Note that $c > 0$ if $\alpha > 1/2$ (or $c < 0$ if $\alpha < 1/2$). If c is positive, then the wave travels to the right and the equilibrium $u = 0$ "wins." Thus whether or not the species invades, and indeed its speed of invasion, is determined by the sign of c. From a biological point of view, c is thus a crucial quantity.

Before tackling the mathematical background we remark that, strictly speaking, we are considering traveling fronts, but, as in most of the literature, we use the term traveling wave. In some circumstances in multi-species problems there are periodic traveling waves and traveling pulses, but we do not consider these here. For an interesting example concerning impulses in nerves, see Rinzel and Terman (1982).

For the above choice of f the solution could be found explicitly. We suppose next that f is a general function with the same sort of shape, so that 0, 1 are stable points; as previously noted, this is commonly known as the bistable case. To tackle this mathematically, we again introduce the traveling wave variable $z = x - ct$ and substitute $u(x - ct)$ into Equation (22.45).

We obtain

$$u'' + cu' + f(u) = 0 , \tag{22.48}$$

where a prime denotes differentiation with respect to z, and require an increasing solution with $u(-\infty) = \lim_{z \to -\infty} u(z) = 0$, $u(\infty) = 1$. Equation (22.48) is then written as a system by introducing $p = u'$:

$$u' = p , \tag{22.49a}$$

$$p' = -cp - f . \tag{22.49b}$$

The "phase plane," which here is the (u, p)-plane, is used for the analysis. The problem is finding a value of c such that there is an orbit joining $(0, 0)$ to $(1, 0)$ with $p > 0$ everywhere. Thus the PDE problem given by Equation (22.45) has been replaced by an ODE problem in Equation (22.49). First it must be decided whether there is a solution – an existence problem in mathematical terminology. This may at first seem esoteric in the biological context, but one should reflect that *it is by no means obvious* that Equation (22.45) should have a solution of the form $u(x - ct)$ as required, so the existence problem is important. It is in fact not difficult, although not trivial (it involves finding a connecting orbit between two saddle points), to show that a solution does exist. Furthermore, it exists for a unique value of c and the solution obtained is stable (for details see Fife 1979, or Britton 1986). One can easily calculate (using a simple integration) the sign of c; it is in fact opposite to that of $\int_0^1 f(u) \, du$. Thus, in Figure 22.1, if the positive part of f is greater (in area) than the negative part, the species will invade (in the traveling-wave sense). This result fits in precisely with the analysis in the previous section. The criterion is exactly the same as it is for invasion of a spatially finite "blob" of the species. Thus the idea of dominance of the equilibrium 1 over 0 is intuitively rather satisfactory from a biological point of view.

We next consider what may be done to extend these ideas to two "types" of individual. For illustration, one might suppose that type-1 individuals inhabit the region with $x < 0$, type-2 individuals inhabit the region with $x > 0$, and an "iron curtain" separates them. The question is, What happens when the curtain is removed, and in particular which species wins? We illustrate the problem by using a bimatrix game (introduced in the last example in Section 22.4) and study the possible traveling wave from $(0, 0)$ to $(1, 1)$. As both points are stable, the problem is still of bistable type. We are searching for a solution to Equation (22.43) of the form

$(u(x - ct), v(x - ct))$ for some value of c. It is, of course, even less clear now whether such a solution exists. The existence problem is a great deal more difficult, but again it can be shown that there is a unique, stable traveling wave.

What can be said about the issue of chief biological importance, namely, the sign of c? Unfortunately this, too, is much more difficult to answer in this case than in the case of a single species, and there is no general method known that leads to a complete solution. However, a considerable amount of information can be obtained. It turns out that, as far as the sign of c is concerned, it is equivalent to take the reaction terms to be

$$f = u(1 - u)(v - \overline{v}) \quad \text{and} \quad g = v(1 - v)(v - \overline{v}) , \qquad (22.50a)$$

where

$$\overline{u} = \frac{a_1}{a_1 + a_2} \quad \text{and} \quad \overline{v} = \frac{b_1}{b_1 + b_2} . \qquad (22.50b)$$

The analysis is rather technical and we shall merely review some of the results here (for the details see Hofbauer *et al.* 1997). First, if $\overline{u}, \overline{v} < 1/2$ then $(1, 1)$ is the winner for all μ, ν. We have introduced the term "uniformly dominant" to reflect the fact that the winner is independent of the diffusion rates. Second, an asymptotic analysis can be carried out by fixing $\nu = 1$ and letting $\mu \to 0$, and then fixing $\mu = 1$ and letting $\nu \to 0$. This shows that there are regions of the parameters $\overline{u}, \overline{v}$ where sometimes $(0, 0)$ wins and sometimes $(1, 1)$ does. In some parts of the region, $(1, 1)$ is favored by high diffusion; in others $(1, 1)$ is favored by low diffusion. An intuitive guess here seems almost impossible.

We have claimed that the direction of the traveling wave is a key issue. To emphasize this point, it is possibly easier to think of two competing species in the bistable situation. The question of which wins in the traveling-wave sense is crucial in determining the species most likely to prevail. It appears to be extremely difficult to find an intuitive argument suggesting whether or not low diffusion is good for a species. In fact, one can construct interaction terms, satisfying the competitive assumptions, where low diffusion is good, while for others it is bad (see, e.g., Cressman and Vickers 1997). The effect of diffusion and space is subtle. [For a recent discussion of the traveling-wave point of view to the *TFT* problem introduced in Section 22.2, see Chapter 17; Ferrière and Michod (1995, 1996); Hutson and Vickers (1995).]

22.6 The Evolution of Diffusion

For this section we choose an example from a class of selection–mutation problems that bears on adaptive dynamics. The question addressed – that is, the evolution of dispersal itself – is particularly relevant in the context of the present volume. For here a central issue is the effect of spatial structure and dispersal (in one form or another) on various evolutionary models, and the ability of the dispersal rates themselves to evolve thus raises extremely interesting and important issues. The example also illustrates some of the techniques described previously. In this section, more technicalities are included, so that interested readers can get a feel for the mathematics involved in a representative problem.

There have been numerous investigations, involving various mechanisms, into the evolution of diffusion (see Johnson and Gaines 1990). The model we describe here (see Dockery *et al.* 1997) attempts to focus on one of the simplest situations where there is no direct advantage (in the sense of reproductive success) to a change in diffusion. We consider phenotypes of a single species in a spatially inhomogeneous environment that differ in their diffusion rates but in no other respect. We also consider mutations and inquire whether high or low diffusion rates are favored by selection.

For purposes of explanation, it is simplest to consider the case of just two phenotypes, 1 and 2, where 2 has the higher diffusion rate. Let the number densities of phenotypes 1 and 2 be u_1, u_2, respectively. Suppose that the self-interaction term is logistic (this is the simplest assumption; more general forms could be included without making a qualitative difference to the results). We take the spatial region Ω to be finite and to have an impermeable boundary. For simplicity, we also assume Ω to be one-dimensional. With $a(x)$ measuring the effect of the environment upon the reproduction rate, the equations take the form

$$\frac{\partial u_i}{\partial t} = u_i[a(x) - u_1 - u_2] + \mu_i \frac{\partial^2 u_i}{\partial x^2} + \varepsilon \sum_{j=1}^{2} M_{ij} u_j \quad (i = 1, 2) , \quad (22.51)$$

where $0 < \mu_1 < \mu_2$, ε is a small parameter, and the matrix (M_{ij}) gives the mutation rates. Consider first the case of no mutation. Assume that each species on its own has a positive equilibrium $(\tilde{u}_1, 0)$ and $(0, \tilde{u}_2)$; conditions for this assumption may be given as described in Section 22.3, and it is the case that the ith equilibrium state is stable when only the ith phenotype is present. We thus have

$$0 = \tilde{u}_i[a(x) - \tilde{u}_i] + \mu_i \frac{d^2 \tilde{u}_i}{dx^2} \quad (i = 1, 2) . \quad (22.52)$$

We examine the stability of phenotype 1 (the type with the lower diffusion rate). Following the ideas of Section 22.3, we linearize Equation (22.51) with $i = 2$ about $(\tilde{u}_1, 0)$ to obtain

$$\frac{\partial u_2}{\partial t} = u_2[a(x) - \tilde{u}_1] + \mu_2 \frac{\partial^2 u_2}{\partial x^2} .$$ (22.53)

Stability is determined by the principal eigenvalue λ_0 of the equation

$$\mu_2 \frac{d^2\phi}{dx^2} + [a(x) - \tilde{u}_1]\phi = \lambda\phi$$ (22.54)

(with, of course, $d\phi/dx = 0$ on the boundary). At first sight it appears difficult to determine the sign of λ_0 since $\tilde{u}_1$ is not known explicitly – it is the solution of the nonlinear boundary value problem in Equation (22.52) – and even if it were known, there remains the problem of solving Equation (22.54). Even allowing that we are considering a one-dimensional domain, this is no easy matter. Nonetheless, a complete answer is possible. The key is using a variational characterization of the eigenvalue λ_0 mentioned in Section 22.3. As stated there, the principal eigenvalue (i.e., that corresponding to a positive eigenfunction ϕ) of the equation

$$\mu \frac{d^2\phi}{dx^2} + h(x)\phi = \lambda\phi$$ (22.55)

is a decreasing function of μ. However, Equation (22.52) with $i = 1$ and $h(x) = (a - \tilde{u}_1)$ says that zero is an eigenvalue when $\mu = \mu_1$. Because $\mu_2 > \mu_1$, it follows that the principal eigenvalue of Equation (22.54) is negative. Thus $(\tilde{u}_1, 0)$ is stable. Likewise $(0, \tilde{u}_2)$ is unstable. This analysis may be readily extended to any (finite) number of phenotypes. The immediate biological conclusion is that, if the local analysis is a reliable guide, evolution favors the lower diffusion rate.

This initial conclusion must now be examined from two additional points of view. First, is the conclusion globally true? The theory of Hirsch and Smith mentioned previously tells us that the answer is yes. Second, does mutation affect the conclusion? When ε is small and positive (and under reasonable conditions on M), the $(\tilde{u}_1, 0)$ equilibrium moves a small distance inside. A further, rather technical, analysis confirms that it remains globally stable. This shows that the initial guide from the local theory is correct.

We conclude with a further problem raised by this model. Because continuous models are usually easier to treat than discrete ones, it is tempting to try to analyze the case where there are many phenotypes by a continuous model. With s as the phenotypic variable, a suitable model equation for $u \equiv u(x, t, s)$ is

$$\frac{\partial u}{\partial t} = u \left(a(x) - \int_0^1 u(x, t, s') \, ds' \right) + \mu(s) \frac{\partial^2 u}{\partial x^2} + \varepsilon \frac{\partial^2 u}{\partial s^2} . \quad (22.56)$$

The analysis of this equation, even for small ε, appears to be a difficult problem. But it is an interesting and relevant one in the context of selection–mutation models. Perhaps some of our readers will be challenged to investigate it further.

22.7 Concluding Comments

This chapter has focused on models based on reaction–diffusion equations, with the aim of persuading the reader of the following two points. First, theoretical analysis can sometimes lead to important insight – insight that would be difficult to obtain from numerical simulation of individual-based models. Second, the analysis shows that in some very important cases, the behavior of the solution of the reaction–diffusion model is qualitatively different from that of the corresponding ODE model based on spatial averaging. This is clear from several of the examples given, in particular, the *TFT* model in Section 22.2, the Turing instability in Section 22.3, and the invasion analyses in Sections 22.4 and 22.5.

23

The Dynamics of Invasion Waves

Johan A.J. Metz, Denis Mollison, and Frank van den Bosch

23.1 Introduction

In this chapter we concentrate on certain macroscopic patterns in the transient behavior of spatially extended ecological systems. Chapters 17 and 22 on reaction–diffusion equations also deal with the macroscopic perspective, but from a different angle. Those chapters forego realistic movement and life-history detail in order to concentrate on interactions between individuals. In this chapter, we restrict ourselves to phenomena that are, in general, only weakly dependent on those interactions to arrive at robust and simple quantitative population-level predictions based on measurements of behavioral characteristics of individuals. Luckily, as Chapter 16 makes clear, such phenomena are not confined to the realm of mathematics, but commonly occur in real ecological systems as well.

Transient behavior is usually viewed as an effect of a temporary external perturbation of an otherwise stationary situation. From a biological perspective there are two principal types of perturbations. The first type are abiotic perturbations, such as an unusually severe drought; these usually affect large regions, leaving the spatial distributions of species macroscopically homogeneous. The other type of perturbation is the introduction of a new species or the occurrence of an advantageous mutation in an already established species. Such perturbations originate locally and from the initial inoculum spread over space in a wavelike manner. It is the second type of transient behavior that we consider here.

An invasion generally starts with the arrival of a small number of individuals of a new species or a mutation in a single individual. Thus the initial phase of an invasion is dominated by demographic stochasticity. If the invading species survives this phase, it starts to spread from its center of origin until it runs into the boundary of the spatial domain, be it a meadow or a continent. Our main concern is calculating the speed of this spatial spread from life-history data.

Section 23.2 is devoted to delineating the class of biological systems amenable to our modeling framework. Section 23.3 contains "do-it-yourself" recipes. Box 23.1 brings to the fore some special parametric model families with the double virtue that they often efficiently graduate empirical reproductive data and allow efficient numerical calculation of the wave speeds. In Boxes 23.2 to 23.4, we consider various special cases for which explicit formulas for the wave speed exist. Box 23.5 indicates by means of two real-life examples the care that must be taken in matching kernels to field simulations. In Section 23.4 we discuss the consequences of transgressing the class boundaries. The link with reaction–diffusion models is made in Section 23.5, where we treat the special conditions under which such models provide useful approximations to our more detailed ones. Finally, Section 23.6, on stratified dispersal, shows how the straightforward application of the simple recipes from Section 23.3 can still lead to unexpected discoveries.

23.2 Relative Scales of the Process Components

Individual population dynamical behavior has at least four natural length scales set by (1) the dispersal distance; (2) the distance over which the sexes still can attract each other, or over which microgametes (sperm or pollen) are transported; and the distances over which individuals interact ecologically, either (3) directly, by helping each other or fighting, or (4) indirectly, by depleting common resources or boosting local predator densities.

- *Dispersal scale.* With certain provisos (discussed at the end of the next section), the dispersal scale corresponds to the root mean square distance between the birth place of a representative individual and that of its parent. In Section 23.6 we also consider multiple dispersal mechanisms, each with its own dispersal scale. Treating dispersal as stratified is only necessary when these scales differ by more than one order of magnitude, as is the case for, say, clonal growth and dispersal by seed. In all other sections, we assume that there is only one such mechanism.
- *Sexual scale.* Usually, reproductive output saturates very quickly with increasing mating opportunities. Thus the sexual scale can be defined by the distance between individuals below which offspring production does not suffer from the lack of such opportunities. This distance is infinite for clonally reproducing organisms, self-fertilizers, and organisms that mate before dispersing. From an abstract viewpoint, the spread of a mutant is simply the spread of a clone of the new allele in the genetic and biotic environment set by the resident population. If the sexual attraction

distance is short relative to the dispersal distance, individuals in the front of the wave suffer from a lack of mating opportunities. Luckily, sexual attraction distances are often relatively long, so that this effect can be neglected and we can treat the spatial spread of sexual populations in the female-dominant tradition of classical demography.

Although we often can neglect sex during the later phases of an invasion process, we should not forget that an obligately outcrossing population starting from a few individuals has a considerably smaller chance of surviving the initial stochastic phase than a clonal or facultatively self-fertilizing one.

Whereas for our purpose the sexual scale must be large relative to the dispersal scale, exactly the opposite must be true for the ecological interaction scales.

- *Direct interaction scale.* In nature, direct interactions often take place only over distances that are short compared with the dispersal distance. This means that individuals in the front of the dispersing wave have few direct interactions.
- *Indirect interaction scale.* Indirect interactions are more complicated than direct ones. It is possible to dream up scenarios in which predators, feeding on the new invader, increase in numbers beyond proportion and move out from the initial invasion area so quickly as to hamper population increase of the invader at distances far beyond its dispersal distance. Such situations can be treated in the simplified context of reaction–diffusion equations (see, e.g., Hosono 1998), but they fall outside the scope of this chapter. Here, we simply assume that both the direct and indirect interactions act on scales much shorter than the dispersal scale.

In this chapter we confine ourselves to cases without a generalized Allee effect – that is, cases where the ecological interactions between individuals have either no or only detrimental effects on their offspring production at all combinations of age and distance, and females never suffer from a lack of mating opportunities. This restriction allows us to handle relatively realistic life-history patterns.

The requirement to mate is so ubiquitous that we cannot offhandedly dismiss the associated Allee effect. However, it often occurs at such low population densities that its effect on the wave speed should be negligible (see our discussion of the sexual scale above and in Section 23.4, Interactions between individuals).

The assumption that interactions play only a small role in the forward tail of the wave and that any Allee-type effects, whether direct or indirect, are negligible implies that the wave speed is determined only by the linear population dynamics in that forward tail.

A final restriction on the applicability of our framework relates to the scale of any inhomogeneities in the spatio-temporal substrate for the invasion. We proceed as if space and time are homogeneous and infinitely extended, and dispersal is always the same in all directions. In practice, this means that any spatio-temporal inhomogeneities should be either very localized in space compared with the dispersal distance, or have a scale large enough to allow the wave to develop before a region with different properties is reached. In particular, the distance between the location of the inoculum and the boundary of the spatial domain in the direction of the wave movement should be much larger than the dispersal distance, preferably by more than one order of magnitude. When the domain does not have a reflecting boundary, its width in the direction orthogonal to the wave should be considerably larger than the dispersal distance.

A spatially restricted inoculum, rotational symmetry, and translation invariance together lead to asymptotically circular spread at a constant speed (provided no symmetry-breaking destabilization of the wave front occurs; the previous assumptions were rigged to preclude such a destabilization). The next section reviews recipes for calculating this speed for those population systems that comply with the restrictions outlined above. We believe that the simplifying assumptions underlying our calculations are fairly harmless. The most important exceptions occur (1) when the interactions between individuals are already felt at such low densities that there is interference with the demographic stochasticity, and (2) when space is inhomogeneous. In Section 23.4, we provide some hints as to how those complications may affect the results. We start in the next section with the simple case of individuals that reproduce and disperse independently in a homogeneous space.

23.3   Independent Spread in Homogeneous Space:
A Natural Gauging Point

We have a good grasp of the case of independent spread, down to the level of the full individual-based stochastic process. Biggins (1997) provides a nice survey at a level of biological generality comparable with that of this chapter [with proofs given in Biggins (1995)] and also discusses the intimate relationship between the stochastic and deterministic results. The older deterministic tradition starting with Kolmogorov *et al.* (1937) and Fisher (1937), followed by Kendall (1957, 1965), Mollison (1972a,

1972b, 1977), who also considers the stochastic case, Atkinson and Reuter (1976), Barbour (1977), Brown and Carr (1977), Aronson and Weinberger (1975), Aronson (1977), Diekmann (1978, 1979), Thieme (1977a, 1977b, 1979a, 1979b), Weinberger (1978, 1982), Radcliffe and Rass (1983, 1984a, 1984b, 1984c, 1984d, 1985, 1986, 1991, 1993, 1995a, 1995b, 1996, 1997, 1998, book in preparation), Lui (1983, 1989a, 1989b), Creegan and Lui (1984), and Kot (1992), immediately took on board some mild forms of nonlinearity. Recent surveys with a focus on applications, and a corresponding stress on the linear deterministic theory, are given by van den Bosch et al. (1990a), Mollison (1991), and Metz and van den Bosch (1995). Here, we also primarily follow the deterministic tradition, since the arguments are easy to convey. In addition, we stick to the case where newborns are stochastically equal – that is, they may differ in some stochastic characteristic, but this characteristic is in no way tied to their parents' birth characteristic or to their space–time coordinates. Analogous results for the case with a Markovian relation between the birth states of parents and offspring can be found in Biggins (1995, 1997), Lui (1983, 1989a, 1989b), and the numerous papers by Radcliffe and Rass cited above.

Model description

Let $b(t, x)$ denote the local birth rate at time t and position x. Then, if the inoculum was put in place at $t = 0-$ and there is no further immigration,

$$
\begin{aligned}
\text{Local birth rate} \; = \; & \text{Cumulative local birth rates from} && (23.1) \\
& \text{parents from all places and of all ages,} \\
& \text{born after the moment of inoculation} \\
& + \\
& \text{Local birth rate from parents in the inoculum ,}
\end{aligned}
$$

or,

$$
b(t, x) = \int_0^t \int_{\mathbb{R}^n} b(t - \tau, x - y) A(\tau, y) \, dy d\tau + h(t, x) , \tag{23.2}
$$

where n is the dimension of the spatial domain under consideration (in practical applications $n = 1, 2,$ or 3, think of a river bank, a field, or some flask with a protozoan culture in a viscous culture fluid); $A(\tau, y)$ denotes the rate at which a mother aged τ places daughters at a position that is a vectorial distance y from her place of birth; and $h(t, x)$ is the birth rate at x from mothers older than t (i.e., the mothers in the inoculum). In the tradition of the theory of Volterra integral equations, we refer to A as the birth kernel and to h as the initial condition.

As an example, we show how the "reaction"–diffusion model

$$\frac{\partial n}{\partial t} = D\frac{\partial^2 n}{\partial x^2} + r_0 n \tag{23.3}$$

fits into this scheme. The simplest individual-based model giving rise to Equation (23.3) is one in which individuals diffuse at a rate D, die at rate δ, and give birth in a Poisson process with rate β, so that

$$r_0 = \beta - \delta . \tag{23.4}$$

Under these assumptions, the probability density that an individual survives to age τ and then resides at a position that is distance y from its place of birth is

$$P(\tau, y) = \exp(-\delta\tau)(4\pi D\tau)^{n/2}\exp[-|y|^2/(4D\tau)] , \tag{23.5}$$

and

$$A(\tau, y) = \beta P(\tau, y) . \tag{23.6}$$

For later reference, we introduce the additional terminology

$$R_0 = \int_0^\infty \int_{\mathbb{R}^n} A(\tau, y)\,dy d\tau , \tag{23.7}$$

the average lifetime number of offspring, or reproduction ratio, and, in the case where $R_0 < \infty$,

$$a(\tau, y) = A(\tau, y)/R_0 , \tag{23.8}$$

the birth distribution; we refer to the marginal distributions of the latter as the age-at-birth and displacement distributions. Box 23.1 gives some examples of such distributions that have proved their worth in adapting the theory to practical applications. The mean and variance of the age-at-birth distribution are denoted by μ and v^2, respectively; the variance of the displacement distribution is denoted by σ^2. For the reaction–diffusion model $R_0 = \beta/\delta$, $\mu = \delta^{-1}$, $v^2 = \delta^{-2}$, and $\sigma^2 = 2D/\delta$.

Calculating the wave speed

The various theorems concerning the development of waves are rather intricate. As a first step in our arguments, we consider for $n = 1$ the existence of exponential wave-type solutions, with λ representing the steepness of the wave front,

$$b(t, x) = \alpha\exp[-\lambda(x - ct)] , \tag{23.9}$$

Box 23.1 Examples of kernels

To apply the theory from Section 23.3, we need submodels for the birth kernel. Preferably, such submodels should have a mechanistic basis and (1) be sufficiently flexible when it comes to fitting observed life-history data, (2) have a limited number of parameters, and (3) have simple Laplace transforms. Property (3) greatly facilitates solving Equation (23.14).

In this box, we concentrate on situations where all movement precedes reproduction. This results in birth kernels that can be written as $R_0 a_1(\tau) a_2(y)$, with a_1 being the age-at-birth and a_2 the displacement distribution. The product form is inherited by the Laplace transform. In general, it is expedient to determine R_0 from population observations under the same circumstances as those under which the wave is observed, and a_1 and a_2 from observations of individuals and/or mechanistic submodels.

Age-at-birth distributions. It is rarely possible to find good mechanistic submodels for the age-at-birth distribution. At best we can consider models that do a fair job in the sense of properties (1) to (3) above. A first useful candidate is the block distribution, expressing the assumption that individuals pass through a maturation period of duration p, after which they are fertile at a constant level for a period of duration i. The advantage of block distributions is that the corresponding models can often be rephrased as delay-differential equations. An example is Vanderplank's equation from phytopathology (see Chapter 16). The mean and variance are respectively $\mu = p + i/2$ and $v^2 = i^2/12$. The Laplace transform is

$$\tilde{a}_1(s) = \exp(-ps)(is)^{-1}[1 - \exp(-is)] .$$

A considerably more flexible family is the delayed gamma densities

$$a_1(\tau) = \begin{cases} 0 & \text{for } 0 < \tau \leq p \\ \alpha(\alpha(\tau - p))^{\beta-1} \exp(-\alpha(\tau - p))/\Gamma(\beta) & \text{for } p \leq \tau \end{cases}$$

(see Figure 16.13), which have mean $\mu = p + \beta/\alpha$, variance $v^2 = \beta/\alpha^2$, and Laplace transform

$$\tilde{a}_1(s) = \exp(-ps)[\alpha/(\alpha + s)]^{\beta} .$$

Together these two families adequately approximate most age-at-birth distributions.

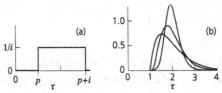

Some age-at-birth distributions. (a) Block and (b) delayed gamma [with $\beta = 2$ (the lowest peak), 4, 10, and $p = \beta/\alpha = 1$].

continued

Box 23.1 *continued*

Displacement distributions. Formulas purportedly describing displacement distributions abound in the literature. We give three such distributions that we have often found to be good descriptors of empirical data and that moreover can be derived from mechanistic considerations. We gear our discussion to two dimensions, but we only give the Laplace transforms of the one-dimensional marginal distribution used in Equation (23.14) [see formulas (23.15) and (23.17)]. We parameterize with the displacement variance σ^2, which can be estimated by averaging the observed variances in the y_1 and y_2 directions, or, equivalently, as half the observed mean square displacement. Below, "transect distribution" refers to the distribution of offspring over a line transect through the parent.

The occupation of home ranges is often well described by a Gaussian distribution. Therefore, a Gaussian is a good descriptor of the transmission of many animal diseases. A Gaussian also results if individuals move for a fixed time according to a (driftless) Brownian motion. Brownian motion is a good description of any continuous movement with little dependence between the displacements in subsequent time intervals, such as in transport by turbulent water or air. The transect and marginal distributions are again Gaussian, with Laplace transform

$$\tilde{a}_2(\lambda, 0) = \exp[(\sigma\lambda)^2] \ .$$

If individuals move for an exponentially distributed time according to a driftless Brownian motion, a Bessel distribution results (Broadbent and Kendall, 1953; Williamson, 1961). The transect distribution is

$$a_2(y_1, 0) = \sqrt{2}(\pi\sigma)^{-1} K_0(\sqrt{2}|y_1|/\sigma)$$

[K_0 is the modified Bessel function of the second kind of order zero; see Abramowitz and Stegun (1965)], see Figure 16.14. It has variance $\frac{1}{2}\sigma^2$. The marginal distribution is a double exponential one. Its Laplace transform is

$$\tilde{a}_2(\lambda, 0) = [1 - (\sigma\lambda)^2/2]^{-1} \ .$$

If individuals move for an exponentially distributed time in a straight line, a rotated exponential distribution results. The marginal distribution has Laplace transform

$$\tilde{a}_2(\lambda, 0) = [1 - (\sigma\lambda)^2]^{-1/2} \ .$$

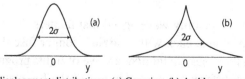

Some marginal displacement distributions: (a) Gaussian, (b) double exponential.

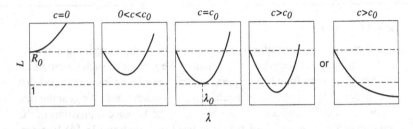

Figure 23.1 Shape of the function L defined by Equations (23.12) and (23.13).

of the time invariant form of Equation (23.2),

$$b(t, x) = \int_0^\infty \int_{\mathbb{R}} b(t - \tau, x - y) A(\tau, y) \, dy d\tau \; . \qquad (23.10)$$

When we substitute (23.9) into (23.10) and rearrange the result, we end up with the characteristic equation

$$L(c, \lambda) = 1 \; , \qquad (23.11)$$

with

$$L(c, \lambda) = \tilde{A}(\lambda c, \lambda) \; , \qquad (23.12)$$

and

$$\tilde{A}(s, \lambda) = \int_0^\infty \int_{-\infty}^\infty \exp(-s\tau - \lambda y) A(\tau, y) \, dy d\tau \qquad (23.13)$$

the Laplace transform of the birth kernel, one-sided in time and two-sided in space.

Using the properties of the Laplace transform, it is easy to show that $L(c, \lambda)$ has the properties depicted in Figure 23.1. From this we deduce that there exists a c_0 such that Equation (23.11) allows real solutions for all $c \geq c_0$ and no solutions for $c < c_0$, where c_0 can be calculated from

$$\frac{\partial L}{\partial \lambda}(c_0, \lambda_0) = 0 \; , \qquad L(c_0, \lambda_0) = 1 \; . \qquad (23.14)$$

In practice, c_0 is the only wave speed that matters. To understand why, consider a row of flares, each with a slightly longer fuse than the one preceding it. When the fuses are ignited, a wave of lights progresses at a speed dependent on the differences in the lengths of the fuses attached to subsequent flares. However, when the flares can also be ignited by a neighbor,

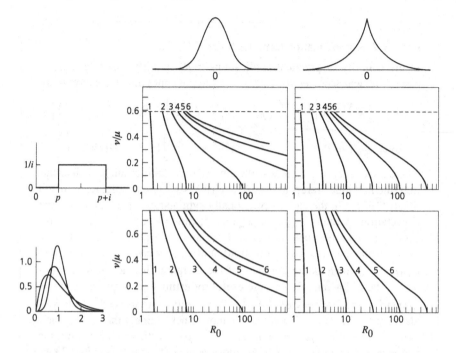

Figure 23.2 Contour plots of the scaled wave speeds $c_0^* = \mu c_0/\sigma$, for the Gaussian and double exponential displacement distributions, in combination with the block and gamma age-at-birth distributions; μ is the mean age-at-birth; ν is its standard deviation, and σ is the standard deviation of the displacement distribution.

a minimum speed exists. Any slower wave will be overtaken by the autonomous one generated by the mutual ignition. This slowest wave speed is the one that is realized when the process is started up with a localized inoculum.

In general, Equation (23.14) has to be solved numerically. In the reaction–diffusion case, $c_0 = 2\sqrt{r_0 D} = (\sigma/\mu)\sqrt{2(R_0 - 1)}$. Figure 23.2 gives contour plots of $c_0^* = \mu c_0/\sigma$ for some other special families of kernels. From these graphs it can be seen that

- for small R_0 the scaled wave speed $c_0^* \approx \sqrt{2 \ln R_0}$, with little dependence on either the type of the displacement distribution or the type or the coefficient of variation ν/μ of the age-at-birth distribution;
- for small ν/μ and a Gaussian displacement distribution $c_0^* \approx \sqrt{2 \ln R_0}$ for all values of R_0, whereas for small ν/μ and a double exponential displacement distribution $c_0^* \approx \sqrt{1/2} \ln R_0$ for large values of R_0.

These observations form the basis for some useful approximation formulas, discussed in Boxes 23.2 and 23.3.

Box 23.2 Approximation formulas for small $\ln R_0$

In the majority of practical applications $\ln R_0$ is fairly small. For these cases, van den Bosch *et al.* (1990a) derived the approximative expressions

$$c_0 = \frac{\sigma}{\mu} \sqrt{2 \ln R_0} \left\{ 1 + \left[\left(\frac{\nu}{\mu} \right)^2 - \frac{\kappa_{1,2}}{\sigma^2 \mu} + \frac{1}{12} \frac{\kappa_{0,4}}{\sigma^4} \right] \ln R_0 + O\left((\ln R_0)^2 \right) \right\} ,$$

$$\lambda_0 = \frac{1}{\mu} \sqrt{2 \ln R_0} \left\{ 1 - \left[\left(\frac{\nu}{\mu} \right)^2 - 2\frac{\kappa_{1,2}}{\mu \sigma^2} + \frac{1}{4} \frac{\kappa_{0,4}}{\sigma^4} \right] \ln R_0 + O\left((\ln R_0)^2 \right) \right\} ,$$

where the $\kappa_{i,j}$ represent the so-called mixed cumulants of the birth distribution (the first index refers to the age at birth); see, for example, Kendall and Stuart (1958). In the case of rotationally symmetric dispersal (with $\mathbb{E}$ the expectation operator), $\kappa_{1,0} = \mu$, $\kappa_{2,0} = \nu^2$, $\kappa_{0,2} = \sigma^2$, $\kappa_{1,2} = \mathbb{E}\tau y_1^2 - \mu\sigma^2$ (the covariance between the age at birth and the square of the displacement component in a given direction), and $\kappa_{0,4} = \mathbb{E}y_1^4 - 3\sigma^4$. The ratio $\kappa_{0,4}/\sigma^4$ is known as the kurtosis (of the marginal displacement distribution). The following relations are useful for estimation purposes: $\mathbb{E}\tau y_1^2 = \frac{1}{2}\mathbb{E}\tau(y_1^2 + y_2^2)$, $\mathbb{E}y_1^4 = \frac{1}{2}\mathbb{E}(y_1^4 + y_2^4)$. When only prereproductive individuals disperse, $\kappa_{1,2} = 0$. If individuals move unchangingly throughout their lives, $\kappa_{1,2} = \nu^2\sigma^2/\mu$, $\sigma^2 = \mu\acute{\kappa}_2$, $\kappa_{0,4} = \mu\acute{\kappa}_4$, with $\acute{\kappa}_2$ representing the variance and $\acute{\kappa}_4$, the kurtosis of the movement per time unit (see Box 23.4). The figure shows how well the approximation for c_0 performs for the models of Figure 23.2.

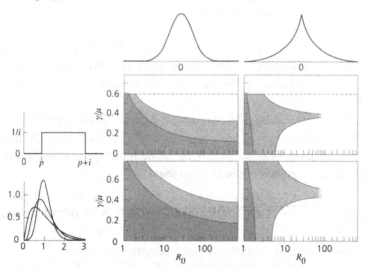

The gray region indicates where the approximation for c_0 differs by less than 10% from the numerical results for the models from Figure 23.2. The darker gray indicates where the lowest-order term on its own already differs by less than 10%.

Box 23.3 Approximations for concentrated reproduction

In this box we give approximation formulas for the case where reproduction takes place in one very narrow pulse, that is, for small v/μ. These formulas also can be used to calculate the wave speeds for integro-difference equation models for the spread of a population of annuals (in a seasonal environment) over a continuous spatial domain. If reproduction occurs in a delta peak in time, we can write the birth distribution as a product of that delta peak and the displacement distribution. The Laplace transform of the delta peak at μ is $\tilde{a}_1(s) = \exp(-\mu s)$. We treat only the three diplacement distributions from Box 23.1. For a Gaussian displacement distribution,

$$c_0 = \frac{\sigma}{\mu}\sqrt{2\ln R_0} \ .$$

This formula equals the lowest-order term of the formula for c_0 from Box 23.2. The figure in that box shows how well the approximation predicts the model results.

No explicit formulas are available for the Bessel and rotated exponential distributions. For (very) large R_0 we have the following asymptotic formulas: Bessel: $c_0 \approx \ln R_0/\sqrt{2}$; exponential: $c_0 \approx \ln R_0$. In both cases the R_0 dependence over ranges of one order of magnitude is well described by $c_0 \approx a+b\ln R_0$, even for fairly small R_0 and fairly large v/μ. This relation was used to good effect by van den Bosch *et al.* (1990b) in a study of the spread of a fungal pathogen in a wheat field sowed with different mixtures of susceptible and resistant cultivars. The right-hand column of Figure 23.2 gives an indication how fast a and b change with other model parameters. For $R_0 > 5$ and small v/μ, the following formulas, with $\rho = \ln R_0$, give values for c_0 that differ by less than 10% from the exact ones:

$$c_0 = \frac{\sigma}{\mu} \frac{2\sqrt{(2\rho + \theta)(2 + \rho)}}{4 - \theta} \ , \quad \theta = \frac{3.2 \ \ln(1 + \rho/2)}{1 + \rho} \quad \text{(Bessel)} ,$$

$$c_0 = \frac{\sigma}{\mu} \frac{\sqrt{(\rho + \theta)(1 + \rho)}}{1 - \theta} \ , \quad \theta = \frac{0.83 \ \ln(1 + \rho)}{1 + 2\rho} \quad \text{(Exponential)} .$$

"Run for your life" theorems

The one-dimensional result immediately generalizes to plane or space waves in two and three dimensions, respectively – that is, waves with a profile that is constant in all but one orthogonal direction, for which the speed is measured in the unique remaining direction. All we have to do is replace the birth kernel with the marginal kernel that results from integrating out all but one of the space directions and we are back at the one-dimensional problem. Thus, for $n = 2$ and $y = (y_1, y_2)$, we can calculate the relevant wave speed from Formula (23.14) together with

$$L(c, \lambda) = \tilde{A}(\lambda c, \lambda, 0) ,\tag{23.15}$$

with

$$\tilde{A}(s, \eta_1, \eta_2) =$$
$$\int_0^\infty \int_{-\infty}^\infty \int_{-\infty}^\infty \exp(-s\tau - \eta_1 y_1 - \eta_2 y_2) A(\tau, y) \, dy_1 dy_2 d\tau ,\tag{23.16}$$

so that

$$\tilde{A}(s, \lambda, 0) = \int_0^\infty \int_{-\infty}^\infty \exp(-s\tau - \lambda y_1) \int_{-\infty}^\infty A(\tau, y) dy_2 \, dy_1 d\tau .\tag{23.17}$$

Circular or spherical waves are a little more complicated. In an expanding wave the curvature changes. As this curvature necessarily affects the local arrival rate of propagules, we can only expect to see a constant speed emerge after the circle, or sphere, has locally become essentially flat when looked at on the dispersal scale. By the same argument, in combination with those used before, the asymptotic speed of spatial expansion from a localized inoculum should be the same as that of the slowest plane or space waves.

Mathematically the convergence to wavelike behavior from some initial condition h is problematic even in the one-dimensional case, both conceptually (What sort of convergence should we think of?) and technically (How should one go about proving it?). Therefore, research has concentrated on the weakest possible form of convergence. Here, we describe the result for $n = 2$. Imagine that you are sitting in a helicopter ascending at speed one, positioned in the sky above $x = 0$. All features on the ground are shrinking in proportion to your altitude. A position x on the ground appears in your field of vision at x/t. Now imagine that every x on the ground for which the local birth rate $b(t, x)$ is between two arbitrary threshold values $b_- < b_+$ is colored black, and every other x is colored white. The black set in your field of vision converges to a circle of radius c_0 around the origin, and everywhere inside that circle the birth rate appears to go to infinity and everywhere outside, to zero. This result is also called the "run for your life" theorem after the following metaphor. Imagine that instead of entering the helicopter you stay on the ground, and that you start running in a straight line at speed c. If $c > c_0$ you will outrun the population front, that is, you will see the birth rate dwindle to zero in an exponentially bounded manner. If $c < c_0$, the birth rates in your neighborhood will keep increasing exponentially [i.e., $t^{-1} \ln b(t, ct)$ converges to some constant]. If the movement, survival, and growth of the individuals satisfy some very mild conditions (e.g., postreproductive individuals should not live forever or move faster than reproductive ones, nor should individuals grow at too high an exponential rate), the statements made about the birth rate apply equally to population densities, biomass densities, etc.

More precise forms, as well as proofs, of these rather loose statements can be found in the papers by Diekmann, Thieme, Lui, Radcliffe and Rass, and Biggins, cited at the start of this section.

The "run for your life" theorem has been proved under very general conditions, and it suffices for many practical purposes. But it is also a rather weak result, since it tells nothing about what happens around you when you run at a speed that keeps the birth rate around you more or less constant. Will you see the birth profile around you converge to a constant form? For a few explicitly solvable cases, such as simple diffusion models, one easily obtains the expected result: the profile becomes proportional to $\exp(-\lambda_0 z)$, where z is the distance from your present position and becomes constant in the orthogonal direction. But it is unknown how far this result extends.

In the stochastic case we have to be a little more circumspect. Concepts like the local birth density or the local density of individuals only make sense for means or for deterministic approximations. Our discussion is therefore largely phrased in terms of the numbers $N(t, r)$ of individuals outside a circle of radius r around the origin (an interval when $n = 1$, a sphere when $n = 3$). If we know $N(t, r)$, we can calculate the number of individuals in a circular band between r_1 and r_2 (where $r_1 < r_2$) by simple subtraction. To get any results, we have to assume that the same innocent conditions needed before to translate results about b into results about numbers of individuals still apply. We only consider the linear case here – that is, we assume that individuals reproduce and disperse independent of each other. In that case, the stochastic mean satisfies the deterministic equations. But, in addition, we have the stronger result that $N(t, ct)$ will almost surely dwindle to zero in an exponentially bounded manner whenever $c > c_0$, and will grow asymptotically like a deterministic exponential, in the sense that $t^{-1} \ln N(t, ct)$ converges to some constant, whenever $c < c_0$. Moreover, the (stochastic) distance r_{max} of the individual farthest from the origin almost surely satisfies $t^{-1} r_{max} \to c_0$ (see Biggins 1995, 1997). Again, notwithstanding the beauty and strength of these results, they are in a certain sense rather weak. We look at the data through an ever-more minifying macro-scope. The stochastic structure that we may encounter in our immediate surroundings while moving at speed c_0 might be rather subtle. A first hint in this direction can be found in the physics-style results about the asymptotic decomposition of the covariance structure in local and global components presented by Lewis and Pacala (in press).

Spatial scales

The approach followed here is based on the assumption that the integral in Equations (23.13) and (23.17) converges for at least some pairs (s, λ). If the far tail of the displacement distribution is so fat that convergence fails, there also is no convergence to a constant wave speed. Instead, in the deterministic case the expanding wave seems to increase ever more in speed (Kot *et al.* 1996; Clark 1998). And in the stochastic model, any convergence is thwarted by occasional large jumps forward, ruining the development of a deterministic-looking spatial configuration (Mollison 1972b, 1977; Lewis, 1997; Lewis and Pacala, in press). In other words, there is no clear dispersal scale.

In practice, it is nearly impossible to obtain good information about the tails of the displacement distribution. Indeed, there is no far tail because any real system is bounded. What really matters is how well the model under consideration captures the dispersal phenomenon in which we are interested. In that sense, the big jumps forward seen in models with really fat-tailed displacement distributions seem to do a fair job. In Section 23.5, we discuss another modeling approach to this phenomenon which has the advantage that we can wring some interesting results from it using nothing but Equation (23.14). Neither approach has yet been shown to be markedly inferior as far as the observations are concerned, and it will be difficult to come to a final verdict (see Appendix 16.A). The types of model considered in this chapter are by necessity rough approximations to reality, and we are stressing phenomena near the verge of the associated observational resolution. Those interested in the practical aspects are referred to Turchin (1998, Section 6.3), and to Clark (1998) and Clark *et al.* (1998); but see also Appendix 16.A.

This is also the place to delve a little deeper into the subtleties of the concept of spatial scale. Since the convergence of the integral in Equation (23.13) is intimately tied to the fact that the tails of the birth kernel are bounded by a negative exponential in the displacement distance (and by some, positive or negative, exponential in time), the relevant dispersal scale in this context is $(\lambda_{\max})^{-1}$, where $\lambda_{\max}$ is the lower bound of those $\lambda > 0$ for which the integral in definition (23.13) converges. According to our previous discussion, there is no clear dispersal scale if $(\lambda_{\max})^{-1} = \infty$. In Section 23.1 we identified the dispersal scale with the root mean square displacement σ of the birthplace of the offspring from that of the mother. This mean square displacement is the appropriate gauge for the dispersal scale when it comes to the scaling of the wave profile λ_0 and speed c_0. Finiteness of $(\lambda_{\max})^{-1}$ implies that σ is finite. For most model families used in practice, the two measures are of the same order of magnitude. Therefore, the precise choice rarely matters, particularly in a heuristic discussion; however, a narrow area of uncertainty remains.

23.4 Complications

In this section, we briefly consider how the results from the previous section are modified by either spatial and temporal inhomogeneities in the substrate for the invasion, or by interactions between individuals.

Spatial and temporal inhomogeneities, non-isotropy

In practice, our assumption of spatio-temporal homogeneity means that any inhomogeneities should have a very fine spatial grain and be effectively invisible on the dispersal scale. Individuals should be independent; if this is the case, nothing counts but the average of the individual reproductive output over the possible environments in which an individual can find itself, independent of, for example, the temporal scale of the local environmental fluctuations. With temporally fine-grained but spatially widespread inhomogeneities, we have no such luck. Our idealized individuals are assumed to have negligible spatial but considerable temporal extension. Therefore, individuals become effectively independent at a sufficiently fine spatial grain, but remain stochastically dependent when we decrease only the temporal grain.

The heuristic arguments above are supported by analytical results for the diffusion case from Shigesada *et al.* (1986, 1987). For spatial but no temporal fluctuations, $c_0 \geq 2\sqrt{\langle r_0 \rangle_{\mathrm{A}} \langle D \rangle_{\mathrm{H}}}$, where $\langle \cdot \rangle_{\mathrm{A}}$ denotes the arithmetic and $\langle \cdot \rangle_{\mathrm{H}}$, the harmonic spatial mean, with equality when there are no fluctuations or in the limit of zero environmental grain size. Unfortunately, the figures in Shigesada *et al.* (1986) show that the speed of convergence to the limit depends rather intricately on the nature of the environmental fluctuations.

Direct calculations for the diffusion case show that for temporal fluctuations alone, $c_0 = 2\sqrt{\langle r_0 \rangle_A \langle D \rangle_H}$, where the averages are now taken in time. In other words, the purely temporal analogue of the spatial limit result is exact.

This last result immediately extends to any model with individuals that diffuse throughout their lives at an age-independent, though time-dependent, rate if we identify $\langle r_0 \rangle_A$ with the overall exponential growth rate of the population. This result is an almost immediate extension of the results in Box 23.4. Unfortunately, there are no easy recipes for determining $\langle r_0 \rangle_A$ from life-history data, except in the special case where the individual birth and death rates depend solely on time, and not on, for example, age. But the result is useful in situations where for other reasons we want to take recourse to a field estimate of $\langle r_0 \rangle_A$ as part of a scheme to estimate the parameters of the birth kernel.

We know from experience that invasion waves may change direction as a result of large-scale spatial inhomogeneities. Therefore, the theory developed in Section 23.3 is of use only when these inhomogeneities occur on a very large scale, so that the wave has time to relax to its asymptotic speed before the next change in terrain is encountered. Unfortunately, the diffusion theory shows that even the "run for your life" type of convergence only happens at the slow speed of $\ln(t)/t$ (Bramson 1983; Ebert and van Saarloos 1998, unpublished). Some pertinent discussions of the consequences of large-scale spatial inhomogeneity for the analysis of empirical spatial spread patterns can be found in Lubina and Levin (1988) and Andow *et al.* (1990, 1993). In addition, there is the interesting open theoretical problem of how to transform data about the local values of c_0, derived from life-history data, into statements about changing directions of the wave front (see also Section 8 of Metz and van den Bosch 1995).

In the anisotropic case the linear wave speed changes with the direction of the wave front. The direction-dependent speeds are readily calculated from Equation (23.14) together with Equations (23.15) and (23.16) or (23.17), with the y_2 direction chosen parallel to the wave front and the y_1 direction, orthogonal to it. The bigger problem is to transform the direction-dependent speed data into a contour that can replace the circle in the "run for your life" theorem. This problem was solved *in abstracto* by Weinberger (1978, 1982) and translated into down-to-earth calculations by van den Bosch *et al.* (1990a); a more extensive explanation of the recipe can be found in Metz and van den Bosch (1995).

Box 23.4 Continuous movement

When individuals move in exactly the same manner throughout their lives, the equation for the wave speed can be solved using a two-step procedure.

We first consider the example where individuals move continuously at a constant rate. In that case,

$$A(\tau, y) = B(\tau)(4\pi D\tau)^{n/2}\exp[-|y|^2/(4D\tau)] ,$$

where B is the average rate of offspring production at different ages (or R_0 times the age-at-birth distribution), leading to

$$L(c, \lambda) = \tilde{B}(c\lambda - D\lambda^2) = 1 ,$$

where $\tilde{B}$ is the Laplace transform of B. A well-mixed population growing according to the same birth regime will, in the long run, grow exponentially at a relative rate r_0 determined by

$$\tilde{B}(r_0) = 1 .$$

The combination of the second and third equations above tells us that

$$c\lambda - D\lambda^2 = r_0 .$$

The minimum value of c_0 for which this equation still allows a solution for λ can again be found by setting the differentiated left-hand side equal to zero:

$$c_0 = 2D\lambda_0 .$$

The combination of the last two equations gives

$$c_0 = 2\sqrt{r_0 D} .$$

Thus the square-root formula applies not only to the simple reaction–diffusion case, but also to any model in which individuals diffuse at a constant rate. With a little creative interpretation of the various terms we can also turn this argument on its head: the only non-contrived models that have the same dependence of wave speed on the population-dynamical parameters as reaction–diffusion models assume that individuals diffuse at a constant rate over their entire reproductive lives.

Many observations on real animals suggest that their dispersal is more leptokurtic than purely diffusive. The easiest way to model this is by assuming that they move according to a more general process with independent increments. Biologically, this means that movement rates are highly variable on a very short time scale so that, on the time scale of interest here, movements at different ages are effectively independent. Well-behaved processes of this type can be characterized by the fact that the Laplace transform of the displacement at age τ can be written as $\exp(k(\eta)\tau)$, where k is the so-called infinitesimal cumulant generating function. The

continued

Box 23.4 *continued*

coefficients in the Taylor series of k correspond to the cumulants of the distribution of the displacement y at age 1. For such processes,

$$L(c, \lambda) = \tilde{B}(c\lambda - k(\lambda)) .$$

[For simplicity, we assume that we are dealing with spread in one dimension, otherwise we have to consider the marginal distribution of k, in the same manner as in Equation (23.17).] Going through the same motions as before, we find that the wave speed satisfies

$$c_0\lambda_0 - k(\lambda_0) = r_0 \qquad \text{and} \qquad c_0 = k'(\lambda_0) .$$

If r is not too large, it suffices to approximate k using estimates for the first terms of its Taylor expansion, which can, for example, be obtained from the estimated cumulants of the dispersal in one direction over one year. If these cumulants are written as $\acute{\kappa}_i$, $\acute{\kappa}_2$ is the variance and $\acute{\kappa}_4$ is the kurtosis (in the diffusion case $\acute{\kappa}_2 = 2D$ and $\acute{\kappa}_4 = 0$),

$$k(\lambda) = \tfrac{1}{2}\acute{\kappa}_2\lambda^2 + \tfrac{1}{24}\acute{\kappa}_4\lambda^4 + \cdots .$$

If we stick to the first two terms of the last equation, Equation (23.14) can be solved explicitly. If the restriction of constant movement applies, the result provides a good approximation for c_0 using only the relatively accessible quantities r_0, $\acute{\kappa}_2$, and $\acute{\kappa}_4$.

Interactions between individuals

In reality, individuals do not remain independent at ever larger population densities. In the language of deterministic modelers, interactions are equivalent with nonlinearities. Below, we assume that the interactions merely induce a relatively harmless sort of nonlinearity. More particularly, we assume that individuals can never do better reproductively, at any age or distance from their place of birth, than a lonely immigrant just arrived on the scene. We refer to the violation of this assumption as the presence of a generalized Allee effect. However, before we assume that generalized Allee effects are absent, we delve a little deeper into one particular Allee effect that is practically ubiquitous, although mathematically somewhat special.

The Allee effect commonly dealt with in the literature corresponds to an absence of births at low population densities, or at least to such a dearth of births that starting with a uniformly low population density leads to sure extinction. We call this a strict Allee effect, as opposed to the much weaker generalized Allee effect introduced in the previous paragraph. A good discussion of the consequences of a strict Allee effect in the context of reaction–diffusion models can be found in Lewis and Kareiva (1993). Veit and Lewis (1996) analyze a discrete-time model with a strict Allee effect for the spread of the house finch in eastern North America. With a strict Allee effect in place, any waves that arise are pushed by a spillover of individuals from the more crowded regions, whereas in the absence of a generalized Allee effect the waves are pulled by the growth at low population densities in the forward tail of the wave. Because the strict Allee effect "cuts the fuses from our flares," no wave speeds exist other than that corresponding to neighbor ignition. A side effect is that the convergence to the asymptotic wave speed and shape is much faster for pushed waves than for pulled ones. Moreover, a strict Allee effect induces a well-defined asymptotic wave speed even for fat-tailed displacement distributions, since it effectively deactivates any individuals that have moved too far beyond the wave front. The down side is that there is no quick route to calculating the velocity of pushed waves. In principle, all interactions between individuals matter, regardless of whether they come into play at low or at high densities only. However, not all situations are equally dire: the numerical results in Cruickshank *et al.* (unpublished) show that, at least for some reaction–diffusion models, if a strict Allee effect exists but only plays a role at population densities below those at which the other, detrimental, nonlinearities kick in, we can obtain a good estimate of the wave speed from a linear model in which both types of nonlinearities are ignored. The question is how far this result can be generalized. The displacement kernels corresponding to reaction–diffusion models have about the slimmest tails encountered in any serious model. We have already seen that any strict Allee effect dramatically alters the conclusions for the fat-tailed case. For exponentially bounded tails, we expect the wave speed to converge to that of the limiting model when the density range over which the Allee effect operates is pushed toward zero, but to converge more slowly for kernels satisfying only weaker exponential bounds.

At this point a warning is in order: generalized Allee effects may spring to life unexpectedly. For example, Hosono (1998) found that the wave speed with which a superior invader takes over in a reaction–diffusion version of the classical Lotka–Volterra competition model is not always equal to $2\sqrt{r_{0i}(1 - \alpha_{ir} K_r) D_i}$, where the indices i and r refer to invader and resident, respectively, α denotes the competition coefficient, and K_r is the equilibrium density of the resident when it is on its own. This result goes directly against the accepted wisdom. The discrepancy arises in cases where the invader so successfully outcompetes the resident that locally the growth rate of the invader can rise well above the depressed value $r_{0i}(1 - \alpha_{ir} K_r)$.

From now on, we assume that Allee effects are absent. This does not necessarily mean that all the effects individuals have on each other are detrimental. It only means that the positive effects can never more than compensate for the detrimental ones if we gauge the results against the situation of an as yet uninvaded environment.

Without an Allee effect, the solution to any deterministic equation for the birth rate is necessarily bounded from above by that of Equation (23.2), where A is the birth kernel in the uninvaded environment and h is the birth rate from the inoculum as calculated from the full nonlinear model. [For many special models, incorporating the feedback from its fellows into the birth rate of an individual in an equation like Equation (23.2) is a daunting task. The only advantage of sticking to the integral equation formalism is that this strategy permits some sweeping generalizations.] The wave speed for the nonlinear case is then bounded from above by that of its linearization around the uninvaded state. The fact that the wave speed is determined essentially in the forward tail makes it plausible that the two wave speeds are actually equal, provided that the detrimental effects from higher population densities stay sufficiently localized for the forward tail to be unaffected by the higher population densities further on in the wave. This so-called linear conjecture has been proved for a number of special models (see the references at the start of Section 23.3). Upon closer inspection, all the apparent exceptions that we know of turn out to violate at least one of the following two assumptions: (1) absence of an Allee effect, or (2) sufficient localization of the interactions. So the linear conjecture forms an excellent first basis for tackling practical problems.

The linear conjecture only refers to the asymptotic speed, and shape, of the far front of the wave. The densities of individuals at any fixed point in space generally increase during the passing of the farthest front. But there is nothing to prevent the initial increase from being followed by crashes as a result of direct or indirect interactions. A second point is that even when the shape of the far front always stabilizes over time, it is not generally true that the population fluctuations in the wake of the wave can be predicted by calculating a unique wave moving at speed c_0. Stabilization of the front does not guarantee the stability of the wake. The difference between the two sorts of stability is demonstrated in the work of Dunbar (1986) and Sherratt et al. (1997) on predator–prey models. Dunbar achieved the mathematical feat of proving the existence of a unique wave with an exponential front and a regularly oscillating rear in a predator–prey model. Ten years later there was a somewhat surprising twist to this result when Sherratt et al. produced numerical simulations indicating that the stable development of the wave front is remarkably robust to even gross changes in the model specification, but that the rear of the wave is unstable, giving way to irregularly fluctuating spatial patterns. Probably the strongest limitation on the extent of the linear conjecture is found in the work of Bramson (1983) and Ebert and van Saarloos (1998; unpublished), who show that for various reaction–diffusion models there exist asymptotic wave shapes $\hat{b}$ and displacement functions v such that $b(t, z + v(t)) \rightarrow \hat{b}(z)$, with $v(t)/c_0 - t = -\frac{3}{4}\ln(t) + O(1)$ in the nonlinear and $v(t)/c_0 - t = -\frac{1}{4}\ln(t) + O(1)$ in the corresponding linear cases.

So far we have concentrated on deterministic models. However, any deterministic population model ultimately has to be justified by its connection to an underlying individual-based model, which in almost all cases has to be stochastic. The usual route from stochastic to deterministic population models is to assume that population numbers are uniformly large. This assumption cannot hold in the extreme exponential front of a population wave. Those small numbers do not cause a problem because in the front of the wave the individuals are effectively independent. In the independent, or linear, case the association between stochastic and deterministic population models is much stronger, since in that case the deterministic model faithfully represents the mean behavior of the stochastic model at all densities, including arbitrarily low ones. This relation lies at the basis of the strong connection between the stochastic and deterministic "run for your life" results discussed in Section 23.3. The upshot is that the connection between the stochastic and deterministic results breaks down only when the nonlinearities kick in at densities that are so low that the stochastic effects are not yet negligible.

Nonlinear stochastic "run for your life" results, called "shape theorems" in the stochastic literature, so far have only been proved for models where space is discretized to a square grid. (A good survey can be found in Durrett 1988a, 1988b; see also Cox and Durrett 1988; Zhang 1993.) As yet, no better methods are available for approximating the exact speed than running an efficient simulation of the full spatial stochastic process. Some approximate methods for lattice-based models using a heuristically adapted pair-approximation technique as their main ingredient can be found in Ellner et al. (1998). Lewis (unpublished) has derived adapted moment expansion plus moment closure methods for the continuous-space case by expanding in the size of the neighborhood over which the nonlinear interaction occurs. This procedure has the considerable advantage of allowing the derivation of direct error estimates, but at present the technique remains tied to some very specific assumptions. Overall, both methods perform well in a comparison with simulation results.

Many nonlinear stochastic models can be coupled to a majorating linear model through a thought experiment in which we selectively remove individuals and all their descendants from the output (in mathematical lingo, the sample function) of the linear model in a manner mimicking the stochastic structure of the feedback loop of the nonlinear model (from the population history and/or the present population composition to the death and reproductive events). At all times and places, the population numbers of the linear model are above those of the coupled nonlinear model. This argument proves that the speeds for those nonlinear models are necessarily below that of the associated linear model. Non-Allee nonlinearities only diminish the speed of spatial spread. This effect is clearly demonstrated in Figure 23.3. The cases considered in this figure are as extreme as possible, since displacement occurs only to the four nearest neighbors on a square grid. For more extended displacement regimes, the speed rapidly converges to that of the associated linear model. Cases where the nonlinearity leads to a speed decrease of more than 50% are rare indeed. However, deviations from the linear results are prominent for the customarily small displacement neighborhoods of grid-based simulations.

23.5 The Link with Reaction–Diffusion Models

In Section 23.3 we discuss one link between our integral-kernel-based models and reaction–diffusion models: the latter can be considered a rather special case of the former. Notwithstanding the somewhat stringent underlying assumptions, reaction–diffusion models often do a good job of capturing the phenomena in which we are interested. Moreover, they are much more accessible mathematically. Reaction–diffusion models are the only models for which results about the convergence to a well-defined wave shape are available, and the numerical calculation of their full solution is relatively easy. The main disadvantage of reaction–diffusion models is their largely phenomenological character. Tying their ingredients to all the wonderful biological detail observable in the field is not always straightforward, or even possible. In this respect, the integral-kernel formalism does a better job.

The main reason reaction–diffusion models do such a good job overall is that they are not just any special case, they are a very special case in that they provide good approximations for large classes of more general models.

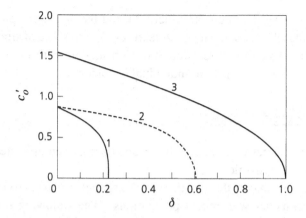

Figure 23.3 Scaled wave speeds $c_0' = c_0/(\beta\sigma)$ for three nearest-neighbor processes on a square lattice. Individuals are supposed to die at rate δ and give birth at rate β. The offspring are randomly dropped into one of the four nearest cells, so that the displacement variance $\sigma^2 = 1/2$. In the epidemic process (curve 1) offspring dropped into cells that are or have been occupied die. In the contact process (curve 2) only offspring that land in occupied cells die. In the spatial birth and death process (curve 3), the occupation history of a cell has no effect. (The reason for the unusual scaling is that this allows the inclusion of the case $\delta = 0$, for which $\mu = 1/\delta$ is infinite.)

The remainder of this section contains a heuristic introduction to the inner workings of this approximation.

A first clue about the the way reaction–diffusion models approximate more general population models can be found by comparing our general approximation $c_0 \approx (\sigma/\mu)\sqrt{2\ln R_0}$ for the wave speeds with the speed for the reaction–diffusion model $c_0 = 2\sqrt{r_0 D} = (\sigma/\mu)\sqrt{2(R_0 - 1)}$. This tells us that we can only expect the approximation to work for small R_0. If we also take into account the constraints imposed by their dimensions, we get the following identification of parameters: $r_0 \approx \mu^{-1}\ln R_0 \approx \mu^{-1}(R_0 - 1)$ and $2D \approx \sigma^2/\mu$. (We cheat a little here by already putting the factor 2 in the right place.)

There are two clues for allocating the factor 2. For $R_0 = 1$, the reaction term drops out and we are back to an ordinary diffusion equation. This equation also crops up in the theory of stochastic process as a description of the change over time of a particle undergoing Brownian motion. There, $2D$ corresponds to the variance increment per unit of time. The closest counterpart to such a quantity in our framework is σ^2/μ. The second clue is that the overall exponential growth rate r_0 for general structured populations is well approximated by $\mu^{-1}\ln R_0$ at small values of $\ln R_0$ or v/μ. So the 2 should not be allocated to r_0.

Thus we have the connection between the parameters in place for the linear regime. What about any nonlinearities? For the sake of simplicity, we concentrate on nonlinear reaction–diffusion models with direct feedback from the population density to individual behavior

$$\frac{\partial n}{\partial t} = D\frac{\partial^2 n}{\partial x^2} + r(n)n \ . \tag{23.18}$$

What is the link between the function r and the more complicated underlying population dynamics?

The fuction r has nothing to do with the spatial structure, so for the time being we consider well-mixed populations. The following observations come to mind. The linear considerations told us that we should primarily look at cases where the population density changes only very slowly. We are considering only direct feedback. Therefore, a slowly changing population density means slowly changing circumstances for our individuals. So what guidance can we get from the theory of population growth under constant conditions? In unchanging environments populations will eventually grow exponentially. The relative growth rate can be determined from the nonspatial analogue of Equation (23.2). Moreover, the population composition stabilizes at an exponential rate, which stays bounded away from zero when we approach the limit of zero growth. If we consider the case of a very small growth rate, we have to rescale time if we still wish to see changes in the population density. In this new time scale, the time in which population structure relaxes to its stable form is very short. The stable form of the population composition is only slightly dependent on the growth rate. In the limit, it is the stable population composition for zero growth rate that, together with the overall population density, enters into the feedback to the reproductive behavior of the individuals. The latter determines the overall population growth rate r. Since the population composition is constant, r effectively only depends on n. Thus we can write $r = \mu^{-1} \ln R(n)$, where $R(n)$ is the average lifetime offspring production of individuals surrounded for their entire lives by conspecifics at a density n in relative frequencies corresponding to the stable population composition (formal details can be found in Greiner et al. 1994, and Metz and van den Bosch 1995).

After this excursion to the general theory of physiologically structured population models, we go a little further into the origin of the diffusion term. In the limit of zero population growth and corresponding constant environment, the basic deterministic equations for our problem become the same as those for a random walk moving to the beat of a renewal process (Cox 1962), because of our assumption that the movement of the individuals is independent across the generations. If we wish to keep the population growth in view, we have to rescale time while taking the limit. The average number of displacement steps made in one time unit by a single line of descent equals μ^{-1}. If we rescale time, this number goes up. Therefore, we have to rescale space to compensate for this increase. However, this combination of time and space scaling is precisely what transforms a random walk into a diffusion (see, e.g., van Kampen 1981).

The upshot is that reaction–diffusion models generically derive from much more general population models through a combination of robust limit procedures. But not every concrete population problem has parameter values putting it close to a reaction–diffusion model!

23.6   Dispersal on Different Scales

Recently, an increasing number of papers have suggested that dispersal often occurs on two (or more) disparate scales, with most displacement taking place over a fairly short scale, and a small part occurring on a longer scale. The effect of such double dispersal is a wave front consisting of clusters of individuals amid a largely empty space – hence the joy of suddenly discovering a meadow full of a plant species until then only known from some distant country. The down side is that, because long-distance dispersers are few and far between, it is difficult to measure the parameters of the far-dispersal process. Yet it is this dispersal that largely determines the speed of spatial spread. The amplification effect of biological reproduction opens the door to the rare but important events that are the bane of experimental population dynamics (and that, on another time scale, drive evolution).

The combination of a double dispersal mechanism and local population dynamics has two accessible extremes, depending on the interaction of the smallest-scale dispersal and the population dynamical nonlinearity. At the first extreme, the nonlinearity comes into play only late in the passing of the wave front, after the various local clusters have merged into a more or less continuous-looking population. At the other extreme, the local nonlinearities already dominate far out in the wave front. We start with the case where the effect of the nonlinearities stays negligible. A different approach has been taken by Shigesada *et al.* (1995; see also Shigesada and Kawasaki 1997).

The fully linear case

If the dispersal scales are sufficiently disparate we can concentrate on the larger scale, making the simplifying assumption that all reproduction over the smaller dispersal scale occurs at an individual's place of birth.

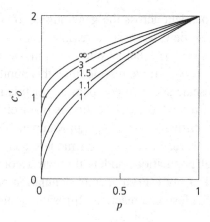

Figure 23.4 Scaled wave speeds $c_0' = c_0/\sqrt{D(\beta - \delta)}$ for a model in which individuals die at rate δ and give birth at rate β to offspring that either stay at home for their whole life with probability p or diffuse at rate D. (The scaling in this figure differs from the usual scaling since we also want to accommodate the case $\delta = 0$, for which $R_0 = \infty$.) The different curves correspond to the values of $R_0 = \beta/\delta$ indicated on the curve.

This leads to reproduction kernels that can be written as the sum of two terms – the term for the far dispersal plus a kernel that is the product of a delta function in space times some time kernel. It turns out that the wave speeds for such models have unexpected properties, which were first brought to light by Cook (unpublished; see also Goldwasser *et al.* 1994). We demonstrate this phenomenon with the following example.

Imagine a situation in which individuals give birth and die randomly at rates β and δ, respectively, and either stay at home for their whole life with probability p or diffuse at rate D. The birth kernel for this model is, for $n = 1$,

$$A(\tau, y) = \beta \exp(-\delta\tau)\big[p\delta_0(y) \qquad\qquad \\ + (1 - p)(4\pi D\tau)^{1/2} \exp[-y^2/(4D\tau)]\big] , \tag{23.19}$$

with δ_0 the delta function at zero. Its Laplace transform is

$$\tilde{A}(s, \lambda) = \beta[p(s + \delta)^{-1} + (1 - p)(s + \delta - D\lambda^2)^{-1}] . \tag{23.20}$$

An explicit formula for c_0 cannot be found, but the two equations for λ_0 and c_0 can be reduced to one equation in one unknown, which allows a quick numerical solution. Figure 23.4 shows the result. Surprisingly, the wave speed does not converge to zero when an increasingly larger fraction of the individuals stay at home. A perturbation expansion gives the following approximate expression for c_0 at small values of p:

$$c_0 \approx \sqrt{1 - R_0^{-1}} + \sqrt{p(1 + R_0^{-1})} \,. \tag{23.21}$$

What better illustration of the importance of rare events could one wish for?

Other mixtures of a dispersal kernel and a kernel expressing the fact that some individuals never disperse give the same sort of result. As one more example, consider individuals that reproduce only at age μ and disperse in two dimensions according to the Bessel displacement distribution from Box 23.1, if they disperse at all. Then, $c_0 = (\sigma/\mu)[\ln(R_0)/\sqrt{2} + O(\sqrt{p})]$. The unpleasant message is that, clearly, the rare and therefore difficult to observe far dispersers dominate the picture. The good message is that the customarily low quality of our estimates for the parameter p has less influence on our predictions than we might expect.

When the nonlinearity dominates at the small scale

When new long-distance arrivals are rare and local population growth is fast, we have a situation where the local population dynamical hot-spots (or foci, see Chapter 16) develop essentially independent of the large-scale structure of the developing wave. In that case we can treat the hot-spots as individuals in a model that concentrates on the lowest resolution. By concentrating only on the higher resolution we can model the development of the hot-spots as just another nonlinear spatial expansion problem. To simplify the discussion, from now on we assume that $n = 2$. In the simplest case, the local population density inside a hot-spot just grows to a fixed ceiling, so that the output of far dispersers from the hot-spot overall grows like a quadratic function of its age. Another possibility, seen *inter alia* in epidemics, is that the population growth leaves in its wake a burnt-out zone that no longer produces any propagules. In that case the output of far dispersers overall grows like a linear function of age. Although the assumption of a linear or quadratic growth of the output of far dispersers is only a rough approximation, since it does not account for the initial buildup of a hot-spot, we stick to this simplification for the remainder of this section.

The reproduction kernel for the hot-spots can be written as a product of a time kernel, for which we have already chosen $\alpha\tau$ or $\beta\tau^2$, and a displacement distribution. For this we take either the Bessel distribution or the rotated exponential from Box 23.1. The wave speeds for the four possible combinations are as follows:

Box 23.5 Dependence among parameters: A lesson from rabies

On a number of occasions in this chapter we have broken the birth kernel into components. In this box we wish to introduce a caveat. One should be careful not to let a decomposition that is mathematically convenient guide one's thoughts when it comes to connecting a model to reality. For example, for a fungal disease, sowing a mixture of resistant and susceptible wheat only changes R_0, whereas changing the sowing density changes both R_0 and the dispersal distribution. The following example shows that paying close attention to interpretational matters sometimes allows one to derive interesting conclusions by very modest means.

In 1940 at the Polish–Russian border, a wave of feral rabies began sweeping over Europe. The veterinary and medical importance of rabies has led to considerable modeling effort and the availability of a lot of good data. The primary host for rabies in Europe is the fox. One of the intriguing observations is that the wave speed seems to be largely independent of fox density. When the local fox density is below a certain threshold, rabies does not enter a region at all; in all other cases the wave proceeds at a speed that shows little dependence on fox density (Moegle *et al.* 1974; Bögel and Moegle 1980). For fox densities above the threshold the earlier models for rabies spread showed a gradual increase in the speed. Van den Bosch *et al.* (1990a) provided a simple explanation for the discrepancy. Although their model is no doubt wrong in its details, the overall message is robust. Transmission of rabies takes place only through direct contact. Foxes have roughly fixed home ranges. Radio-tracking data suggest that home range use by rabid and uninfected foxes does not differ much (Andral *et al.* 1982). Therefore, we can write $A(\tau, y) = R_0 a_1(\tau)a_2(y)$. It is usual to assume R_0 to be proportional to the local fox density at the start of the epidemic. The argument from Box 23.1 suggests taking a Gaussian density for a_2. Rabies has a long latent period and a relatively short infectious period. Rough estimates of the mean and standard deviation of the age-at-birth kernel a_1 are respectively $\mu = 35$ days and $\nu = 5$ days (Berger, 1976). The main interest is in the dependence of c_0 on the properties of the local fox community. Lower fox densities correspond both to smaller R_0 and to larger home ranges. More particularly, $\sigma^2 \propto$ [area of home range] $\propto$ [susceptible density]$^{-1} \propto R_0^{-1}$. If we substitute this into the first formula from Box 23.3, we get

$$c_0 \propto \sqrt{\frac{\ln R_0}{R_0}} .$$

To arrive at a quantitative formula, van den Bosch *et al.* (1990a) used two additional pieces of data gleaned from the literature. At a fox density of 1 fox per km^2, the approximate radius of a home range is 2.3 km

continued

Box 23.5 *continued*

(Lambinet *et al.* 1978), corresponding to $\sigma \approx \sqrt{2.3^2 + 2.3^2} = 3.2$ km, and the observed threshold density for the occurrence of rabies (i.e., $R_0 = 1$) is 0.4 fox per km^2 (Steck and Wandeler 1980). The resulting relation between the predicted speed of the rabies wave and fox densities is depicted in the figure. Effectively, there is either an epidemic wave traveling at a more or less constant velocity, or no epidemic at all. Beyond the threshold density the predicted speed decreases from 45 to 30 km per year for the largest feasible fox densities, which compares well with a reported range of speeds of between 30 and 60 km per year.

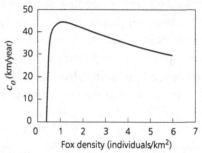

Predicted dependence of speed of rabies spread on fox density.

■ Linear × Bessel: $c_0 = 2^{1/2}\sigma\alpha^{1/2}$
■ Linear × exponential: $c_0 = (27/4)^{1/4}\sigma\alpha^{1/2}$
■ Quadratic × Bessel: $c_0 = (5/6)^{1/2}5^{1/3}\sigma\beta^{1/3}$
■ Quadratic × exponential: $c_0 = (4/3)^{1/2}4^{1/3}\sigma\beta^{1/3}$

On the mathematical side we are in the best possible position. On the empirical side it is a different story. Although we have gone out of our way to produce a parameter-scarce family of models, it is generally impractical to estimate even those few parameters. The theory does, however, provide a basis for organizing comparisons between related species and for studying the potential impact of enforced changes of the parameter values, for example, resulting from a government's agricultural policy. Some initial explorations in the latter direction can be found in van den Bosch *et al.* (1994, 1999); these authors even went so far as to consider disperal on three scales, with the smallest scale dominated by the nonlinearities and the next two scales being treated in the manner outlined in the previous subsection.

23.7 Concluding Comments

In this section we change the perspective a little and put slightly more emphasis on the inherent stochastic nature of the population processes that underlie the occurrence of invasion waves. Such waves appear when a macroscopically homogeneous system of largish spatial extent is seeded with a concentrated invader mass. By macroscopic spatial homogeneity, we mean that the environment, averaged over the scales at which the individuals interact or disperse, should change little at the spatial scale over which the invasion wave develops. To study the basic properties of invasion waves, we have concentrated on infinitely extended systems. Seen from afar, such waves looks like circles (spheres) extending at an asymptotically constant speed. This speed has been our main object of study. To obtain simple results, we have concentrated on systems without an Allee effect, since for such systems the wave speed is bounded from above by the wave speed of the local linearization around the uninvaded case. However, we have also given arguments for why we expect such systems to be good approximations to systems with a slight strict Allee effect (and no other Allee effect). From the stochastic viewpoint, the main distinction that we have to make is between the locally finite and infinite cases. Systems are effectively locally infinite when the numbers of individuals potentially contained in the dispersal or interaction ranges are sufficiently large and the effect of each separate interaction is correspondingly small. No general quantitative results are available for the locally finite case; all we are able to do is summarize some qualitative trends. The fortunate conclusion is that infinity tends to be rather nearby.

In the locally infinite case, the so-called linear conjecture says that the wave speed of the full model and the wave speed of its local linearization around the uninvaded case coincide. The latter wave speed can be calculated from two equations in two unknowns, expressed in terms of the Laplace transform of the average distribution over time and space of the offspring produced by a single individual. These observations provide the technical background for discussing a number of phenomena, such as the approximation of more complicated population processes by a reaction–diffusion equation, and the curious fact that, in general, the wave speed approaches a nonzero limit when the fraction of individuals who never disperse approaches 1.

24

Epilogue

Johan A.J. Metz, Ulf Dieckmann, and Richard Law

We would like to close this volume with a look at the future of mechanistic modeling in spatial ecology. We hope that at least some of the optimistic views sketched below come true. No doubt others, hopefully fewer, will turn out to be mirages.

Just as theory is at its best when it is demonstrably applicable to real ecological systems, field research is most important when it addresses questions that clearly transcend a particular study system. Few researchers, however, have sufficient command of both theory and experiment to actively participate at the two research fronts. It is therefore essential to extend chains of collaboration between empiricists and theorists. These chains should not become too long lest they break or the message passed along becomes too garbled. If such collaborative chains are to work effectively, each partner must have a good understanding of the others' vocabulary, basic concepts, and techniques.

One of this book's objectives is to foster dialogue between those researchers with empirical competence and those with theoretical skills in the field of spatial ecology. In practice, there is still an appreciable distance between the detailed investigations of plant interactions reported in Part A of this volume and the mathematical methods advanced in Part D. However, ecological theory is making great strides toward integrating more ecological realism into manageable models. Theorists and empiricists alike are searching for new kinds of models that are better able to account for the complex implications of spatial heterogeneity. In this context, we clearly discern an increasing appreciation for the importance of constructing ecological theory from the bottom up, starting from the level of the individual and working up to the dynamics of populations and communities. We believe that this is why there already is a close conceptual correspondence between the interaction neighborhoods considered by the plant ecologists in Part A and those used by the mathematicians in Part D (notwithstanding the distance between the two research fields as practiced today). At a larger spatial scale, the good match between observed epidemiological foci and the theory of invasion waves is also encouraging. We may not yet have all the connections in place, but contacts are being made.

Whereas Parts A and D show some correspondence in their concepts and methods, there is substantial diversity in the intermediate realm, covered in Parts B and C of this book. Here, scientific imagination is roaming more freely, spurred by the availability of powerful computers and the multitude of novel phenomena to be discovered. Humans are visual creatures, with a fascination for spatial pattern. It is becoming increasingly clear that many important ecological phenomena can only be understood in terms of the self-generated spatial patterns found in ecosystems.

Computer simulations and resulting spatial patterns have considerable, and rightful, appeal. In the process of covering new ground in ecological understanding, simulation studies often take on the role of pioneers. In the long run, however, we hope to see the former shoals transformed into safe, fertile ground through systematic consolidation during later successional stages. Here, mathematics can provide the necessary infrastructure. Such consolidation is currently under way in spatial ecology and may eventually help us to see the pattern in the patterns.

At the moment, the pattern is clearest at the extremes of spatial scale. Current mathematical techniques are most successful at the relatively large and the relatively small scales. Moment methods and pair-approximation techniques help us to unravel some of the intricate consequences of small-scale spatial structure. Yet, these methods fail, or need to be extended, in the presence of long-range heterogeneities. On the conceptual side, moment methods alert us to the critical importance of adopting an "individual's-eye view": it is the (necessarily local) environments experienced by individuals that shape a population's response. And in these local environments, neighbors are not always abundant enough to permit us to neglect sampling variance; fluctuation corrections are thus important complements of the correlation corrections that take care of small-scale structure. Reaction–diffusion and integral equation methods, on the other hand, are tailored to describe large-scale heterogeneities. Models of these types are typically derived from so-called rapid-stirring limits, ensuring the local equivalent of mean-field conditions; for this reason, correlation and fluctuation corrections are rarely considered in such models.

Models in which space and populations are discrete (as in cellular automata) and those where they are continuous (as in partial differential equations) are often viewed as interchangeable descriptions that can be appropriately applied to the same kinds of systems. This view obscures the fact that matches between real systems and their simplified mathematical descriptions are only as good as the assumptions under which those simplifications have been derived from individual-based considerations. Putting more emphasis on formal derivations therefore is not just pedantry. The derivations unveil assumptions and help eliminate misunderstandings that otherwise would soon permeate spatial ecology.

Recent studies underline that it is the discreteness of individuals in particular that has unexpected consequences. In continuum-based descriptions, this discreteness can to some extent be fudged by applying ad hoc threshold rules, but more rigorous ways of handling are needed.

For intermediate-scale patterns, no obvious solution is on the horizon. This is mainly because, for such patterns, clear mathematical limits from which to derive suitably simplified descriptions do not seem to be available. We are more hopeful about the problem of coping with the simultaneous presence of small- and large-scale spatial structure. Here, it seems worth aiming at a merger of moment methods and reaction–diffusion techniques. One step in this direction has already been taken by incorporating fluctuation corrections into reaction–diffusion models (based on so-called hydrodynamic limits of interacting particle systems). Yet, the real challenge remains: to systematically incorporate correlation corrections into reaction terms. We expect some rapid progress on this problem will be possible in the near future.

We believe that all of these developments together will lead to a third generation of models in spatial ecology. After the drastic oversimplification that has compromised mean-field models and, to a lesser extent, reaction–diffusion models, and following the bewildering intractability of many of the contemporary individual-based simulation models, a bridge will be established, constructed from elaborate but manageable models of intermediate complexity.

We expect that this third generation of models will have the following features:

- They will be intimately linked to individual-based models by sound approximation schemes that make explicit the underlying assumptions.

- In particular, new approaches will respect the discreteness of individuals by putting into place fluctuation corrections that go beyond current threshold heuristics.
- They will entail the merging of the insights and techniques already available for small- and large-scale patterns.
- Some will address spatial heterogeneity at intermediate scales. Development of a suite of spatial statistics, geared to particular types of intermediate-scale structure, will allow more systems to be approximated by relaxation projections.
- Based on such advances, evolutionary implications of spatial structure will receive more attention. A theory of spatial adaptive dynamics, in which descriptions of local mutant growth are translated into predictions of phenotypic change, is in the making.

Yet there are many reasons for modesty. More powerful methods generally impose steeper learning curves on their practitioners. Unfortunately, no amount of effort will ever result in models of spatial complexity possessing the simplicity that mean-field approximations once offered. Nonetheless, we believe that the new methods presented in Parts C and D of this book have enticing cost-to-benefit ratios, and we hope that this volume makes those benefits accessible. If the new methods can successfully be applied to an increasing number of relevant ecological questions, then some of them, stripped down to their essentials, may eventually become part of the standard ecological repertoire.

One should be aware, though, that spatial processes have an inexhaustible potential for dynamical complications, and that it will never be possible to deal with this complexity through just one method. Instead, we need inspired combinations of a range of techniques for constructing helpful simplifications. Spatial systems of realistic complexity need to be approached from many angles to achieve the greatest understanding.

References

Page numbers of reference citations in this volume are given in square brackets.

Aarssen LW (1983). Ecological combining ability and competitive combining ability in plants: Towards a general evolutionary theory of coexistence in systems of competition. *The American Naturalist* **22**:707–731 [*41*]

Aarssen LW & Epp GA (1990). Neighbor manipulations in natural vegetation: A review. *Journal of Vegetation Science* **1**:13–30 [*13*]

Abramowitz M & Stegun IA, eds. (1965). *Handbook of Mathematical Functions, with Formulas, Graphs, and Mathematical Tables.* New York, NY, USA: Dover [*392, 489*]

Adams J, Kinney T, Thompson S, Rubin L & Helling RB (1979). Frequency-dependent selection for plasmid-containing cells of *Escherichia coli*. *Genetics* **91**:627–637 [*243*]

Albano AM, Muench J, Schwartz C, Mees AI & Rapp PE (1988). Singular value decompostion and the Grassberger–Procaccia algorithm. *Physical Review A* **38**:3017–3026 [*225*]

Allouche J & Reder C (1984). Spatio-temporal oscillations generated by cellular automata. *Discrete Applied Mathematics* **8**:215 [*95*]

Ammerman AJ & Cavalli-Sforza LL (1973). A population model for the diffusion of early farming in Europe. In *The Explanation of Culture Change*, ed. Renfrew C, pp. 343–357. London, UK: Duckworth [*316*]

Ammerman AJ & Cavalli-Sforza LL (1984). *The Neolithic Transition and the Genetics of Populations in Europe.* Princeton, NJ, USA: Princeton University Press [*316*]

Anderson RM & May RM (1991). *Infectious Diseases of Humans: Dynamics and Control.* Oxford, UK: Oxford University Press [*250, 271, 412*]

Andow DA, Kareiva PM, Levin SA & Okubo A (1990). Spread of invading organisms. *Landscape Ecology* **4**:177–188 [*316, 388, 498*]

Andow DA, Kareiva PM, Levin SA & Okubo A (1993). Spread of invading organisms: Pattern of spread. In *Evolution of Insect Pests: Patterns of Variation*, eds. Kim KC & McPheron BA, pp. 219–242. New York, NY, USA: Wiley [*316, 498*]

Andral L, Artois M, Aubert MFA & Blancou J (1982). Radio-pistage de renards enragés. *Comparative Immunology, Microbiology and Infectious Diseases* **5**:284–291 [*316, 510*]

Anonymous (1953). Some further definitions of terms used in plant pathology. *Transactions of the British Mycological Society* **36**:267 [*292*]

Aronson DG (1977). The asymptotic speed of propagation of a simple epidemic. In *Nonlinear Diffusion, Research Notes in Mathematics*, Vol. 14, eds. Fitzgibbon WE & Walker HF, pp. 1–23. London, UK: Pitman [*486*]

Aronson DG & Weinberger HF (1975). Nonlinear diffusion in population genetics, combustion and nerve-pulse propagation. In *Partial Differential Equations and Related Topics, Lecture Notes in Mathematics, 446*, ed. Goldstein JA, pp. 5–49. Berlin, Germany: Springer [*475, 486*]

Atkinson C & Reuter GE (1976). Deterministic epidemic waves. *Mathematical Proceedings of the Cambridge Philosophical Society* **80**:315–330 [*486*]

Atsatt PR & O'Dowd DJ (1976). Plant

defense guilds. *Science* **193**:24–29 [*33*]

Augspurger CK (1984). Pathogen mortality of tropical tree seedlings: Experimental studies of the effects of dispersal distance, seedling density, and light conditions. *Oecologia* **61**:211–217 [*18*]

Axelrod R (1981). The emergence of cooperation among egoists. *American Political Science Review* **75**:306–318 [*329*]

Axelrod R (1984). *The Evolution of Cooperation.* New York, NY, USA: Basic Books [*136, 138, 147, 332, 382*]

Axelrod R (1997). *The Complexity of Cooperation: Agent-Based Models of Competition and Collaboration.* Princeton, NJ, USA: Princeton University Press [*342*]

Axelrod R & Hamilton WD (1981). The evolution of cooperation. *Science* **211**:1390–1396 [*135, 318, 324*]

Bacon PJ, ed. (1985). *Population Dynamics of Rabies in Wildlife.* London, UK: Academic Press [*316*]

Bailey NTJ (1975). *The Mathematical Theory of Infectious Diseases and Its Applications.* London, UK: Griffin [*412*]

Bak P, Chen K & Tang C (1990). A forest fire model and some thoughts on turbulence. *Physics Letters A* **147**:297–300 [*275*]

Ballaré CL, Scopel AL & Sanchez RA (1990). Far-red radiation reflected from adjacent leaves: An early signal of competition in plant canopies. *Science* **247**:329–332 [*14*]

Barbour AD (1977). The uniqueness of Atkinson's and Reuter's epidemic waves. *Mathematical Proceedings of the Cambridge Philosophical Society* **82**:127–130 [*486*]

Bazzaz FA (1990). Plant–plant interactions in successional environments. In *Perspectives on Plant Competition,* eds. Grace JB & Tilman D, pp. 239–263. San Diego, CA, USA: Academic Press [*12, 14*]

Been TH & Schomaker CH (1998). *Quantitative Studies on the Management of Potato Cyst Nematodes (Globodera spp) in the Netherlands.* PhD dissertation. Agricultural University Wageningen, Netherlands [*295–296, 313*]

Belitsky P, Ferrari N, Konno N & Liggett T (1997). A strong correlation inequality for contact processes and oriented percolation. *Stochastic Processes and their Applications* **67**:213–225 [*357*]

Bell AD (1984). Dynamic morphology: A contribution to plant population ecology. In *Perspectives on Plant Population Ecology,* eds. Dirzo R & Sarukhan J, pp. 48–65. Sunderland, MA, USA: Sinauer Associates [*15*]

Bell AD (1986). The simulation of branching patterns in modular organism. *Philosophical Transactions of the Royal Society of London B* **313**:143–159 [*24*]

Bergé P, Pomeau Y & Vidal C (1984). *Order within Chaos.* Chichester, UK: Wiley [*197*]

Berger J (1976). Model of rabies control. In *Mathematical Models in Medicine, Lecture Notes in Biomathematics,* Vol. 11, eds. Berger J, Buhler W, Repges R & Tautu P, pp. 75–88. Berlin: Springer [*510*]

Berger RD & Luke HH (1979). Spatial and temporal spread of oat crown rust. *Phytopathology* **69**:1199–1201 [*299*]

Berkowitz AR, Canham CD & Kelly VR (1995). Competition versus facilitation of tree seedling growth and survival in early successional communities. *Ecology* **76**:1156–1168 [*13*]

Bertero M & Pike ER (1982). Resolution in diffraction-limited imaging, a singular value analysis. I. The case of coherent illumination. *Optica Acta* **29**:727 [*221*]

Bertness MD & Shumway SW (1993). Competition and facilitation in marsh plants. *The American Naturalist* **124**:718–724 [*13*]

Besag J (1974). Spatial interaction and the statistical analysis of lattice systems. *Journal of the Royal Statistical Society B* **367**:192–236 [*86*]

Besag J (1975). Statistical analysis of non-lattice data. *The Statistician* **24**:179–195 [*87*]

Biggins JD (1995). The growth and spread of the general branching random walk. *Annals of Applied Probability* **5**:1008–1024 [*485–486, 495*]

Biggins, JD (1997). How fast does a branching random walk spread? In *Classical and Modern Branching Processes*, eds. Athreya KB & Jagers P, pp. 19–39. New York, NY, USA: Springer [*485–486, 495*]

Blackman GE (1935). A study by statistical methods of the distribution of species in grassland association. *Annals of Botany* **49**:749–777 [*48*]

Bleasdale JKA & Nelder JA (1960). Plant population and crop yield. *Nature* **188**:342 [*18*]

Blomberg C, von Heijne G & Leimar O (1981). Competition, coexistence and irreversibility in models of early evolution. In *Origin of Life*, ed. Wollman E, pp. 385–392. Dordrecht, Netherlands: D. Reidel [*157*]

Böddeker B & McCaskill J. Self-replicating spots as a basis for the spatial stabilisation of catalytic function. Unpublished [*180–181*]

Boerlijst MC (1994). *Selfstructuring: A Substrate for Selection*. PhD dissertation, University of Utrecht, Netherlands [*172*]

Boerlijst MC & Hogeweg P (1991a). Selfstructuring and selection: Spiral waves as a substrate for evolution. In *Artificial Life*, Vol. 2, eds. Langton CG, Taylor C, Farmer JD & Rasmussen S, pp. 255–276. Redwood City, CA, USA: Addison-Wesley [*152, 160, 172, 178–179*]

Boerlijst MC & Hogeweg P (1991b). Spiral wave structure in pre-biotic evolution: Hypercycles stable against parasites. *Physica D* **48**:17–28 [*116–118, 128, 152, 160, 172, 181*]

Boerlijst MC & Hogeweg P (1995a). Attractors and spatial patterns in hypercycles with negative interactions. *Journal of Theoretical Biology* **176**:199–210 [*227*]

Boerlijst MC & Hogeweg P (1995b). Spatial gradients enhance persistence of hypercycles. *Physica D* **88**:29–39 [*172*]

Boerlijst M, Lamers ME & Hogeweg P (1993). Evolutionary consequences of spiral waves in a host-parasitoid system. *Proceedings of the Royal Society of London B* **253**:15–18 [*181, 334, 368, 382, 385*]

Bögel K & Moegle H (1980). Characteristics of the spread of a wildlife rabies epidemic in Europe. *Biogeographica* **8**:251–258 [*510*]

Bolker B (1995). Group report: Spatial dynamics of infectious diseases in natural populations. In *Ecology of Infectious Diseases in Natural Populations*, eds. Grenfell B & Dobson A, pp. 399–420. Cambridge, UK: Cambridge University Press [*271*]

Bolker B & Grenfell B (1995). Space, persistence and dynamics of measles epidemics. *Philosophical Transactions of the Royal Society of London B* **348**:309–320 [*271*]

Bolker B & Pacala SW (1997). Using moment equations to understand stochastically driven spatial pattern formation in ecological systems. *Theoretical Population Biology* **52**:179–197 [*252, 442*]

Bonan GB (1993). Analysis of neighborhood competition among annual plants: Implications of a plant growth model. *Ecological Modelling* **65**:123–136 [*21*]

Bonde R & Schultz ES (1942/1943). Potato refuse piles as a factor in the dissemination of late blight. *Maine Agricultural Experiment Station Bulletin* **416**:229–246 [*296*]

Bourke PMA (1964). Emergence of potato blight, 1843–46. *Nature* **203**:805–808 [*300*]

Bradburg R, van der Laan J & MacDonald B (1990). Modelling the effects of predation and dispersal on the generation of waves of starfish outbreaks. *Mathematical and Computer Modelling* **13**:61–67 [*95*]

Bradshaw AD (1965). Evolutionary significance of phenotypic plasticity in plants. *Advanced Genetics* **13**:115–155 [*14*]

Bramson M (1983). Convergence of

solutions of the Kolmogorov equation to travelling waves. *AMS Memoirs*, Vol. 44, No. 285. Providence, RI, USA: American Mathematical Society [*498, 503*]

Brenchley GH & Dadd CV (1962). Potato blight recording by aerial photography. *NAAS Quarterly Review* **14**:21–24 [*292, 295, 299, 311*]

Bresch C, Niesert U & Harnasch D (1980). Hypercycles, parasites and packages. *Journal of Theoretical Biology* **85**:399–405 [*171*]

Britton N (1986). *Reaction–Diffusion Equations and Their Application to Biology*. New York, NY, USA: Academic Press [*458, 471, 477*]

Broadbent SR & Kendall DG (1953). The random walk of *Trichostrongylus retortaeformis*. *Biometrics* **9**:460–465 [*489*]

Brockwell P & Davis R (1991). *Time Series: Theory and Methods*, 2nd edn. New York, NY, USA: Springer [*68*]

Broomhead DS & King GP (1986). Extracting qualitative dynamics from experimental data. *Physica D* **20**:217–236 [*221, 225*]

Broomhead DS, Jones R & King GP (1987). Topological dimension and local coordinates from time-series data. *Journal of Physics A* **20**:L563–L569 [*225*]

Broomhead DS, Jones R & King GP (1988). Singular value decomposition and embedding dimension – Comment. *Physical Review A* **37**:5004–5005 [*225*]

Broomhead DS, Indik R, Newall AC & Rand DA (1991). Local adaptive Galerkin bases for large dimension dynamical systems. *Nonlinearity* **4**:159–197 [*225*]

Brown J, Sanderson M & Michod R (1982). Evolution of social behavior by reciprocation. *Journal of Theoretical Biology* **99**:319–339 [*327*]

Brown KJ & Carr J (1977). Deterministic epidemic waves of critical velocity. *Mathematical Proceedings of the Cambridge Philosophical Society* **81**:431–433 [*486*]

Brown W (1993). Cheat thy neighbour, a recipe for success. *New Scientist* **189**:19 [*272*]

Bullock JM, Hill BC, Dale MP & Silvertown J (1994). An experimental study of the effects of sheep grazing on vegetation change in a species-poor grassland and the role of seedlings recruitment into gaps. *Journal of Applied Ecology* **31**:493–507 [*34*]

Burgman MA (1987). An analysis of the distribution of plants on granite outcrops in southern Western Australia using Mantel tests. *Vegetatio* **71**:79–86 [*53*]

Cain ML, Pacala SW, Silander JA Jr & Fortin MJ (1995). Neighborhood models of clonal growth in the white clover *Trifolium repens*. *The American Naturalist* **145**:888–917 [*24*]

Caldwell MM, Manwaring JH & Durham SL (1991). The microscale distribution of neighboring plant roots in fertile soil microsites. *Functional Ecology* **5**:765–772 [*17*]

Caldwell MM, Manwaring JH & Durham SL (1996). Species interactions at the level of fine roots in the field: Influence of soil nutrient heterogeneity and plant size. *Oecologia* **106**:440–447 [*17*]

Callaghan TV (1988). Physiological and demographic implications of modular construction in cold environments. In *Plant Population Ecology*, eds. Davy AD, Hutchings MJ & Watkinson AR, pp. 111–136. Oxford, UK: Blackwell [*228*]

Callaway RM (1995). Positive interactions among plants. *Botanical Review* **61**:306–349 [*32*]

Callaway RM & King L (1996). Temperature-driven variation in substrate oxygenation and the balance of competition and facilitation. *Ecology* **77**:1189–1195 [*13*]

Capy P, Bazin C, Higuet D & Langin T (1998). *Dynamics and Evolution of Transposable Elements*. New York, NY, USA: Springer [*313–314*]

Carey JR (1996). The incipient Mediterranean fruit fly population in California:

Implications for invasion biology. *Ecology* **77**:1690–1697 [*72*]

Carlile DW, Skalski JR, Batker JE, Thomas JM & Cullinan VI (1989). Determination of ecological scale. *Landscape Ecology* **2**:203–213 [*210*]

Casdagli M, Eubank S, Farmer JD & Gibson J (1991). State space reconstruction in the presence of noise. *Physica D* **51**:52–98 [*218*]

Casparie WA (1969). Bult- und Schlenkenbildung in Hochmoortorf. *Vegetatio* **19**:146–180 [*59, 61*]

Casparie WA (1972). Bog development in Southeastern Drente, The Netherlands. *Vegetatio* **25**:1–271 [*59*]

Caswell H (1989). *Matrix Population Models: Construction, Analysis, and Interpretation*, Sunderland, MA, USA: Sinauer Associations [*228*]

Caswell H & Etter RJ (1992). Ecological interactions in patchy environments: From patch-occupancy models to cellular automata. In *Patch Dynamics*, eds. Levin SA, Powell TM & Steele JH, pp. 93–109. New York, NY, USA: Springer [*227*]

Chacón P & Nuño J (1995). Spatial dynamics of a model for prebiotic evolution. *Physica D* **81**:398–410 [*152*]

Chandler RE (1996). The second-order spectral analysis of spatial-temporal rainfall models. Research Report 158. University College London, UK: Department of Statistical Science [*84*]

Chandler RE (1997). A spectral method for estimating parameters in rainfall models. *Bernoulli* **3**:301–322 [*84*]

Chao L & Levin BR (1981). Structured habitats and the evolution of anticompetitor tòxins in bacteria. *Proceedings of the National Academy of Sciences of the USA* **78**:6324–6328 [*243*]

Charles AH (1964). The differential survival of plant types in swards. *Journal of the British Grassland Society* **19**:198–204 [*41*]

Claessen D & de Roos AM (1995). Evolution of virulence in a host-pathogen system with local pathogen transmission. *Oikos* **74**:401–413 [*385*]

Clark JS (1990). Disturbance and population structure on the shifting mosaic landscape. *Ecology* **72**:1119–1137 [*95*]

Clark JS (1998). Why trees migrate so fast: Confronting theory with dispersal biology and the paleorecord. *The American Naturalist* **15**:202–224 [*317, 496*]

Clark JS, Fastie C, Hurt G, Jackson C, Johnson C, King GA, Lewis M, Lynch J, Pacala S, Prentice C, Scupp EG, Webb T III & Wyckoff P (1998). Reid's paradox of rapid plant migration. *Bioscience* **48**:13–24 [*317, 496*]

Cliff A (1995). Incorporating spatial components into models of epidemic spread. In *Epidemic Models: Their Structure and Relation to Data*, ed. Mollison D, pp. 119–149. Cambridge, UK: Cambridge University Press [*271*]

Cliff A, Haggett P, Ord J & Versey G (1981). *Spatial Diffusion: An Historical Geography of Epidemics in an Island Community*. Cambridge, UK: Cambridge University Press [*271*]

Collins SL & Glenn SM (1988). Disturbance and community structure in North American prairies. In *Diversity and Pattern in Plant Communities*, eds. During HJ, Werger MJA & Willems JH, pp. 131–143. The Hague, Netherlands: SPB Academic Publishing [*53*]

Colwell RN (1956). Determining the prevalence of certain cereal crop diseases by means of aerial photography. *Hilgardia* **26**:223–286 [*292, 299*]

Comins HN & Blatt DWE (1974). Prey–predator models in spatially heterogeneous environments. *Journal of Theoretical Biology* **48**:75–83 [*188*]

Comins HN, Hassell MP & May RM (1994). Species coexistence and self-organizing spatial dynamics. *Nature* **370**:290–292 [*209*]

Condit R, Hubbell SP & Foster RB (1996). Assessing the response of plant functional types to climatic change in tropical forests. *Journal of Vegetation Science* **7**:405–416 [*40*]

Cook J. Biological Invasions: The Effect of Long-Range Dispersal. Unpublished [*508*]

Cook RE (1983). Clonal plant populations. *American Scientist* **71**:244–253 [*228*]

Cornet AF, Delhoume JP & Montana C (1988). Dynamics of striped vegetation patterns and water balance in the Chihuahuan desert. In *Diversity and Pattern in Plant Communities*, eds. During HJ, Werger MJA & Willems JH, pp. 221–231. The Hague, Netherlands: SPB Academic Publishing [*60*]

Cox DR (1958). *Planning of Experiments*. New York, NY, USA: Wiley [*65*]

Cox DR (1962). *Renewal Theory*. London, UK: Methuen [*506*]

Cox DR (1984). Long-range dependence: A review. In *Statistics: An Appraisal*, eds. David HA & David HT, pp. 55–74. Ames, IA, USA: Iowa State University Press [*67*]

Cox DR & Isham V (1980). *Point Processes*. London, UK: Chapman and Hall [*72, 74*]

Cox JT & Durrett R (1988). Limit theorems for the spread of epidemics and forest fires. *Stochastic Processes and their Applications* **30**:171–191 [*277, 503*]

Crank J (1975). *The Mathematics of Diffusion*. Oxford, UK: Clarendon Press [*156*]

Crawley MJ & May RM (1987). Population dynamics and plant community structure: Competition between annuals and perennials. *Journal of Theoretical Biology* **125**:475–489 [*11, 227, 385*]

Creegan P & Lui P (1984). Some remarks about the wave speed and travelling wave solutions of a nonlinear integral operator. *Journal of Mathematical Biology* **20**:59–68 [*486*]

Cressie NAC (1991). *Statistics for Spatial Data*. New York, NY, USA: Wiley [*49, 64, 66, 72, 75–76, 393–394*]

Cressman R & Dash A (1987). Density dependence and evolutionarily stable strategies. *Journal of Theoretical Biology*

126:393–406 [*320, 333*]

Cressman R & Vickers G (1997). Spatial and density effects in evolutionary game theory. *Journal of Theoretical Biology* **184**:359–369 [*318, 320, 333, 478*]

Cronhjort M (1995). *Models and Computer Simulations of Origins of Life and Evolution*. PhD dissertation, Kungliga Tekniska Högskolan, Stockholm, Sweden [*172*]

Cronhjort M & Blomberg C (1995). Hypercycles versus parasites in a two-dimensional partial differential equations model. *Journal of Theoretical Biology* **169**:31–49 [*118, 128, 152, 159, 172*]

Cronhjort M & Blomberg C (1997a). Chasing: A mechanism for resistance against parasites in self-replicating systems. In *Artificial Life V: Proceedings of the Fifth International Workshop on the Synthesis and Simulation of Living Systems*, ed. Weinstein J, pp. 413–417. Cambridge, MA, USA: MIT Press [*152, 165, 167*]

Cronhjort M & Blomberg C (1997b). Cluster compartmentalization may provide resistance to parasites for catalytic networks. *Physica D* **101**:289–298 [*152, 165*]

Cronhjort M & Nyberg A (1996). 3D hypercycles have no stable spatial structure. *Physica D* **90**:79–83 [*162*]

Cruickshank I, Gurney WSC & Veitch AR. The characteristics of epidemics and invasions with thresholds. Unpublished [*501*]

Czárán T (1997). *Spatiotemporal Models of Population and Community Dynamics*. London, UK: Chapman and Hall [*128, 450*]

Czárán T & Bartha S (1992). Spatiotemporal dynamical models of plant populations and communities. *Trends in Ecology and Evolution* **7**:38–42 [*96, 128*]

Daley DJ & Vere-Jones D (1988). *An Introduction to the Theory of Point Processes*. New York, NY, USA: Springer [*72*]

Dawson TE (1993). Hydraulic lift and water use by plants: Implications for water balance, performance and plant–plant interactions. *Oecologia* **95**:565–574 [*13*]

DeAngelis DL & Gross LJ (1992). *Individual-based Models and Approaches in Ecology: Populations Communities and Ecosystems.* New York, NY, USA: Chapman and Hall [*1*]

de Roos AM, McCauley E & Wilson WG (1991). Mobility versus density-limited predator–prey dynamics of different spatial scales. *Proceedings of the Royal Society of London B* **246**:117–122 [*184–187, 210, 227, 361, 385*]

Dieckmann U (1994). *Coevolutionary Dynamics of Stochastic Replicator Systems.* Juelich, Germany: Central Library of the Research Center Juelich [*423*]

Dieckmann U (1997). Can adaptive dynamics invade? *Trends in Ecology and Evolution* **12**:128–131 [*319*]

Dieckmann U & Law R (1996). The mathematical theory of coevolution: A derivation from stochastic ecological processes. *Journal of Mathematical Biology* **34**:579–612 [*285*]

Dieckmann U, Herben T & Law R (1997). Spatio-temporal processes in plant communities. In *Yearbook 1995/96*, Institute for Advanced Study Berlin, ed. Lepenies W, pp. 296–326. Berlin, Germany: Nicolaische Verlagsbuchhandlung [*252, 360, 423, 442*]

Dieckmann U, Herben T & Law R. Spatio-temporal processes in ecological communities. *CWI Quarterly.* In press [*423, 442*]

Diekmann O (1977). Limiting behaviour in an epidemic model. *Nonlinear Analysis, Theory, Methods and Applications* **1**:459–470 [*305*]

Diekmann O (1978). Thresholds and travelling waves for the geographical spread of infection. *Journal of Mathematical Biology* **6**:109–130 [*305, 486*]

Diekmann O (1979). Run for your life. *Journal of Differential Equations* **33**:58–73 [*305, 486*]

Diekmann O, Christiansen F & Law R (1996). Evolutionary dynamics, Editorial. *Journal of Mathematical Biology* **34**:483 [*319*]

Diggle PJ (1983). *Statistical Analysis of Spatial Point Patterns.* London, UK: Academic Press [*66, 388, 393, 429*]

Diggle PJ (1990). *Time Series: A Biostatistical Introduction.* Oxford, UK: Oxford University Press [*66, 68*]

Diggle PJ & Gratton RJ (1984). Monte Carlo methods of inference for implicit statistical models. *Journal of the Royal Statistical Society B* **46**:193–227 [*84*]

Dirac PAM (1926). The physical interpretation of the quantum dynamics. *Proceedings of the Royal Society of London A* **113**:621–641 [*419*]

Dirac PAM (1958). *The Principles of Quantum Dynamics*, 4th edn. Oxford, UK: Clarendon Press [*419, 423*]

Dockery J, Hutson V, Mischaikow K & Pernarowski M (1997). The evolution of slow dispersal rates: A reaction–diffusion model. *Journal of Mathematical Biology* **37**:61–83 [*479*]

Doebeli M & Knowlton N (1998). The evolution of interspecific mutualism. *Proceedings of the National Academy of Sciences of the USA*, **95**:8676–8680 [*149*]

Dugatkin L & Wilson D (1991). Rover: A strategy for exploiting cooperators in a patchy environment. *The American Naturalist* **138**:687–701 [*332*]

Dunbar SR (1986). Traveling waves in diffusive predatory–prey equations: Periodic orbits and point-to-periodic heteroclinic orbits. *SIAM Journal of Applied Mathematics* **46**:1057–1078 [*502*]

Dunning JB Jr, Stewart DJ, Danielson BJ, Noon BR, Root TL, Lamberson RH & Stevens EE (1995). Spatially explicit population models: Current forms and future uses. *Ecological Applications* **5**:3–11 [*252*]

During HJ & Lloret F (1996). Permanent grid studies in bryophyte communities. I. Pattern and dynamics of individual species. *Journal of the Hattori Botanical Laboratory* **79**:1–41 [*52*]

During HJ & Ter Horst B (1983). The diaspore bank of bryophytes and ferns in chalk grassland. *Lindbergia* **9**:57–64 [*57*]

During HJ, Brugues M, Cros RM & Lloret F (1988). The diaspore bank of bryophytes and ferns in the soil in some contrasting habitats around Barcelona, Spain. *Lindbergia* **13**:137–149 [*57, 60–61*]

Durrett R (1982). On the growth of one dimensional contact processes. *Annals of Probability* **8**:890–907 [*248*]

Durrett R (1988a). Crabgrass, measles and gypsy moths: An introduction to interacting particle systems. *The Mathematical Intelligencer* **10**:37–47 [*271, 503*]

Durrett R (1988b). *Lecture Notes on Particle Systems and Percolation*. Pacific Grove, CA, USA: Wadsworth and Brooks/Cole [*503*]

Durrett R (1995a). Spatial epidemic models. In *Epidemic Models: Their Structure and Relation to Data*, ed. Mollison D, pp. 187–201. Cambridge, UK: Cambridge University Press [*271, 276*]

Durrett R (1995b). *Ten Lectures on Particle Systems*. New York, NY, USA: Springer [*357*]

Durrett R & Levin S (1994a). Stochastic spatial models: A user's guide to ecological applications. *Philosophical Transactions of the Royal Society of London B* **343**:329–350 [*271*]

Durrett R & Levin S (1994b). The importance of being discrete (and spatial). *Theoretical Population Biology* **46**:363–394 [*20, 119, 129, 149, 248, 271, 342, 359–360, 388, 405, 450*]

Durrett R & Levin S (1997). Allelopathy in spatially distributed populations. *Journal of Theoretical Biology* **185**:165–171 [*244, 247*]

Durrett R & Neuhauser C (1994). Particle systems and reaction–diffusion equations. *Annals of Probability* **22**:289–333 [*248, 405*]

Durrett R & Schinazi R (1993). Asymptotic critical value for a competition model. *Annals of Applied Probability* **3**:1047–1066 [*357*]

Durrett R & Swindle G (1991). Are there

bushes in a forest? *Stochastic Processes and their Applications* **37**:19–31 [*357*]

Dushoff JG (1997). *Modeling the Effects of Host Heterogeneity on the Spread of Human Diseases*. PhD dissertation, Princeton University, Princeton, NJ, USA [*390*]

Dwyer G (1992a). Density dependence and spatial structure in the dynamics of insect pathogens. *The American Naturalist* **143**:533–562 [*316*]

Dwyer G (1992b). On the spatial spread of insect pathogens – Theory and experiment. *Ecology* **73**:479–494 [*271, 316*]

Dwyer G (1994). On the spatial spread of insect pathogens: Theory and experiment. *Ecology* **73**:479–494 [*316*]

Dwyer G & Elkinton JS (1995). Host dispersal and the spatial spread of insect pathogens. *Ecology* **76**:1262–1275 [*316*]

Dzhaparidze K & Yaglom A (1983). Spectrum parameter estimation in time series analysis. In *Developments in Statistics*, Vol. 4, ed. Krishnaiah P, pp. 1–96. San Diego, CA, USA: Academic Press [*84*]

Eberhart K & Woodard P (1987). Distribution of residual vegetation associated with large fires in Alberta. *Canadian Journal of Forest Research* **17**:1207–1212 [*110, 112*]

Ebert U & van Saarloos W (1998). Universal algebraic relaxation of fronts propagating into an unstable state and implications for moving boundary approximations. *Physics Review Letters* **80**:2350–2353 [*498, 503*]

Ebert U & van Saarloos W. Fronts propagating uniformly into unstable states: Universal algebraic rate of convergence of pulled fronts. Unpublished [*498, 503*]

Eckmann J P & Ruelle D (1985). Ergodic theory of chaos and strange attractors. *Reviews of Modern Physics* **57**:617–656 [*218, 220*]

Eckmann J-P, Kamphorst SO, Ruelle D & Ciliberto S (1986). Liapunov exponents from time series. *Physical Review A* **34**:4971–4979 [*220*]

Edelstein-Keshet L (1988). *Mathematical*

Methods in Biology. New York, NY, USA: Random House [*470*]

Eigen M (1971). Self-organization of matter and the evolution of biological macromolecules. *Naturwissenschaften* 10:465–523 [*116, 171*]

Eigen M (1992). *Steps Towards Life – A Perspective on Evolution.* Oxford, UK: Oxford University Press [*171*]

Eigen M & Schuster P (1978). The hypercycle: A principle of natural self-organization. Part C. The realistic hypercycle. *Naturwissenschaften* 65:341–569 [*119*]

Eigen M & Schuster P (1979). *The Hypercycle: A Principle of Natural Self-organisation.* Berlin, Germany: Springer [*153, 161, 171*]

Eigen M & Schuster P (1982). Stages of emerging life – Five principles of early organization. *Journal of Molecular Evolution* 19:47–61 [*171*]

Eigen M, Schuster P, Gardiner W & Winkler Oswatitsch R (1981). The origin of genetic information. *Scientific American* 244:78–94 [*117*]

Eissenstat DM & Caldwell MM (1988). Seasonal timing of root growth in favorable microsites. *Ecology* 69:870–873 [*16*]

Ellison AM, Dixon PM & Ngai J (1994). A null model for neighborhood models of plant competitive interactions. *Oikos* 71:225–238 [*20–21*]

Ellner SP, Sasaki A, Haraguchi Y & Matsuda H (1998). Speed of invasion in lattice population models: Pair-edge approximation. *Journal of Mathematical Biology* 36(5):469–484 [*248, 405, 426, 452, 503*]

Enquist M & Leimar O (1993). The evolution of cooperation in mobile organisms. *Animal Behaviour* 45:747–757 [*332*]

Epstein JM & Axtell R (1996). *Growing Artificial Societies.* Boston, MA, USA: The Brookings Institution Press [*150*]

Eriksson J & Henning F (1896). *Die Getreideroste, Ihre Geschichte und Natur sowie Massregeln gegen dieselben.* Stockholm, Sweden: Norstedt and Söner [*292, 299*]

Eriksson O (1986). Mobility and space capture in the stoloniferous plant *Potentilla anserina.* *Oikos* 46:82–87 [*11*]

Ermentrout G & Edelstein-Keshet L (1992), Cellular automata approaches to biological modelling. *Journal of Theoretical Biology* 160:97–133 [*95, 271*]

Eshel I & Cavalli-Sforza L (1982). Assortment of encounters and the evolution of cooperativeness. *Proceedings of the National Academy of Sciences of the USA* 79:1331–1335 [*329*]

Eshel I, Sansone E & Shaked A. Evolutionary dynamics of populations with a local interaction structure. Sonderforschungsbereich Bonn. Unpublished [*148*]

Etter RJ & Caswell H (1994). The advantages of dispersal in a patch environment: Effects of disturbance in a cellular automata model. In *Reproduction, Larval Biology and Recruitment in the Deep-Sea Benthos*, eds. Eckelbarger KJ, Young CM, pp. 93–109 New York, NY, USA: Columbia University Press [*227*]

Evans JP (1991). The effect of resource integration on fitness related traits in a clonal dune perennial, *Hydrocotyle bonariensis.* *Oecologia* 86:268–275 [*15*]

Evans JP (1992). The effect of local resource availability and clonal integration on ramet functional morphology in *Hytrocotyle bonariensis.* *Oecologia* 89:265–276 [*15*]

Fahrig L & Paloheimo J (1988). Effect of spatial arrangement of habitat patches on local population size. *Ecology* 69:468–475 [*209*]

Feldman B & Nagel K (1993). Lattice games with strategic takeover. In *Lectures in Complex Systems*, SFI Studies in the Sciences of Complexity, eds. Nadel L & Stein D, Redwood City, CA, USA: Addison-Wesley [*144*]

Ferrière R & Michod R (1995). Invading wave of cooperation in a spatial iterated Prisoner's Dilemma. *Proceedings of the Royal Society of London B* 259:77–83 [*149, 318, 328–329, 478*]

Ferrière R & Michod R (1996). The evolution of cooperation in spatially heterogeneous populations. *The American Naturalist* **147**:692–717 [*251, 318, 325–327, 478*]

Fife P (1979). *Mathematical Aspects of Reacting–Diffusing Systems. Lecture Notes on Biomathematics 28*. Berlin, Germany: Springer [*404, 456, 458, 460, 470, 473, 477*]

Fisher R (1937). The wave of advance of advantageous genes. *Annals of Eugenics* **7**:353–369 [*475, 485*]

Fitter AH (1986). Acquisition and utilization of resources. In *Plant Ecology*, ed. Crawley MJ, pp. 375–405. Oxford, UK: Blackwell [*14*]

Fitter AH & Stickland TR (1991). Architectural analysis of plant root systems. II. Influence of nutrient supply on architecture in contrasting plant species. *New Phytologist* **118**:383–389 [*16*]

Fitter AH, Stickland TR, Harvey ML & Wilson GW (1991). Architectural analysis of plant root systems. I. Architectural correlates of exploitation efficiency. *New Phytologist* **118**:375–382 [*16*]

Forrest S & Jones T (1994). Modeling complex adaptive systems with eco. In *Complex Systems: Mechanism of Adaptation*, eds. Stonie R & Yu X, pp. 3–21. Amsterdam, Netherlands: IOS Press [*342*]

Fraedrich K (1986). Estimating the dimensions of weather and climate attractors. *Journal of Atmospheric Science* **43**:419–432 [*225*]

Franco M (1986). The influence of neighbors on the growth of modular organisms with an example from trees. *Proceedings of the Royal Society of London B* **313**:209–225 [*25*]

Franco M & Harper JL (1988). Competition and the formation of spatial pattern in spacing gradients: An example using *Kochia scoparia*. *Journal of Ecology* **76**:959–674 [*25*]

Franco M & Silvertown J (1996). Life history variation in plants: An exploration of the fast-slow continuum hypothesis. *Philosophical Transactions of the Royal Society of London B* **351**:1341–1348 [*40*]

Franco Pizana JG, Fulbright TE, Gardiner DT & Tipton AR (1996). Shrub emergence and seedling growth in microenvironments created by *Prosopis glandulosa*. *Journal of Vegetation Science* **7**:257–264 [*13*]

Frisch T & Rica S (1992). Dynamics of spiral rings in the three-dimensional Ginzburg-Landau equation. *Physica D* **61**:155 [*152*]

Gandhi A, Levin S & Orszag S (1998). "Critical slowing down" in time-to-extinction: An example of critical phenomena in ecology. *Journal of Theoretical Biology* **192**:363–376 [*388*]

Gauch HG (1982). *Multivariate Analysis in Community Ecology*. Cambridge, UK: Cambridge University Press [*223*]

Gäumann E (1946). *Pflanzliche Infektionslehre*. Basel, Switzerland: Birkhäuser [*298*]

Gause GF (1969). Reprint. *The Struggle for Existence*. New York: Hafner Publishing Company. Original edition, Williams and Wilkins (1934) [*183*]

Gause GF, Smaragdova P & Witt AA (1936). Further studies of interaction between predators and prey. *Journal of Animal Ecology* **5**:1–18 [*183*]

Geman S & Geman D (1984). Stochastic relaxation, Gibbs distributions and the Bayesian restoration of images. *IEEE Transactions* PAMI-**6**:721–741 [*86*]

Geritz S, Metz J, Kisdi E & Mesézna G (1997). The dynamics of adaptation and evolutionary branching. *Physical Review Letters* **78**:2024–2027 [*319*]

Gilbert W (1986). The RNA world. *Nature* **319**:618 [*153*]

Gilks WR, Richardson S & Spiegelhalter DJ (1996). Introducing Markov chain Monte Carlo. In *Markov Chain Monte Carlo in Practice*, eds. Gilks WR, Richardson S & Spiegelhalter DJ, pp. 1–19. London, UK: Chapman and Hall [*86*]

Gilpin M & Hanski I (1991). *Metapopulation Dynamics: Empirical and Theoretical Investigations*. London, UK: Academic Press [*209*]

Goldberg DE (1987). Neighborhood competition in an old-field plant community. *Ecology* **68**:1211–1223 [*22*]

Goldberg DE (1990). Components of resource competition in plant communities. In *Perspectives on Plant Competition*, eds. Grace JB & Tilman D, pp. 27–47. New York, NY, USA: Academic Press [*14*]

Goldberg DE (1995). Generating and testing predictions about community structure: Which theory is relevant and can it be tested with observational data? *Folia Geobotanica et Phytotaxonomica* **30**:511–518 [*58*]

Goldberg DE (1997). Competitive ability: Definitions, contingency and correlated traits. In *Plant Life Histories*, eds. Silvertown J, Franco M & Harper JL, pp. 283–306. Cambridge, UK: Cambridge University Press [*39, 40*]

Goldberg DE & Barton AM (1992). Patterns and consequences of interspecific competition in natural communities: A review of field experiments with plants. *The American Naturalist* **139**:771–801 [*32, 34, 40*]

Goldberg DE & Werner PA (1983). Equivalence of competitors in plant communities: A null hypothesis and a field experimental approach. *American Journal of Botany* **70**:1098–1104 [*23*]

Goldwasser L, Cook J & Silverman ED (1994). The effects of variability on metapopulation dynamics and rates of invasion. *Ecology* **75**:40–47 [*508*]

Gordon D (1997). The population consequences of territorial behavior. *Trends in Ecology and Evolution* **12**:63–66 [*342*]

Grassberger P, Schreiber T & Schaffrath C (1991). Non-linear time series anlaysis. *International Journal of Bifurcation and Chaos in Applied Sciences and Engineering* **1**:521–547 [*218, 225*]

Gray LF (1982). The positive rates problem for attractive nearest neighbor spin system on Z. *Zeitung für Wahrscheinlichkeitstheorie* **61**:389–404 [*248*]

Green D (1989). Simulated effects of fire, dispersal and spatial pattern on competition within forest mosaics. *Vegetatio* **82**:139–154 [*95, 227*]

Gregory PH (1968). Interpreting plant disease dispersal gradients. *Annual Review of Phytopathology* **6**:189–212 [*306, 310*]

Greig-Smith P (1964). *Quantitative Plant Ecology*, 2nd edn. London, UK: Butterworths [*66*]

Greig-Smith P (1983). *Quantitative Plant Ecology*, 3rd edn. Oxford, UK: Blackwell [*48–49, 255, 270*]

Greiner G, Heesterbeek J & Metz J (1994). A singular perturbation theorem for evolution equations and time-scale arguments for structured population models. *Canadian Applied Mathematics Quarterly* **2**:435–459 [*506*]

Grenfell B, Kleczkowski A, Gilligan C & Bolker B (1995a). Spatial heterogeneity, nonlinear dynamics and chaos in infectious diseases. *Statistical Methods in Medical Research* **4**:160–183 [*271*]

Grenfell B, Wilson K, Isham V, Boyd H & Dietz K (1995b). Modeling patterns of parasite aggregation in natural populations: Trichostrongylid nematode–ruminant interactions as a case-study. *Parasitology* **111**:S135–S151 [*390*]

Grim P (1995). The greater generosity of the spatialized Prisoner's Dilemma. *Journal of Theoretical Biology* **173**:353–359 [*148, 227*]

Grime JP, Thompson K, Hunt R, Hodgson JG, Cornelissen JHC, Rorison IH, Hendry GAF, Ashenden TW, Askew AP, Band SR, Booth RE, Bossard CC, Campbell BD, Cooper JEL, Davison AW, Gupta PL, Hall W, Hand DW, Hannah MA, Hillier SH, Hodkinson DJ, Jalili A, Liu Z, Mackey JML, Matthews N, Mowforth MA, Neal AM, Reader RJ, Reiling K, Ross Fraser W, Spencer RE, Sutton F, Tasker DE, Thorpe PC & Whitehouse

J (1997). Integrated screening validates primary axes of specialisation in plants. *Oikos* **79**:259–281 [*40*]

Grimm V (1994). Mathematical models and understanding in ecology. *Ecological Modelling* **75/76**:641–651 [*26*]

Grindrod P (1991). *Patterns and Waves: The Theory and Applications of Reaction–Diffusion Equations.* Oxford, UK: Oxford University Press [*458, 462*]

Grindrod P & Sleeman B (1987). Weak travelling fronts for population models with density dependent dispersion. *Mathematical Methods in the Applied Sciences* **9**:576–586 [*462*]

Grünbaum D (1994). Translating stochastic density-dependent individual behavior with sensory constraints to an Eulerian model of animal swarming. *Journal of Mathematical Biology* **33**:139–161 [*404*]

Gurevitch J, Morrow LL, Wallace A & Walsh JS (1992). Meta analysis of field experiments. *The American Naturalist* **140**:539–572 [*22*]

Habtu A, van den Bosch F & Zadoks JC (1995). Focus expansion of bean rust in cultivar mixtures. *Plant Pathology* **44**:503–509 [*299, 309, 317*]

Hadeler K (1981). Diffusion in Fisher's population model. *Rocky Mountain Journal of Mathematics* **11**:39–45 [*320*]

Halley J, Comins H, Lawton J & Hassell M (1994). Competition, succession and pattern in fungal communities: Towards a cellular automaton model. *Oikos* **70**:435–442 [*95*]

Hamilton WD (1964). The genetic evolution of social behavior. *Journal of Theoretical Biology* **7**:1–52 [*148, 327, 361*]

Hammerstein P & Hoekstra R (1995). Mutualism on the move. *Nature* **376**:121–122 [*116*]

Hansen LR (1982). Large sample properties of generalized method of moments estimators. *Econometrica* **50**:1029–1054 [*83*]

Hara T, Wyszomirski T (1994). Competitive asymmetry reduces spatial effects on size-

structure dynamics in plant populations. *Annals of Botany* **73**:285–297 [*21*]

Hara T, Nishimura N & Yamamoto S (1995). Tree competition and species coexistence in a cool-temperate old-growth forest in southwestern Japan. *Journal of Vegetation Science* **6**:565–574 [*63*]

Harada Y & Iwasa Y (1994). Lattice population dynamics for plants with dispersing seeds and vegetative propagation. *Researches on Population Ecology* **36**:237–249 [*227–234, 252, 343, 361*]

Harada Y & Iwasa Y (1996). Analyses of spatial patterns and population processes of clonal plants. *Researches on Population Ecology* **38**:153–164 [*228, 236–237*]

Harada Y, Ezoe H, Iwasa Y, Matsuda H & Satō K (1995). Population persistence and spatially limited social interaction. *Theoretical Population Biology* **48**:65–91 [*228, 348, 352, 355–357, 361, 373, 377, 386*]

Harper J & White J (1974). The demography of plants. In *Annual Review of Ecology and Systematics*, Vol. 5, pp. 419–463 [*400*]

Harper JL (1977). *Population Biology of Plants.* London, UK: Academic Press [*12, 14–15, 50*]

Harrison GW (1995). Comparing predator–prey models to Luckinbill's experiment with *Paramecium* and *Didinium*. *Ecology* **76**:357–374 [*183*]

Harsanyi J & Selten R (1988). *A General Theory of Equilibrium Selection in Games.* Cambridge, MA, USA: MIT Press [*474*]

Hartman P (1964). *Ordinary Differential Equations.* New York, NY, USA: Wiley [*193*]

Hassell M, Comins HN & May RM (1991a). Spatial structure and chaos in insect population dynamics. *Nature* **353**:255–258 [*209, 227*]

Hassell M, May R, Pacala S & Chesson P (1991b). The persistence of host–parasitoid associations in patchy environments. I. A general criterion. *The American Naturalist* **138**:568–583 [*390*]

Hassell M, Comins HN & May RM (1994). Species coexistence and self-organizing spatial dynamics. *Nature* **370**:290–292 [*227*]

Hastings A (1994). Conservation and spatial structure: Theoretical approaches. In *Frontiers in Mathematical Biology*, ed. Levin SA, pp. 494–503. Berlin, Germany: Springer [*3*]

Hastings A & Wolin CL (1989). Within-patch dynamics in a metapopulation model. *Ecology* **70**:1261–1266 [*209*]

Heald FD & Studhalter RA (1914). Birds as carriers of the chestnut blight fungus. *Journal of Agricultural Research* **2**:405–422 [*298, 308*]

Hediger T, Passamante A & Farrell ME (1990). Characterising attractors using local intrinsic dimensions calculated by singular value decomposition and information theortic criteria. *Physical Review A* **41**:5325–5332 [*225*]

Heesterbeek JAP & Zadoks JC (1987). Modelling pandemics of quarantine pests and diseases: Problems and perspectives. *Crop Protection* **6**:211–231 [*293, 300–301, 304, 314*]

Hegselmann R (1996). Solidarität unter Ungleichen. In *Modelle sozialer Dynamiken*, eds. Hegselmann R & Peitgen HO, pp. 105–128. Vienna, Austria: Hölder-Pichler-Tempski [*150*]

Hendry R, McGlade J & Weiner J (1996). A coupled map lattice model of the growth of plant monocultures. *Ecological Modelling* **84**:81–90 [*96*]

Hengeveld R (1989). *Dynamics of Biological Invasions*. London, UK: Chapman and Hall [*308, 312*]

Henry D (1981). *Geometric Theory of Semilinear Parabolic Equations. Lecture Notes in Mathematics, 840*. New York, NY, USA: Springer [*458*]

Herben T, Krahulec F, Hadincova V & Kovarova M (1993). Small scale spatial dynamics of plant species in a grassland community during six years. *Journal of Vegetation Science* **4**:171–178 [*49, 52–53*]

Herben T, During HJ & Krahulec F (1995). Spatio-temporal dynamics in mountain grasslands: Species autocorrelations in space and time. *Folia Geobotanica et Phytotaxonomica* **30**:185–196 [*52, 60*]

Hermansen JE (1968). Studies on the spread and survival of cereal rust and mildew diseases in Denmark. *Friesia* **8**:1–206 [*296, 299*]

Herz AVM (1994). Collective phenomena in spatially extended evolutionary games. *Journal of Theoretical Biology* **169**:65–87 [*149, 227*]

Hiebeler D (1997). Stochastic spatial models: From simulations to mean field and local structure approximations. *Journal of Theoretical Biology* **187**:307–319 [*405, 452*]

Hill MO & Radford GL (1986). *Register of Permanent Vegetation Plots*. Abbots Ripton, UK: Institute of Terrestrial Ecology [*50*]

Hirst JM & Stedman OJ (1960). The epidemiology of *Phytophthora infestans*. II. The source of inoculum. *Annals of Applied Biology* **48**:489–517 [*299*]

Hjalten J, Danell K & Lundberg P (1993). Herbivore avoidance by association – Vole and hare utilization of woody-plants. *Oikos* **68**:125–131 [*34*]

Hochberg M, Hassell M & May R (1990). The dynamics of host–parasitoid–pathogen interactions. *The American Naturalist* **135**:74–94 [*390*]

Hofbauer J (1998). Equilibrium selection via travelling waves. In *Game Theory, Experience, Rationality*, eds. Leinfellner W & Köhler E, pp. 245–259. Dordrecht, Netherlands: Kluwer Academic [*474*]

Hofbauer J & Sigmund K (1988). *Dynamical Systems and the Theory of Evolution*. Cambridge, UK: Cambridge University Press [*474*]

Hofbauer J & Sigmund K (1998). *Evolutionary Games and Population Dynamics*.

Cambridge, UK: Cambridge University Press [318–319]

Hofbauer J, Hutson V & Vickers G (1997). Travelling waves in economics and biology. *Nonlinear Analysis: Theory, Methods, and Applications* **30**:1235–1244 [474, 478]

Hoffman R & Waring N (1996). The simulation of localised interaction and learning in artificial adaptive agents. In *Evolutionary Computing*, ed. Fogarty TC, New York, NY, USA: Springer [148]

Hogeweg P (1988). Cellular automata as a paradigm for ecological modeling. *Applied Mathematics and Computation* **27**:81–100 [95, 412]

Hogeweg P (1994). Multilevel evolution: Replicators and the evolution of diversity. *Physica D* **75**:275–291 [152]

Holling CS (1965). The functional response of predators to prey density and its role in mimicry and population regulation. *Memoirs of the Entomological Society of Canada* **45**:5–60 [185]

Holmes EE & Wilson HB (1998). Running from trouble: Long-distance dispersal and the competitive coexistence of inferior species. *The American Naturalist* **151**:578–586 [402]

Holsinger KE & Roughgarden J (1985). A model for the dynamics of an annual plant-population. *Theoretical Population Biology* **28**:288–313 [412]

Hook PB & Lauenroth WK (1994). Root system response of a perennial bunchgrass to neighborhood-scale soil water heterogeneity. *Functional Ecology* **8**:738–745 [17]

Hoppensteadt FC (1982). *Mathematical Methods of Population Biology*. Cambridge, UK: Cambridge University Press [459]

Horn H & MacArthur R (1972). Competition among fugitive species in a harlequin environment. *Ecology* **53**:749–752 [390]

Hosono Y (1998). The minimal speed of traveling fronts for a diffusive Lotka–Volterra competition model. *Bulletin of Mathematical Biology* **60**:435–448 [484, 501]

Houston A (1993). Mobility limits cooperation. *Trends in Ecology and Evolution* **8**:194–196 [332]

Hubbell SP & Foster RB (1986). Canopy gaps and the dynamics of a neotropical forest. In *Plant Ecology*, ed. Crawley MJ, pp. 77–96. Oxford, UK: Blackwell [237–238, 240]

Huber-Sannwald E, Pyke DA & Caldwell MM (1996). Morphological plasticity following species-specific recognition and competition in two perennial grasses. *American Journal of Botany* **83**:919–931 [17, 24]

Huberman BA & Glance NS (1993). Evolutionary games and computer simulations. *Proceedings of the National Academy of Sciences of the USA* **90**: 7716–7718 [123, 140–141]

Huffaker CB (1958). Experimental studies on predation: Dispersion factors and predator–prey oscillations. *Hilgardia* **27**:343–383 [183]

Huston MA & DeAngelis DL (1994). Competition and coexistence: The effects of resource transport and supply rates. *The American Naturalist* **144**:954–977 [16]

Huston MA, DeAngelis DL & Post W (1988). New computer models unify ecological theory – Computer simulations show that many ecological patterns can be explained by interactions among individual organisms. *Bioscience* **38**:682–691 [1]

Hutchings MJ (1986). The structure of plant populations. In *Plant Ecology*, ed. Crawley MJ, pp. 97–136. Oxford, UK: Blackwell [11]

Hutchings MJ (1988). Differential foraging for resources and structural plasticity in plants. *Trends in Ecology and Evolution* **3**:200–204 [15]

Hutchings MJ & de Kroon H (1994). Foraging in plants: The role of morphological plasticity in resource acquisition. In *Advances in Ecological Research*, Vol. 25, eds. Begon M & Fitter AH, pp. 159–238. London, UK: Academic Press [*15, 25*]

Hutson V (1986). Stability in a reaction-diffusion model of mutualism. *SIAM Journal of Mathematical Analysis* **77**:58–66 [*474*]

Hutson V & Vickers GT (1992). Travelling waves and dominance of ESSs. *Journal of Mathematical Biology* **30**:457–471 [*318, 321–322, 333*]

Hutson V & Vickers GT (1995). The spatial struggle of *Tit-For-Tat* and defect. *Philosophical Transactions of the Royal Society of London B* **348**:393–404 [*149, 321, 329–331, 463, 466, 471, 478*]

IITA (1996). Update on "Mighty Mite," *T. aripo*. Biocontrol of cassava green mite gives African farmers a bonanza. Annual Report, pp. 31–32. [*308*]

Isham V (1981). An introduction to spatial point processes and Markov random fields. *International Statistical Review* **49**:21–43 [*78–79*]

Ives A & May R (1985). Competition within and between species in a patch environment. *Journal of Theoretical Biology* **115**:65–92 [*343*]

Iwao S (1968). A new regression method for analyzing the aggregation pattern of animal populations. *Researches in Population Ecology* **10**:1–20 [*231*]

Iwao S & Kuno E (1971). An approach to the analyzing the aggregation pattern of animal populations. In *Statistical Ecology*, Vol. 1, pp. 461–513. Philadelphia, PA, USA: Philadelphia University Press [*231*]

Iwasa Y, Satō K & Nakashima S (1991). Dynamic modeling of wave regeneration (Shimagare) in subalpine Abies forests. *Journal of Theoretical Biology* **152**:143–158 [*227, 341*]

Iwasa Y, Nakamaru M & Levin SA (1998). Allelopathy of bacteria in a lattice population: Competition between colicin-sensitive and colicin-producing strains. *Evolutionary Ecology* **12**:785–802 [*232, 244, 246–248, 358*]

Jackson RB & Caldwell MM (1993). The scale of nutrient heterogeneity around individual plants and its quantification with geostatistics. *Ecology* **74**:612–614 [*54*]

James J (1995). *A Student's Guide to Fourier Transforms with Applications in Physics and Engineering*. Cambridge, UK: Cambridge University Press [*399*]

Jammer M (1966). *The Conceptual Development of Quantum Mechanics*. New York, NY, USA: McGraw-Hill [*419*]

Jansen VAA (1994). On the bifurcation structure of two diffusively coupled, predator–prey systems. In *Theoretical Aspects of Metapopulation Dynamics*, PhD dissertation, pp. 93–115, Leiden University, Leiden, Netherlands [*196–197*]

Jansen VAA (1995). Regulation of predator–prey systems through spatial interactions: A possible solution to the paradox of enrichment. *Oikos* **74**:384–390 [*196*]

Janzen DH (1970). Herbivores and the number of tree species in tropical forests. *The American Naturalist* **104**:501–529 [*18*]

Jeger MJ (1983). Analysing epidemics in time and space. *Plant Pathology* **32**:5–11 [*293, 302–303*]

Jeger MJ, ed. (1989). *Spatial Components of Plant Disease Epidemics*. Englewood Cliffs, NJ, USA: Prentice Hall [*293*]

Jeltsch F & Wissel C (1994). Modelling dieback phenomena in natural forests. *Ecological Modelling* **75/76**:111–121 [*96, 341*]

Jeltsch F, Wissel C, Eber S & Brandl R (1992). Oscillating dispersal patterns of tephritid fly populations. *Ecological Modelling* **60**:63–75 [*96*]

Jeltsch F, Milton S, Dean W & van Rooyen N (1996). Tree spacing and coexistence in semiarid savannas. *Journal of Ecology* **84**:583–595 [*96*]

Jeltsch F, Mueller MS, Grimm V, Wissel C & Brandl R (1997). Pattern formation triggered by rare events: Lessons from the spread of rabies. *Proceedings of the Royal Society of London B* **264**:495–503 [*96, 98, 101, 359*]

Johnson C & Seinen I. Growth and competitive ability in spatial competition systems. Unpublished [182]

Johnson M & Gaines M (1990). Evolution of dispersal: Theoretical models and empirical tests using birds and mammals. Annual Reviev of Ecology and Systemics 21:449–480 [479]

Jones M (1985). Modular demography and form in silver birch. In Studies on Plant Demography: A Festschrift for John L. Harper, ed. White J, pp. 223–237. London, UK: Academic Press [25]

Jones M & Harper JL (1987a). The influence of neighbors on the growth of trees. I. The demography of buds in Betula pendula. Proceedings of the Royal Society of London B 232:1–18 [25]

Jones M & Harper JL (1987b). The influence of neighbors on the growth of trees. II. The fate of buds on long and short shoots in Betula pendula. Proceedings of the Royal of Society London B 232:19–33 [25]

Joshi LM & Palmer LT (1973). Epidemiology of stem, leaf and stripe rusts of wheat in Northern India. Plant Disease Reporter 57:8–12 [299]

Judson O (1994). The rise of the individual-based model in ecology. Trends in Ecology and Evolution 9:9–14 [1, 342]

Kallén A, Arcuri P & Murray JD (1985). A simple model for the spatial spread and control of rabies. Journal of Theoretical Biology 123:377–393 [316]

Kampmeijer P & Zadoks JC (1974). EPIMUL, A Simulator of Foci and Epidemics in Mixtures, Multilines and Mosaics of Resistant and Susceptible Plants. Simulation Monographs, Wageningen, Netherlands: Pudoc [302, 309]

Kaneko K (1992). Overview of coupled map lattices. Chaos 2:279–282 [210]

Kaneko K (1993). Theory and Applications of Coupled Map Lattices. Chichester, UK: Wiley [197]

Kanzaki M, Yoda K & Dhanmaonda P (1994). Mosaic structure and tree growth pattern in a monodominant tropical seasonal evergreen forest in Thailand. In Vegetation Science in Forestry, eds. Peet RK & Box EO, pp. 499–517. Dortrecht, Netherlands: Kluwer [237–238]

Kareiva P & Wennergren U (1995). Connecting landscape patterns to ecosystem and population processes. Nature 173:299–302 [209]

Karlson RH & Jackson JBC (1981). Competitive networks and community structure: A simulation study. Ecology 62:670–678 [39]

Karr AF (1991). Point Processes and Their Statistical Inference. New York, NY, USA: Dekker [72]

Kato H & Kozaka T (1974). Effect of temperature on lesion enlargement and sporulation of Pyricularia oryzae in rice leaves. Phytopathology 64:828–830 [307]

Katori M & Konno N (1990). Correlation inequalities and lower bounds for the critical value λ_c of contact processes. Journal of the Physics Society of Japan 59:877–887 [352]

Kawano K & Iwasa Y (1993). A lattice structured model for beech forest dynamics: The effect of understory dwarf bamboo. Ecological Modelling 66:261–275 [227]

Kawano S, Takada T, Nakayama S & Hiratsuka A (1987). Demographic differentiation and life-history evolution in temperate woodland plants. In Differentiation Patterns in Higher Plants, ed. Urbanska KM, pp. 153–181. New York, NY, USA: Academic Press [228]

Keddy P (1991). Plant competition and resources in oldfields. Trends in Ecology and Evolution 6:235–237 [14]

Keddy PA & Shipley B (1989). Competitive hierarchies in herbaceous plant communities. Oikos 54:234–241 [23, 39]

Keeling M (1995). Dynamics and Evolution of Spatial Host–parasite Systems. PhD dissertation, Warwick University, Warwick, UK [273, 361, 373, 377, 386]

Keeling M & Rand D (1995). A spatial

mechanism for the evolution and maintenance of sexual reproduction. *Oikos* **74**:414–424 [*181, 273*]

Keeling M & Rand D. Space and fluctuations in the dynamics of infection. In *From Finite to Infinite Dimensional Dynamical Systems*, ed. Glendinning P. Amsterdam, Netherlands: Kluwer. In press [*272–273, 281, 288*]

Keeling M, Rand DA & Morris A (1997a). Correlation models for childhood epidemics. *Oikos* **264**:1149–1156 [*376, 387*]

Keeling M, Mezic I, Hendry R, McGlade J & Rand D (1997b). Characteristic length scales of spatial models in ecology. *Philosophical Transactions of the Royal Society of London B*. **352**:1589–1601 [*225, 274*]

Keizer J (1985). Theory of rapid bimolecular reactions in solution and membranes. *Accounts of Chemical Research* **18**:235–241 [*390*]

Kendall DG (1957). In discussion on Bartlett, MS: Measles periodicity and community size. *Journal of the Royal Statistical Society A* **120**:48–70 [*305, 485*]

Kendall DG (1965). Mathematical models of the spread of infection. In *Mathematics and Computer Science in Biology and Medicine*, pp. 213–225. London, UK: HMSO [*305, 485*]

Kendall MG (1980). *A Course in Multivariate Analysis*. London, UK: Griffin [*223*]

Kendall MG & Stuart A (1958). *The Advanced Theory of Statistics*, Vol. 1. London, UK: Griffin [*492*]

Kermack WO & McKendrick AG (1927). A contribution to the mathematical theory of epidemics. *Proceedings of the Royal Society of London A*, **115**:700–721 [*312*]

Kershaw KA & Looney HH (1985). *Quantitative and Dynamic Plant Ecology*, 3rd edn. London, UK: Edward Arnold [*48–49*]

Killingback T & Doebeli M (1996). Spatial evolutionary game theory: Hawks and Doves revisited. *Proceedings of the Royal Society of London B* **263**:1135–1144 [*116, 144*]

Killingback T & Doebeli M (1998). Self-organized criticality in spatial evolutionary game theory. *Journal of Theoretical Biology* **191**:335–340 [*144*]

Kirchkamp O. Spatial evolution of automata in the Prisoner's Dilemma game. Unpublished [*149*]

Kirkwood JG (1935). Statistical mechanics of fluid mixtures. *Journal of Chemical Physics* **3**:300–313 [*442*]

Kis T. Predictable properties of the spatial iterated Prisoner's Dilemma game. Unpublished [*149*]

Kiyosawa S & Shiyomi M (1974). A theoretical evaluation of mixing resistant variety with susceptible variety for controlling plant disease. *Annals of the Phytopathological Society of Japan* **38**:41–51 [*302*]

Kocks CG & Zadoks JC (1996). Cabbage refuse piles as sources of inoculum for black rot epidemics. *Plant Disease* **80**:789–792 [*295–296, 313*]

Kolmogorov A, Petrovsky N & Piskounov NS (1937). A study of the equation of diffusion with increase in the quantity of matter, and its application to a biological problem. *Moscow University Bulletin of Mathematics* **1**:1–25 [*485*]

Konno N & Katori M (1990). Applications of the CAM based on a new decoupling procedure of correlation functions in the one-dimensional contact process. *Journal of the Physics Society of Japan* **59**:1581–1592 [*352*]

Kot M (1992). Discrete-time travelling waves: Ecological examples. *Journal of Mathematical Biology* **30**:413–436 [*486*]

Kot M, Lewis MA & van den Driesche P (1996). Dispersal data and the spread of invading organisms. *Ecology* **77**:2027–2042 [*463, 496*]

Krannitz PG & Caldwell MM (1995). Root growth responses of three Great Basin perennials to intra- and interspecific contact with other roots. *Flora* **190**:161–167 [*17*]

Kubo T, Iwasa Y & Furumoto N (1996).

Forest spatial dynamics with gap expansion: Total gap area and gap size distribution. *Journal of Theoretical Biology* **180**:229–246 [*235–237, 239–243, 341*]

Küppers M (1994). Canopy gaps: Competitive light interception and economic space filling – A matter of whole plant allocation. In *Exploitation of Environmental Heterogeneity by Plants*, eds. Caldwell MM & Pearcy RW, pp. 111–144. San Diego, CA, USA: Academic Press [*25*]

Lambert EB (1929). The relation of weather to the development of stem rust in the Mississippi Valley. *Phytopathology* **19**:1–71 [*294*]

Lambinet D, Boisvieux JF, Mallet A, Artois M & Andral L (1978). Modèle mathématique de la propagation d'une épizootie de rage vulpine. *Revue d'Epidémiologie et de Santé Publique* **26**:9–28 [*511*]

Lannou C & Mundt CC (1996). Evolution of a pathogen population in host mixtures: Simple race–complex race competition. *Plant Pathology* **45**:440–453 [*302*]

Lannou C, de Vallavieille-Pope C & Goyeau H (1994). Host mixture efficacy: Effect of lesion growth analysed through computer simulated epidemics. *Plant Pathology* **43**:651–662 [*302*]

Launchbaugh KL & Provenza FD (1993). Can plants practice mimicry to avoid grazing by mammalian herbivores? *Oikos* **66**:501–504 [*34*]

Law R & Dieckmann U. A dynamical system for neighbourhoods in plant communities. *Ecology*. In press [*252, 256, 258, 260–262*]

Law R, McLellan A & Mahdi AS (1993). Spatio-temporal processes in a calcareous grassland. *Plant Species Biology* **8**:175–193 [*52, 60*]

Law R, Herben T & Dieckmann U (1997). Non-manipulative estimates of competition coefficients in a montane grassland community. *Journal of Ecology* **85**:505–518 [*30, 51*]

Lawlor LR (1979). Direct and indirect

effects of N-species competition. *Oecologia* **43**:355–364 [*46*]

Lawton RO & Putz FE (1988). Natural disturbance and gap-phase regeneration in a wind-exposed tropical cloud forest. *Ecology* **69**:764–777 [*237–238*]

Leadley PW, Reynolds JF & Chapin FS (1997). A model of nitrogen uptake by *Eriophorum vaginatum* roots in the field: Ecological implications. *Ecological Monographs* **67**:1–22 [*16*]

Le Cam L (1961). A stochastic description of precipitation. In *Proceedings of the Fourth Berkeley Symposium on Mathematical Statistics and Probability*, Vol. 3, ed. Neyman J, pp. 165–186. Berkeley, CA, USA: University of California Press [*79*]

Lechowicz MJ & Bell G (1991). The ecology and genetics of fitness in forest plants. II. Microspatial heterogeneity of the edaphic environment. *Journal of Ecology* **79**:687–696 [*19*]

Lee K, McCormick W, Ouyang Q & Swinney H (1993). Pattern formation by interacting chemical fronts. *Science* **261**:192–194 [*164*]

Lee K, McCormick W, Pearson J & Swinney H (1994). Experimental observation of self-replicating spots in a reaction–diffusion system. *Nature* **369**:215–218 [*164, 172*]

Leitner W. From individual life history processes to population dynamics: a derivation of the Ricker map. Unpublished [*450*]

Lensink R (1997). Range expansion of raptors in Britain and the Netherlands since the 1960s: Testing an individual based diffusion model. *Journal of Animal Ecology* **66**:811–826 [*317*]

Leonard DE (1971). Air-borne dispersal of larvae of the gypsy moth and its influence on concepts of control. *Journal of Economic Entomology* **64**:638–641 [*312*]

Leung A (1989). *Systems of Partial Differential Equations*. Boston, MA, USA: Kluwer [*458, 471, 474*]

Levin S & Paine R (1974). Disturbance, patch formation, and community structure. *Proceedings of the National Academy of Sciences of the USA* **71**:2744–2747 [*388*]

Levin SA (1974). Dispersion and population interactions. *The American Naturalist* **108**:207–228 [*390, 412*]

Levin SA (1976). Population dynamical models in heterogeneous environments. *Annual Review of Ecology and Systematics*, Vol. 7, pp. 287–310 [*412*]

Levin SA (1992). The problem of pattern and scale in ecology. *Ecology* **73**:1943–1967 [*210*]

Levin SA, Grenfell B, Hastings A & Perelson AS (1997). Mathematical and computational challenges in population biology and ecosystems science. *Science* **275**:334–343 [*3*]

Lewis MA (1997). Variability, patchiness, and jump dispersal in the spread of an invading population. In *Spatial Ecology: The Role of Space in Population Dynamics and Interspecific Interactions*, eds. Tilman D & Kareiva P, pp. 46–69. Princeton, NJ, USA: Princeton University Press [*496*]

Lewis MA. Spread rate for a nonlinear stochastic invasion. Unpublished [*503*]

Lewis MA & Kareiva P (1993). Allee dynamics and the spread of invading organisms. *Theoretical Population Biology* **43**:141–158 [*501*]

Lewis MA & Pacala SW. Modeling and analysis of stochastic invasion processes. *Journal of Mathematical Biology*. In press [*404, 496*]

Lewontin RC (1974). *The Genetic Basis of Evolutionary Change*. New York, NY, USA: Columbia University Press [*413*]

Liggett T (1978). Attractive nearest neighbor spin systems on the integers. *Annals of Probability* **6**:629–636 [*248*]

Liggett T (1985). *Interacting Particle Systems*. New York, NY, USA: Springer [*342*]

Lindgren K (1991). Evolutionary phenomena in simple dynamics. In *Artificial Life II*, eds. Langton CG *et al.*, pp. 295–312. Redwood City, CA, USA: Addison-Wesley [*147*]

Lindgren K & Nordahl MG (1994). Evolutionary dynamics of spatial games. *Physica D* **75**:292–309 [*141, 145–147, 318, 334*]

Lloyd M (1967). Mean crowding. *Journal of Animal Ecology* **36**:1–30 [*231*]

Lovett Doust L (1981). Population dynamics and local specialization in a clonal perennial (*Ranunculus repens*). I. The dynamics of ramets in contrasting habitats. *Journal of Ecology* **69**:743–755 [*37*]

Lubina J & Levin SA (1988). The spread of a reinvading organism: Range expansion in the Californean sea otter. *The American Naturalist* **131**:526–543 [*498*]

Luckinbill L (1974). The effects of space and enrichment on a predator prey system. *Ecology* **55**:1142–1147 [*183*]

Lui R (1983). Existence and stability of a non-linear integral operator. *Journal of Mathematical Biology* **16**:199–220 [*486*]

Lui R (1989a). Biological growth and spread modeled by systems of recursions. I. Mathematical theory. *Mathematical Biosciences* **93**:267–296 [*486*]

Lui R (1989b). Biological growth and spread modeled by systems of recursions. II. Biological theory. *Mathematical Biosciences* **93**:297–312 [*486*]

Mahall BE & Callaway RM (1992). Root communication mechanisms and intracommunity distributions of two Mojave Desert shrubs. *Ecology* **73**:2145–2151 [*17*]

Mahdi A & Law R (1987). On the spatial organization of plant species in a limestone grassland community. *Journal of Ecology* **75**:459–476 [*270, 426–427*]

Maillette L (1982a). Structural dynamics of Silver birch. I. The fate of buds. *Journal of Applied Ecology* **19**:203–218 [*25*]

Maillette L (1982b). Structural dynamics of Silver birch. II. A matrix model of the bud population. *Journal of Applied Ecology* **19**:219–238 [*25*]

Maly EJ (1969). A laboratory study of the interaction between the predatory rotifer *Asplancha* and *Paramecium*. *Ecology* **50**:59–73 [*183*]

Mar G & St Denis P (1994). Chaos in cooperation: Continuous-valued Prisoner's Dilemmas in infinite-valued logic. *International Journal of Bifurcation and Chaos* **4**:943–958 [*148*]

Mar G & St Denis P. Chaos in cooperation: Two-dimensional Prisoner's Dilemmas in infinite-valued logic. Unpublished [*148–149*]

Mariau D, van de Lande HL, Renard JL, Dollet M, Rocha de Souza R, Rios R, Orellana F & Corrado F (1992). Les maladies de type pourriture du coeur sur le palmier à huile en amérique latine: Symptomatologie-epidémiologie–incidence. *Oléagineux* **47**:605–618 [*313*]

Marshall EJP (1990). Interference between sown grasses and the growth of rhizome of *Elymus repens* (couch grass). *Agriculture, Ecosystems and Environment* **33**:11–22 [*35*]

Martinez-Sanchez JJ, Casares-Porcel M, Guerra J, Gutierrez-Carretero L, Ros RM, Hernandez-Bastida J & Cano MJ (1995). A special habitat for bryophytes and lichens in arid zones of Spain. *Lindbergia* **19**:116–121 [*56, 61*]

Masaki T, Suzuki W, Niiyama K, Iida S, Tanaka H & Nakashizuka T (1992). Community structure of a species-rich temperate forest, Ogawa forest reserve, central Japan. *Vegetatio* **98**:97–111 [*237*]

Matérn B (1960). Spatial variation. *Meddelanden från Statens Skogsforskningsinstitut* **49**:5 [*66*]

Matsuda H (1987). Condition for the evolution of altruism. In *Animal Societies: Theories and Facts*, eds. Itô Y, Brown J & Kikkawa J, pp. 67–80. Tokyo, Japan:

Japan Scientific Society Press [*244, 251, 343*]

Matsuda H, Tamachi N, Ogita N & Sasaki A (1987). A lattice model for population biology. In *Mathematical Topics in Biology: Lecture Notes in Biomathematics*, Vol. 71, eds. Teramoto E & Yamaguti M, pp. 154–161. New York, NY, USA: Springer [*244, 251, 343*]

Matsuda HN, Ogita A, Sasaki A & Satō K (1992). Statistical mechanics of population: The lattice Lotka-Volterra model. *Progress in Theoretical Physics* **88**:1035–1049 [*228, 232, 252, 345, 347, 361, 365–366, 372, 382, 384–385*]

May RM (1991). Hypercycles spring to live. *Nature* **353**:607–608 [*172*]

Maynard Smith J (1979). Hypercycles and the origin of life. *Nature* **280**:445–446 [*117, 171*]

Maynard Smith J (1982). *Evolution and the Theory of Games*. Cambridge, UK: Cambridge University Press [*318–319*]

Maynard Smith J & Szathmáry E (1993). The origin of chromosomes. I. Selection for linkage. *Journal of Theoretical Biology* **164**:437–446 [*116*]

Maynard Smith J & Szathmáry E (1995). *The Major Transitions in Evolution*. Oxford, UK: Freeman [*116, 133*]

Maynard Smith J, Burian R, Kauffman S, Alberch P, Campbell J, Goodwin B, Lande R, Raup D & Wolpert L (1985). Developmental constraints and evolution. *The Quarterly Review of Biology* **60**:265–287 [*182*]

McCauley E, Wilson WG & de Roos AM (1993). Dynamics of age-structured and spatially structured predator–prey interactions: Individual based models and population level formations. *The American Naturalist* **142**:412–442 [*184–186, 209, 388*]

McLellan AJ (1995). *Fine-scale Spatiotemporal Dynamics of a Limestone Grassland Community*. PhD dissertation, University of York, York, UK [*52*]

Mees AI & Rapp PE (1988). Singular value

decomposition and embedding dimension – Reply. *Physical Review A* **37**:5006 [*225*]

Mees AI, Rapp PE & Jennings LS (1987). Singular value decomposition and embedding dimension. *Physical Review A* **36**:340–346 [*225*]

Mehta YR & Zadoks JC (1970). Uredospore production and sporulation period of *Puccinia recondita* f. sp. *triticina* on primary leaves of wheat. *Netherlands Journal of Plant Pathology* **76**:267–276 [*307*]

Meinhardt H (1982). *Models of Biological Pattern Formation*. London, UK: Academic Press [*470*]

Metz JAJ & van den Bosch F (1995). Velocities of epidemic spread. In *Epidemic Models: Their Structure and Relation to Data*, ed. Mollison D, pp. 150–186. Cambridge, UK: Cambridge University Press [*271, 293, 311, 486, 498, 506*]

Metz JAJ, Nisbet R & Geritz S (1992). How should we define "fitness" for general ecological scenarios? *Trends in Ecology and Evolution* **7**:198–202 [*319, 382*]

Michalakis Y, Olivieri I, Renaud F & Raymond M (1992). Pleiotropic action of parasites: How to be good for the host. *Trends in Ecology and Evolution* **7**:59–62 [*133*]

Michod R (1983). Population biology of the first replicators: On the origin of genotype, phenotype, and organism. *American Zoologist* **23**:5–14 [*116–117*]

Mikhailov AS, Davydov VA & Zykov V (1994). Complex dynamics of spiral waves and motion of curves. *Physica D* **70**:1–39 [*174*]

Miller R & Michael A (1982). *Ordinary Differential Equations*. New York, NY, USA: Academic Press [*471*]

Miller TE & Werner PA (1987). Competitive effects and responses between plant species in a first-year old-field community. *Ecology* **68**:1201–1210 [*29, 38*]

Minogue KP & Fry WE (1983a). Models for the spread of disease: Model description. *Phytopathology* **73**:1168–1173 [*301*]

Minogue KP & Fry WE (1983b). Models for the spread of disease: Some experimental results. *Phytopathology* **73**:1173–1176 [*301*]

Mitchell-Olds T (1987). Analysis of local variation in plant size. *Ecology* **68**:82–87 [*19*]

Mitchley J (1988). Control of relative abundance of perennials in chalk grassland in Southern England. III. Shoot phenology. *Journal of Ecology* **76**:607–616 [*52*]

Mitchley J & Willems JH (1995). Vertical canopy structure of Dutch chalk grasslands in relation to their management. *Vegetatio* **117**:17–27 [*52*]

Mochizuki A, Iwasa Y & Takeda Y (1996). A stochastic model for cell sorting and measuring cell–cell adhesion. *Journal of Theoretical Biology* **179**:129–146 [*234*]

Moegle H, Knorpp F, Bögel K, Arata A, Dietz N & Dietzhelm P (1974). Zur Epidemiologie der Wildtollwut. Untersuchungen im südlichen Teil der Bundesrepublik Deutschland. *Zentralblatt für Veterinär Medizin B* **21**:647–59 [*316, 510*]

Mollison D (1972a). Possible velocities for a simple epidemic. *Advances in Applied Probability* **4**:233–257 [*305, 485*]

Mollison D (1972b). The rate of spatial propagation of simple epidemics. In *Proceedings of the Sixth Berkeley Symposium on Mathematical Statistics and Probability* Vol. 3, pp. 579–614. Berkeley, CA, USA: University of California Press [*305, 486, 496*]

Mollison D (1977). Spatial contact models for ecological and epidemic spread. *Journal of the Royal Statistical Society B* **39**:283–326 [*271, 486, 496*]

Mollison D (1991). Dependence of epidemic and population velocities on basic parameters. *Mathematical Biosciences* **107**:255–287 [*360, 486*]

Mollison D & Levin S (1995). Spatial dynamics of parasitism. In *Ecology of Infectious Diseases in Natural Populations*, eds. Grenfell B & Dobson A. Cambridge,

UK: Cambridge University Press   [*271*]

Moloney K & Levin SA (1996). The effects of disturbance architecture on landscape-level population dynamics. *Ecology* **77**:375–394   [*29*]

Moloney K, Morin A & Levin S (1991). Interpreting ecological patterns generated through simple stochastic processes. *Landscape Ecology* **5**:163–174   [*96*]

Moloney K, Levin S, Chiarello N & Buttel L (1992). Pattern and scale in a serpentine grassland. *Theoretical Population Biology* **41**:257–276   [*96*]

Molski A & Keizer J (1995). Spatially nonlocal fluctuation theory of rapid chemical reactions. *Journal of Chemical Physics* **104**:3567–3578   [*390*]

Moriya S, Inoue K & Mabuchi M (1989). The use of *torymus sinensis* to control chestnut gall wasp, *Dryocosmus kuriphilus*, in Japan, Taipei. *ASPAC Technical Bulletin* **118**:12 pp, and loose data sheet on 1991 situation   [*308*]

Morris A (1997). *Representing Spatial Interactions in Simple Ecological Models*. PhD dissertation, University of Warwick, Coventry, UK   [*363, 377, 379, 387*]

Mukherji A, Rajan V & Slagle JR (1995). Robustness of cooperation. *Nature* **379**:125–126   [*140–141*]

Mundt CC & Leonard KJ (1986). Analysis of factors affecting disease increase and spread in mixtures of immune and susceptible plants in computer simulated epidemics. *Phytopathology* **76**:832–840   [*302, 310*]

Mundt CC, Leonard KJ, Thal WM & Fulton JH (1986). Computerized simulation of crown rust epidemics in mixtures of immune and susceptible oat plants with different genotype unit areas and spatial distribution of initial disease. *Phytopathology* **76**:590–598   [*302*]

Murray J (1975). Non-existence of wave solutions for the class of reaction–diffusion equations given by the Lotka–Volterra equations with diffusion. *Journal of Theoretical Biology* **52**:459–469   [*188*]

Murray J (1988). How the leopard gets its spots. *Scientific American* **258**:80–87   [*470*]

Murray J (1989). *Mathematical Biology*. Berlin, Germany: Springer   [*316, 458, 460, 469–470, 475*]

Murray J (1990). *Mathematical Biology. Biomathematics*, Vol. 19, 2nd edn. Berlin, Germany: Springer   [*391, 396, 404*]

Nagarajan S & Singh H (1975). The Indian stem rust rules – An epidemiological concept on the spread of wheat stem rust. *Plant Disease Reporter* **59**:133–136   [*303*]

Nagarajan S, Singh H, Joshi LM & Saari EE (1976). Meteorological conditions associated with long distance dissemination and deposition of *Puccinia graminis tritici* uredospores in India. *Phytopathology* **66**:198–203   [*303*]

Nagylaki T (1992). *Introduction to Theoretical Population Genetics. Biomathematics*, Vol. 21. Berlin, Germany: Springer   [*390*]

Nakamaru M, Matsuda H & Iwasa Y (1997). The evolution of cooperation in a lattice-structured population. *Journal of Theoretical Biology* **184**:65–81   [*147–148, 250, 318, 334, 361*]

Nakashizuka T (1987). Regeneration dynamics of beech forests in Japan. *Vegetatio* **69**:169–175   [*237*]

Nakashizuka T (1991). Population dynamics of coniferous and broad-leaved trees in a Japanese temperate mixed forest. *Journal of Vegetation Science* **2**:413–418   [*227, 237*]

Noble JV (1974). Geographic and temporal development of plagues. *Nature* **250**:726–728   [*301*]

Northrop PJ (1996). *Modelling and Statistical Analysis of Spatial-temporal Rainfall Fields*. PhD dissertation, University of London, UK   [*81*]

Novoplansky A, Cohen D & Sachs T (1989). Ecological implications of correlative inhibition between plant shoots. *Physiologia Plantarum* **77**:136–140   [*25*]

Nowak MA & May RM (1992). Evolutionary games and spatial chaos. *Nature* **359**:826–829 [*116, 137, 139, 144, 148–149, 318, 324, 382*]

Nowak MA & May RM (1993). The spatial dilemmas of evolution. *International Journal of Bifurcation and Chaos* **3**:35–78 [*139, 142, 149, 324*]

Nowak MA & Sigmund K (1992). Tit For Tat in heterogeneous populations. *Nature* **355**:250–253 [*324*]

Nowak MA & Sigmund K (1993). A strategy of win–stay, lose–shift that outperforms Tit-for-Tat in the Prisoner's Dilemma game. *Nature* **364**:56–58 [*145, 324*]

Nowak MA, Bonhoeffer S & May RM (1994a). Spatial games and the maintenance of cooperation. *Proceedings of the National Academy of Sciences of the USA* **91**:4877–4881 [*140, 143*]

Nowak MA, Bonhoeffer S & May RM (1994b). More spatial games. *International Journal of Bifurcation and Chaos* **4**:33–56 [*140, 142, 144*]

Nowak MA, May RM & Sigmund K (1995a). The arithmetics of mutual help. *Scientific American* **272**:76–81 [*137*]

Nowak MA, Sigmund K & El-Sedy E (1995b). Automata, repeated games, and noise. *Journal of Mathematical Biology* **33**:703–732 [*146*]

Nychka DW, Ellner S, Gallant AR & McCaffrey D (1992). Finding chaos in noisy systems. *Journal of the Royal Statistical Society B* **54**:1–27 [*220*]

Ohsawa M & Yamane M (1988). Pattern and population dynamics in patchy communities on a maritime rock outcrop. In *Diversity and Pattern in Plant Communities*, eds. During HJ, Werger MJA & Willems JH, pp. 209–220. The Hague, Netherlands: SPB Academic Publishing [*53*]

Ohtsuki T & Keyes T (1986). Kinetic growth percolation, epidemic processes with immigration. *Physics Review A* **33**:1223–1232 [*227*]

Okubo A (1980). *Diffusion and Ecological Problems: Mathematical Models*. Berlin, Germany: Springer [*5, 460*]

Okubo A (1986). Dynamical aspects of animal grouping: Swarms, schools, flocks and herds. *Advances in Biophysics* **22**:1–94 [*461*]

Okubo A, Maini PK, Williamson MH & Murray JD (1989). On the spatial spread of the grey squirrel in Britain. *Proceedings of the Royal Society of London B* **238**:113–125 [*316*]

Oort AJP (1940). De verspreiding van de sporen van tarwestuifbrand (*Ustilago tritici*) door de lucht. [The dissemination of the spores of loose smut of wheat (*Ustilago tritici*) through the air]. *Tijdschrift over Plantenziekten* **46**:1–18 [*294*]

Osae-Danso F. *Invasions of pests and diseases in Africa*, Department of Phytopathology, Wageningen. Unpublished [*305, 308*]

Pacala SW (1986). Neighborhood models of plant population dynamics. II. Multispecies models of annuals. *Theoretical Population Biology* **29**:262–292 [*20, 412*]

Pacala SW (1987). Neighborhood models of plant population dynamics. III. Models with spatial heterogeneity in the physical environment. *Theoretical Population Biology* **31**:359–392 [*20*]

Pacala SW & Deutschman D (1995). Details that matter: The spatial distribution of individual trees maintains forest ecosystem function. *Oikos* **74**:357–365 [*20, 388*]

Pacala SW & Silander JA Jr (1985). Neighborhood models of plant population dynamics. I. Single-species models of annuals. *The American Naturalist* **125**:385–411 [*20, 395, 412*]

Pacala SW & Silander JA Jr (1987). Neighborhood interference among velvet leaf, *Abutilon theophrasti*, and pigweed, *Amaranthus retroflexus*. *Oikos* **48**:217–224 [*20, 395*]

Pacala SW & Silander JA Jr (1990). Field tests of neighborhood population dynamic models of two annual weed species. *Ecological Monographs* **60**:113–134 [20]

Pacala SW & Tilman D (1994). Limiting similarity in mechanistic and spatial models of plant competition in heterogeneous environments. *The American Naturalist* **143**:222–257 [360]

Pacala SW, Canham CD, Saponara J, Silander JA Jr, Kobe RK & Ribbens E (1996). Forest models defined by field measurements: Estimation, error analysis and dynamics. *Ecological Monographs* **66**:1–43 [1, 18, 23, 54, 256]

Packard NH, Crutchfield JP, Farmer JD & Shaw RS (1980). Geometry from a time series. *Physics Review Letters* **45**:712–716 [218]

Paczuski M & Bak P (1993). Theory of the one-dimensional forest fire model. *Physical Review E* **48**:3214–3216 [275]

Panfilov A & Hogeweg P (1993). Spiral breakup in a modified FitzHugh–Nagumo model. *Physics Letters A* **176**:295–299 [181]

Pao C (1992). *Nonlinear Parabolic and Elliptic Equations*. New York, NY, USA: Plenum [458, 471, 474]

Parrish JAD & Bazzaz FA (1976). Underground niche separation in successional plants. *Ecology* **57**:1281–1288 [16]

Parrish JAD & Bazzaz FA (1985). Ontogenetic niche shifts in old-field annuals. *Ecology* **66**:1296–1302 [16]

Pärtel M & Zobel M (1995). Small-scale dynamics and species richness in successional alvar plant communities. *Ecography* **18**:83–90 [52]

Pearcy RW, Chazdon RL, Gross LJ & Mott KA (1994). Photosynthetic utilization of sunflecks: A temporally patchy resource on a time scale of seconds to minutes. In *Exploitation of Environmental Heterogeneity by Plants*, eds. Caldwell MM & Pearcy RW, pp. 175–208. San Diego, CA, USA: Academic Press [54]

Pearson JE (1993). Complex patterns in a simple system. *Science* **261**:189–192 [164, 172, 174]

Pfeifer D, Bäumer HP, Ortleb H, Sach G & Schleier U (1996). Modeling spatial distributional patterns of benthic meiofauna species by Thomas and related processes. *Ecological Modelling – Modelling of Geo-Biosphere Processes* **87**:285–294 [77]

Pfister CA & Hay ME (1988). Associational plant refuges – Convergent patterns in marine and terrestrial communities result from differing mechanisms. *Oecologia* **77**:118–129 [33]

Phillips DL & MacMahon JA (1981). Competition and spacing in desert shrubs. *Journal of Ecology* **69**:97–115 [50]

Pickett STA & White PS (1985). *The Ecology of Natural Disturbance and Patch Dynamics*. New York, NY, USA: Academic Press [19]

Pielou EC (1968). *An Introduction to Mathematical Ecology*. New York, NY, USA: Wiley InterScience [48–49, 255]

Pielou EC (1977). *Mathematical Ecology*, 2nd edn. New York, NY, USA: Wiley [394]

Populer C (1964). Le comportement des épidémies de mildiou du tabac *Peronospora tabacina*. I. La situation en Europe. *Bulletin de l'Institut agronomique et des Stations de Recherches de Gembloux* **32**:339–378 [298, 301]

Preston C (1974). *Gibbs States on Countable Sets*. Cambridge, UK: Cambridge University Press [79]

Priestley MB (1981). *Spectral Analysis and Time Series*. London, UK: Academic Press [66, 69, 84]

Radcliffe J & Rass L (1983). Wave solutions for the deterministic non-reducible N-type epidemic. *Journal of Mathematical Biology* **17**:45–66 [486]

Radcliffe J & Rass L (1984a). The spatial spread and the final size of models for the deterministic host-vector epidemic. *Mathematical Biosciences* **70**:123–146 [486]

Radcliffe J & Rass L (1984b). The uniqueness of wave-solutions for the deterministic non-reducible N-type epidemic. *Journal of Mathematical Biology* **19**:303–308 [*486*]

Radcliffe J & Rass L (1984c). The spatial spread and final size of the deterministic non-reducible N-type epidemic. *Journal of Mathematical Biology* **19**:309–327 [*486*]

Radcliffe J & Rass L (1984d). Saddle point approximations in N-type epidemics and contact-birth processes. *Rocky Mountain Journal of Mathematics* **14**:599–617 [*486*]

Radcliffe J & Rass L (1985). The rate of spread of infection in models for the deterministic host–vector epidemic. *Mathematical Biosciences* **74**:257–273 [*486*]

Radcliffe J & Rass L (1986). The asymptotic spread of propagation of the deterministic non-reducible N-type epidemic. *Journal of Mathematical Biology* **23**:341–359 [*486*]

Radcliffe J & Rass L (1991). Effect of reducibility on the deterministic spread of infection in a heterogeneous population. In *Mathematical Population Dynamics: Proceedings of the Second International Conference*, eds. Arino O, Axelrod DE & Kimmel M, pp. 93–114. New York, NY, USA: Marcel Dekker [*486*]

Radcliffe J & Rass L (1993). Reducible epidemics: Choosing your saddle. *Rocky Mountain Journal of Mathematics* **14**:599–617 [*486*]

Radcliffe J & Rass L (1995a). Spatial branching and epidemic processes. In *Mathematical Population Dynamics: Analysis and Heterogeneity*, eds. Arino O, Langlais M & Kimmel M, pp. 147–170. Winnipeg, Canada: Wuertz [*486*]

Radcliffe J & Rass L (1995b). Multitype contact branching processes. *Proceedings of First World Conference on Branching Processes, Springer Lecture Notes in Statistics* **99**:169–179 [*486*]

Radcliffe J & Rass L (1996). The asymptotic behavior of a reducible system of non-

linear integral equations. *Rocky Mountain Journal of Mathematics* **26**:731–752 [*486*]

Radcliffe J & Rass L (1997). Discrete time spatial models arising in genetics, evolutionary game theory, and branching processes. *Mathematical Biosciences* **140**:101–129 [*486*]

Radcliffe J & Rass L (1998). Spatial mendelian games. *Mathematical Biosciences* **151**:191–218 [*486*]

Radcliffe J & Rass L. *Spatial Deterministic Epidemics*. American Mathematical Society Surveys and Monographs. Book in preparation [*486*]

Rand DA (1999). Correlation equations and pair approximations for spatial ecologies. In *Advanced Ecological Theory: Advances in Principles and Applications*, ed. McGlade JM. Oxford, UK: Blackwell [*363, 376, 387*]

Rand DA & Wilson HB (1995). Using spatio-temporal chaos and intermediate-scale determinism to quantify spatially extended ecosystems. *Proceedings of the Royal Society of London B* **259**:111–117 [*3, 187, 215, 225*]

Rand DA, Wilson HB & McGlade JM (1994). Dynamics and evolution: Evolutionarily stable attractors, invasion exponents and phenotype dynamics. *Philosophical Transactions of the Royal Society of London B* **343**:261–283 [*382*]

Rand DA, Keeling M & Wilson HB (1995). Invasion, stability and evolution to criticality in spatially extended, artificial host-pathogen ecologies. *Proceedings of the Royal Society of London B* **259**:55–63 [*181, 273, 281, 361*]

Ratz A (1995). Long term spatial pattern created by fire: A model oriented towards boreal forests. *International Journal of Wildland Fire* **5**:25–34 [*109*]

Ratz A (1996). *A Generic Forest Fire Model: Spatial Patterns in Forest Fire Ecosystems*. PhD dissertation, University of Marburg, Germany [*109*]

Raven PH, Evert RF & Curtis H (1981).

Biology of Plants, 3rd edn. New York, NY, USA: Worth [228]

Rees M, Grubb PJ & Kelly D (1996). Quantifying the impact of competition and spatial heterogeneity on the structure and dynamics of a four-species guild of winter annuals. *The American Naturalist* **147**:1–32 [29]

Reeve JD (1990). Stability, variability, and persistence in host-parasitoid systems. *Ecology* **71**:422–426 [209]

Remmert H (1991). The mosaic-cycle concept of ecosystems – An overview. *Ecological Studies* **85**:1–21 [55]

Renshaw E (1991). *Modelling Biological Populations in Space and Time.* Cambridge, UK: Cambridge University Press [66, 392]

Rhodes C & Anderson R (1996). Evaluation of epidemic thresholds in a lattice model of disease spread. *Physics Letters A* **210**:183–188 [271]

Ribbens E, Pacala S & Silander J Jr (1994). Seedling recruitment in forests: Calibrating models to predict patterns of seedling dispersion. *Ecology* **75**:1794–1806 [392]

Rice EL (1984). *Allelopathy*, 2nd edn. Orlando, FL, USA: Academic Press [14]

Ricker WE (1954). Stock and recruitment. *Journal of the Fisheries Research Board of Canada* **11**: 559–623 [450]

Ricklefs RE & Schluter D (1993). *Species Diversity: Historical and Geographical Patterns.* Chicago, IL, USA: University of Chicago Press [18]

Rinzel J & Terman D (1982). Propagation phenomena in a bistable reaction-diffusion system. *SIAM Journal of Applied Mathematics* **42**:1111–1137 [476]

Ripley BD (1981). *Spatial Statistics.* New York, NY, USA: Wiley [66, 69, 75, 429]

Ripley BD (1988). *Statistical Inference for Spatial Processes.* Cambridge, UK: Cambridge University Press [71, 85–86]

Robertson GP & Gross KL (1994). Assessing the heterogeneity of the belowground resources: Quantifying pattern and scale. In *Exploitation of Environmental Hetero-*

geneity by Plants, eds. Caldwell MM & Pearcy RW, pp. 237–254. San Diego, CA, USA: Academic Press [54]

Robinson RA (1976). *Plant Pathosystems.* Berlin, Germany: Springer [294]

Roelfs AP (1972). Gradients in horizontal dispersal of cereal rust uredospores. *Phytopathology* **62**:57–62 [299]

Room PM & Thomas PA (1985). Nitrogen and establishment of a beetle for biological control of the floating weed *Salvinia* in Papua New Guinea. *Journal of Applied Ecology* **22**:139–156 [313]

Rosén E (1995). Periodic droughts and long-term dynamics of alvar grassland vegetation on Öland, Sweden. *Folia Geobotanic et Phytotaxonomica* **30**:131–140 [63]

Rosenzweig ML & MacArthur RH (1963). Graphical representation and stability conditions of predator–prey interactions. *The American Naturalist* **97**:209–223 [185]

Ross J, Müller S & Vidal C (1988). Chemical waves. *Science* **240**:460 [151]

Roughgarden J (1997). Production functions from ecological populations: A survey with emphasis on spatially implicit models. In *Spatial Ecology: The Role of Space in Population Dynamics and Interspecific Interactions*, eds. Tilman D & Karieva P, pp. 296–317. Princeton, NJ, USA: Princeton University Press [267]

Roxburgh SH, Watkins AJ & Wilson JB (1993). Lawns have vertical stratification. *Journal of Vegetation Science* **4**:699–704 [46]

Runkle JR (1981). Gap regeneration in some old-growth forests of the eastern United States. *Ecology* **62**:1041–1051 [238]

Runkle JR (1984). Development of woody vegetation in treefall gaps in a beech-sugar maple forest. *Hoarctic Ecology* **7**:157–164 [237–238]

Ryan CC (1988). Epidemiology and control of Fiji disease virus of sugarcane. *Advances in Disease Vector Research* **5**:163–176 [306, 308, 313]

Sache I & Zadoks JC (1995). Effect of rust (*Uromyces viciae-fabae*) on yield components of faba bean. *Plant Pathology* **44**:675–685 [*295, 298–299, 301–303, 311*]

Sakoda JM (1971). The checkerboard model of social interaction. *Journal of Mathematical Sociology* **1**:119–132 [*150*]

Sano M & Sawada Y (1985). Measurement of the Lyapunov spectrum from a chaotic time series. *Physical Review Letters* **10**:1082–1085 [*220*]

Satō K & Iwasa Y (1993). Modeling of wave regeneration (shimagare) in subalpine Abies forests: Population dynamics with spatial structure. *Ecology* **74**:1538–1550 [*227, 341*]

Satō K & Konno N (1995). Succesional dynamical models on the 2-dimensional lattice space. *Journal of the Physics Society of Japan* **64**:1866–1869 [*357–358*]

Satō K, Matsuda H & Sasaki A (1994). Pathogen invasion and host extinction in lattice structured populations. *Journal of Mathematical Biology* **32**:251–268 [*227–228, 250, 347–351, 353–354, 357, 361, 377, 386, 388, 405*]

Satulovsky J & Tomé T (1994). Stochastic lattice gas model for a predator–prey system. *Physics Review E* **49**:5073–5079 [*358*]

Sauer T, Yorke JA & Casdagli M (1991). Embedology. *Journal of Statistical Physics* **65**:579–616 [*218, 225*]

Savary S (1986). The effect of age of the groundnut crop on the development of primary gradients of *Puccinia arachidis* foci. *Netherlands Journal of Plant Pathology* **93**:15–24 [*293, 299*]

Savary S & van Santen G (1991). Effet de l'age de la culture sur les gradients primaires de cercosporiose tardive de l'arachide. In *Approches de la pathologie des cultures tropicales. L'Exemple de l'arachide en Afrique de l'Ouest*, ed. Savary S, pp. 65–90. Paris, France: Karthala [*294*]

Savill NJ (1997). *Eco-Evolutionary Con-*

sequences of Spatial Pattern Formation. PhD dissertation, University of Utrecht, Netherlands [*181*]

Savill NJ, Rohani P & Hogeweg P (1997). Self-reinforcing spatial patterns enslave evolution in a host-parasitoid system. *Journal of Theoretical Biology* **188**:11–20 [*181*]

Sayers B, Ross JA & Saencharoenrat P (1985). Pattern analysis of the case occurrences of fox rabies in Europe. In *Population Dynamics of Rabies in Wildlife*, ed. Bacon P, pp. 235–254. London, UK: Academic Press [*97–98*]

Schelling TC (1971). Dynamic models of segregation. *Journal of Mathematical Sociology* **1**:143–186 [*150*]

Schinazi R (1996). On an interacting particle system modeling an epidemic. *Journal of Mathematical Biology* **34**:915–925 [*353*]

Schlesinger WH, Raikes JA, Hartley AE & Cross AF (1996). On the spatial pattern of soil nutrients in desert ecosystems. *Ecology* **77**:364–374 [*54*]

Schmid B (1985). Clonal growth in grassland perennials. II. Growth form and fine-scale colonizing ability. *Journal of Ecology* **73**:809–818 [*38*]

Schmid B (1990) Some ecological and evolutionary consequences of modular organization and clonal growth in plants. *Evolutionary Trends in Plants* **4**:25–34 [*15, 24*]

Schmid B & Harper JL (1985). Clonal growth in grassland perennials. I. Density- and pattern-dependent competition between plants with different growth forms. *Journal of Ecology* **73**:793–808 [*37*]

Schmitt CG, Kingsolver CH & Underwood JF (1959). Epidemiology of stem rust of wheat: I. Wheat stem rust development from inoculation foci of different concentration and spatial arrangement. *Plant Disease Reporter* **43**:601–606 [*303*]

Schmitt J (1993). Reaction norms of morphological and life-history traits to light availability in *Impatiens capensis*.

Evolution **47**:1654–1668   [*14*]

Schmitt J & Wulff RD (1993). Light spectral quality, phytochrome and plant competition. *Trends in Ecology and Evolution* **8**:47–51   [*14*]

Schouten HJ (1991). *Studies on Fire Blight*. PhD dissertation, Agricultural University, Wageningen, Netherlands   [*313*]

Schwartz L (1950). *Théorie des distributions*. Paris, France: Hermann   [*419*]

Schwinning S & Parsons AJ (1996a). Analysis of the coexistence mechanisms for grasses and legumes in grazing systems. *Journal of Ecology* **84**:799–813   [*46*]

Schwinning S & Parsons AJ (1996b). A spatially explicit population model of stoloniferous n-fixing legumes in mixed pasture with grass. *Journal of Ecology* **84**:815–826   [*46*]

Shaw MW (1995). Simulation of population expansion and spatial pattern when individual dispersal distributions do not decline exponentially with distance. *Proceedings of the Royal Society of London B* **259**:243–248   [*311*]

Sherratt JA (1996). Periodic travelling waves in a family of deterministic cellular automata. *Physica D* **95**:329–335   [*457, 461*]

Sherratt JA, Eagan BT & Lewis MA (1997). Oscillations and chaos behind predator–prey invasion: Mathematical artifact or ecological reality? *Philosophical Transactions of the Royal Society of London B* **352**:21–38   [*457, 461, 502*]

Shigesada N & Kawasaki K (1997). *Biological Invasions: Theory and Practice* Oxford, UK: Oxford University Press   [*459, 507*]

Shigesada N, Kawasaki K & Teramoto E (1986). Traveling periodic waves in heterogeneous environments. *Theoretical Population Biology* **30**:143–160   [*497*]

Shigesada N, Kawasaki K & Teramoto E (1987). The speed of travelling frontal waves in heterogeneous environments. In *Mathematical Topics in Population Biology, Morphogenesis and Neurosciences,*

Lecture Notes in Biomathematics, Vol. 71, eds. Teramoto E & Yamaguti M, pp. 88–97. Berlin, Germany: Springer   [*497*]

Shigesada N, Kawasaki K & Takeda Y (1995). Modeling stratified diffusion in biological invasions. *The American Naturalist* **146**:229–251   [*507*]

Shipley B (1993). A null model for competitive hierarchies in competition matrices. *Ecology* **74**:1693–1699   [*39*]

Shmida A & Ellner S (1984). Coexistence of plant species with similar niches. *Vegetatio* **58**:29–55   [*402*]

Sigmund K (1992). On prisoners and cells. *Nature* **359**:774   [*137*]

Sigmund K (1995). *Games of Life*. Harmondsworth, UK: Penguin   [*135*]

Silander JA Jr & Pacala SW (1985). Neighborhood predictors of plant performance. *Oecologia* **66**:256–263   [*20–21*]

Silvertown J (1980). The dynamics of a grassland ecosystem: Botanical equilibrium in the Park Grass Experiment. *Journal of Applied Ecology* **17**:491–504   [*63*]

Silvertown J (1987). *Introduction to Plant Population Ecology*, 2nd edn. London, UK: Longman Group UK Ltd   [*228*]

Silvertown J (1992). Cellular automaton models of interspecific competition for space – The effect of pattern process. *Journal of Ecology* **80**:527–534   [*227*]

Silvertown J & Law R (1987). Do plants need niches? *Trends in Ecology and Evolution* **2**:24–26   [*39*]

Silvertown J & Lovett Doust J (1993). *Introduction to Plant Population Biology*, 3rd edn. Oxford, UK: Blackwell   [*12*]

Silvertown J & Smith B (1989). Mapping the environment for seed germination in the field. *Annals of Botany* **63**:163–167   [*54*]

Silvertown J, Holtier S, Johnson J & Dale P (1992). Cellular automaton models of interspecific competition for space – The effect of pattern on process. *Journal of Ecology* **80**:527–534   [*34, 42, 45, 95*]

Silvertown J, Franco M, Pisanty I & Mendoza A (1993). Comparative plant demography – Relative importance of life-cycle components to the finite rate of increase in woody and herbaceous perennials. *Journal of Ecology* **81**:465–476 [*40*]

Silvertown J, Lines CEM & Dale MP (1994). Spatial competition between grasses – Rates of mutual invasion between four species and the interaction with grazing. *Journal of Ecology* **82**:31–38 [*30, 34–37, 42*]

Skellam J (1951). Random dispersal in theoretical populations. *Biometrika* **38**:196–218. Reprinted in Real LA, Brown JH, eds. (1991). *Foundations of Ecology: Classic Papers with Commentaries.* Chicago, IL, USA: University of Chicago Press [*308, 312, 315, 388, 392, 402*]

Skellam J (1973). The formulation and interpretation of mathematical models of diffusional processes in population biology. In *The Mathematical Theory of the Dynamics of Biological Populations*, eds. Bartlett MS & Hiorns RW, pp. 63–85. London, UK: Academic Press [*460*]

Slade AJ & Hutchings MJ (1987a). The effects of nutrient availability on foraging in the clonal herb *Glechoma hederacea. Journal of Ecology* **75**:95–112 [*15*]

Slade AJ & Hutchings MJ (1987b). The effects of light intensity on foraging in the clonal herb *Glechoma hederacea. Journal of Ecology* **75**:639–650 [*16*]

Slatkin M (1974). Competition and regional coexistence. *Ecology* **55**:128–134 [*390*]

Smith H (1995). *Monotone Dynamical Systems.* Providence, RI, USA: American Mathematical Society [*475*]

Smith TM & Urban DL (1988). Scale and resolution of forest structural pattern. *Vegetatio* **74**:143–150 [*227*]

Snyder DL & Miller MI (1991). *Random Point Processes in Time and Space.* New York, NY, USA: Springer [*72*]

Sole RV & Bascompte J (1995). Measuring chaos from spatial information. *Journal of Theoretical Biology* **175**:139–147 [*218*]

Sorrensen-Cothern KA, Ford ED & Sprugel DG (1993). A model of competition incorporating plasticity through modular foliage and crown development. *Ecological Monographs* **63**:277–304 [*25*]

Spitters CJT (1983). An alternative approach to the analysis of mixed cropping experiments. I. Estimation of competition coefficients. *Netherlands Journal of Agricultural Science* **31**:1–11 [*31*]

Sprugel DG, Hinckley TM & Schaap W (1991). The theory and practice of branch autonomy. *Annual Review of Ecology and Systematics* **22**:309–334 [*25*]

Stampfli A (1995). Species composition and standing crop variation in an unfertilized meadow and its relationship to climatic variability during six years. *Folia Geobotanica et Phytotaxonomica* **30**:117–130 [*63*]

Stark JM (1994). Causes of soil nutrient heterogeneity at different scales. In *Exploitation of Environmental Heterogeneity by Plants*, eds. Caldwell MM & Pearcy RW, pp. 255–284. San Diego, CA, USA: Academic Press [*53–54*]

Stauffer D (1985). *Introduction to Percolation Theory.* London, UK: Taylor and Francis [*235*]

Stauffer D (1994). *Introduction to Percolation Theory,* London, UK: Taylor and Francis [*276*]

Steck F & Wandeler A (1980). The epidemiology of fox rabies in Europe. *Epidemiological Review* **2**:71–96 [*511*]

Stewart J & Potvin C (1996). Effects of elevated CO_2 on an artificial grassland community: Competition, invasion and neighbourhood growth. *Functional Ecology* **10**:157–166 [*35*]

Stoll P & Schmid B (1998). Plant foraging and dynamic competition between branches of *Pinus sylvestris* in contrasting light environments. *Journal of Ecology* **86**:934–945 [*25*]

Stoll P, Weiner J & Schmid B (1994). Growth variation in a naturally established population of *Pinus sylvestris. Ecology* **75**:660–670 [*24*]

Stone L & Roberts A (1991). Conditions for a species to gain advantage from the presence of competitors. *Ecology* **72**:1964–1972 [*46*]

Stuefer JF (1995). Separating the effects of assimilate and water integration in clonal fragments by the use of steam-girdling. *Abstracta Botanica* **19**:75–81 [*15*]

Stuefer JF, During HJ & deKroon H (1994). High benefits of clonal integration in two stoloniferous species, in response to heterogeneous light environments. *Journal of Ecology* **82**:511–518 [*15*]

Sugden R (1986). *The Economics of Cooperation, Rights and Welfare*. New York, NY, USA: Blackwell [*136*]

Sugihara G, Grenfell BT & May RM (1990). Distinguishing error from chaos in ecological time series. *Philosophical Transactions of the Royal Society of London B* **330**:235–251 [*209, 218*]

Sugiyama S. Genetic variation in the relative dominance of mixed swards: Potential influence and its mechanism. *Grassland Science*. In press [*40*]

Sugiyama S & Nakashima H (1995). Mechanisms responsible for the changes in competitive outcome betweem two cultivars of tall fescue (*Festuca arundinacea schreb.*). *Grassland Science* **41**:93–98 [*41*]

Sutherland WJ & Stillman RA (1988). The foraging tactics of plants. *Oikos* **52**:239–244 [*25*]

Sykes MF & Glen M (1983). Percolation processes in two dimensions. I. Low-density series expansions. *Journal of Physics A* **9**:87–95 [*235–236*]

Sykes MT, van der Maarel E, Peet RK & Willems JH (1994). High species mobility in species rich plant communities: An intercontinental comparison. *Folia Geobotanica et Phytotaxonomica* **29**:439–448 [*52*]

Szathmáry E (1986). The eukaryotic cell as an information integrator. *Endocytobiosis and Cell Research* **3**:113–132 [*116*]

Szathmáry E (1992). Natural selection and the dynamical coexistence of defective and complementing virus segments. *Journal of Theoretical Biology* **157**:383–406 [*116–117, 129*]

Szathmáry E & Demeter L (1987). Group selection of early replicators and the origin of life. *Journal of Theoretical Biology* **128**:463–486 [*116–117*]

Szathmáry E & Maynard Smith J (1995). The major evolutionary transitions, *Nature* **374**:227-232 [*117*]

Tainaka K (1988). Lattice model for the Lotka–Volterra system. *Journal of the Physics Society of Japan* **57**:2588–2590 [*227*]

Tainaka K (1994). Vortices and strings in a model ecosystem. *Physics Review E* **50**:3401–3409 [*358*]

Takenaka Y, Matsuda H & Iwasa Y (1997). Competition and evolutionary stability of plants in a spatially structured habitats. *Researches on Population Ecology* **39**:67–75 [*250*]

Takens F (1981). Detecting strange attractors in turbulence. In *Dynamical Systems and Turbulence*, eds. Rand D & Young L-S, pp. 366–381. Berlin, Germany: Springer [*218*]

Tang Y & Washitani I (1995). Characteristics of small-scale heterogeneity in light availability within a *Miscanthus sinensis* canopy. *Ecological Research* **10**:189–197 [*54*]

Taylor AD (1990). Metapopulations, dispersal and predator–prey dynamics: An overview. *Ecology* **71**:429–433 [*209*]

Taylor GI (1938). Statistical theory of turbulence. *Proceedings of the Royal Society of London A* **164**:476–490 [*70*]

Taylor LR (1984). Assessing and interpreting the spatial distributions of insect populations. In *Annual Review of Entomology*, Vol. 29, pp. 321–357 [*231*]

Taylor P & Jonker L (1978). Evolutionarily stable strategies and game dynamics. *Mathematical Biosciences* **40**:145–156 [*319*]

Taylor PD (1992). Altruism in viscous populations – An inclusive fitness model. *Evolutionary Ecology* 6:352–356 [*251*]

Taylor R & Aarssen LW (1990). Complex competitive relationships among genotypes of three perennial grasses: Implications for species coexistence. *The American Naturalist* 136:305–327 [*41*]

Thieme HR (1977a). A model for the spatial spread of an epidemic. *Journal of Mathematical Biology* 4:337–351 [*305, 486*]

Thieme HR (1977b). The asymptotic behaviour of solutions of nonlinear integral equations. *Mathematische Zeitschrift* 157:141–154 [*486*]

Thieme HR (1979a). Asymptotic estimates of the solutions of nonlinear integral equations and asymptotic speeds for the spread of populations. *Journal für reine und angewandte Mathematik* 306:94–121 [*305, 486*]

Thieme HR (1979b). Density-dependent regulation of spatially distributed populations and their asymptotic speed of spread. *Journal of Mathematical Biology* 8:173–187 [*486*]

Thomas SC & Weiner J (1989). Including competitive asymmetry in measures of local interference in plant populations. *Oecologia* 80:349–355 [*23*]

Thompson L & Harper JL (1988). The effect of grasses on the quality of transmitted radiation and its influence on the growth of white clover *Trifolium repens. Oecologia* 75:343–347 [*30*]

Thompson S (1992). *Sampling.* New York, NY, USA: Wiley [*65*]

Thompson S & Seber G (1996). *Adaptive Sampling.* New York, NY, USA: Wiley [*65*]

Thomson JD, Weiblen G, Thomson BA, Alfaro S & Legendre P (1996). Untangling multiple factors in spatial distributions: Lilies, gophers and rocks. *Ecology* 77:1698–1715 [*68*]

Thórhallsdóttir TE (1990a). The dynamics of a grassland community: A simultaneous investigation of spatial and temporal heterogeneity at various scales. *Journal of*

Ecology 78:884–908 [*49, 52*]

Thórhallsdóttir TE (1990b). The dynamics of five grasses and white clover in a simulated mosaic sward. *Journal of Ecology* 78:909–923 [*35*]

Thresh JM (1958). The spread of virus disease in cacao. *West African Cocoa Research Institute, Technical Bulletin* 5:36 [*294–295, 299, 310, 313*]

Tilman D (1982). *Resource Competition and Community Structure.* Princeton, NJ, USA: Princeton University Press [*14, 16*]

Tilman D (1988). *Plant Strategies and the Dynamics and Structure of Plant Communities.* Princeton, NJ, USA: Princeton University Press [*14*]

Tilman D (1994). Competition and biodiversity in spatially structured habitats. *Ecology* 75:2–16 [*402*]

Tilman D (1997). Community invasibility, recruitment limitation and grassland biodiversity. *Ecology* 78:81–92 [*18*]

Tilman D & Karieva P (1997). *Spatial Ecology: The Role of Space in Population Dynamics and Interspecific Interactions.* Princeton, NJ, USA: Princeton University Press [*3*]

Toffoli T & Margolus N (1987). *Cellular Automata Machines: A New Environment for Modeling.* Cambridge, MA, USA: MIT Press [*122*]

Tong H (1995). A personal overview of nonlinear time series from a chaos perspective (with discussion and comments). *Scandinavian Journal of Statistics* 22:399–445 [*68*]

Tremmel DC & Bazzaz FA (1993). How neighbor canopy architecture affects target plant performance. *Ecology* 74:2114–2124 [*14*]

Trivers R (1971). The evolution of reciprocal altruism. *Quarterly Review of Biology* 46:35–57 [*135, 318*]

Turchin P (1998). *Quantitative Analysis of Animal Movement: Measuring and Modeling Population Redistribution in Animals and Plants.* Sunderland, MA, USA: Sinauer [*496*]

Turelli M & Barton N (1994). Genetic and statistical analyses of strong selection on polygenic traits: What, me normal? *Genetics* **138**:913–941   [*390*]

Turing A (1952). The chemical basis of morphogenesis. *Philosophical Transactions of the Royal Society of London B* **237**:37–72   [*323, 457, 469*]

Turkington R (1994). Effect of propagule source on competitive ability of pasture grasses – Spatial dynamics of six grasses in simulated swards. *Canadian Journal of Botany – Revue Canadienne de Botanique* **72**:111–121   [*35*]

Turkington RA & Harper JL (1979). The growth, distribution and neighbour relationships of *Trifolium repens* in a permanent pasture. I. Ordination, pattern and contact. *Journal of Ecology* **67**:201–218   [*270, 426*]

Tyler CM & D'Antonio CM (1995). The effects of neighbors on the growth and survival of shrub seedlings following fire. *Oecologia* **102**:255–264   [*18*]

Tyson JJ & Keener JP (1988). Singular pertubation theory of traveling waves in excitable media (a review). *Physica D* **32**:327–361   [*177*]

Underwood JF, Kingsolver CH, Peet CE & Bromfield KR (1959). Epidemiology of stem rust of wheat. III. Measurements of increase and spread. *Plant Disease Reporter* **43**:1154–1159   [*303*]

Urban D, Bonan G, Smith R & Shugart H (1991). Spatial applications of gap models. *Forest Ecology and Management* **42**:95–110   [*96*]

van Andel J & Nelissen HJM (1981). An experimental approach to the study of species interference in a patchy vegetation. *Vegetatio* **45**:155–163   [*35*]

van Baalen M & Rand DA (1998). The unit of selection in viscous populations and the evolution of altruism. *Journal of Theoretical Biology* **193**:631–648   [*334, 361, 366, 383–384, 386*]

van de Lande H & Zadoks JC. Spatial patterns of spear rot in oil palm plantations in Suriname. *Plant Pathology* **48**. In press   [*297, 299, 313*]

van den Bergh JP (1979). Changes in the composition of mixed populations of grassland species. In *The Study of Vegetation*, ed. Werger MJA, pp. 57–80. The Hague, Netherlands: Junk   [*63*]

van den Bosch F (1990). *The Velocity of Spatial Population Expansion*. PhD dissertation. University of Leiden, Netherlands   [*306*]

van den Bosch F, Zadoks JC & Metz JAJ (1988a). Focus expansion in plant disease. I. The constant rate of focus expansion. *Phytopathology* **78**:54–58   [*317*]

van den Bosch F, Zadoks JC & Metz JAJ (1988b). Focus expansion in plant disease. II. Realistic parameter-sparse models. *Phytopathology* **78**:59–64   [*307, 310, 317*]

van den Bosch F, Frinking HD, Metz JAJ & Zadoks JC (1988c). Focus expansion in plant disease. III. Two experimental examples. *Phytopathology* **78**:919–925   [*293, 297, 299–300, 307–308, 310–311, 317, 404*]

van den Bosch F, Metz JAJ & Diekmann O (1990a). The velocity of spatial population expansion. *Journal of Mathematical Biology* **28**:529–565   [*317, 392, 404, 486, 492, 498, 510*]

van den Bosch F, Verhaar MA, Buiel AAM, Hoogkamer W & Zadoks JC (1990b). Focus expansion in plant disease. IV. Expansion rates in mixtures of resistant and susceptible hosts. *Phytopathology* **80**:598–602   [*293, 299, 309, 317, 493*]

van den Bosch F, Hengeveld R & Metz JAJ (1992). Analyzing the velocity of animal range expansion. *Journal of Biogeography* **19**:135–150   [*308, 317*]

van den Bosch F, Zadoks JC & Metz JAJ (1994). Continental expansion of plant disease: A survey of some recent results. In *Predictability and Nonlinear Modelling in Natural Sciences and Economics*, eds. Grasman J & van Straten G, pp. 274–281. Dordrecht, Netherlands:

Kluwer Academic Publishers [*312, 511*]

van den Bosch F, Metz JAJ & Zadoks JC (1999). Pandemics of focal plant disease. A model. *Phytopathology* **89** [*311–313, 511*]

van der Maarel E (1996). Pattern and process in the plant community: Fifty years after A.S. Watt. *Journal of Vegetation Science* **7**:19–28 [*52, 55, 57*]

van der Maarel E & Sykes MT (1993). Small scale plant species turnover in a limestone grassland: The carousel model and some comments on the niche concept. *Journal of Vegetation Science* **4**:179–188 [*52, 56–57, 60–61*]

van der Maarel E, Noest V & Palmer MW (1995). Variation in species richness on small grassland quadrats: Niche structure or small-scale plant mobility? *Journal of Vegetation Science* **6**:741–752 [*57–58*]

Vandermeer J (1984). Plant competition and the yield-density relationship. *Journal of Theoretical Biology* **109**:393–399 [*18*]

Vanderplank JE (1963). *Plant Diseases: Epidemics and Control.* New York, NY, USA: Academic Press [*293–294, 302*]

Vanderplank JE (1975). *Principles of Plant Infection.* New York, NY, USA: Academic Press [*300, 311*]

van der Werf W (1988). *Yellowing Viruses in Sugarbeet: Epidemiology and Damage.* PhD dissertation, Agricultural University Wageningen, Netherlands [*292, 295, 299, 313*]

van der Zaag DA (1956). Overwintering en epidemiologie van *Phytophthora infestans*, tevens enige nieuwe bestrijdingsmogelijkheden, *Tijdschrift over Plantenziekten* **62**:89–165 [*294*]

van Doorn AM (1959). Investigations on the occurrence and the control of downy mildew (*peronospora destructor*) in onions. [In Dutch with English summary.] *Tijdschrift over Plantenziekten* **65**:193–255 [*300*]

van Kampen NG (1981). *Stochastic Processes in Physics and Chemistry.* Amsterdam, Netherlands: North-Holland [*507*]

van Kampen NG (1992). *Stochastic Processes in Physics and Chemistry.* Amsterdam, Netherlands: North-Holland [*421–422, 433*]

van Voorst G, Vos EA & Zadoks JC (1987). Dispersal of *Phytophthora nicotianae* on tomatoes grown by nutrient film technique in a greenhouse. *Netherlands Journal of Plant Pathology* **93**:195–199 [*301*]

Vautard R & Ghil M (1989). Singular spectrum analysis in nonlinear dynamics, with applications to paleoclimatic time series. *Physica D* **35**:395–424 [*225*]

Veit RR & Lewis MA (1996). Dispersal, population growth and the Allee effect: Dynamics of the house finch invasion of North America. *The American Naturalist* **148**:255–274 [*317, 501*]

Vickers GT (1989). Spatial patterns and ESSs. *Journal of Theoretical Biology* **140**:129–135 [*318, 320–321*]

Vickers GT, Hutson V & Budd C (1993). Spatial patterns in population conflicts. *Journal of Mathematical Biology* **31**:411–430 [*318, 322–323*]

Wächtershäuser G (1988). Before enzymes and templates: Theory of surface metabolism. *Microbiological Reviews* **52**:452–484 [*133*]

Wächtershäuser G (1992). Groundworks for an evolutionary biochemistry: The iron–sulphur world. *Progress in Biophysics and Molecular Biology* **58**:85–201 [*134*]

Waggoner PE (1968). Weather and the rise and fall of fungi. In *Biometeorology*, ed. Lowry W. Corvallis, OR, USA: Oregon State University Press [*302*]

Wälder O & Stoyan D (1996). On variograms in point process statistics. *Biometrical Journal* **38**:895–905 [*72*]

Walker MD, Webber PJ, Arnold EH & Ebert-May D (1994). Effects of interannual climate variation on aboveground phytomass in alpine vegetation. *Ecology* **75**:393–408 [*63*]

Watkins AJ & Wilson JB (1992). Fine-scale

community structure of lawns. *Journal of Ecology* **80**:15–24 [*29*]

Watkinson AR (1980). Density-dependence in single-species populations of plants. *Journal of Theoretical Biology* **83**:345–357 [*18*]

Watt AS (1919). On the causes of failure of natural regeneration in British Oakwoods. *Journal of Ecology* **7**:173–203 [*34*]

Watt AS (1947). Pattern and process in the plant community. *Journal of Ecology* **35**:1–22 [*1, 50, 55, 59–60, 415*]

Watt AS (1960). Population changes in acidiphilous grass-heath in Breckland, 1936–57. *Journal of Ecology* **48**:605–629 [*50, 61*]

Watts DJ & Strogatz SH (1998). Collective dynamics of 'small-world' networks. *Nature* **393**:440–442 [*363*]

Waymire E, Gupta VK & Rodriguez-Iturbe I (1984). A spectral theory of rainfall intensity at the meso-beta scale. *Water Resources Research* **20**:1453–1465 [*79*]

Wedin D & Tilman D (1993). Competition among grasses along a nitrogen gradient – Initial conditions and mechanisms of competition. *Ecological Monographs* **63**:199–229 [*35*]

Weinberger HF (1978). Asymptotic behaviour of models in population genetics. In *Nonlinear Partial Differential Equations and Applications, Lecture Notes in Mathematics*, Vol. 648, ed. Chadam JM, pp. 47–98. Berlin, Germany: Springer [*486, 498*]

Weinberger HF (1982). Long-time behaviour of a class of biological models. *SIAM Journal of Mathematical Analysis* **13**:353–396 [*486, 498*]

Weiner J (1982). A neighborhood model of annual-plant interference. *Ecology* **63**:1237–1241 [*412*]

Weiner J (1984). Neighborhood interference amongst Pinus rigida individuals. *Journal of Ecology* **72**:183–195 [*21*]

Weiner J (1995). On the practice of ecology. *Journal of Ecology* **83**:153–158 [*26*]

Weiner J & Conte PT (1981). Dispersal and neighborhood effects in an annual plant competition model. *Ecological Modelling* **13**:131–147 [*412*]

Weiss MP (1907). L'hypothèse du champ moléculaire et la propriété ferromagnétique. *Journal de Physique* **6**:661–690 [*4*]

Wheater HS, Isham V, Cox DR, Chandler RE, Kakou A, Northrop PJ, Oh L, Onof C & Rodriguez-Iturbe I (1996). Spatial-temporal rainfall fields: Modelling and statistical aspects. Research Report 76. Department of Statistical Science, University College London, UK [*79*]

Whitmore TC (1975). *Tropical Rain Forest in the Far East*. Oxford, UK: Clarendon Press [*236, 238*]

Wiegand T & Milton S (1996). Vegetation change in semiarid communities, simulating probabilities and time scales. *Vegetatio* **125**:169–183 [*104*]

Wiegand T, Milton S & Wissel C (1995). A simulation model for a shrub-ecosystem in the semi-arid Karoo, South Africa. *Ecology* **76**:2205–2221 [*96, 104–106*]

Wiegand T, Dean W & Milton S (1997). Simulated plant population responses to small disturbances in semi-arid shrublands. *Journal of Vegetation Science* **8**:163–176 [*104*]

Wiens JA, Stenseth NC, van Horne B & Ims RA (1993). Ecological mechanisms and landscape ecology. *Oikos* **66**:369–380 [*3, 252*]

Willems JH, Peet RK & Bik L (1993). Changes in chalk-grassland structure and species richness resulting from selective nutrient additions. *Journal of Vegetation Science* **4**:203–212 [*52*]

Williamson EJ (1961). The distribution of larvae of randomly moving insects. *Australian Journal of Biological Sciences* **14**:598–604 [*489*]

Williamson MH & Brown KC (1986). The analysis and modelling of British invasions. *Philosophical Transactions of the Royal Society of London B* **314**:505–522 [*312*]

Wilson DS (1980). *The Natural Selection of Populations and Communities*. Menlo Park, CA, USA: Benjamin Cummings [*116, 126, 129*]

Wilson JB & Roxburgh SH (1994). A demonstration of guild-based assembly rules for a plant community, and determination of intrinsic guilds. *Oikos* **69**:267–276 [*58, 61*]

Wilson DS, Pollock GB & Dugatkin LA (1992). Can altruism evolve in purely viscous populations? *Evolutionary Ecology* **6**:331–341 [*141, 251, 329*]

Wilson W, de Roos A & McCauley E (1993). Spatial instabilities within the diffusive Lotka–Volterra system: Individual-based simulation results. *Theoretical Population Biology* **43**:91–127 [*184, 227, 388*]

Wilson JB, Sykes MT & Peet RK (1995a). Time and space in the community structure of a species-rich limestone grassland. *Journal of Vegetation Science* **6**:729–740 [*57–58, 60*]

Wilson W, McCauley E & de Roos A (1995b). Effect of dimensionality on Lotka–Volterra predator–prey dynamics: Individual based simulation results. *Bulletin of Mathematical Biology* **57**:507–526 [*184, 388*]

Winfree A (1974). Rotating chemical reactions. *Scientific American* **230**:82 [*151*]

Wissel C (1992). Modelling the mosaic-cycle of a middle European beech forest. *Ecological Modelling* **63**:29–43 [*96*]

Wolf A, Swift JB, Swinney HL & Vastano JA (1985). Determining Lyapunov exponents from a time series. *Physica D* **16**:285–317 [*220*]

Wolfram S (1986). *Theory and Applications of Cellular Automata*. Singapore: World Scientific Publishing [*95, 155*]

Wright AJ (1981). The analysis of yield-density relationships in binary mixtures using inverse polynomials. *Journal of Agricultural Science* **96**:561–567 [*31*]

Wu JG & Levin SA (1994). A spatial patch dynamic modeling approach to pattern and process in an annual grassland. *Ecological Monographs* **64**:447–464 [*29, 96*]

Yamamoto S (1992). The gap theory in forest dynamics. *Botanical Magazine of Tokyo* **105**:375–383 [*236–237*]

Yamamoto S (1993). Gap characteristics and gap regeneration in a subalpine coniferous forest on Mt. Ontake, central Honshu, Japan. *Ecological Research* **8**:277–285 [*237–238*]

Yang X, Madden LV & Brazee RD (1991). Application of the diffusion equation for modelling splash dispersal of point-source pathogens. *New Phytologist* **118**:295–301 [*295, 313*]

Zadoks JC (1961). Yellow rust on wheat, studies in epidemiology and physiologic specialization. *Tijdschrift over Plantenziekten (European Journal of Plant Pathology)* **67**:69–256 [*292–295, 299, 303, 311, 313*]

Zadoks JC (1968). The epidemiology of yellow rust. In *Abstracts of Papers, 1st International Congress of Plant Pathology, London*, p. 224 [*302*]

Zadoks JC (1981). Mr. Duhamel's 1728 treatise on the violet root rot of saffron crocus: "Physical explanation of a disease that perishes several plants in the Gastinois, and saffron in particular." *Mededelingen Landbouwhogeschool, Wageningen* 81-7 [*292, 313*]

Zadoks JC (1988). Twenty five years of botanical epidemiology. *Philosophical Transactions of the Royal Society of London B* **321**:377–387 [*301, 303*]

Zadoks JC & Kampmeijer P (1977). The role of crop populations and their deployment, illustrated by means of a simulator EPIMUL 76. *Annals New York Academy of Science* **287**:164–190 [*301, 303–304, 308, 312*]

Zadoks JC & Schein RD (1979). *Epidemiology and Plant Disease Management*. New York, NY, USA: Oxford University Press [*292, 295, 299*]

Zadoks JC & van den Bosch F (1994). On the spread of plant disease: A theory on foci. *Annual Review of Phytopathology* **32**:503–521   [*293, 312*]

Zandvoort R (1968). Wind dispersal of *Puccinia horiana*. *Netherlands Journal of Plant Pathology* **74**:124–127   [*303*]

Zawolek MW (1993). Shaping a focus: Wind and stochasticity. *Netherlands Journal of Plant Pathology* **93**, Suppl. 3:241–255   [*292, 295, 310, 313*]

Zawolek MW & Zadoks JC (1989). A physical theory of focus development in plant disease. *Wageningen Agricultural University Papers* 89-3. Wageningen, Netherlands: Agricultural University [*303, 309–311*]

Zeeman E (1980). Population dynamics from game theory. In *Global Theory of Dynamical Systems. Lecture Notes in Mathematics 819*, eds. Nitecki A & Robinson C, pp. 471–497. Berlin, Germany: Springer [*319*]

Zhang Y (1993). A shape theorem for epidemics and forest fires with finite range interactions. *Annals of Probability* **21**:1755–1781   [*277, 503*]

Ziman JM (1979). *Models of Disorder: The Theoretical Physics of Homogeneously Disordered Systems.* Cambridge, UK: Cambridge University Press [*442*]

Zwankhuizen MJ, Govers F & Zadoks JC (1998). Development of potato late blight epidemics: Disease foci, disease gradients, and infection sources. *Phytopathology* **88**:754–763   [*296, 299, 303*]

Index

AD *see* Always Defect
age-at-birth distribution, *487*
   see also birth kernel
aggregation, *216–217*
   see also clustering
Allee effect, *317, 484, 500–502, 504, 512*
allelopathy, *12, 14, 17, 50*
altruism, *171, 334, 342, 361, 366, 368*
Always Cooperate, *146*
Always Defect, *137, 145–147, 208, 251,*
   323–332, 463–466
anisotropy, *69, 294, 296–297, 313, 498*
Anti-Tit For Tat, *146*
artificial ecologies, *205, 361, 364, 366,*
   370–371, 376, 380, 386, 418
   see also cellular automaton models
assessment of fit, *87–88*
autocatalysis, *119, 134*
auto-correlation, *255, 266, 427, 429, 436,*
   444, 448
   see also correlation density; covariance
      function; doublet density; pair density

bacteria, *154, 243–249, 295, 313, 315, 457*
Bessel density, distribution, kernel, *306,*
   310, 317, 397, 409, 489, 493, 509, 511
Bessel function, *409*
bimatrix games, *474–475, 477*
birth–death–movement process
   continuous space, *256–258, 263, 391,*
      418–425, 486–487, 495–496, 503–504
   lattice, *365–366, 370*
birth distribution, *487, 488, 492*
   see also birth kernel; reproduction kernel
birth kernel, *486, 488–490, 493, 496, 498,*
   502, 508, 510
   see also birth distribution; reproduction
      kernel
Bully, *144, 146*

c_0, *300, 308, 316, 490–500, 502*
   see also wave speed
CA *see* cellular automaton models

carousel model, *10, 55–61, 64*
catalytic support, *153–154, 160, 165-167,*
   171–172, 174, 178-179
Cayley tree, *346, 347*
cellular automaton models
   colicin bacteria, *243–244*
   and diffusion, *122, 155–157*
   dwarf shrub community, *104–109*
   forest fires, *109–114*
   forest gaps, *237–238*
   game theory, *138–148*
   general specification, *155*
   grassland communities, *43–46*
   host–parasite, *271–274, 283–284, 288*
   metabolic replication, *118–123*
   plant reproduction strategies, *229*
   predator–prey, *212–213, 184–186*
   see also artificial ecologies;
      birth–death–movement process,
      lattice; grid-based models; lattice
      logistic model; lattice models
chaos, *220–221*
chemical spatial systems, *154*
Chicken game, *136, 139, 144*
   see also Hawk–Dove game
clumping *see* clustering
clustering, *49–50, 76–77, 165–168,*
   228–236, 247–248, 342–343, 350–353,
   382, 386, 396, 400–405
   see also foci; hot-spots
clusters, *163–169*
   see also scroll rings; spiral waves;
      self-replicating spots
cluster sizes, *234–236*
   see also aggregation
coexistence
   and community size, *126–127*
   and diffusion, *124–126*
   game theory, *139, 141, 146*
   grasses, *39, 40–41*
   macromolecular replicators, *116, 119,*
      121, 123–129
   and metabolic coupling, *119*

molecular replicators, *116, 119, 121,*
128–129
and neighborhood size, *117, 126–127,*
131
and parasites, *131, 166–169*
primitive genomes, *116*
species, *71, 107, 342, 424*
and surface metabolism, *133*
colicin bacteria, *206, 243–249, 358*
comparison methods in diffusion equations,
471–475
competition
altruists versus non-altruists, *334*
birth–death–movement process
representation, *256–258, 261–263,*
366, 388–411, 421, 424
clonally versus sexually reproducing
plants, *227, 228, 341, 361*
coefficients, *28, 31–32*
colicin versus non-colicin bacteria,
243–245, 248, 358
in coral reefs, *182*
in dwarf shrub community, *104–106*
among genes, *210–211*
in grassland communities, *28–47, 50*
influence of self-generated pattern,
260–261, 265–269, 287, 386
measurement, *30, 31, 33*
neighborhood structure, *12–25, 263–269,*
334, 385–386, 448, 450
in replicator molecules, *129, 178*
scale, *396, 405, 409*
within species modulated by other
species, *149*
and species turnover, *57*
as source of spatial separation, *50, 77*
transitivity of, *36, 39, 41, 42*
wavelike takeover, *316, 478, 501*
competition kernels, *257, 391, 392, 411*
competitive hierarchy, *36–37, 39, 42, 46*
conditional intensity, *75*
contact process, *344, 352, 358, 505*
see also lattice logistic model
continuous movement, *489, 499–500*
convolution, *394, 395, 399, 420*
Conway's Game of Life, *139*
corrected correlation density, *430–431*
corrected pair density *see* corrected

correlation density
correlation corrections, *340, 449–451,*
514–515
correlation density
advantages, *429–430*
auto-correlation, *255, 266, 427, 429, 436,*
444, 448
corrected, *430–431*
cross-correlation, *255, 427, 429, 444, 448*
definition, *254, 428*
dynamics, *260, 265, 432–438, 452–455*
examples, *254, 427, 428*
as expression of pattern, *255*
further reduction of, *451–452*
higher order, *428*
as measure of local environment,
426–427
measurement, *428*
and relaxation projections, *417, 430–431,*
451–452
see also conditional intensity; covariance
function; doublet density; moment
methods; pair-approximation methods;
pair density; spatial moments
coupled map lattice, *197, 210–211*
covariance function
average, *394*
definition, *393–394*
dynamics, *396, 408*
measure of local crowding, *394*
quasi-equilibrium, *398–400*
see also conditional intensity; correlation
density; doublet density; moment
methods; pair-approximation methods;
pair density; spatial moments
critical transmissibility, *275–278, 281–283,*
286–288
cross-correlation, *255, 429, 444*
see also correlation density; covariance
function; doublet density; pair density

data analysis, *65–88, 209–226*
Bayesian methods, *85–86*
correlation density, measurement of, *428*
correlation structures, *69, 88*
data configurations, *66, 70*
descriptive methods, *66–70*
fluctuation diagram, *216–217*

Gibbs sampler, *86*
image restoration, *85*
Markov random fields, *77–79, 84–86*
maximum likelihood estimation, *81–83,
    84–87*
method of moments, *83*
model adequacy, *87–88, 94–95, 111*
model fitting, *80–88*
pseudo-likelihood estimation, *87*
simulation methods, *84*
spatial scales, determining, *213–216*
spectral methods, *84*
stochastic models, *70–80*
time series analysis, *67–68, 70, 213–225*
trend, *67, 69–70*
variogram, *72*
Whittle's method, *84*
degrees of freedom, *413–417, 425–426,
    451–452*
delta function
    Dirac, *253, 255, 407, 416–419, 422–423,
        460*
    generalized, *422–423, 433, 438*
demographic stochasticity, *93, 199–200,
    258, 269, 342, 358, 381, 385–386, 482,
    485*
deterministic approximation, *259, 264, 267,
    268*
deterministic dynamics, embedded, *71,
    170, 209, 219–224, 226*
deterministic models, *170, 187, 340, 404,
    500–503*
diffusion, *122, 155–157, 459–463*
    see also reaction–diffusion models,
        systems
diffusion operator, *156, 174, 462*
direct interaction scale, *484*
dispersal kernel, *257, 302, 306, 310–312,
    391–392, 397, 410, 421, 425, 509*
    see also displacement distribution;
        movement kernel
dispersal scale, *312, 396, 405, 409,
    483–484, 496–497, 507–511*
displacement distribution, *317, 487, 489,
    491–493, 496, 501, 509*
    see also dispersal kernel; movement
        kernel
displacibility, *32*

disturbance, *18, 45, 57–58, 236–237, 242,
    400, 462*
    fire, *18, 109–114*
    and resources, *18*
    see also perturbation
doublet density, *205–206, 231–232,
    345–347, 351, 429*
    see also correlation density; pair density
doubly stochastic Poisson process, *76*
dwarf shrub community dynamics,
    104–108, 109
dynamic simulation in plant pathology, *302*
    EPIMUL, *302–303, 305, 309*
    PODESS, *303, 309–312*
dynamic sufficiency, *413, 415, 417*
    see also degrees of freedom; relaxation
        projections

ecological signal, *2, 205, 256, 269, 412*
embedding dimension, *218, 219, 221, 222*
epidemics
    first-order, *293–298, 300–303, 310–313*
    integral equation models, *304–306*
    on a lattice structure, *250, 271–288, 349*
    plant, *292–315, 317*
    rabies, *91, 97–104, 114, 316–317,
        510–511*
    second-order, *293, 298, 301, 304–306,
        308, 312–315*
    zero-order, *293–300, 302, 309, 312–313,
        315; see also* foci
    see also foci; hot-spots
EPIMUL spatio-temporal simulation
    model, *302–303, 305, 309*
equilibrium density, *31, 234, 247, 263, 265,
    267–268, 501*
ergodicity, *60–63, 64, 273, 352, 394, 396*
    see also stationarity
ESS, *144, 172, 250, 319, 321–322, 323,
    324, 327, 333*
evolution
    cooperation, *463–466, 474–475*
    diffusion, *479–480*
    games, *135–150, 318, 335, 463–466,
        474–475*
    prebiotic, *116–134, 153–169, 171–188*
    transmissibility, *282–287*
evolutionarily stable strategies *see* ESS

evolutionary game theory *see* evolution
external heterogeneity, externally generated
    spatial patterns, *52–54, 448–449,
    497–500*

fat-tailed kernels, *311, 317, 496, 501*
Fick's law, *460–461*
fit, assessment of, *87–88*
Floquet multiplier, *190, 192–193*
fluctuation corrections, *449–451, 514–516*
fluctuation diagram, *196, 216–217, 225*
foci, *300, 305, 307, 311–315, 317*
    lattice, *362, 385*
    plant disease, *292–315, 317*
    phytopathology, *315, 488*
    and wave front, *340*
    see also epidemics, zero-order; hot-spots;
        wave speed
focus *see* foci
foraging, by plants, *15–16, 25*
forest dynamics
    fires, *109–111, 113–114*
    gaps, *236–243*
    herbivory, and oaks, *34*
    oaks, spread of after ice age, *315–316*
    regeneration waves, *341*
forest gaps, *236–243*
fungal diseases of plants *see* epidemics

Game of Life, *139*
game theory, *135–156, 474–475, 463–466,
    318–335*
gene competition, *210–211*
genets, *16, 24–25*
grassland communities, *38–47, 48–64*
    carousel models, *56–58, 59–60, 61*
    cellular automaton model, *42–46*
    competition, *28–47*
    competition coefficients determined from
        traits, *40–41*
    ergodicity, *60–61, 62–63, 64*
    grazing, *36–37, 46, 53*
    guild proportionality, *58, 59–60, 61*
    herbivory, *32–34*
    invasion, *30, 31, 34–36, 37, 42*
    measuring competition, *29–38*
    mosaic cycle, *55–56, 59, 60–61*
    mutual invasion, *34–36, 42*

    neighbor effects, *32–38*
    patch dynamics, *1, 19, 28–29*
    space preemption, *58–59, 60, 61*
    spatial interaction outcomes, *42–46*
    spatial-temporal patterns, *48–54*
    transition matrices, *38–42*
    transitivity, *36, 39, 41–42*
    see also carousel model; ergodicity;
        guild proportionality; mosaic cycle;
        plant interactions; space preemption
grazing
    and competition, *36–37, 46, 53*
    dwarf shrub community, *109*
    and grassland patterns, *53*
grid-based models
    bacterial competition, *243–247*
    descriptors, *96–97*
    dwarf shrub community, *104–109*
    and ecological systems, *95–96, 114–115*
    forest fires, *109–114*
    forest gaps, *236–243*
    games, *135–150*
    gene competition, *210–211*
    host–parasite, *272–277*
    and mathematics, *97, 114–115*
    pair approximations, *227–247, 341–357,
        374–377*
    versus partial differential equations,
        153–169
    prebiotic evolution, *120–134*
    predator–prey, *184–187, 212–213*
    procedures, *96*
    rabies, *97–104*
    rules and algorithms, *94–96, 114–115*
    seeds versus clonal growth, *228–236*
    see also cellular automaton models;
        lattice models
Grim, *145*
guerilla species, *37*
guild proportionality, *58, 59–61*

Hamilton rule, *327–328, 334*
Hawk–Dove game, *136, 144*
    see also Chicken game
herbivory, *32–34*
host–parasite models
    critical transmissibility, *275, 277–278,
        282–284, 297*

dynamics, *272–277*
and mean-field assumptions, *277, 285*
modeling, *271–276*
pair approximation, *301*
PATCH model, *279–282, 287, 289–291*
scale, *274*
spatial dynamics, *273–274, 275*
hot-spots, *509*
see also foci
hypercycles, *92–93, 117, 156–163,
171–182*
and parasites, *116, 154–155, 160–161,
171–172, 178–180*
scroll rings, *162–163*
spirals, *159–162, 174, 178–179*
spots, *174, 179*

indirect interaction scale, *484*
individual-based model components,
18–19, 487–489, 499–500, 509–512
individual-based models, *1–3*
birth–death–movement process, *418, 425*
and cellular automaton models, *151–152*
continuous space, *253, 391–394,
418–424*
discrete-entity simulations, *184–187*
for fat-tailed kernels, *311*
and lattice models, *342*
limitations, *2, 5, 412–413, 515*
predator–prey models, *184–187, 196,
199–200*
and reaction–diffusion models, *152, 457*
spatial structure, *3*
see also artificial ecologies; cellular
automaton models
insect diseases in plants, *292, 295, 308,
312, 313, 315, 316*
insects, *292, 295, 308, 313, 315*
integral equations, *486–497, 507–511*
and epidemic spread, *304–306*
input variables, *306, 308*
methods, *514*
as rapid-stirring limits, *514*
and reaction–diffusion models, *504–507*
integral kernel *see* birth distribution; birth
kernel; competition kernels; dispersal
kernel; displacement distribution
integro-difference models, *317*

intensity, *75*
interacting particle systems, *271, 342, 388,
515*
see also lattice models
interaction kernels, *420–421, 425, 434,
436–437*
interaction neighborhood, *24–25, 46–47*
invasibility
calculating, *247*
competitive hierarchy, *46*
criteria, *319, 333*
and evolutionary stability, *321–323*
games, *139*
and invasiveness, *31–32, 35, 36*
pair-approximation analysis, *250*
see also invasiveness
invasion conditions, *246–247, 363,
381–385, 386*
invasion of neighboring space, *30, 31,
34–37, 42*
invasion waves
and Allee effects, *317, 484–485, 501, 512*
circular, *494*
dispersive, *293*; *see also* kernels,
fat-tailed
independent spread, *485–500*
linear models, *485–500, 504, 507–509*
nonlinear models, *504*
pushed, *501*
rabies, *99–101, 510–511*
in reaction–diffusion models, *321–322,
331–332, 315–316, 475–478, 504–507*
run for your life theorem, *493–498, 503*
shape theorems, *503*
spatial inhomogeneity, *497–499*
spherical, *494*
temporal inhomogeneity, *497–499*
see also traveling wave front; wave
speed; wavelike spatial spread
invasiveness
competitive hierarchy, *37, 39, 42, 46*
and invasibility, *31–32, 35, 46*
see also invasibility
isopath map, *296–297, 299–301, 304, 305,
314*
isotropy,
correlation structure by assumption, *255,
344, 394, 407*

deviations from, *312, 497*
observed, *300*
use of, *255, 265, 345, 407, 427–428, 451*
see also anisotropy; rotational symmetry
Iterated Prisoner's Dilemma game,
  *136–137, 145–150, 250–251, 323–335,
  463–466*

Kermack–McKendrick equation, *312, 359*
kernel-based models
  continuous-time, *252–270, 317,
    388–411, 482–512*
  discrete-time, *293, 317*
kernels
  birth, *486, 488–490, 493, 496, 498, 502,
    508, 510*; *see also* birth distribution;
    reproduction kernel
  competition, *257, 391–392, 410, 411*
  discrete-time, *293, 317*
  dispersal, *257, 302, 306, 310–312, 391,
    392, 397, 421, 425, 509*; *see also*
    displacement distribution; movement
    kernel
  displacement, *317*; *see also* displacement
    distribution
  fat-tailed, *311, 317*
  interaction, *420, 421, 425, 434, 436, 437*
  marginal, *493*
  movement, *420, 438*
  normalized, *307*
  reproduction, *508, 509*
  spatial, *392, 395*
Kirkwood superposition approximation,
  442
Kronecker symbol, *254, 260, 420, 422, 431,
  436,*
kurtosis, *492, 500*

Laplace transform, *419, 488–489, 490, 493,
  499, 508, 512*
lattice logistic model, *344–349, 355*
  see also contact process
lattice models
  artificial societies, *150*
  birth–death–movement process,
    365–366, 370
  complexity, *2*
  coupled map lattice, *197, 210–211*

invasion dynamics, *382–385*
  limitations, *361, 362–363*
  logistic models, *344–349, 355*
  and moment methods, *404–406*
  and pair approximations, *344–358*
  pair-dynamics models, *359–387*
  random lattices, *362, 373–374, 380*
  square lattices, *376–377*
  superimposed lattices, *149*
  triangular lattices, *374–376, 380*
  see also artificial ecologies; cellular
    automaton models; grid-based models
learning strategies, *149*
life-history parameters, *398*
life-history traits in plants, *31, 37, 40–41,
  228–234, 379*
linear conjecture, *502, 503*
linearized stability analysis, *145, 466–471*
local configuration dynamics, *279–282, 452*
local extinction, *160–162, 169, 175–177,
  180–181*
local mean field, *335*
  see also rapid-stirring limit
local stability analysis *see* linearized
  stability analysis
locally infinite systems, *512*
logistic growth, *359, 363, 381, 383, 391,
  465*
  see also spatial logistic model
Lotka–Volterra models
  competition, *39, 389, 400*
  and moment approximation, *256, 258,
    434–435*
  multi-patch, *193–196*
  predator–prey, *187, 359*
  spatial analogues, *263, 421, 449*
  spatial dynamics, *187–190, 199*
  two-patch, *187–193, 197*
  see also predator–prey models;
Lyapunov exponents, *174, 218, 219,
  220–221*
Lyapunov function, *188*

Markov process, *68, 77, 229, 286, 350, 391,
  421–422, 486*
Markov random fields, *77–79, 84–86*
  clique, *78*
  hard-core process, *78*

homogeneous pairwise-interaction process, *78*
master equation, *421, 422*
maximum likelihood estimation, *81–84, 87*
mean crowding, *231*
mean density
  comparison of models, *256–267, 445–446*
  correlation corrections, *449–451*
  definition, *254, 393, 426, 428*
  dynamics, *258, 394–395, 406–407, 433–434*
  fluctuation corrections, *449–451*
  see also intensity; singlet density
mean-field approximation
  assumptions, *4, 231*
  as basis of ecological ideas, *18, 270, 412*
  colicin bacteria, *245–246, 248, 358*
  correlation corrections to, *450–451*
  epidemics, *354–355*
  and evolution rate, prediction of, *207, 287*
  fluctuation corrections to, *450–451*
  host–parasite, *277, 285, 287*
  lattice logistic model, *348, 380*
  local mean field, *335*
  Lotka–Volterra model, *261–262*
  mutant parasite, *285*
  and pair-dynamics models, *387*
  and parasites, *160*
  plants reproducing by seeds and growth, *231, 236*
  and predator–prey modeling, *183*
  selective pressure, *285*
  shortcomings, *1–4, 90–91, 148, 160, 207, 231, 252, 275, 343, 350, 358, 412, 424*
  and transmission rate, *275*
metabolic replication, *119–123, 124*
method of moments, *82*
  see also moment methods; pair-approximation methods
model adequacy, *87–88, 94–95, 111*
model fitting, *80–88*
moment methods, *83*
  accuracy, *259, 261–262, 264, 267, 398–399, 402–404, 442–447, 503*
  closures, *371–379, 363, 389–390, 396, 438–447, 503*

competition model, spatial, *250–261, 400–403*
  correlation corrections, *450–451*
  equations, *258, 260, 261, 263, 265, 394, 396, 406–411, 432–438, 452–455*
  fluctuation corrections, *450–451*
  and lattice models, *388–389, 392, 404–406*
  logistic model, spatial, *261–269, 391–400*
  and pair approximations, *252, 254–255, 258, 260, 265, 404–405, 424–431, 434–449, 451–455*
  and reaction–diffusion models, *252, 269, 359–360, 404–405, 515*
  and relaxation projections, *440*
  spatial moments, definition, *254–255, 392–394, 426–432*
  waves, *503–504*
  see also corrected correlation density; correlation density; pair-approximation methods
Moore neighborhood, *120, 128–129, 155, 243*
mosaic cycle, *55–56, 59, 60–61*
movement kernel, *420, 438*
  see also dispersal kernel; displacement distribution

neighbor effect *see* neighborhood
neighborhood
  and cooperation, *149–150*
  Markov random field, *77–78*
  Moore, *120, 128–129, 155, 243*
  nearest-neighbor disturbances, *75*
  neighbor effects, *3, 12–14, 18–19, 21, 23, 32–34, 229, 238, 257, 379*
  plants, *4–26*
  refuge, *33–34*
  size, *20, 119, 123–129, 263, 265, 267, 362, 398, 503*
  spatial structures, *23–25, 50*
  von Neumann, *43, 135, 155, 235, 243, 272, 462*
  see also patch dynamics, in plant communities; plant interactions
nematodes, *292, 295–296, 313, 315*
von Neumann neighborhood, *43, 135, 155, 235, 243, 272, 462*

Neyman–Scott Poisson cluster process, *74, 77*
non-isotropy *see* anisotropy; isotropy

oak trees *see* forest dynamics
origin of life, *116, 133, 152, 153–154, 157*
    see also prebiotic evolution
overdispersion, *49, 75, 76, 444*

pair-approximation methods
    accuracy, *234, 237, 240–243, 250,*
        348–349, 355, 357, 379–383
    approximation step *see* closure
    basic contact process *see* contact process
    birth–death–movement process,
        365–368, 370–371, 380–382
    closure, *232, 346–347, 351–353, 363,*
        371–379
    cluster sizes, *234–236*
    colicin bacteria, *243–247*
    contact process, *344–349*
    definitions, *228–232, 343–346, 350,*
        355–356, 364, 371
    epidemics, *250, 349–355, 357–358*
    equations, *230–232, 238–239, 345–348,*
        351–352, 354, 360, 364, 367,
        373–377, 384
    forest gaps, *236–243*
    improved, *349–357*
    invasion conditions, dynamics, *246–247,*
        382–385
    lattice logistic model, *344–349; see also*
        birth–death–movement process;
        contact process
    pair-edge approximation, *248–250, 452,*
        503
    predator–prey, *358*
    Prisoner's Dilemma, *250–251*
    and reaction–diffusion models, *359–360*
    relation to moment methods, *252, 343,*
        346, 372, 405
    seed versus clonal representation,
        228–234
    socially influenced mortality, *355–357*
    waves, *503*
    see also moment methods
pair density
    closure, continuous space, *434–444*
    closure, lattice, *363, 372, 377, 386–387*

corrected, *430*
definition, continuous space, *427–429*
definition, lattice, *364–365*
dynamics, continuous space, *260, 339,*
    434–435, 447, 452–455
example, *472*
as expression of spatial pattern, *360,*
    427–430, 434
simplified characterization, *451*
see also correlation density; doublet
    density
pair-edge approximation, *248–250, 452, 503*
pandemics *see* epidemics, second-order
parasite–host *see* host–parasite models
parasites, metabolic *116, 119, 129–133,*
    154, 160–162, 165–170, 171–172,
    178–181
    see also hypercycles
partial differential equations *see*
    reaction–diffusion equations
patch dynamics, in plant communities, *19,*
    28–29, 48–53
PATCH model *see* host–parasite models
patchiness index, *231*
pattern dynamics, *421, 423–425*
Pavlov strategy, *145, 146, 147, 149*
PCA *see* cellular automaton models
PDE *see* reaction–diffusion equations
perimeter polynomials, *235, 236*
perturbation, *482*
    see also disturbance
phalanx species, *37*
plant interactions
    clonal growth, *15*
    competition, measurement of, *29–38*
    competition coefficients determined from
        traits, *40–41*
    competition mechanisms, *12–18*
    foraging, *15*
    life-history traits in plants, *31, 40–41,*
        228–234, 379
    neighbor effects, *32–38*
    neighborhood, *19–26*
    patch dynamics, *19, 28–29*
    positive neighbor effects, *13*
    resources, *14–17*
    roots, *17*
    seed dispersal, *17*

size plasticity, *21, 23*
see also dwarf shrub community
    dynamics; grassland communities;
    foci, phytopathology
PODESS spatio-temporal simulation
    model, *303, 309–310*
Poincaré map, *192–193*
point processes, *72–75, 77–79, 338–339,*
    388–389, 391, 405–406, 418–425
see also Poisson processes
Poisson processes, *10, 49, 70, 72–80,*
    82–83, 253, 310, 407, 450, 487
polycyclic fungal diseases *see* foci
prebiotic evolution, *116–134, 153–169,*
    171–182
see also hypercycles; parasites,
    metabolic; replicator macromolecules;
    surface metabolism
predator–prey models
    demographic stochasticity, *93, 199–200,*
        258, 269, 342, 358, 381, 385–386,
        482, 485
    differential equation models, *187–198*
    embedding dimension, *218, 221–222*
    individual-based, *184–188, 196,*
        199–200, 212–213
    lattice, *358*
    Lyupanov exponents, *219, 220–221*
    multi-patch, *199*
    spatial Lotka–Volterra, *187–196*
    spatial Rosenzweig–MacArthur,
        185–186, 196–199
    spatial scales, *183–184, 213, 216, 221*
    statistical stabilization, *200*
    time series, *214–215, 219, 221*
    see also host–parasite models;
        Lotka–Volterra models
principal components analysis, *223*
Prisoner's Dilemma, *135–150, 250–251,*
    323–335, 463–466
probabilistic cellular automaton models *see*
    cellular automaton models
pseudo-likelihood estimation, *87*

quasi-monotone systems, *473–475*

rabies, *91, 97–104, 114, 316–317, 510–511*
ramets, *24, 29*

random drift, *342*
see also demographic stochasticity
random lattices
    foci, *362*
    invasion dynamics, *385*
    and pair approximation, *373–374, 378,*
        380–382
    simulation, *380–381*
rapid-stirring limit, *514*
see also local mean field
reaction–diffusion equations
    catalytic networks *see* hypercycles
    versus cellular automaton models, *91,*
        118, 151–153, 155–157, 169–170
    clusters *see* spots
    comparison methods, *471–475*
    cut-offs, *159, 161–163, 165–167, 169,*
        177, 515
    derivation, *156, 459–566*
    evolution of diffusion, *479–481*
    and fluctuation corrections, *515*
    games, *318–335, 463–466, 474–475*
    Hamilton rule, *327–328, 334*
    Hutson–Vickers model, *329–331, 333*
    hypercycles, *157–159, 173–174*
    and integral equation models, *504–507*
    invasion waves, *315–316, 322–323,*
        327–329, 331–332, 475–478, 491,
        497–498, 501–503, 504, 508
    linearized stability, *466–471*
    literature on, *458–459*
    numerical methods, *156, 174, 459*
    prebiotic evolution, *116–118, 153–154,*
        171–174
    predator–prey models, *187–201,*
        502–503
    Prisoner's Dilemma, *149, 321–335,*
        463–466
    as rapid-stirring limits, *514*
    scroll rings, waves, *162–163*
    spatial population expansion *see* invasion
        waves
    spirals, spiral patterns, spiral waves,
        161–162, 174–175, 178–179
    spots, *163–169, 174–175, 179–180*
    Turing instability, *323, 466–471*
    waves *see* invasion waves; scroll rings;
        spiral waves

reaction–diffusion models, systems
116–134, 151–170, 171–182, 187–201,
309, 315–316, 318
see also cellular automaton models;
reaction–diffusion equations
relaxation projections, 414–415, 417
alternative methods, 451–452
and correlation densities, 425
and moment closure, 440
as simplification, 414–415, 425
and spatial scale, 426
see also degrees of freedom; dynamic
sufficiency
repeated games see Iterated Prisoner's
Dilemma game
replication–diffusion equations see
reaction–diffusion equations
replication–diffusion models see
reaction–diffusion models, systems
replicator macromolecules
"advantage of the rare" effect, 126, 130
as cellular automata, 119–123
characteristics, 118–119
clusters, 163–169
coexistence, 116, 123–129
community size, 126–127, 129–132
and hypercycle theory, 116–117,
153–154, 171–172
and parasites, 116, 129–133, 163–169
persistence, 126–128
surface metabolism, 120–121, 133–134
see also hypercycles; parasites,
metabolic; reaction–diffusion
equations; surface metabolism
reproduction kernel, 508–509
see also birth distribution, birth kernel
resistance to parasites, 116–118, 129–133,
160–163, 165–168, 171–172, 178–182
see also evolution, transmissibility
Retaliator, 144
Rock–Scissors–Paper game, 144
Rosenzweig–MacArthur models, 185–186,
196–198
see also Lotka–Volterra models;
predator–prey models
rotational symmetry, 375, 485, 492
see also isotropy
run for your life theorem, 493–498, 503

sandpile avalanches, 113–114
scale
competition, 396, 405, 409
direct interaction scale, 484
dispersal, 312, 396, 405, 409, 483–484,
496–497, 507–511
indirect interaction scale, 484
sexual scale, 483–484
spatial heterogeneity, 425–426, 497–498
spatial scales, 496–497
scroll rings, 155–156, 160–163
see also self-replicating spots; spiral waves
second moment, 206, 254–256, 260–265,
389, 396, 429, 439
see also correlation density; covariance
function; doublet density
second-order models, 389–391, 394, 396,
403, 405, 407–408, 439–440
selective pressure, 282, 284–287
self-organized criticality, 113–114
self-replicating entities see replicator
macromolecules
self-replicating spots, 172, 174, 177
resistance to parasites, 178–180
see also clusters; scroll rings; spiral waves
self-structuring patterns and evolution, 118,
128, 181–182
see also scroll rings; self-replicating
spots; spiral waves
sexual scale, 483–484
shape theorems, 503
SI model see host–parasite models
singlet density, 205–206
definition, 345, 426, 428
dynamics, 231, 251, 345, 347, 350–351,
363, 380, 385, 426, 428, 443
see also mean density
singular value decomposition, 225–226
SIR model, 349–355, 357–358
Snowdrift game, 136, 144
see also Hawk–Dove game
solidarity game, 150
space preemption, 58–61
spatial competition experiments, 29–40,
46–47
spatial competition models, 361
colicin bacteria, 243–250
genes, 210–213

grassland plants, *43–46*
Lotka–Voltera, *257–261, 400–403, 420–421, 446*
plant reproductive strategies, *228–237*
spatial density function, *418*
spatial evolutionary game theory *see* evolution
spatial heterogeneity, inhomogeneity, *52–54, 448–449, 497–500*
spatial hypercycle model *see* hypercycles
spatial logistic model
continuous, *261–269, 391–400, 445*
lattice, *344–349, 355, 360, 370–371, 381, 383*
spatial moments, *253–256, 392–394*
see also correlation density; moment methods
spatial population expansion *see* invasion waves; wave form; wavelike spatial spread
spatial scales, *213–218, 425–426, 496–497*
species traits and competition matrices, *40–41*
spectral methods, *66, 68, 84, 399–400*
spiral waves, *117–118, 128, 159–162, 174–175*
bounding conditions, *155*
cellular automaton models, *160–161*
and parasites, *160, 172, 178–179, 180*
and partial differential equation models, *161–162, 174–177*
and self-replication, *174–175*
see also clustering; scroll rings; self-replicating spots
spots *see* self-replicating spots
square lattices, *376–377*
see also cellular automaton models; grid-based models; lattice models
stability analysis, *195, 200–201, 330–331, 466–471*
stationarity, *394, 407*
see also data analysis, trend; ergodicity
statistical stabilization, *196, 198–200*
stochastic models and data analysis, *70–80, 84–86, 88*
surface metabolism, *133–134*
susceptible–infected *see* host–parasite models

susceptible–infected–recovered model *see* SIR model

temporal inhomogeneity, *497–499*
TFT *see* Tit For Tat
third moments, *260, 377–379, 403–404*
see also triplet densities
time series analysis, *67–68, 70, 84, 213–225*
Tit For Tat, *137–138, 145–149, 208, 251, 323–332, 463–466, 471, 478, 481*
transition matrices, estimation from invasion experiments, *42*
transition rates, *370, 422–433*
birth–death–movement process in continuous space, *257, 263, 420–421*
colicin bacteria model, *244*
forest gaps model, *237–238*
lattice birth–death–movement process, *344*
plant reproduction model, *229*
SIR model, *350*
socially changed mortality model, *355*
spatial logistic model, *344*
spiral logistic model, *263, 391–392*
transition rules, *96*
dwarf shrub community model, *105*
forest fire model, *110*
games, *137–138, 140–142, 145–146*
gene competition, *210*
grassland competition model, *42–43*
host–parasite model, *272–273*
metabolic replication model, *119–121*
PATCH model, *279–281, 289–291*
predator–prey models, *184–185*
predator–prey–resource models, *212–213*
rabies model, *101*
reaction–diffusion, *155–157*
transitive competition, *39*
traveling wave front
games, *321–322, 329, 331, 333–334*
plant diseases, *292–315*
rabies, *99–102*
in reaction–diffusion models, *475–478*
and sexual interaction, *384*
shape, *476, 495–476, 502–503*
see also foci; invasion waves; scroll rings; spiral waves; wave speed

trend, *67, 69–70, 200, 258, 512*
triangular lattice models, *374–376, 380, 382, 383, 385*
triplet densities
  closure assumptions, *260, 346, 351, 372–379, 386–387, 439–444*
  definition, *231–232, 345, 429*
  effect on pair dynamics, *232, 251, 366, 369, 371, 431, 435, 436–438*
  examples, *383, 431, 443*
  see also moment methods, closure; pair-approximation methods, closures
Turing
  bifurcation, *466–470*
  instability, *323, 458, 470*
  pattern, *340*
two-patch models
  Lotka–Volterra, *187–193*
  Rosenzweig–MacArthur, *196–197*

underdispersion *see* clustering

Vanderplank's equation, *301, 311, 488*
viruses, *295, 299, 313*

wave form, *60*
waves
  fir forest regeneration, *341*
  heathlands, *60*
  see also invasion waves; scroll rings; spiral waves; traveling wave front; wave speed
wave speed, *481–493, 497–501, 505, 507–512*
  Allee effect, *317, 501, 512*
  amplification effect, *507*
  concentrated reproduction, *493*
  continuous movement, *499–500*
  experimental results, *299, 303, 308*
  home ranges, *489*
  hot-spots, *509*
  quantitative applications, *315–317*
  see also c_0
wavelike spatial spread
  approximation formulas, *492, 493*
  in reaction–diffusion models, *487, 491*
  run for your life theorem, *493–498, 503*
  wave speed, *487, 490–491, 493–494*
Whittle's method, *84*

International Institute for Applied Systems Analysis

IIASA is an interdisciplinary, nongovernmental research institution founded in 1972 by leading scientific organizations in 12 countries. Situated near Vienna, in the center of Europe, IIASA has been for more than two decades producing valuable scientific research on economic, technological, and environmental issues.

IIASA was one of the first international institutes to systematically study global issues of environment, technology, and development. IIASA's Governing Council states that the Institute's goal is: *to conduct international and interdisciplinary scientific studies to provide timely and relevant information and options, addressing critical issues of global environmental, economic, and social change, for the benefit of the public, the scientific community, and national and international institutions.* Research is organized around three central themes:

- Energy and Technology;
- Environment and Natural Resources;
- Population and Society.

The Institute now has national member organizations in the following countries:

Austria
The Austrian Academy of Sciences

Bulgaria*
The Bulgarian Committee for IIASA

Finland
The Finnish Committee for IIASA

Germany**
The Association for the
Advancement of IIASA

Hungary
The Hungarian Committee for
Applied Systems Analysis

Japan
The Japan Committee for IIASA

Kazakstan*
The Ministry of Science –
The Academy of Sciences

Netherlands
The Netherlands Organization for
Scientific Research (NWO)

Norway
The Research Council of Norway

Poland
The Polish Academy of Sciences

Russian Federation
The Russian Academy of Sciences

Slovak Republic*
The Slovak Committee for IIASA

Sweden
The Swedish Council for Planning and
Coordination of Research (FRN)

Ukraine*
The Ukrainian Academy of Sciences

United States of America
The American Academy of
Arts and Sciences

*Associate member
**Affiliate